ISBN 978-1-5282-0231-2
PIBN 10922466

English
Français
Deutsche
Italiano
Español
Português

www.forgottenbooks.com

Mythology Photography **Fiction**
Fishing Christianity **Art** Cooking
Essays Buddhism Freemasonry
Medicine **Biology** Music **Ancient**
Egypt Evolution Carpentry Physics
Dance Geology **Mathematics** Fitness
Shakespeare **Folklore** Yoga Marketing
Confidence Immortality Biographies
Poetry **Psychology** Witchcraft
Electronics Chemistry History **Law**
Accounting **Philosophy** Anthropology
Alchemy Drama Quantum Mechanics
Atheism Sexual Health **Ancient History**
Entrepreneurship Languages Sport
Paleontology Needlework Islam
Metaphysics Investment Archaeology
Parenting Statistics Criminology
Motivational

NBS SPECIAL PUBLICATION

U.S. DEPARTMENT OF COMMERCE/National Bureau of Standards

Wind and Seismic Effects

Proceedings of the Eleventh Joint UJNR Panel Conference

NATIONAL BUREAU OF STANDARDS

The National Bureau of Standards[1] was established by an act of Congress on March 3, 1901. The Bureau's overall goal is to strengthen and advance the Nation's science and technology and facilitate their effective application for public benefit. To this end, the Bureau conducts research and provides: (1) a basis for the Nation's physical measurement system, (2) scientific and technological services for industry and government, (3) a technical basis for equity in trade, and (4) technical services to promote public safety. The Bureau's technical work is performed by the National Measurement Laboratory, the National Engineering Laboratory, and the Institute for Computer Sciences and Technology.

THE NATIONAL MEASUREMENT LABORATORY provides the national system of physical and chemical and materials measurement; coordinates the system with measurement systems of other nations and furnishes essential services leading to accurate and uniform physical and chemical measurement throughout the Nation's scientific community, industry, and commerce; conducts materials research leading to improved methods of measurement, standards, and data on the properties of materials needed by industry, commerce, educational institutions, and Government; provides advisory and research services to other Government agencies; develops, produces, and distributes Standard Reference Materials; and provides calibration services. The Laboratory consists of the following centers:

Absolute Physical Quantities[2] — Radiation Research — Chemical Physics — Analytical Chemistry — Materials Science

THE NATIONAL ENGINEERING LABORATORY provides technology and technical services to the public and private sectors to address national needs and to solve national problems; conducts research in engineering and applied science in support of these efforts; builds and maintains competence in the necessary disciplines required to carry out this research and technical service; develops engineering data and measurement capabilities; provides engineering measurement traceability services; develops test methods and proposes engineering standards and code changes; develops and proposes new engineering practices; and develops and improves mechanisms to transfer results of its research to the ultimate user. The Laboratory consists of the following centers:

Applied Mathematics — Electronics and Electrical Engineering[2] — Manufacturing Engineering — Building Technology — Fire Research — Chemical Engineering[2]

THE INSTITUTE FOR COMPUTER SCIENCES AND TECHNOLOGY conducts research and provides scientific and technical services to aid Federal agencies in the selection, acquisition, application, and use of computer technology to improve effectiveness and economy in Government operations in accordance with Public Law 89-306 (40 U.S.C. 759), relevant Executive Orders, and other directives; carries out this mission by managing the Federal Information Processing Standards Program, developing Federal ADP standards guidelines, and managing Federal participation in ADP voluntary standardization activities; provides scientific and technological advisory services and assistance to Federal agencies; and provides the technical foundation for computer-related policies of the Federal Government. The Institute consists of the following centers:

Programming Science and Technology— Computer Systems Engineering.

[1] Headquarters and Laboratories at Gaithersburg, MD, unless otherwise noted; mailing address Washington, DC 20234.
[2] Some divisions within the center are located at Boulder, CO 80303.

Wind and Se

Proceedings of the 11th Joint
Panel Conference of the U.S.-Japan
Cooperative Program in
Natural Resources

NBS Special publication

H.S. Lew, Editor
Center for Building Technology
National Engineering Laboratory
National Bureau of Standards
Washington, DC 20234

Library of Congress Catalog Card Number: 83-600550

National Bureau of Standards Special Publication 658
Natl. Bur. Stand. (U.S.), Spec. Publ. 658, 755 pages (July 1983)
CODEN: XNBSAV

PREFACE

The Eleventh Joint Meeting of the U.S. - Japan Panel on Wind and
Seismic Effects was held in Tokyo, Japan on September 4-7, 1979. This
panel is a part of the U.S. - Japan Cooperative Program in Natural
Resources (UJNR). The UJNR was established in 1964 by the U.S. - Japan
Cabinet-level Committee on Trade and Economic Affairs. The purpose of
the UJNR is to exchange scientific and technological information which
will be mutually beneficial to the economics and welfare of both
countries.

These proceedings include the program, the formal resolutions, and
the technical papers presented at the Joint Meeting. The texts of the
papers, all of which were prepared in English, have been edited for
obvious errors and clarity.

It should be noted that throughout the proceedings certain commercial equipment, instruments or materials are identified in order to
specify adequately experimental procedure. In no case does such identification imply recommendation or endorsement by the National Bureau of
Standards, nor does it imply that the material or equipment identified
is necessarily the best for the purpose.

Preparation of the proceedings was partially supported by funds
from the Department of Housing and Urban Development

H. S. Lew, Secretary
U.S. Panel on Wind and
Seismic Effects

SI CONVERSION UNITS

In view of the present accepted practice for wind and seismic tech
of measurements were used throughout this publication. In recognition
United States as a signatory to the General Conference on Weights and M
official status to the International System of Units (SI) in 1960, the
sented to facilitate conversion to SI Units. Readers interested in mak
coherent system of SI units are referred to: NBS SP 330, 1977 Edition,
System of Units; and ASTM Standard for Metric Practice.

Table of Conversion Factors to SI Units

	Customary Units	International (SI) UNIT	
Length	inch (in)	meter (m)[a]	1
	foot (ft)	meter (m)	1
Force	pound (lbf)	newton (N)	1
	kilogram (kgf)	newton (N)	1
Pressure	pound per square		
Stress	inch (psi)	newton/meter^2	1
	kip per square		
	inch (ksi)	newton/meter^2	1
Energy	inch·pound (in-lbf)	joule (J)	1
	foot·pound (ft-lbf)	joule (J)	1
Torque or	pound·inch	newton-meter (N·m)	1
Bending Moment	pound·foot (lbf-ft)	newton-meter (N·m)	1
Weight	pound (lb)	kilogram (kg)	1
Unit Weight	pound per cubic foot (pcf)	kilogram per cubic meter (kg/m^3)	1
Velocity	foot per second (ft/sec)	meter per second (m/s)	1
Acceleration	foot per second second (ft/sec^2)	meter per second per second (m/s^2)	1

---

[a] Meter may be subdivided. A centimeter (cm) is 1/100 m and a millimete

* Exactly

ABSTRACT

The Eleventh Joint Meeting of the U.S. - Japan Panel on Wind and
Seismic Effects was held in Tsukuba, Japan on September 4-7, 1979. The
proceedings of the Joint Meeting include the program, the formal resolution
and the technical papers. The subjects covered in the paper include (1)
the engineering characteristics of wind, (2) the characteristics of earthquake ground motions, (3) the earthquake response of structures, (4) the
wind response of structures, (5) recent design criteria against wind and
earthquake disturbances, (6) the design and analysis of special structures,
(7) the evaluation, repairing, and retrofitting for wind and earthquake
disaster, (8) earthquake disaster prevention planning, (9) storm surge and
tsunamis, (10) and technical cooperation with developing countries.

KEYWORDS: Accelerograph; codes; design criteria; disaster; earthquakes;
           hazards; ground failures; seismicity; solids; standards;
           structural engineering; structural responses; tsunamis;
           wind loads; and winds.

CONTENTS

Themes and Technical Papers

Opening Themes:

Theme I: ENGINEERING CHARACTERISTICS OF WIND

Theme II: CHARACTERISTICS OF EARTHQUAKE GROUND MOTIONS (Including Engineering
           Seismology)

Theme III: EARTHQUAKE RESPONSE OF STRUCTURES (Including Hydraulic Structures and
            Earth Structures)

THEME X: TECHNICAL COOPERATION WITH DEVELOPING COUNTRIES

OF THE
U.S.-JAPAN PANEL ON WIND AND SEISMIC EFFECTS
SEPTEMBER 4-7, 1979
AT THE
PUBLIC WORKS RESEARCH INSTITUTE
TSUKUBA SCIENCE CITY, JAPAN

TUESDAY –   September 4   OPENING SESSION

11:00 a.m.   Call to order by Dr. Ryuichi Iida, Secretary, Japanese Panel, Director,
Planning and Research Administration Department, Public Works Research
Institute

Remarks by Mr. Shin-ichiro Asai, Engineer General, Ministry of Construction

Remarks by Mr. Justin L. Bloom, Counselor for Scientific and Technological
Affairs, Embassy of the United States of America

Remarks by Mr. Yoshijiro Sakagami, Chairman, Japanese Panel, Director-General,
Public Works Research Institute, Ministry of Construction

Remarks by Dr. E. O. Pfrang, Chairman, U.S. Panel, Chief, Structures and
Materials Division, Center for Building Technology, National Engineering
Laboratory, National Bureau of Standards

Introduction of Japan Panel Members by Japanese Chairman and U.S. Panel
Members by U.S. Chairman

Election of Conference Chairman

Adoption of Agenda

1:20 p.m.   New Initatives in Earthquake Hazards Mitigation — Charles C. Thiel,
E. O. Pfrang

1:35   Earthquake Hazard Reduction Research at the National Bureau of Standards --
Edward O. Pfrang, Edgar V. Leyendecker, James R. Harris, Richard N. Wright

1:50   Outline of Research on Wind and Earthquake Effects in Japan -- Yoshijiro
Sakagami

2:05   Discussion

2:40   Modeling the 1978 Tokyo Tornado that Overturned in the Tozai Subway Train —
Eiji Uchida, Ryozo Tatehira, Ichiro Tabe, Kazuyuki Ohtsuka, Shigemi Fujiwhara

3:00   Extreme Wind Speeds at 129 Stations in the Contiguous United States --
Michael J. Changery, Emil Simiu, James J. Filliben, Celso Barrientos

3:20   Analysis of High Wind Observations from Very Tall Towers -- Keikichi Naito,
Isao Tabata, Noboru Banno, Katsumi Takahashi

3:40   Discussion

4:10   Wind and Structure Motion Study for Pasco-Kennewick Bridge -- M. C. C. Bampton,
Harold Bosch, David H. Cheng, Charles F. Scheffey

4:30   Calculation of the Gust Responses of Long-Span Bridges -- Nobuyuki Narita, Hiroshi Sato

4:50   Discussion

WEDNESDAY - September 5

9:00 a.m.   A Proposal for a New Parameter in Assessing Seismic Disaster -- Michio Otsuka

9:20   Strong-Motion Data Exchange Through National Data Bases -- A. Gerald Brady,

9:40   Discussion

9:55   Simulation of Earthquake Ground Motion and Its Application to Dynamic Response Analysis -- Tetsuo Kubo, Makoto Watabe

10:15   Comparison of Vertical Component of Strong-Motion Accelerograms for Western United States and Japan -- Tatsuo Uwabe

10:35   Earthquake Observation System in and around Structure in Japan -- Kiyoshi Nakano, Yoshikazu Kitagawa

10:55   Discussion

11:25   Single-story Residential Masonry Construction in Uniform Building Code Seismic Zone 2 -- G. Robert Fuller

11:45   The Performance of Lapped Splices in Reinforced Concrete Under High-Level Repeated Loading -- Peter Gergely, Fernando Fagundo, Richard N. White

12:05 p.m.   Discussion

1:20   Dynamic Behavior of Reinforced Concrete Frame Structures -- Keiichi Ohtani, Chikahiro Minowa

             Seismic Response of Precast Concrete Walls -- James M. Becker, Carlos Llorente, Peter Mueller

2:00   California School and Hospital Ceilings -- John F. Meehan

2:20   Seismic Response Analysis of the Itajima Bridge with Use of Strong Motion Acceleration Records -- Masamitsu Ohashi, Toshio Iwasaki, Kazuhiko Kawashima

2:40   Discussion

3:00   Field and Laboratory Determination of Soil Moduli -- William F. Marcuson III, Joseph R. Curro, Jr., G. Robert Fuller

3:20   An Experimental Study on Liquefaction of Sandy Soils on a Cohesive Soil Layer -- Tatsuo Asama, Yukitake Shioi, Hideya Asanuma

3:40   Stress-strain Behavior of Dry Sand and Normally Consolidated Clay by Inter-laboratory Cooperative Cyclic Shear Tests -- Hiroshi Oh-oka, Kohjiroh Itoh, Yoshihiro Sugimura, Masaya Hirosawa

4:00   Discussion

4:30   Discussion of Task Committee Reports (* Chairmen of Task Committees)

             a) Strong-Motion Arrays and Data -- *Jerry S. Dodd, *Hajime Tsuchida, Toshio Iwasaki, Keiichi Ohtani, Michio Ohtsuka, Makoto Watabe

b) Large-Scale Testing Programs -- *H. S. Lew, *Keiichi Ohtani, Tatsuor Murota, Nobuyuki Narita, Setsuo Noda, Yashushi Sasaki, Makota Watabe

c) Repair and Retrofit of Existing Structures -- *John B. Scalzi, *Kiyoshi Nakano, Tatsuo Asama, Eiichi Kuribayashi

d) Structural Performance Evaluation -- *G. Robert Fuller, *Makoto Watabe, Tatsuo Asama, Yukitake Shioi

f) Disaster Prevention Methods for Lifeline Systems -- *Charles F. Scheffey, *Tadayoshi Okubo, Shigemi Fujiwhara, Toshio Iwasaki, Eiichi Kuribayashi, Tatsuro Murota, Setsuo Noda

g) High Wind -- *Celso Barrientos, *Shigemi Fujiwhara, Tatsuro Murota, Keikichi Naito, Nobuyuki Narita

THURSDAY - September 6

9:00 a.m.   Draft Seismic Design Guidelines for Highway Bridges -- James D. Cooper, Charles F. Scheffey, Roland L. Sharpe, Ronald L. Mayer

9:20   Discussion

9:30   Analysis and Design of Cracked Reinforced Concrete Nuclear Containment Shells for Earthquakes -- Peter Gergely, Richard N. White

9:50   Comparison of Measured and Computed Response of Yuda Dam during the July 6, 1976 and June 12, 1978 Earthquakes -- Ryuichi Iida, Norihisa Matsumoto, Satoru Kondo

10:10   Discussion

10:35   Report on the 1978 Miyagi-ken-oki Earthquake -- Makoto Watabe, Yutaka Matsushima, Yuji Ishiyama, Tetsuo Kubo, Yuji Ohashi

10:50   Disastrous Ground Failure in a Residential Area of Large-Scale Cut-and-Fill in the Sendai Region caused by the Earthquake of 1978 -- Hitoshi Haruyama, Motoo Kobayashi

11:05   Damage Features of Civil Engineering Structures due to the Miyagi-ken-oki Earthquake of 1978 -- Tadayoshi Okubo, Masamitsu Ohashi, Toshio Iwasaki, Kazuhiko Kawashima, Ken-ichi Tokida

11:20   Damage to River Dykes caused by the Miyagi-ken-oki Earthquake of June, 1978 -- Kazuya Yamamura, Yasushi Sasaki, Yasuyuki Koga, Eiichi Taniguchi

11:35   Damage to Port Structures by the 1978 Miyagi-ken-oki Earthquake -- Hajime Tsuchida, Setsuo Noda

11:50   Discussion

1:15 p.m.   Wood Diaphragms Masonry Buildings -- Mihran S. Aghabian

1:30   Mitigation of Seismic Hazards in Existing Unreinforced Masonry Buildings -- M. S. Agbabian

1:45   Development of a Universal Fastener for Wooden Buildings Roof Frames -- Tatsuo Murota, Yuji Ishiyama

2:05   Discussion

2:20   The NASA/MSFC Experimental Facilities at Huntsville, Alabama -- John B. Scalzi, George F. McDonough, Jr., Nicholas C. Costes

2:40	Earthquake Hazards Reduction Research Supported in 1978 -- Charles C. Thiel, William A. Anderson, Michael P. Gaus, William Hakala, Frederick Krimgold, Shih Chi Liu, John B. Scalzi
3:00	Functional Damage and Rehabilitation of Lifelines in the Miyagi-ken-oki Earthquake of 1978 -- Kazuto Nakazawa, Eiichi Kuribayashi, Tadayuki Tazaki, Takayuki Hadate, Ryoji Hagiwara
3:20	Discussion
3:50	Social Aspects of Earthquake Mitigation and Planning in the United States -- William A. Anderson
4:10	Discussion
4:20	Seismic Risk Maps in the Southeast Asian Countries (Maximum Acceleration and Maximum Particle Velocity) - Phillippines, Indonesia and Indo-China -- Sadaiku Hattori
4:50	Discussion of Task Committee Reports (* Chairmen of Task Commitees)
	e) Land Use Programs for Controlling Natural Hazard Effects -- *G. Robert Fuller, *Eiichi Kuribayashi, Hitoshi Haruyama, Sadaiku Hattori
	h) Soil Behavior and Stability during Earthquakes -- *Jerry S. Dodd, *Toshio Iwasaki, Yasushi Sasaki, Yukitake Shioi, Hajime Tsuchida
5:30	Adjourn

FRIDAY - September 7

9:00 a.m.	Orientation of Tsunami Research in Japan -- Hiroshi Takahashi, Yukio Fujinawa
9:20	Wave Setup Caused by Typhoon 7010 -- Hiroshi Hashimoto, Taka-aki Uda
9:40	Specification and Prediction of Surface Wind Forcing for Ocean Current and Storm Surge Models -- Celso S. Barrientos, Kurt W. Hess
9:55	A Dynamic Model to Predict Storm Surges and Overland Flooding in Bays and Estuaries -- Chester P. Jelesnianski, Celso S. Barrientos, Jye Chen
10:10	Topics on Tsunami Protection along the Port Areas in Japan -- Yoshimi Goda
10:30	Discussion
11:05	Task Committee Reports
11:05	a) Strong-Motion Instrumentation Arrays and Data
11:15	b) Large-Scale Testing Programs
11:25	c) Repair and Retrofit of Existing Structures
11:35	d) Structural Performance Evaluation
11:45	e) Land Use Programs for Controlling Natural Hazard Effects
12:55 p.m.	Task Committee Reports
12:55	f) Disaster Prevention Methods for Lifeline Systems

U.S. PANEL ON WIND AND SEISMIC EFFECTS

MEMBERSHIP LIST

JANUARY 1980

Dr. Edward O. Pfrang (CHAIRMAN)
Chief, Structures and Materials Division
Center for Building Technology, NEL
National Bureau of Standards
Washington, DC 20234

Dr. H. S. Lew (SECRETARIAT)
Structures and Materials Division
Center for Building Technology, NEL
National Bureau of Standards
Washington, DC 20234
(301) 921-2647

Dr. S. T. Algermissen
Office of Earthquake Studies
Denver Federal Center
Branch of Earthquake Tectonics, USGS
Stop 978, Box 25046
Denver, CO 80225
(303) 234-4014

Dr. Celso Barrientos
Research Meteorologist
NOAA
8060 13th Street
Silver Spring, MD 20910
(301) 427-7613

Mr. Billy Bohannan
Assistant Director
Wood Engineering Research
Forest Products Laboratory, USDA
P.O. Box 5130
Madison, WI 53705
(608) 257-2211

Dr. Roger D. Borcherdt
Chief, Branch of Ground Motion
and Faulting
Office of Earthquake Studies, USGS
345 Middlefield Road
Menlo Park, CA 94025
(415) 323-8111

Dr. A. Gerald Brady
Physical Scientist
Office of Earthquake Studies, USGS
345 Middlefield Road
Menlo Park, CA 94025
(415) 323-8111

Mr. James D. Cooper
Structures and Applied Mechanics Division
Federal Highway Administration
Office of Research HRS-11
Washington, DC 20590
(202) 557-4315

Mr. Jerry Dodd
Bureau of Reclamation
P.O. Box 25007
Denver Federal Center
Denver, CO 80225
(303) 234-3089

Mr. G. Robert Fuller
Architectural and Engineering Division
Department of Housing and Urban Development
Washington, DC 20411
(202) 755-5924

Dr. Michael P. Gaus
Division of Problem-Focused Research
  Applications
National Science Foundation
1800 G Street, NW
Washington, DC 20550
(202) 632-5700

Mr. John J. Healy
Chief, Research and Development Office
Department of the Army
DAEN-RDM
Washington, DC 20314
(202) 693-7287

Dr. William B. Joyner
Office of Earthquake Studies
Branch of Ground Motion and Faulting, USGS
345 Middlefield Road
Menlo Park, CA 94025
(415) 323-8111

Mr. John W. Kaufman
Atmospheric Sciences Division
National Aeronautics and Space Administration
Marshall Space Flight Center, AL 35812
(205) 453-3104

Mr. James Lander
National Oceanic and Atmospheric
  Administration
Environmental Data Service
National Geophysical and Solar-Terrestrial
  Data Center
Boulder, CO 80302
FTS 323-6474 or (303) 449-1000, ext. 6474

Dr. E. V. Leyendecker
Leader, Earthquake Hazards Reduction Group
Structures and Materials Division
Center for Building Technology, NEL
National Bureau of Standards
Washington, DC 20234
(301) 921-3471

Dr. Shih C. Liu
Program Manager
Division of Problem-Focused Research
  Applications
National Science Foundation
Washington, DC 20550

Dr. Richard D. McConnell
Office of Construction
Veterans Administration
811 Vermont Avenue, NW
Washington, DC 20420
(202) 389-3103

Dr. William F. Marcuson, III
Research Civil Engineer, WES-SH
Department of the Army
Waterways Experiment Station,
  Corps of Engineers
P.O. Box 631
Vicksburg, MS 39180
FTS 542-2202 or (601) 636-3111, ext. 2202

Dr. Richard D. Marshall
Leader, Structural Engineering
Structures and Materials Division
Center for Building Technology, NEL
National Bureau of Standards
Washington, DC 20234
(301) 921-3471

Mr. John F. Meehan
Department of General Services
Office of Architectural Construction
Sacramento, CA 95805
(916) 445-8730

Mr. Howard L. Metcalf
Deputy Director for Construction Standards
  and Design
Office of the Assistant Secretary of Defense
(MRA&L) Room 3E763
The Pentagon
Washington, DC 20301
(202) 695-2713

Dr. Thomas D. Potter
Deputy Director
Environmental Data Service, NOAA
3300 Whitehaven Street, NW
Washington, DC 20235
(202) 634-7319

Mr. Ronald J. Morony
Program Manager
Building Technology Research Staff
Department of Housing and Urban Development
Washington, DC 20410
(202) 755-0640

Mr. Drew A. Tiedemann
Bureau of Reclamation
Engineering and Research Center
Denver Federal Center
Denver, CO 80225
(303) 234-3029

NATIONAL EARTHQUAKE HAZARDS REDUCTION PROGRAM

The National Earthquake Hazards Reduction Act [1] established the purpose "to
risks of life and property from future earthquakes in the United States through th
lishment of an effective earthquake hazards reduction program." The Act specified
objectives of the program shall include:

1) The development of technologically and economically feasible design and co
methods and procedures to make new and existing structures, in areas of seismic ris
quake resistant, giving priority to the development of such methods and procedures
nuclear power generating plants, dams, hospitals, schools, public utilities, public
structures, high occupancy buildings, and other structures which are especially nee
time of disaster;

2) The implementation in all areas of high or moderate seismic risk, of a sys
(including personnel, technology, and procedures) for predicting damaging earthqual
identifying, evaluating, and accurately characterizing seismic hazards;

3) The development, publication, and promotion, in conjunction with State and
officials and professional organizations, of model codes and other means to coordin
mation about seismic risk with land-use policy decisions and building activity;

4) The development, in areas of seismic risk, of improved understanding of,
bility with respect to, earthquake-related issues, including methods of controllin
from earthquakes, planning to prevent such risks, disseminating warnings of earthq
organizing emergency services, and planning for reconstruction and redevelopment a
earthquake;

5) The education of the public, including State and local officials as to ea
phenomena, the identification of locations and structures which are especially sus
earthquake damage, ways to reduce the adverse consequences of an earthquake, and r
matters;

6) The development of research on--

(a) Ways to increase the use of existing scientific and engineering know
mitigate earthquake hazards;

(b) The social, economic, legal, and political consequences of earthqual
prediction; and

Dr. Nobuyuki Narita
Head, Structure Division
Structure and Bridge Department
Public Works Research Institute
Ministry of Construction
Asahi 1-banchi, Toyosato-machi, Tsukuba-gun
Ibaraki-ken 305

Mr. Setsuo Noda
Chief, Earthquake Disaster Prevention
   Laboratory
Structure Division
Port and Harbour Research Institute
Ministry of Transport
3-1-3, Nagase, Yokosuka-shi,
Kanagawa-ken 239

Dr. Tadayoshi Okubo
Deputy Director-General
Public Works Research Institute
Ministry of Construction
Asahi 1-banchi, Toyosato-machi, Tsukuba-gun
Ibaraki-ken 305

Mr. Keiichi Ohtani
Chief, Earthquake Engineering Laboratory
Second Research Division
National Research Center for Disaster
   Prevention
Science and Technology Agency
3-1, Tennodai, Sakura-mura, Niihari-gun,
Ibaraki-ken 305

Mr. Yasushi Sasaki
Head, Soil Dynamics Division
Construction Method and Equipment Department
Public Works Research Institute
Ministry of Construction
Asahi 1-banchi, Toyosato-machi, Tsukuba-gun
Ibaraki-ken 305

Mr. Yukitake Shioi
Head, Foundation Engineering Division
Structure and Bridge Department
Public Works Research Institute
Ministry of Construction
Asahi 1-banchi, Toyosato-machi, Tsukuba-gun
Ibaraki-ken 305

Mr. Hiroshi Takahashi
Head, Second Research Division
National Research Center for Disaster
   Prevention
Science and Technology Agency
3-1, Tennodai, Sakura-mura, Niihari-gun
Ibaraki-ken 305

Mr. Hajime Tsuchida
Chief, Earthquake Resistant Structures
   Laboratory
Port and Harbour Research Institute
Ministry of Transport
3-1-1, Nagase, Yokosuka-shi,
Kanagawa-ken 239

Mr. Osamu Ueda
Head, Earthquake Engineering Division
Earthquake Disaster Prevention Department
Public Works Research Institute
Ministry of Construction
Asahi 1-banchi, Toyosato-machi, Tsukuba-gun
Ibaraki-ken 305

Dr. Makoto Watabe
Head, Structure Division
Building Research Institute
Ministry of Construction
Tatehara 1-banchi, Oh-ho-machi, Tsukuba-gun
Ibaraki-ken 305

SECRETARIAT:

Dr. Ryuichi Iida, Secretary-General
Director
Planning and Research Administration
   Department
Public Works Research Institute
Ministry of Construction
Asahi 1-banchi, Toyosato-machi, Tsukuba-gun
Ibaraki-ken 305

RESOLUTIONS ON THE ELEVENTH JOINT MEETING

U.S.-JAPAN PANEL ON WIND AND SEISMIC EFFECTS

U.J.N.R

September 4 - 7, 1979

The following resolutions for future activities of this Panel are hereby adopted:

1. The Eleventh Joint Meeting was an extremely valuable exchange of technical informatic
    which was beneficial to both countries. In view of the importance of cooperative
    programs on the subject of wind and seismic effects, the continuation of Joint Panel
    Meetings is considered essential.

2. The exchange of technical information, especially revised codes and specifications
    relevant to wind and seismic effects, and the promotion of research programs includin
    exchange of personnel and available equipment should be strengthened. The secretaria
    of both sides shall be kept informed of such exchanges of information by individual
    members.

3. The Panel concurs in the provisions described in the "Implementing Arrangement" signe
    by the Science and Technology Agency and the Ministry of Construction for Japan and t
    National Science Foundation for the U.S. on August 10th, 1979. To insure successful
    implementation of the Joint Research Program of Large-scale Testing of Structures, th
    Panel should expand its existing Task Committee on the Large-scale Testing Program fo
    providing scientific and technical advice to participating institutions in the Progra
    The co-chairmen of this Task Committee shall be appointed by the co-chairmen of the
    Panel. Additional Task Committee members shall also be appointed by the Panel co-cha
    men. The Task Committee should consider recommendations from other Task Committees c
    this Panel for expansion of the joint research program such as life line componennts,
    repair and retrofit and soil-structure interaction.

4. Considerable advancement has been made by various Task Committees. Cooperative resea
    programs should be promoted through the Task Committees, specifically in the area of
    repair and retrofit of structures.

5. A new Task Committee on Tsunami and Storm Surge is hereby established. Co-chairmen a
    members of this Task Committee will be appointed by the co-chairmen of the Panel.

6. The date and location of the Twelveth Joint Meeting on Wind and Seismic Effects will
    May 1980 in Washington, D.C. Specific dates and itinerary will be determined by the
    U.S. Panel with concurrence by the Japanese Panel.

NEW INITIATIVES IN EARTHQUAKE HAZARDS MITIGATION

Charles C. Thiel

Earthquake Hazards Reduction Coordination Group

Office of Science and Technology Policy

Executive Office of the President

Washington, D.C.

ABSTRACT

Under the National Earthquake Hazards Reduction Program a number of significant steps have been taken to improve national earthquake mitigation policies and practices. The recently established Earthquake Hazards Reduction Coordination Group within the Executive Office of the President, is coordinating a number of efforts to improve building practices, land use, insurance, preparedness, emergency response and post-earthquake recovery activities. An "Interagency Committee on Seismic Safety in Construction" has been established to:

-- develop seismic design and construction standards for Federal projects;

-- develop guidelines to ensure serviceability following an earthquake of vital facilities constructed or financed by the Federal government; and

-- develop guidelines that provide independent and State and local review of seismic considerations in the construction of critical facilities constructed and financed by the Federal government, where appropriate.

In a collateral activity a "Building Seismic Safety Council" has been formed within the private sector to enhance the public safety by providing a national forum to foster improved seismic safety provisions for use by the building community. The scope of the Council's activities encompasses seismic safety of building-type structures with explicit consideration and assessment of the social, technical, administrative, political, legislative, and economic implications of its deliberations and recommendations.

NATIONAL EARTHQUAKE HAZARDS REDUCTION PROGRAM

The National Earthquake Hazards Reduction Act [1] established the purpose "to
risks of life and property from future earthquakes in the United States through the
lishment of an effective earthquake hazards reduction program." The Act specified
objectives of the program shall include:

1) The development of technologically and economically feasible design and co
methods and procedures to make new and existing structures, in areas of seismic ris
quake resistant, giving priority to the development of such methods and procedures
nuclear power generating plants, dams, hospitals, schools, public utilities, public
structures, high occupancy buildings, and other structures which are especially nee
time of disaster;

2) The implementation in all areas of high or moderate seismic risk, of a sys
(including personnel, technology, and procedures) for predicting damaging earthqual
identifying, evaluating, and accurately characterizing seismic hazards;

3) The development, publication, and promotion, in conjunction with State and
officials and professional organizations, of model codes and other means to coordi
mation about seismic risk with land-use policy decisions and building activity;

4) The development, in areas of seismic risk, of improved understanding of,
bility with respect to, earthquake-related issues, including methods of controllin
from earthquakes, planning to prevent such risks, disseminating warnings of earthq
organizing emergency services, and planning for reconstruction and redevelopment a
earthquake;

5) The education of the public, including State and local officials as to ea
phenomena, the identification of locations and structures which are especially sus
earthquake damage, ways to reduce the adverse consequences of an earthquake, and r
matters;

6) The development of research on—

   (a) Ways to increase the use of existing scientific and engineering know
   mitigate earthquake hazards;

   (b) The social, economic, legal, and political consequences of earthqual
   prediction; and

(c) Ways to assure the availability of earthquake insurance or some functional substitute; and

7) The development of basic and applied research leading to a better understanding of the control or alternation of seismic phenomena.

The Act further specified that an implementation plan be prepared to achieve these objectives. In response, the National Earthquake Hazards Reduction Program [2] (NEHRP) was prepared and transmitted by the President to the Congress. This program sets forth a wide ranging, comprehensive set of activities. Of the tasks outlined, the highest priorities for immediate action are:

-- The establishment of a focus--a lead agency--to provide national leadership and to guide and coordinate Federal activities;

-- The determination of the interest of States for the development of State and local strategies and capabilities for earthquake hazards reduction;

-- The completion of Federal, State and local contingency plans for responding to earthquake disasters in the densely populated areas of highest risk;

-- The development of seismic resistant design and construction standards for application in Federal construction and encouragement for adoption of improved seismic provisions in State and local building codes;

-- The estimation of the hazard posed to life by possible damage to existing Federal facilities from future earthquakes;

-- The maintenance of a comprehensive program of research and development for earthquake prediction and hazards mitigation.

The Earthquake Hazards Reduction Group (EHRG) was established in September 1978 as an adjunct to the Office of Science and Technology Policy, Executive Office of the President, to serve as the interim lead agency for coordination of the NEHRP. The President's Reorganization Plan No. 3 of 1978, June 19, 1978, specifies that the Federal Emergency Management Agency (FEMA) will assume lead agency responsibilities upon its activation. The balance of this paper describes several initiatives taken under the NEHRP.

INTERAGENCY COMMITTEE ON SEISMIC SAFETY IN CONSTRUCTION

The Interagency Committee on Seismic Safety in Construction was formed in October 1978. It is chaired by the head of EHRG and has participation from over 130 individuals from 17 Federal agencies and departments. The Committee has three principle objectives:

-- develop seismic design and construction standards for Federal projects;

-- develop guidelines to ensure serviceability following an earthquake of vital facilities constructed or financed by the Federal government;

-- develop guidelines that provide for independent and State and local review of seismic considerations in the construction of critical facilities constructed and financed by the Federal government, where appropriate.

The detailed work of preparing specific provisions for consideration by the Committee is underway in ten subcommittees and one task group. Since these groups are pivotal to the Federal Program, their objectives are described below:

1) Format and Notation

The Subcommittee is to develop an overall plan for the organization, arrangement, and terminology for standards for all types of Federal facilities. Specific areas of activity are to assure that scope and organization of subcommittees' reports and standards are consistent with their mission; definitions and symbols are consistent for reports and standards of the Interagency Committee; levels of technology, resulting levels of risk, and uses of reference standards are consistent; and, identify research needs, receive and synthesize research needs recommendations of other subcommittees, recommend or endorse research programs of Federal agencies to Interagency Committee.

2) Standards for Buildings

The Subcommittee is to recommend seismic design and construction standards for Federal buildings, their appurtenances and nonstructural components. Recommendations received from other subcommittees will be reviewed for incorporation into the standards developed by this Subcommittee. The assigned target date for completion of this task is 1980. After appropriate testing and evaluation by Federal construction agencies, the results will be reviewed, and the standards will be revised as necessary and presented to the full Interagency Committee for adoption. A procedure will be developed for periodically reviewing the adopted standards for the purpose of improvement.

4

3) Existing Hazardous Buildings

The Subcommittee is to develop a strategy and technique for the identification of
existing seismically hazardous buildings and to develop hazard mitigation techniques and procedures to reduce these hazards.

4) Lifelines

This Subcommittee is to identify and recommend Federal standards for seismic design,
construction and retrofit of energy, transportation, water and telecommunication systems.
The Subcommittee will study strategies for evaluating the seismic vulnerability of existing
lifelines and for improving their resistance to seismic effects and ease of repair. These
strategies will permit identification of those lifeline facilities important in the emergency, immediate recovery, and long-term economic recovery periods, and provide guidance for
appropriate levels of seismic protection for each type. The Subcommittee will establish
liaison with existing professional and industrial groups active in the seismic design. The
Subcommittee will classify lifeline components into three groups: 1) those elements which
behave seismically in a manner similar to buildings and installations already adequately
covered by codes and standards; 2) those unique facilities where existing lifeline engineering information exists to prepare a standard; and 3) those where new technology or
research will be required to provide a technically sound recommendation.

5) Risk Analysis

The Subcommittee is to recommend a suitable risk analysis methodology for policy and
technical decisions on seismic protection in Federal design and construction programs and in
capital investment decisions. This Subcommittee will recommend a risk analysis methodology
that is sufficiently general to include consideration of existing structures and new construction. Various levels of risk will also be considered. The methodology of risk analysis is to permit joint consideration of the seismic risk at a specific site with other types
of risk for the same facility such as damage from fire, floods or high winds.

6) Federal Grant, Lease and Assistance Programs

The Subcommittee is to develop strategies for implementation of the Interagency
Committee's aseismic standards for Federal programs such as those involving grants, loans,
leasing, aid, assistance or regulatory programs. The Subcommittee will deal with policy
matters necessary to enable application of direct aseismic standards to these programs,
in conjunction with State and local codes. The Subcommittee will develop recommendations

for implementation and for regulatory procedures which can be adopted by in

Proper coordination will be maintained between all related regulatory organ

Federal, State and local agencies involved in construction under Federal as

to assure resolution of conflicts prior to adoption of mandatory aseismic s

implementation procedure will be developed for periodically reviewing aseis

for the purpose of revising and upgrading the standards, and for adopting f

standards.

7) Evaluation of Site Hazards

The Subcommittee is to establish guidelines, procedures, and criteria
of seismic risk and seismically induced geologic hazards to Federally funde
regulated construction sites. Areas of concern include procedures for site
analyses, identification of special study zones, and seismic risk maps. Ha
considered include: seismicity, ground shaking, liquefaction, landslides,
and subsidence.

8) Tsunami and Flood Waves

This Subcommittee is to develop standards for the hydrodynamic aspects
construction against tsunami, seiche, and flood wave threats. The mission
opment of procedures for tsunami risk assessment and proposal of methods fo
possible hazards associated with tsunamis, seiches, and flood waves.

9) Post-Earthquake Serviceability

This Subcommittee is to develop guidelines to ensure serviceability fo
quake of vital community facilities constructed or financed by the Federal
guidelines are to be directed toward those functions within the facilities
which are vital during the post-earthquake period. These facilities inclu
installations such as hospitals, communication centers, police and firefig
systems, and computer centers and disbursement centers whose loss could ha
impact on the community.

10) Evaluation of Seismic Safety of Critical Facilities

The Subcommittee is to develop guidelines for involving Federal, Stat
mental bodies in the evaluation of seismic safety of critical facilities t

operated, regulated, or financed by the Federal government. Critical facilities are those whose seismically induced failure might increase the destructive impact and public consequences of earthquakes by accidental release of contained materials.

Task Group 1. Unregulated Federal Nuclear Facilities

The mission of this Task Group is to assess whether standards for unlicensed and unregulated Federal nuclear facilities should be developed by the Interagency Committee.

BUILDING SEISMIC SAFETY COUNCIL

A key element in the achievement of increased public earthquake safety is the continued development, evaluation and improvement of model seismic design provisions suitable for incorporation into local building codes and practices. A major impediment to achieving this objective has been that the diverse elements of the building community (professional societies, and labor, trade, model code, voluntary standards, public interest and public agency organizations) have not had a forum to develop such standards. This has been remedied through the formation, in April, 1979, of the Building Seismic Safety Council as an independent, voluntary body to enhance the public's safety by fostering improved seismic safety provisions. The Council was formed through the efforts of organizations representing all segments of the building industry, including ASCE, AIA, SEAOC, NCSBCS, CABO, NBS, PC, AGC, NIBS, ATC and EHRG. The Council's objectives are to:

a) promote the development of seismic safety provisions suitable for use throughout the United States;

b) recommend, encourage, and promote the adoption of appropriate seismic safety provisions in voluntary standards and model codes;

c) assess implementation progress by Federal, State and local regulatory and construction agencies;

d) identify opportunities for the improvement of seismic regulations and practices and encourage public and private organizations to effect such improvements;

e) promote the development of training and educational courses and materials for use by design professionals, builders, building regulatory officials, elected officials, industry representatives, other members of the building community and the public;

f) provide advice to governmental bodies on their programs of research, development, and implementation; and

g) periodically review and evaluate research findings, practice, and experience and make recommendations for incorporation into seismic design practices.

The first specific task undertaken by the Council is to extensively review the Tentative Provisions for the Development of Seismic Regulations of Buildings, [3] prepared by the Applied Technology Council under contract to the National Bureau of Standards with the National Science Foundation's support. This document contains tentative seismic design provisions for use in the development of seismic code regulations for design and construction of buildings. The provisions represent the result of a concerted effort by a multidisciplinary team of nationally recognized experts in earthquake engineering. Design professionals, researchers, Federal agency representatives, staffs from the model code organizations and representatives from State and local governments throughout the United States were involved. The provisions are comprehensive in nature and deal with earthquake resistant design of the structural system, architectural and non-structural elements and mechanical-electrical systems in buildings. Both new and existing buildings are included. They embody several new concepts which are significant departures from existing seismic design provisions. An extensive commentary documenting the basis for the provisions is included.

The Council has developed a review procedure depicted in figure 1 that is quite extensive to assure that the resulting procedures are widely considered on their technical, administrative and applications merit. The NBS has agreed to provide technical management of some of the tasks as indicated. BSSC will maintain responsibility for the overall effort as well as the tasks indicated. After completion of this review, modification of the provisions, and test by trial designs the member organizations will be encouraged to adopt the resultant provisions. It is important to note that the Council has not been formed to replace the initiatives of its member organizations, but is to provide a national forum where its members can develop seismic design provisions that are responsive to the public's interest. This is a unique undertaking in the building community for which there is no precedent within the U.S. The opportunity for different groups (e.g., public officials, owners, architects) to consider the merits of proposed provisions before they are incorporated into a code or national standard without the usual heated advisory climate should substantially improve the process of reaching consensus views which provide appropriate and effective public earthquake building regulations.

8

EARTHQUAKE RESPONSE PLANNING

From 1971 to 1976, the Federal Disaster Assistance Administration (FDAA) and its prede-
cessor agency had conducted earthquake response planning studies [4] to respond to the special needs of earthquake victims. The NEHRP has reinstated that program. Concentrated in a
limited number of heavily populated areas in highly seismic zones, it consists of two phases:
a) an estimate of casualties and physical damage to critical facilities that might occur as
a result of one or more maximum credible earthquakes in the area; and b) the development of
response plans by the regional Federal agencies--coordinated by FDAA--to provide assistance
to supplement local and State resource needs identified by companion State and local plans.

In providing new emphasis to this program, a significant shift in approach was also
made. The affected States have been asked to assume a leadership role in the preparation of
both damage estimates and State and local plans, rather than just the latter. Considerable
effort will now be expended in marshalling the support of local jurisdictions and identifying
more specifically the resources that State and local organizations will provide--so that any
shortfall potentially can be filled by supplemental Federal assistance.

Tables 1 and 2 summarize the programs progress to date and expected future activities.
FY 79 activities have concentrated on starting a loss study for the Anchorage, Alaska area
and communities on Prince William Sound (containing critical facilities), under the leadership of the Alaska State Emergency Services. The initial phase of the effort, stressing the
geotechnical aspects and the encouragement of participation of local jurisdictions is
expected to be completed in early 1980. A somewhat similar effort, directed at planning for
coping with an earthquake and tsunami that might affect Honolulu, Hawaii, is expected to be
started this summer, with active participation of the State.

EARTHQUAKES AND FINANCIAL INSTITUTIONS

A study of earthquakes and their impacts on financial institutions has been undertaken
with three related objectives by EHRG with the cooperation of affected Federal agencies:

1) The viability of the financial system in the aftermath of a truly catastrophic
earthquake in a densely populated area. The system is known to have resilience, but the
degree of resilience in such an eventuality has never been examined.

9

2) The effects of a credible earthquake prediction upon the behavior of the public a
its consequent demands on the financial system. Preliminary indications point to some pos
bly heavy burdens to be placed on the system, particularly if the prediction time window i
of considerable length (i.e., more than six months).

3) The feasibility of effecting mitigation of the earthquake hazards through the fin
cial institutions, without causing undue dislocations in the process. As in the case of f
hazards--where financial institutions have played a significant role in encouraging better
codes and prevention programs--it may be possible to identify non-disruptive mechanisms to
cope with earthquake hazards reduction.

There are many kinds of financial institutions, both private and public. The private
institutions are financial intermediaries; the public ones, regulatory and subsidizing age
cies as well as financial intermediaries. For private institutions, especially deposit-ty
institutions, earthquakes exacerbate three ordinary financial concerns--liquidity, solvenc
and records maintenance and replication. These concerns impact on the public institutions
whose roles require careful analysis to determine what public policy problems, if any, exi
and what action may be indicated. The study is expected to be completed by the end of 197
with policy recommendations by March 1980.

EARTHQUAKE PREDICTION EVALUATION COUNCIL

The Council has been formed, as specified in the National Earthquake Hazards Reductic
Program. It is made up of members appointed by the Director of United States Geological
Survey. The Council is to advise the Director on issuing predictions as to the completeness and scientific validity of the available data. In evaluating predictions, the Counc:
objectives are:

-- to recommend to the appropriate scientists any actions that might be desirable
   or required to clarify or verify the basis for a prediction;

-- to maintain an accurate record of predictions evaluated and evidence pertinent
   to them; and

-- to provide the Director a timely and concisely written review of the evidence
   relevant to a prediction of any potentially damaging earthquake and a written

10

recommendation as to whether the evidence is sufficiently clear that an official prediction by the Director should be issued or, if not, what if any other
official position the Director should take.

As of this date no predictions have been submitted for their evaluation.

RESEARCH PROGRAM

FY 1979 is the second year of substantial research under the National Earthquake
Hazards Reduction Program. PL 95-124 and the report Earthquake Prediction and Hazard
Mitigation Option for USGS and NSF Programs, [5] 1979 (Newmark Report) provide the basis for
these agencies programs. The main elements of the program paraphrased from the Newmark
Report are:

I   Fundamental studies--research into the basic causes and mechanisms of earthquakes.

II   Prediction--forecasting the time, place, magnitude and effects of an earthquake.

III Induced Seismicity--prevention or modification of an inadvertently induced or
     natural earthquake.

IV   Hazard Assessment--identification and analysis of the potential for earthquakes within
     a region, their frequency and their effects.

V   Engineering--design and construction of structures for acceptable performance during
     and after an earthquake.

VI   Policy Research--impacts of earthquakes on the community and options for dealing with
     them.

EHRG is currently leading an interagency effort among NSF, USGS and NBS to develop
program options for Fiscal years 1981 through 1984. It is expected that FEMA once it is
fully operational will assume an active role in supporting research, particularly on related
natural disasters, which will strengthen the earthquake effort.

REFERENCES

[1] U.S. Congress, Public Law 95-124, October 7, 1977. Also printed as Appendix I of
     Earthquake Hazards Reduction: Issues for an Implementation Plan, (Office of Science
     and Technology Policy, Executive Office of the President, Washington, D.C., 1978).

[2] National Earthquake Hazards Reduction Program (Office of Science and Technology Policy,
     Executive Office of the President, June 22, 1978).

11

[3] Applied Technology Council, _Tentative Provisions for the Development of Seismic Regulations for Buildings_, U.S. Government Printing Office, 1978, Number 003-003-01939-9.

[4] Department of Commerce, "A Study of Earthquake Losses in the Los Angeles, California, Area," U.S. Government Printing Office, 1973, Number 0319-00026.

Department of Commerce, "A Study of Earthquake Losses in the San Francisco Bay Area," U.S. Government Printing Office, 1972, Number 4101-00011.

Department of Interior, "A Study of Earthquake Losses in the Puget Sound, Washington, Area," U.S. Government Printing Office, 1975, Number 681-256/29.

Department of Interior, "A Study of Earthquake Losses in the Salt Lake City, Utah, Area," U.S. Government Printing Office, 1975, Number 677-530/292.

[5] National Science Foundation and Department of the Interior, _Earthquake Prediction and Hazard Mitigation Options for USGS and NSF Programs_, U.S. Government Printing Office, 1976, Number 038-000-00332-1.

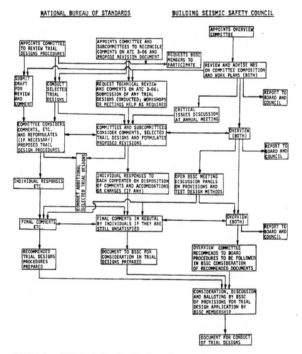

FIGURE 1: BSSC Workplan for Review of Seismic Design Provisions

TABLE 1

EARTHQUAKE LOSS STUDIES AND CONTINGENCY PLANS:
PROGRAM PLAN

Legend: ▓▓▓ Loss Studies (risk analyses, damage estimates); ▧▧▧ Response Plans (Federal, State, local)

Region/State	Population Area	1971-76	1979	1980	Fiscal Years 1981	1982	1983	1984
9 California	San Francisco Bay	▓▧▧	▧▧▧▧		Loss study completed 1972, followed by F-S-L plng; Fed plng resumed 1978			
9 California	Los Angeles Area	▓▧		▧▧▧	Study completed 1973 followed by S&L plng; Fed plan to follow SF plan			
10 Washington	Puget Sound Area	Study 1975 ▓▧	▧▧ Federal Plan 12/78		▧▧▧▧			
8 Utah	Salt Lake City Area	Study 1976 ▓	▧▧▧		Federal Plan 8/78, State Emergency Plan 7/77 Local plans 1977-79 under FDAA preparedness gnt			
10 Alaska	Anchorage Area			▓▓▓	▧▧▧▧			
9 Hawaii	Honolulu Area			▓▓▓	▧▧▧▧			
4 S. Carolina	Charleston Area			▓▓▓	▧▧▧▧▧		Order of these may be reversed	
1 Massachusetts	Boston Area					▧▧▧		
*Central U.S.	Mississippi Valley			FY 80 Project Design	▓▓▓	▧▧▧▧	▧▧▧▧	▧▧▧▧
2 New York**	Rochester-Seaway				▓▓▓	▧▧▧		

*At least Regions 4, 5, 6, and 7 (Kentucky, Tennessee, Mississippi, Illinois, Indiana, Missouri and Arkansas are in the Zone 3--major damage--area; major Zone-3 city: Memphis; Zone 2 includes St. Louis, Cincinnati)

**Zone-3 extends from extreme western New York to northern tip of Maine, includes Buffalo and Montreal

TABLE 2: EARTHQUAKE LOSS STUDIES AND CONTINGENCY PLANS: STATUS

	Loss Study	Federal Response	State Response	Local Response	Functional Study
San Francisco	X	X	X	X	
Los Angeles	X	FY 80	X	X	Temporary Housing Underwa
Salt Lake	X	X	X	X	
Puget Sound	X	X	FY 81		
Anchorage	FY 79	FY 80	FY 80		Winter Housing Underway
Honolulu	FY 79	FY 81	FY 80		
Central U.S.*	FY 80	FY 82	FY 83		
Charlestown S. Carolina**	FY 80	FY 81	FY 81		
Boston**	FY 81	FY 82	FY 82		
Upstate New York	FY 81	FY 82	FY 82		

* -- To be preceded by a project design effort in FY 80.   X --
**-- Order of these two items may be reversed

	USGS	NSF	Total	Newmark Repo Recommendati Option B	Opt
1. Fundamental Studies	2.9	6.0	8.9	9.1	
2. Earthquake Prediction	15.9		15.9	17.9	
3. Induced Seismicity	1.2		1.2	2.0	
4. Hazards Assessment	11.1		11.1	14.8	1
5. Engineering		16.6	16.6	20.3	1
6. Research for Utilization		0.8	0.8	6.1	
TOTAL	31.1	23.4	54.5	70.2	5

TABLE 3: FY 79 Research Expenditures (Amounts in Millions of Dollars).

EARTHQUAKE HAZARD REDUCTION RESEARCH AT

THE NATIONAL BUREAU OF STANDARDS

Edgar V. Leyendecker

James R. Harris

Richard N. Wright

Edward O. Pfrang

Center for Building Technology

National Bureau of Standards

U.S. Department of Commerce

ABSTRACT

Current and planned Earthquake Hazard Reduction programs at the National Bureau of Standards in Research and Standards Development are being conducted in order to achieve the responsibilities assigned to NBS under the President's National Earthquake Hazards Reduction Program. These responsibilities, to:

1) provide technical support to the building community in the development of seismic design and construction provisions for building codes and national standards

2) provide technical support to the Federal agencies in development of seismic design and construction provisions for Federal programs and

3) perform research on performance criteria and supporting measurement technology for earthquake-resistant construction,

are being carried out in cooperation with the Federal and private sectors. This research is complementary to that of the National Science Foundation (NSF) and U.S. Geological Survey (USGS).

INTRODUCTION

The National Bureau of Standards (NBS) Earthquake Hazards Reduction Prog
the national effort to reduce the risks to life and property from future eart
United States. Under the President's National Earthquake Hazards Reduction P
issued June 22, 1978, to implement the Earthquake Hazards Reduction Act of 19
been assigned the following continuing responsibilities:

° provide technical support to the building community, including the Fe

   agencies involved in construction, "in continuing the development,

   and improvement of model seismic design and construction provisions

   for incorporation in local codes, standards (including Federal stan

   practices."

° perform "research on performance criteria and supporting measurement

   for earthquake resistant construction."

NBS has participated in the planning of Federal programs in earthquake h
since 1971 and played a major role in providing input which was used in draft
Earthquake Hazards Reduction Program.

OBJECTIVES

In order to meet its responsibilities under the President's National Ea
Reduction Program, NBS is concentrating its efforts in (1) Research for Stand
(2) Standards Development. The objective of these areas are:

Research for Standards - Improve building design and construction pract

   conducting fundamental research necessary to establish consistent sta

   include reliable measures of the performance of buildings and their e

   during an earthquake.

Standards Development - Provide technical support in the "development o

   design and construction standards for consideration and subsequent ap

   in Federal Construction, and encouragement for the adoption of improv

   provisions in State and local building codes."

---

* Numbers in brackets indicate references.

16

Both of these objectives are consistent with past NBS efforts in the development of technical bases for standards and on measurement technology and with the interagency effort among NSF, USGS, and NBS to develop program options for Fiscal years 1981-1984 [3].

RESEARCH FOR STANDARDS

A research program at NBS to develop technical bases for performance criteria and the supporting measurement technology is a new element of the national earthquake research program. The research is complementary to the U.S. Geological Survey (USGS) and the National Science Foundation (NSF).

Earth sciences research led by the USGS and NSF provides greatly improved understanding of ground shaking and displacement. To be suitable for use in building design, the knowledge must be further developed to loading criteria for the various types of buildings and components. Structural, geotechnical, mechanical, architectural and economic studies supported by NSF have provided substantial advances in abilities to model and predict the earthquake responses of various built elements and their consequences. This knowledge must be synthesized to develop design criteria that will provide consistent safety among the various built elements and various materials or technologies for each. Evaluation and measurement methods consistent with practices of design, manufacture, construction and inspection, must be developed to implement the design criteria.

The NBS research program is essential to the NBS standards development role and will complement the research conducted or funded by USGS and NSF and contribute substantially to the implementation of the latter research. Also the NBS research program provides for an enhanced Federal expertise in the area of earthquake engineering research. This expertise is vital for effective Federal participation with the building community and during the establishment of public policy.

Research for standards are divided into four tasks:

Task 1 — Research for Earthquake Loading Standards —

Develop methods for use in standards to characterize ground motion and the acceleration, velocity, and displacement time-histories of this motion for use in engineering analysis, planning and design. The effect of the earthquake source, transmission path, amplification caused by local site conditions, and soil-structure interaction will be considered.

17

Task 2 - Research for New Building Standards -

Develop methods for use in standards to characterize the performance of buildings in four areas:

i) Develop laboratory and in-situ techniques for characterizing the dynamic properties of soils, develop analytical methods to evaluate soil failures, and investigate the design of various types of foundations in order to develop criteria for selection and design of appropriate foundations.

ii) Develop procedures for characterizing the dynamic behavior of structures and components up to the ultimate limit state and provide a basis for the formulation and validation of nonlinear and inelastic methods of analysis and design.

iii) Develop methods to characterize the response of non-structural systems to earthquake loadings and develop test methods for evaluating the characteristics. Methods to obtain loadings for use in evaluating non-structural systems as a transfer from the structural and site response information will be developed. Critical response parameters will be formulated and test methods will be developed for evaluating the characteristics of non-structural systems and components such as machines, luminaires, piping, elevators, etc. System design conflict arising from non-structural response (secondary hazards, such as fire) will also be examined.

iv) Develop the basis for performance criteria for occupant safety and building functional requirements during and after earthquakes by relating casualties to both structural and non-structural response to earthquake motion and by determining functionality requirements both during and after earthquakes for various types of building uses.

Task 3 - Research for Existing Building Standards -

Develop nondestructive methods to determine the properties of older buildings and components in place, for use in determining realistic values of structural parameters and evaluation of the strength of existing buildings and repairs. Analytical and laboratory studies on components will be followed by field studies on whole buildings, taking advantage of buildings scheduled for demolition where possible.

Task 4 - Research for Assessment of Standards -

Develop methods for assessing the impact of standards for improved seismic resistance. Methodologies will be developed for assessing the benefits of earthquake hazard reduction

18

and defining risk for both building and community scales. The influence of building type and configuration on serviceability and costs will be considered. Costs and benefits of improved practices for seismic resistance associated with planning, design, construction and regulation will be included. Methods for assessing the clarity, consistency, and completeness of standards will also be improved.

Research in each of these tasks is planned to begin in Fiscal year 1981, depending on approved funding. At the current level of funding, research is concentrating on Task 2 in the development of procedures for characterizing the dynamic behavior of structures, particularly for masonry construction.

STANDARDS DEVELOPMENT

A program for the development of standards for buildings is an important element for the implementation of the substantial research programs at USGS and NSF. Criteria for the earthquake-resistant design of new construction used in many current Federal, State, and local building codes, standards and practices do not reflect the current state of the art and should be updated. These codes and the standards and professional practices underlying them should not only represent our best knowledge, but be adaptable to different areas of the United States according to differing seismic risks and the costs and benefits they entail.

Under the leadership of the Office of Science and Technology Policy, an Interagency Committee on Seismic Safety in Construction was organized in 1978. All Federal agencies involved in construction, working through the Interagency Committee on Seismic Safety in Construction, will develop seismic design standards for Federal building construction. NBS is providing technical support to several of the subcommittees of the Interagency Committee in the development of unified Federal standards, including chairing the subcommittee on Building Standards [3].

The vast majority of the construction in the United States is undertaken by the private sector and regulated by local government. To assist State and local governments, industry, and the public in developing construction standards, criteria, and practices, the National Bureau of Standards will work with other Federal agencies, the National Institute of Building Sciences, professional organizations, model code groups, State and local building departments, and the newly formed Building Seismic Safety Council. The Bureau will assist and cooperate

19

with these groups in continuing the development, evaluation, and improvement of model design provisions suitable for incorporation into local codes and practices. Incorp(of these seismic design provisions into local codes is, of course, voluntary, but the sions must be flexible and give consideration to costs and benefits, regional variati seismic hazard, and adaptation to local conditions. They must also be adequately tes

The Building Seismic Safety Council formed in April 1979, when more than fifty | sional societies, trade associations, model code organizations and public interest o1 tions met under the auspices of the National Institute of Building Sciences. The Cou provides a national forum to foster improved seismic provisions for use by the buildi community. NBS will conduct technical studies in support of the Council. In turn, 1 Council will base its recommendations for provisions in building standards and codes these studies.

Standards Development is divided into three tasks:

Task 1 — Seismic Design Provisions for Consensus Standards and Model Codes —

Provide technical studies and draft provisions to assist the building community the development and implementation of seismic design provisions in State and local b codes. This will be accomplished by working through the existing systems of volunta dards and model building codes, making appropriate use of the assessments of the Fed building standards in Task 2.

Task 2 — Seismic Design Standards for Federal Building Construction —

Develop seismic design standards for Federal building construction, test and as safety and economic consequences of their use, improve the standards following asses implement in Federal programs.

Task 3 — Standards for Evaluation and Strengthening of Hazardous Existing Buil

Provide technical assistance in the development of methods for the evaluation safety of existing buildings and in the development of practices and criteria for s ing existing buildings or removing them from service.

Research in each of these tasks is planned to begin in Fiscal year 1981, depen approved funding. At the current level of funding NBS is working with the Interage mittee on Seismic Safety in Construction to develop uniform Federal criteria.

20

Efforts are now under way to assess the "Tentative Provisions for the Development of Seismic Regulations for Buildings [4]," proposed by the Applied Technology Council under contract to the NBS with NSF funding. This effort will be in cooperation with the Building Seismic Safety Council, and will involve both the Federal and private sectors.

CONCLUSION

The role of the NBS under the National Earthquake Hazards Reduction Program includes the responsibility to:

1) provide technical support to the building community in the development of seismic design and construction provisions for building codes and national standards,

2) provide technical support to the Federal agencies in development of seismic design and construction provisions for Federal programs, and

3) perform research on performance criteria and supporting measurement technology for earthquake-resistant construction.

These responsibilities are being met by conducting (1) Research for Standards and (2) Standards Development. Current efforts are concentrating on (1) developing procedures for characterizing the dynamic behavior of structures, particularly masonry construction, and (2) working with the Interagency Committee on Seismic Safety in Construction to develop uniform Federal criteria and with the Building Seismic Safety Council to assess the newly prepared Tentative Provisions for the Development of Seismic Regulations for Buildings.

Current and future plans are consistent with the interagency effort among NSF, USGS, and NBS to develop program options for Fiscal years 1981-1984.

REFERENCES

[1] National Earthquake Hazards Reduction Program, Office of Science and Technology Policy, Executive Office of the President, June 22, 1978.

[2] U.S. Congress, Public Law 95-124, October 7, 1977, Also printed as Appendix I of Earthquake Hazards Reduction: Issues for an Implementation Plan, (Office of Science Technology Policy, Executive Office of the President, Washington, D.C., 1978.

[3] Thiel, C. C., New Initiatives in Earthquake Hazards Mitigation, 11th Joint UJNR Conference, 1979.

[4] Applied Technology Council, Tentative Provisions for the Development of Seismic Regulations for Buildings, SP-510, U.S. Government Printing Office, Washington, D.C., 1978.

OUTLINE OF RESEARCH ON WIND AND EARTHQUAKE EFFECTS IN JAPAN

Yoshijiro Sakagami

Public Works Research Institute

Note: Though Mr. Sakagami presented the research outline
        as cited above, no written version could be obtained
        for inclusion in these proceedings.

ON THE U.S.-JAPAN COOPERATIVE PROGRAM OF LARGE-SCALE TESTING

Tadayoshi Okubo

Public Works Research Institute

Ministry of Construction

1. INTRODUCTION

The Implementing Arrangement between the Science and Technology Agency (STA) and the Ministry of Construction (MOC) of Japan and the National Science Foundation (NSF) of the U.S.A. for cooperation in the U.S.-Japan Joint Earthquake Research Program involving Large-Scale Testing, under the auspices of the Panel on Wind and Seismic Effects of the UJNR was signed by Mr. Nobuo Kozu (STA), Mr. Hidenobu Takahide (MOC) and Dr. Jack T. Sanderson (NSF) on August 10, 1979. (Appendix I).

The U.S.-Japan Cooperative Program on Large-Scale Testing was initiated in 1974 by this Panel and has begun implementation. In this presentation, the author outlines the Program.

2. BACKGROUND

At the Sixth Joint Meeting of the UJNR Panel on Wind and Seismic Effects, held in Washington, D.C. May 15-17, 1974, the necessity of a joint research program which includes large scale testing of structures was emphasized and the Panel produced the following resolution:

"3) that increased effort should be made in the near future to encourage joint research programs, especially in the area of the mutual utilization of research facilities and the exchange of researchers."

At the Seventh Joint Meeting of our Panel held in Tokyo, May 20-23, 1975, a special session, "Theme IX: Joint Research Program Utilizing Large-Scale Testing Facilities" was held and the exchange of information and opinion was carried out with two papers submitted to the session: "A Proposal to the Japan - U.S. Joint Earthquake Engineering Research Project Utilizing Large-Scale Testing Facilities," by M. Watabe, M. Hirosawa, and S. Nakata, and "Earthquake Disaster Mitigation: A Joint Research Approach," by Charles T. Thiel. The Panel concluded with the following resolution:

"7) Cooperative research programs including exchange of personnel and equipment should be undertaken by both governments to address the following problems of mutual interest;

23

a)   Strong motion instrumentation arrays, at selected sites throughout

b)   Large-scale testing programs,

c)   Repair and retrofit of existing structures, i.e., buildings, bridg

d)   Structural performance evaluation,

e)   Land use programs for controlling natural hazard effects,

f)   Disaster prevention methods for lifeline systems."

Based upon the above resolution, six Task Committees were established Eighth Joint Meeting of the Panel and the Task Committees energetically sta ities.

After the 1968 Tokachi-oki earthquake in Japan, a U.S.-Japan Cooperati on Earthquake Engineering which emphasized the Safety of School Buildings w under the U.S.-Japan Cooperative Science Program for May 1973-October 1975. meeting of the Cooperative Research Program, held in Hawaii, August 18-20, set of recommendations was drafted and signed by all official participants

"Recommendation No. 2.   Establish a Cooperative Research Program in Ea ing with Emphasis on Large-Scale Testing of Structural Systems under the Sp U.S.-Japan Panel on Wind and Seismic Effects, UJNR Program.   This program w on controlled dynamic testing of full-scale buildings in the field and labo of large-scale building systems using the available specially designed test associated reaction walls and using a large-sized shaking table.   These tes designed to provide needed information on force-deformation, energy absorpt characteristics of such systems.   Correlative and analytical studies should improve mathematical modeling and computer-oriented dynamic analysis capabi

Recommendation No. 3.   Establish a Task Committee under the U.S.-Japan Seismic Effects, UJNR Program.   The assignment given to this Task Committee detailed plans and recommendations for implementing Recommendation No. 2 wh effective research program of maximum benefit to both countries within the sponsoring government agencies.

In view of the time scale for constructing the presently planned test the major effort required to plan the program, the Task Committee should be U.S.-Japan Panel on Wind and Seismic Effects at the earliest possible date. one Japanese should be appointed as co-chairman to head this Task Committee

24

At the Eighth Joint Meeting of our Panel, held in Washington, D.C., May 18-21, 1976, the Task Committee on Large-Scale Testing Program concluded the following.

"1) Planning activities should be initiated through the combined efforts of U.S. Federal Government and the Government of Japan for programs in large-scale testing of structures under the administration of the United States-Japan Panel on Wind and Seismic Effects.

2) Collaborative programs should be initiated in the areas of wind and seismic effects so that technological advances can be utilized to best advantage. Excellent large-scale test facilities in operation or proposed in Japan will provide a vehicle for advancement in technology. These large-scale test facilities should be combined with technological advances in instrumentation, data acquisition and analysis, and measurements on structures in service so that solutions of programs of mutual interest will be optimized.

3) The Task Committee exchanges the list of plans, including organization equipment and status, on wind and seismic effects concerning these large-scale testing programs. It is anticipated that the Task Committee can update the list periodically using inputs provided by members of the U.S.-Japan Panel and individuals aware of large-scale testing programs underway."

In succession, the Eighth Joint Meeting of the Panel adopted following resolution:

"2) The resolutions of the Seventh Joint Panel Meeting are important to the continued cooperative program and are hereby incorporated as a part of these resolutions:

e) Promote cooperative research programs through the Task Committees including exchange of personnel and available equipment. Publish reports of these Task Committees."

Meanwhile, based on a proposal, "Planning a Cooperative Research Program in Earthquake Engineering with Emphasis on Large-Scale Testing of Structural Systems," submitted by the University of California-Berkeley to the NSF which called for planning effort to be conducted under the auspices of the U.S.-Japan Panel on Wind and Seismic Effects, the Planning Group was established and an adjustment of activities between our Panel and the Group was necessitated. (The activity of the Planning Group is introduced later.)

Therefore, an unofficial meeting for the adjustment of the Panel and the Group activities was held on May 23, 1977, the day before the Ninth Joint Meeting of the Panel. Professor H. Umemura, Japanese co-chairman of the Planning-Group, Dr. E. O. Pfrang, U.S. co-chairman of the Panel, Mr. K. Ohtani, Dr. M. Watabe and the author participated in this meeting and concurred on the following.

25

"1) The cooperative research program on large-scale testing should be carried out under the auspices of the UJNR Panel on Wind and Seismic Effects.

2) UJNR members of the Task Committee on Large-Scale Testing Programs should officially participate with the Planning Group for Large-Scale Testing.

3) The final report to be prepared by the Planning Group should be submitted to the UJNR Panel for review and suggestions relative to implementation of the cooperative test program."

Following this unofficial meeting, the Task Committee on Large-Scale Testing Programs at the Ninth Joint Meeting of the Panel in Tokyo, May 24-27, 1977, concluded as follows:

"2) Future Programs

1) Official participation of UJNR members on the U.S.-Japan Planning Group for large-scale testing chaired by Dr. J. Pengien and Dr. H. Umemura shall be carried out.

2) The Planning Group and Task Committee shall coordinate their efforts.

3) The first joint meeting of the Planning Group with Task Committees is expected to be held in the fall of 1977, in Japan.

4) The final report to be prepared by the Planning Group should be submitted to the UJNR Panel for review and suggestions relative to implementation of the cooperative test program.

5) Japanese Committee members will endeavor to complete arrangements, in the near future, for the official acceptance of the U.S.-Japan cooperative research programs and participants."

The Ninth Joint Meeting of our Panel endorsed the conclusion of the Task Committee by adopting the following resolution:

"4) Considerable advancement in various Task Committees was observed. These activities should be encouraged and continued. Correspondence between Task Committee chairmen should be encouraged, and additional meetings should be held as required."

At the Tenth Joint Meeting of our Panel, held in Washington, D.C., May 23-26, 1978, summaries of the first and second Joint Meeting of the Planning Group were reported to the Task Committee Meeting and the following was agreed on at the Task Committee Meeting:

"The Task Committee on Large-Scale Testing Programs should keep the UJNR Panel on Wind and Seismic Effects informed of developments in large-scale testing programs. Programs proposed by the current planning group should be circulated to UJNR for review and comment with appropriate recommendations returned to the planning group for consideration."

Also, the Tenth Joint Meeting of our Panel drafted the following resolutions.

"3. The Panel on Wind and Seismic Effects recognizes the importance of the U.S.-Japan Cooperative Program on Large-Scale Testing and it urges early implementation of the program under the auspices of this Panel.

4. Considerable advancement has been made by various Task Committees. Cooperative research programs should be promoted through the Task Committees."

The Planning Group has held four meetings: Tokyo, Japan, September 5-10, 1977; San Francisco, California, May 15-19, 1978; Tokyo, Japan, December 18-23, 1978; and Berkeley-California, July 9-14, 1979.

At the earlier meetings, six types of buildings -- reinforced concrete, steel frame, precast and/or prestressed concrete, steel-reinforced concrete, masonry, and wooden -- were discussed. Reinforced concrete and steel frame buildings, and the pseudo-dynamic test method were selected for cooperative research. The summaries and resolutions of the four meetings are shown in Appendix II.

3. COOPERATIVE RESEARCH PROGRAM RECOMMENDED BY THE PLANNING GROUP

The final meeting of the fourth meeting of the Planning Group recommended three research programs. Details of these research programs are shown in Appendix III.

Recommended research on reinforced concrete building structures consists of a full-scale seven-story reinforced building structure to be tested in the Large-size Structure Laboratory, Building Research Institute, Ministry of Construction, and associated tests to be conducted in Japan and the U.S. on the test specimens of element assemblies, structural walls, planer frames, wall-frame assemblies and scaled models of the full-scale seven-story structures.

Recommended research on steel building structures consists of a full-scale seven-story steel building test and complementary tests on structural components, beam-and-column assemblages, braced and unbraced frame bents and small- and medium-scale models of the prototype structure.

Recommended research on the pseudo-dyamic test method consists of verification

sensitivity studies of the numerical algorithm, verification of accuracy of hydrauli

tor control system and verification of adequacy of basic differential equation of mo

4. IMPLEMENTING ARRANGEMENT

At the occasion of the third meeting of the Planning Group held in December, 19

first draft of the implementing arrangement (at that time it was called a "Memorandu

Understanding") was officially proposed by the U.S. side.

After surveying the first draft, the Japanese side made up a counter proposal (

draft) and sent it to the U.S. side in May, 1979.

On June 19, 1979, members concerned of the Ministry of Construction and the Sci

Technology Agency of Japan and the National Science Foundation of the U.S. held a me

Tokyo. Mr. J. Bloom, the Science Counselor of the U.S. Embassy, also attended at th

ing. The third draft (the official title was "Implementation Protocol") proposed by

side was discussed, however complete concurrence could not be reached.

During and after the fourth meeting of the Planning Group in July 1979, negotia

the third draft was being continued and during the negotiation the draft title was c

"Implementing Arrangement."

Finally in early August, 1979, a draft agreed to by both sides was codified as

Appendix I and was signed on August 10. However, there are many detailed items whic

be agreed to by both sides in the "exchange of letters."

APPENDIX I

IMPLEMENTING ARRANGEMENT
BETWEEN THE
SCIENCE AND TECHNOLOGY AGENCY AND
THE
MINISTRY OF CONSTRUCTION OF JAPAN
AND THE
NATIONAL SCIENCE FOUNDATION OF THE
UNITED STATES OF AMERICA
FOR COOPERATION IN THE U.S.-JAPAN JOINT EARTHQUAKE
RESEARCH PROGRAM INVOLVING LARGE-SCALE TESTING,
UNDER THE AUSPICES OF THE PANEL ON WIND AND SEISMIC
EFFECTS OF THE U.S.-JAPAN NATURAL RESOURCES
DEVELOPMENT PROGRAM (UJNR)

The Science and Technology Agency (STA) and the Ministry of Construction (MOC)

and the National Science Foundation (NSF) of the United States agree to cooperate in implementation of the joint earthquake research program involving large-scale testing (hereinafter referred to as the "Joint Program"), to the extent possible within the existing laws and regulations of the respective countries, under the conditions as set forth below:

1) The purpose of this Joint Program is to improve scientific knowledge and engineering practices in the earthquake-resistant design of structures by conducting cooperating earthquake research involving full-scale building structural tests in Japan, and their components and assembly tests in both countries on the basis of equity and mutual benefit.

2) MOC and NSF agree to serve as the principal points of contact between the Japanese side and the U.S. side and to provide the required mutual liaison on matters of implementation of the Joint Research Program.

3) STA and MOC, the Japanese Party, and NSF, the U.S. Party, (hereinafter referred to as "The Parties") agree that the Joint Program shall be implemented with the endorsement of the UJNR Panel on Wind and Seismic Effects. The Parties may establish, through separate exchange of letters, whenever necessary, appropriate bilateral implementation procedures.

4) The Parties agree that the research tasks to be supported under this Implementing Arrangement shall include: (1) tests of full-scale reinforced concrete building structures and their components and assemblies; and (2) study of pseudo-dynamic test procedures.

5) To assure the success of this Joint Program, the Parties agree to provide access to the Program by all U.S. and Japanese research institutions to the maximum extent possible, in accordance with their standard procedures and with the endorsement of the UJNR Panel on Wind and Seismic Effects. The participating institutions may include, but are not limited to, in Japan, the National Research Center for Disaster Prevention, the Building Research Institute, the Public Works Research Institute, and universities; in the United States, the National Bureau of Standards, U.S. universities, industrial organizations and other governmental or private laboratories or agencies. Participant institutions shall make their facilities, including the Tsukuba Test Facility of MOC, available for use in the Joint Program where such availability and the terms of use are approved by an appropriate authority.

6) The Parties shall make every effort to secure funds necessary to implement the Joint Program. Any obligations under this Implementing Arrangement are subject to the availability of funds to the Parties. The Parties retain the executive and administrative authority and responsibilities for activities under this Joint Program in accordance with the agencies' established policies and normal procedures. The Parties shall also consult

with each other to make necessary adjustments of annual implementation plans in accordance with the laws and procedures of both countries. It is anticipated that the parties shall share support for the total program on an approximately equal basis.

7) The Parties agree that the results of the test program shall be made available to the UJNR Secretariat of each country and to the scientific community through the normal publication channels in each country and in accordance with the normal procedures of each agency.

8) The Parties shall, as mutually agreed, conduct as annual project review, including an assessment of scientific and technical achievements for the past year and plans for the coming year. Additional reviews may be held at the request of either party to address the terms and conditions of this Implementing Arrangement and other matters of mutual interest.

9) Details which are not prescribed in this Implementing Arrangement shall be decided between the Parties.

10) Modifications of this Implementing Arrangement may be made by consultation between the Parties.

11) This Implementing Arrangement will take effect on August 10, 1979, and may be terminated by either party giving the other party advance written notice of at least six months. Such termination shall not affect projects underway at the time of termination. Unless so terminated, this Implementing Arrangement shall be for a term of two years.

FOR JAPAN: FOR THE UNITED STATES OF AMERICA:

_____   _____
Nobuo Kozu   Jack T. Sanderson
Deputy Director-General   Assistant Director for
Research Coordination Bureau   Engineering and Applied Science
Science and Technology Agency   National Science Foundation

_____
Hidenobu Takahide
Counsellor for Engineering Affairs
Minister's Secretariat
Ministry of Construction

APPENDIX II

A. First Planning Group Meeting

The first planning group meeting, U.S.-Japan Cooperative Research Program Utilizing Large-Scale Testing Facilities, was held in Tokyo during the period September 5-10, 1977, to discuss future research programs in earthquake engineering utilizing large-scale testing facilities.

During this meeting, the individuals from academic and government institutions and from private industrial groups exchanged information and views on the following topics:

1) Locations and performance characteristics of existing medium- and large-scale testing facilities in both countries with example tests illustrating capabilities.

2) Locations and performance characteristics of medium- and large-scale testing facilities under construction or planned for future construction in both countries with planned programs.

3) Possible large-scale tests which might be conducted under the recommended cooperative research program including estimates of efforts involved.

4) Implementation of the recommended cooperative research program and level of effort possible considering funding procedures, restrictions, and limitations in both countries, and

5) Selection of large-scale tests and supporting smaller-scale tests to be conducted under the recommended cooperative research program.

At the conclusion of the first planning group meeting, the following resolutions were adopted:

1) The first major test to be carried out under the recommended cooperative research program should be conducted on a full-size multi-story building to determine its seismic behavior utilizing the Large-Size Structures Laboratory, Building Research Institute, Tsukuba New Town for Research and Education. It is intended that the number of stories on the test structure be the maximum permitted by the facility (8-10 stories) unless the type of construction or building practice requires a lower number of stories.

2) Six different types of buildings representative of good current practice should be considered for possible testing, namely, a reinforced concrete building, precast and/or pre-stressed concrete building, a steel frame building, a steel-reinforced concrete building, a masonry building, and a wooden building.

31

3) A preliminary design of each building type should be made prior to the second planning group meeting so that a choice can be made at that time as to which type will be tested first.

4) The building to be tested first should include non-structural elements such as curtain walls, partition walls, piping, etc.

5) The building should be tested by a pseudo-dynamic procedure so as to develop and define realistic seismic behavior.

6) After initial testing, the building should be repaired and retested to assess the effectiveness of repair procedures.

7) Complementary tests, such as component and model tests, should be conducted in various institutions prior to and/or parallel with each major full-scale test.

8) Preliminary long-range plans should be made for a possible full-scale test on the same building type selected for the first major test utilizing the super large-scale two-dimensional shaking table.

9) Long-range plans should include the possibility of testing large-scale structures other than buildings and experiments to study the effect of wind on buildings and civil engineering structures.

10) Concerned government agencies in both countries are urged to complete arrangements for implementing the preceding resolutions. It is recommended that the U.S.-Japan Cooperative Science Program and the U.S.-Japan Panel on Wind and Seismic Effects UJNR Program, coordinate their efforts towards implementing the large-scale testing program.

11) The second planning group meeting, U.S.-Japan Cooperative Research Program Utilizing Large-Scale Testing Facilities, should be held during the period May 15-19, 1978, in San Franscisco, California, USA.

B. Second Planning Group Meeting

The second planning group meeting, U.S.-Japan Cooperative Research Program Utilizing Large-Scale Testing Facilities, was held in San Francisco during the period May 15-19, 1978, to continue planning for future research programs in earthquake engineering utilizing large-scale testing facilities.

After opening remarks by J. Penzien and M. Watabe, preliminary design considerations of full-scale and support tests were presented by H. Aoyama and G. Corley, A. Mattock and

32

T. Okada, R. Hanson and B. Kato, C. Culver and S. Nakata, and A. Mattock, and V. Bertero and M. Ozake for building Types 1-6 (No. 1 - Reinforced Concrete, No. 2 - Precast/prestressed Concrete, No. 3 - Structural Steel, No. 4 - Masonry, No. 5 - Timber, No. 6 - Mixed Steel/ Reinforced Concrete), respectively. Extensive discussion toward developing final designs and testing procedures and toward prioritizing structural types followed these presentations. Considering such factors as (1) need based on understanding seismic performance of current construction, (2) need based on developing improved seismic performance in future construc- tion, (3) economic benefits, (4) safety considerations, (5) test feasibility, (6) mutual benefit to both countries, and (7) cost and manpower advantages of conducting full-scale tests on a cooperative basis, it was decided that the immediate future planning should recognize the following priority listing of structural types: No. 1 - Reinforced Concrete, No. 2 - Structural Steel, No. 3 - Precast/prestressed Concrete, No. 4 - Mixed Steel/ Reinforced Concrete, No. 5 - Masonry, and No. 6 - Timber. It was agreed however that detailed planning of the full-scale tests for both the first and second priorities, namely, reinforced concrete and structural steel, should be completed and that plans for component testing involving these and other types should be developed as deemed appropriate. A general flow chart of the first full-scale test project was developed as shown in Figure 1.

K. Otani and M. Watabe reported on the "Current Status of BRI Facilities" and "Design Earthquake Intensity," respectively, and S. C. Liu and T. Okubo discussed plans for imple- menting the program under the auspices of the U.S.-Japan Panel on Wind and Seismic Effects, UJNR Program. It was proposed that the overall program be implemented in accordance with the flow chart shown in Figure 2, that the UJNR Panel on Wind and Seismic Effects be used as the implementation and coordination mechanisms for the joint program, that the funding agencies have administrative management responsibility for the joint program, and that the program utilize scientists and engineers from academic, government, and industrial organiza- tions in both countries as appropriate to the conduct of specific projects.

At the conclusion of the second planning group meeting, the following resolutions were adopted:

1) Due to safety and economic consideration, it is urgent that full-scale pseudo- dynamic tests be conducted on various building types for the purpose of verifying and improving seismic performance.

33

2) In the interest of improved efficiency and research productivity, it is highly desirable that full-scale pseudo-dynamic tests on buildings be conducted by Japan and the United States on a cooperative basis.

3) Immediate plans for conducting full-scale pseudo-dynamic tests on a cooperative basis should be developed in accordance with the following priority listing of building types: No. 1 - Reinforced Concrete, No. 2 - Structural Steel, No. 3 - Prestressed/Precast Concrete, No. 4 - Mixed Steel/Reinforced Concrete, No. 5 - Masonry, and No. 6 - Timber. This listing should remain flexible, however, to accommodate the availability of test facilities and equipment and to meet other special conditions which may arise prior to and at the time of testing.

4) Complete detailed structural designs of the reinforced concrete and structural steel buildings should be prepared for presentation at the third planning group meeting. The designs of support tests for both types should also be presented at this meeting.

5) Because of their great importance, extensive full-scale tests should be conducted on non-structural members together with the structural full-scale tests. Plans for these tests should be developed prior to and be presented at the third planning group meeting.

6) A subgroup should be established to develop procedures and techniques for laboratory simulation of seismic excitations.

7) The implementing agencies of both governments should proceed immediately to request funding for the procurement of special test equipment (e.g. hydraulic actuators) and instrumentation still required by the overall proposed test program.

8) Due to the importance of a balanced exchange of research personnel prior to and during the testing program, every effort should be made on both sides to bring about such an exchange.

9) Administrative and management arrangements and guidelines for implementing the above resolutions should be completed by the government agencies involved as soon as conveniently possible.

10) The third planning group meeting, U.S.-Japan Cooperative Research Program Utilizing Large-Scale Testing Facilities, should be held during the period December 18-23, 1978, in Japan.

C. Third Planning Group Meeting

The third planning group meeting of the U.S.-Japan Cooperative Research Program Utilizing Large-Scale Testing Facilities was held in Tokyo during the period December 18-23, 1978. The meeting continued discussions to develop future research programs in earthquake engineering utilizing large-scale testing facilities.

The summary and resolutions of the first and the second planning group meetings held in Tokyo in September, 1977, and in San Francisco, in May, 1978, were presented and confirmed.

During the third planning group meeting, representatives from academic institutions, governmental agencies and private industrial firms exchanged information and views. Preliminary design considerations for full-scale tests and their supporting tests were presented by representatives from both U.S. and Japan for four structural types;

No. 1   Reinforced Concrete

No. 2   Structural Steel

No. 3   Precast/Prestressed Concrete

No. 4   Mixed Steel/Reinforced Concrete

Preliminary discussion were also presented on masonry and timber types.

For the first two of these structural types, U.S. and Japan representatives discussed in depth and reached certain conclusions on experiment designs. In addition to working sessions on structural design, a session on loading procedures for full-scale tests was held. A major discussion in the loading procedures session centered on the feasibility of the pseudo-dynamic test method in which sufficient numbers of electro-hydraulic jacks are used.

At the conclusion of the Third Planning Group Meeting in Tokyo, the following resolutions were adopted:

1) The goal of the joint program is to improve seismic safety practices through studies to determine the relationship among full-scale tests, small-scale and component tests, and analytical studies.

2) The joint program shall be designed and conducted to:

a) achieve clearly stated scientific objectives;

b) represent total building systems as realistically as possible;

c) balance the simplicity and economy of test specimens with the need to test structures representing real situations;

d) maintain a balance among small-scale, component, and full-scale tests;

35

e) utilize previously performed experiments and studies to the extent practical;

f) represent the best design and construction practice in use in both countries;

g) check the validity of newly developed earthquake-resistant design procedures;

h) maintain flexibility to accommodate new knowledge and conditions as successive experiments are completed; and

i) assure the practicality of program results.

3) This program should be initiated in 1979 jointly and cooperatively in both and Japan.

4) To implement this program, the establishment of the following committees and subpanels is recommended for inclusion in the governmental MEMORANDUM OF UNDERSTANDII

a) Joint Executive Committee for the purpose of providing scientific advice to participating institutions in this program and to appoint subpanels other tha stated below to perform tasks as agreed necessary;

b) subpanel for execution of the full-scale and supporting tests for each structural type; and

c) subpanel for assessing the feasibility and validity for use in this program of pseudo-dynamic loading techniques.

5) To implement this joint program, quick and positive response by both govern to funding and staff arrangements is requested. Strong emphasis is placed on fundin loading systems needed to assure adequacy of the facilities to perform the planned e ments.

6) The planned order of testing is first the reinforced concrete structures, a the steel structure. Precast-prestressed concrete structure and mixed steel-reinfor crete structure are the next priorities. Masonry and timber structures should be st further for inclusion in this program.

7) Additional tests and analyses found to be required beyond the planned progr be conducted to assure that research results can be applied in the practical design ings.

8) All activities of the joint program (full-scale tests, support tests, analytical studies, etc.) should be conducted cooperatively with balanced participation from both countries to the extent possible.

9) The fourth planning group meeting, U.S.-Japan cooperative research program utilizing large-scale testing facilities, should be held in the U.S. during the period July 9-14, 1979.

D.   Fourth Planning Group Meeting

At the conclusion of the fourth planning group meeting in Berkeley, the Resolutions of the first three meetings were re-confirmed and additional resolutions were adopted as follows:

1) In view of the importance of improving seismic safety practices through studies to determine the relationship among full-scale, small-scale, component tests, and analytical studies, the final report of the planning group should be completed and be transmitted to the appropriate agencies of government as soon as possible.

2) It is recommended that the program of cooperative research as set forth in the planning group final report be initiated in 1979 in both countries.

3) The level of funding provided to the cooperative research program should permit the fulfillment of goals and objectives as defined in the planning group final report.

4) To insure successful execution of the recommended cooperative research program, full coordination of all research activities carried out in both countries is essential; therefore, the following committees are recommended:

a)   Joint Technical Coordination Committee to provide scientific and technical advice to participating institutions in the program. This committee should meet at least once a year during the program.

b)   Joint subcommittees to provide advice on the execution of research related to reinforced concrete buildings, steel buildings, pseudo-dynamic loading techniques, and other major areas of activity. These committees should meet as frequently as needed.

5) To assure and enhance the cooperative effort in this joint research program, adequate exchange of research personnel from both countries is needed. It will be desirable for one exchange researcher to be associated with each participating research organization in each country.

the experience gained in the use of the pseudo-dynamic testing procedure. Appropriate
planning for this activity should be initiated at an early date.

7) Long-range plans should include the possibility of testing large-scale structures
and systems, other than buildings, in this cooperative program.

8) The planning group encourages support of the recommended cooperative large-scale
test program by the UJNR Panel on Wind and Seismic Effects.

APPENDIX III

I. RECOMMENDED RESEARCH ON REINFORCED CONCRETE BUILDING STRUCTURES

It is recommended that a full-scale seven-story reinforced concrete building structure
representing good current practice be tested in the Large-Size Structures Laboratory, Building Research Institute, Tsukuba New Town for Research and Education, Japan. This test structure should have a layout as indicated in the plan and elevation views of Figures 3 and 4,
respectively, and it should include non-structural elements. The tests should be conducted
using a procedure intended to simulate dynamic response to prescribed seismic excitations.
A pseudo-dynamic loading procedure based on actuator on-line system should be considered for
this purpose. As indicated in Figure 3, the structure should be loaded only in the direction
of the shear wall.

After an initial series of tests limiting the response to moderate damage levels, the
full-scale structure should be repaired and then retested under severe simulated seismic
conditions to near collapse conditions. The objectives of this final test are to determine
large deformation and failure characteristics and to assess the effectiveness of repair
procedures.

It is recommended that a series of coordinated experiments associated with the full-
scale tests be conducted in Japan and the United States on the following test specimens:
(1) element assemblies, (2) structural walls, (3) planar frames, (4) wall-frame assemblies,
and (5) scaled models of the full-scale seven-story structure. The purpose of these tests

are to investigate the range of validity of each type in interpreting the seismic response of full-scale buildings. Both quasi-static and dynamic loadings should be used in conducting these experiments.

The correlation of the following three groups of experimental parameters are considered important:

1) Methods of loading such as dynamic (earthquake-simulation and harmonic) and static (pre-programmed reversals and pseudo-dynamic).

2) Scales ranging from smallest practical (around 1/10) to that which uses regular concrete and standard reinforcing bars (say, 1/3 scale to full scale).

3) Complexity of test assemblies required to give reliable information on the performance of full-size structures.

It is suggested that the test specimens include: (1) two- and three-dimensional models of the full-scale structure; (2) two- and three-dimensional simple element assemblies; and (3) simple and complex element assemblies including walls and frames. In conducting this series of tests, the importance of including non-structural elements should be determined.

The results of all associated tests in both countries should be fully correlated with each other and with the results of the tests made in Japan on the full-scale seven-story structure. The objectives of these correlation studies will be to (a) identify the relative merits of static loading and dynamic loading and of component tests, small-scale tests, and medium scale tests in predicting prototype performance under moderate to severe seismic conditions, and (b) to improve structural details for earthquake resistant design.

It should be noted that the overall investigation recommended represents the first attempt anywhere in the world at carrying out a fully integrated test program having both of the above objectives. It, therefore, should prove of great value in setting long-range priorities for future test programs based on benefit-cost considerations.

Prior to testing the full-scale structure, a thorough analysis of its seismic performance should be carried out in both Japan and the United States. The best available mathematical modelling and computational procedures should be used. In developing the mathematical model of the structure, use should be made of existing component test data in both countries supplemented by the results of the associated test previously described. Through these analyses, seismic excitations to be used in the full-scale tests as well as in the associated tests can be selected.

Researchers from both Japan and the United States should participate in the full-scale tests, the associated tests, and the correlation studies.

II. RECOMMENDED RESEARCH ON STEEL BUILDING STRUCTURES

There is a recognized need to establish the relationship among full-scale tests, reduced-scale and component tests, and analytical studies of steel building structures designed to resist earthquake ground motions. In recognition of this need, it is recommended that a full-scale seven-story steel building, a series of components and assemblages, and several scaled models of the building be tested. The full-scale structure should be designed by the latest U.S. and Japanese codes and it should be tested in the Large-Size Structures Laboratory of the Japanese Building Research Institute in Tsukuba New Town. The test structure should be representative of an office building and the structural system should consist of moment-resisting exterior frames with a braced interior core. Plan and elevation views of such a building are shown in Figures 5 and 6, respectively. The interior core should be braced by either concentric or eccentric K braces. It is suggested that the floor system be a metal deck with composite floor slab. Non-structural elements such as curtain walls and partitions should be added to the structural system at some point in the testing program. The structure should be tested in a manner which simulates realistic earthquake conditions.

The recommended test program for the full-scale building should include free vibration tests and simulated seismic loading tests at three intensity levels, namely working stress level, post-buckling of braces level, and post-moment frame yielding level. In order to maximize the amount of knowledge that may be gained, a multi-stage test program including bare frames, frames with interior braces, and frames with non-structural elements should be considered.

As part of the overall investigation, it is recommended that a series of complementary quasi-static tests be conducted on (a) structural components including girder-to-column connections, column-to-footing connections, individual braces, and floor slabs, (b) beam-and-column assemblages, both two- and three-dimensional, and (c) braced and unbraced framed bents with and without slab interconnections. Further, it is recommended that small- and medium-scale models of the prototype structure be tested under simulated earthquake ground motions on shaking tables and with the pseudo-dynamic test procedures. Some of the test specimens may be full-size while others may be half- or third-scale. These associated tests, which are

40

to be carried out by researchers in both countries, should be carefully planned and be made fully complementary. The results of these tests should greatly assist in predicting the response of the prototype test structure at all levels of load application. The various quasi-static, pseudo-dynamic tests and shaking table tests should be correlated with each other and with the prototype structure tests.

The entire program should be well coordinated and researchers from both the U.S. and Japan should cooperate fully during the course of the investigation. It is recommended that researchers from both countries work together in each others' laboratories to the extent possible and that U.S. investigators take part in the testing of full-scale building in Japan.

III. RECOMMENDED RESEARCH ON THE PSEUDO-DYNAMIC TEST METHOD

The most realistic method for assessing the seismic behavior of structures is to test them using earthquake simulators. However, many structures of interest are either too large, strong or massive to be tested on available shaking tables. Two approaches have been suggested for quasi-static testing of such structures. One approach assumes a first mode shape, such as an inverted triangle, and lateral forces are applied in a cyclic manner with the forces at each floor level being proportional to their first mode contribution. Since the actual conditions depend on the type of structure, the size of structure and the selected input earthquake accelerogram, the selected test sequences must be idealized. A second, more sophisticated approach assumes the force-deformation characteristics of the structure and a nonlinear dynamic analysis is performed for a given input earthquake accelerogram to determine the response history of the structure. These computed response histories are used as input data to control displacements, for the subsequent tests of the structure. The major shortcoming of these two types of quasi-static test methods is that the nonlinear force-deformation characteristics of the structure must be known or assumed prior to starting the test.

The computer-actuator on-line (pseudo-dynamic) test procedure has been suggested as the most realistic method for simulating the dynamic earthquake responses of test specimens which are too large for existing shaking tables. In the on-line test procedure, the computer uses the preceding test data to select the next step in the test sequence. Thus, the test data are used in controlling the progression of the test in much the same way that current

41

building damage predicates the response of that building during the remaining portion earthquake. A significant advantage of the on-line procedure is that it is possible during the test to record visual observations as well as to recalibrate or relocate mentation.

Since about 1973, various investigators at the University of Tokyo, in particula Institute of Industrial Science, have been developing and evaluating algorithms for test procedures. Similar efforts in the United States have not maintained the same nuity of effort.

At the Third Planning Group Meeting of the U.S.-Japan Cooperative Research Prog Utilizing Large-Scale Testing Facilities, concerns were expressed on the capability trolling up to eight degrees-of-freedom simultaneously and also on the required accu the prescribed displacements and measured force outputs necessary to maintain accura linear dynamic responses of the building. Based on these discussions, there are thre distinct areas which need additional research and/or verification efforts before the procedure can be used with full confidence. The first of these three areas is verif: and sensitivity studies of the numerical algorithm used to connect the dynamic equat: motion and the test specimen for the computer control. The second major aspect is tl fication that the hydraulic actuator control systems can operate accurately in multi; displacement control operation with velocity control of the actuator movement during taneous motion at each of the load points in order to eliminate inaccurate structure, system feedback to the measured forces. This verification would require a physical which a specimen with force-deformation characteristics similar to those expected in full-size test structure are used. The third aspect which needs verification is tha basic differential equation of motion is an adequate representation of the actual dy; response of the building system. This could be verified only by comparing on-line s test results with shaking-table test results.

Analytical studies and experimental verification tests should be carried out in and the U.S. in a coordinated effort so that the pseudo-dynamic test method can be u with confidence for the full-scale testing.

42

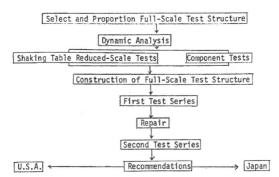

Fig. 1. Flow Chart of First Full-Scale Test Project.

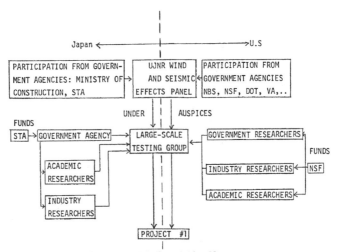

Fig. 2. Flow Chart of Implementation Plan.

43

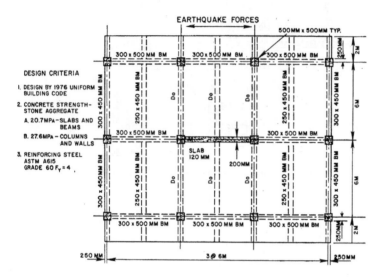

Fig. 3 Plan of Reinforced Concrete Test Structure

44

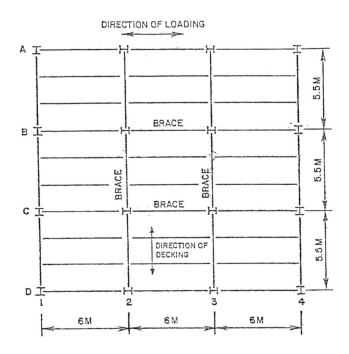

Fig. 4   Seven-story Office Building 18 m X 16.5 m X 22.5 m

MODELING THE 1978 TOKYO TORNADO

THAT OVERTURNED THE TOZAI SUBWAY TRAIN

Eiji Uchida

Shigemi Fujiwhara

Meteorological Research Institute

Ryozo Tatehira

Japan Meteorological Agency

Ichiro Tabe

Kazuyuki Ohtsuka

Teito Rapid Transit Authority

ABSTRACT

A high wind developed in the evening of February 28, 1978, over areas from Kawasaki City (Kanagawa Prefecture) to Kamagaya City (Chiba Prefecture). The wind caused extensive damage (for example, overturning the Tozai subway train, houses, cars, vessels etc.)

Judging from weather conditions which were detailed by data from weather maps, automatic records of wind direction and speed, of air pressure, radar echo patterns, AMeDAS data, and disaster distributions, we envisage that this high wind was associated with a tornado formed within a tornadic cyclone having 5 ~ 6 x 10 km diameter.

We estimate that this tornado was generated within a coverging and unstable area near a squall line preceding a cold front, the diameter of the tornado core is estimated as 100 ~ 200 m (radius 50 ~ 100 m), the maximum wind velocity 60 ~ 80 m/s and the propagation speed 25 m/s.

A numerical simulation experiment was carried out in relation to meteorological elements of the tornado (core radius, the maximum-wind velocity, propagation speed) under an assumption of a single tornado having a straight path and a uniform velocity. The most adaptable values of tornado elements, in regard to the kinematical analysis of the overturned train were located within 75~100 m in core radius, 70~80 m/s in maximum-wind velocity and 25~35 m/s in propagation speed.

These results from the engineering aspect nearly coincide with those of the estimated values and the reliability of this model was thereby roughly verified.

1. INTRODUCTION

In the evening of February 28, 1978, a high wind developed over the areas from Kawasaki City in Kanagawa Prefecture to Kamagaya City in Chiba Prefecture. The wind caused extensive damage (for example, overturning the Tozai subway train at about 21H 34M local time). Since the disaster occurred in a densely populated area in Tokyo, the overturning of the subway train was widely reported. It was a spectacular accident, but no one was killed, and the injured list grew to be no more than twenty-one.

Using the available meteorological data, which consists of weather maps, automatic records of wind directions and speed, radar, and AMeDAS (Automated Meteorological Data Acquisition System) data, the structure of the high wind was analysed.

This analysis is used in a simple mathematical simulation of the engineering aspects for the overturning of the train vehicles.

2. METEOROLOGICAL DATA AND DAMAGES BY THE HIGH WIND

The subway train was composed of ten vehicles, each 20 m in length (200 m in total length). The rear two vehicles were overturned close to 21H 34M local time. The last rear vehicle weights 27.6 ton, and the next to last one weighs 36 ton. The direction of overturning was in the right side (hill side) of the train moving direction, corresponding roughly to a side of shifting direction of the high wind area regarding as the railway as shown in Figure 1 and 2.

The total damage area was extended over about 30 km long and 500 ~ 1500 m wide. Within the area, some extensive damage occurred, including injuries to the 21 people (refer to Figure 3 and reference [1] (1979).

The weather map on that day showed a developing cyclone moved northeastward cross the Japan Sea, and after the passage of its warm front over the accident site, a tornado occurred within the unstable and convergence region of squall line in front of the cold front. The thunderstorms were active in this region (see Figure 4). AMeDAS data showed that high wind belt ran through the damage area during the evening as illustrated in Figure 5. Two typical examples of high wind records were obtained by automatic wind recorders within the disaster area. One of them is shown in Figure 6. In this figure, an instantaneous high wind and a

cyclonic wind direction change were recorded. However, as the cyclonic wind direction
changed during a 30-minute period, the scale of wind direction did not correspond to one of
wind speed. However, the change of wind direction endorses the existence of a tornadic
cyclone.

When these wind records were compared with past high wind data, including tornadoes,
this high wind was viewed to have the typical characteristics of a tornado from the point
of view of an instantaneous high wind, although the maximum pressure drop was recorded as
only about 2 mm within the disaster area (see Figure 7).

On the other hand, Tokyo weather radar detected a hook echo at 21H 50M local time just
after the accident, as shown in Figure 8 schematically. This hook echo is suggested like
to be stemming from the tornadic cyclone, and the existence of tornado itself is also
endorsed.

Fujita (1979) [2] proposed the relationships between downburst, tornado, and hook echo
such as those classified in Figure 9, which includes the probability of tornado generation
about 30% for the occurrence of downburst and about 40% for that of hook echo.

Therefore, the high wind in question was estimated synthetically as a tornado under the
following consideration: the characteristics of wind (the instantaneous high wind, and the
cyclonic change of wind direction suggested as a tornadic cyclone), a hook echo and damage
situations (very narrow regions and damage reports suggesting a tornado flew over). There
was unfortunately no positive, visual observation as to whether the high wind was definitely
a rotating vortex or not because it was night time.

3.   STUDIES RELATING TO THE TOKYO TORNADO

Muramatsu (1979) [3] analyzed a -50°C disturbance areas along the squall line by using
GMS (Geostationary Meteorological Satellite) data. This disturbance, its diameter 40 to
50 km, corresponds to the tornadic cyclone mentioned above (see Figure 10).

Fujita (1978) [4] depicted precipitation areas, stream lines, frontal lines, and a
radar echo pattern through mesoscale analysis as shown in Figure 11 and 12. Soma (1978) [5]
performed a detailed analysis about damages and suggested evidence of a tornado from the
damage features of the Ooi wharf, etc. He also proposed a three tornadoes model from time
sequence verification for disaster generations.

Miyazawa, Mitsuta, Shoji, etc. (1979) [1] also presented reports.

4. MODELING AND RELATING A TORNADO TO THE OVERTURNED VEHICLES

Based on the foregoing observational results, a tornado model concerning the overturned train vehicles is conventionally proposed. It is to be noted that the model would rather stress to simulate its engineering characteristics than physical one. In addition, the tornado model may be considered as a complex of several vortexes or a complex of a normal high wind and a tornado. However, because of the deficit of unique and definite data concerning a complex of several vortexes as a simple step, a single tornado model of a Rankine vortex [6] type is assumed, and its moving state is also assumed in straight forward motion with uniform velocity. This seems to compare reasonably with the observations around the train accident site, especially in regarding to the overturning of the two vehicles.

In order to define the modeling scale, we adjust the following three elements within an applicable range for observations of train overturning:

Core radius of the tornado (A)

    (radius of maximum wind position in a Rankine vortex)

    $\cdots\cdots$50 ~ 100 m estimated by the disaster survey.

Maximum wind velocity of the tornado (V_{max})

    $\cdots\cdots$60 ~ 80 m/s estimated by wind records of Sunamachi drainage office and

    No. 10 signal office.

Propagation speed of the tornado (V_1)

    $\cdots\cdots$25 m/s estimated by damage survey.

The reliability of this model is to be verified by the fact of the overturning of the train.

Using the above three element values with extended ranges, A: 25 ~ 100 m, Vmax: 60 ~ 80 m/s, V_1: 15 ~ 35 m/s, the perpendicular component of wind concerning the train is calculated (as shown in Figure 13).

It took about 7 (± 1) sec until the trains two rear cars were overturned after the driver operated a break facing with some dangers.

On the other hand, calculated wind distributions are composed of a group of circles expanding to both sides, and higher wind areas are located in the right hand side of tornado's trajectory (see Figures 14, 15 and 16).

We can hypothesize the overturning of the vehicles as the following steps in the Figures:

1) The first vehicle received a wind-shock when it contacted the 40 m/s circle which is calculated from the kinematical survey for a train vehicle [1].

2) At that time, the driver applied the brake.

3) Successive relative positions between the train and the tornado were depicted each second.

4) Next, the cases in which the two rear cars were fallen into within more than 80 m/s wind areas were picked up which is the threshold value of overturning of a vehicle from the kinematical survey [1].

The final results are summarized in Table 1. In the table, the optiumum examples are as follows:

Core radius (A): 75 ~ 100 m.

The maximum wind velocity (Vmax): 70 ~ 80 m/s.

Propagation speed (V_1): 25 ~ 35 m/s.

Comparing the observation data, we can verify the tornado model clearly by the overturned train without entering into a more complex tornado model.

ACKNOWLEDGMENT

The authors are greatly indebted to acknowledge Prof. T. T. Fujita of Chicago University, Professor Mitsuta of Kyoto University, Dr. S. Miyazawa (Chief forecaster of JMA), Mr. T. Muramatsu of JMSC, Professor S. Arakawa of Met. College, Dr. M. Shimada of Kofu Local Met. Obs., Mr. S. Ooishi of Tokyo District Met. Obs., Mr. H. Tokuue of Obs. Dept. (JMA) for their valuable suggestions and helpful opinions to our research.

REFERENCES

[1] Report on the Counter-Measure Research for Tozai Subway Train Accident (in Japanese), Committee of the Counter-Measure Research for Tozai Subway Train Accident, March, 1979, Teito Rapid Transit Authority.

[2] Fujita, T. T., 1978: Workbook of Tornadoes and High Winds for Engineering Application, Satellite and Mesometeorology Research Project No. 165, 17-60.

[3] Muramatsu, T., 1979: Analysis for Squall-Line and Tornado by Use of GMS, Radar and AMeDAS System. (in Japanese) Tenki, 26, No. 7, 399-411.

[4] Soma, K., 1978: On Tornadoes and Similar Phenomena (in Japanese), No. 5 Symposium on Wind-Proof Structures.

[5] Ishizaki, H., Mitsuta, Y., Kawamura, S., Murata, T., Kimoto, E., and Tahira, M.: Studies of a Tatsumaki at Toyohashi, December 7, 1969, A, pp. 481-500

Fig. 1 Overturned vehicles viewed from Nishifunabashi side

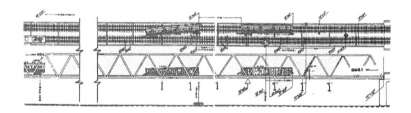

Fig. 2 A schematic sketch of accident scene on Tozai subway bridge across Arakawa river[1].

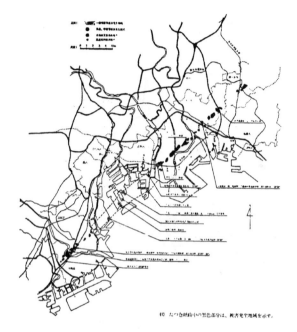

Fig. 3 A schematic sketch of damage distribution caused by a tornado on Feb 28, 1978.
Black spot areas show severe damage areas[1].

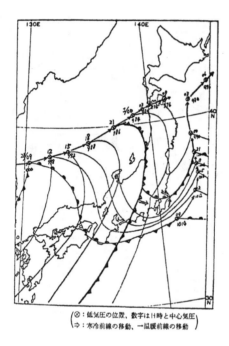

（⊗：低気圧の位置、数字は14時と中心気圧
⇒：寒冷前線の移動、→温暖前線の移動）

Fig. 4 Successive positions of the moving cyclone and accompanying
on Feb. 28, 1978.[1]

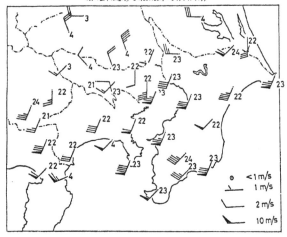

（図の数字は時刻，2月28日20時～3月1日4時）

Fig. 5 Maximum wind distribution during Feb. 28, 20 H～Mar. 1, 04 H (local time) over Kanto district[1]

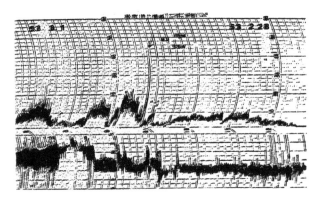

Fig. 6 Automatic records of wind direction and speed (three cup Celsin type instrument, 14.6m in height from the ground) at Sunamachi drainage office[1]

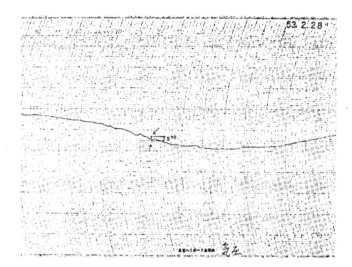

Fig. 7 Automatic air pressure records at Tokyo helicopter airport showing maximum pressure drop of only 2 mm inside the disaster area.[1]

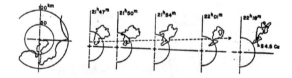

Fig. 8 Movement of echo cell sketchs accompanied with the tornado[1].

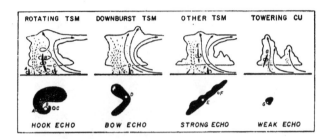

ROTATING TSM	DOWNBURST TSM	OTHER TSM	TOWERING CU
HOOK ECHO	BOW ECHO	STRONG ECHO	WEAK ECHO

Fig. 9 Four types of convective mother (Parent, Ed.) cloud which could start tornadoes[2]

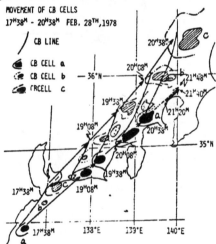

MOVEMENT OF CB CELLS
$17^H38^M - 20^H38^M$ FEB. 28TH,1978

CB LINE

CB CELL a
CB CELL b
CB CELL c

Fig. 10 Movement of squall line and cell[3]

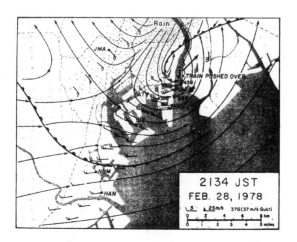

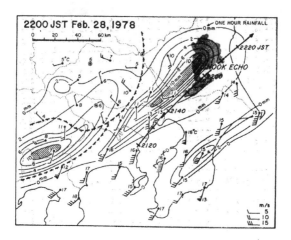

Fig. 12 Surface weather map at 22 H 00 M of Feb. 28, 1978 (local time)[4].

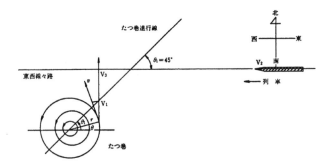

Fig. 13 A schematic figure concerning the tornado and subway.

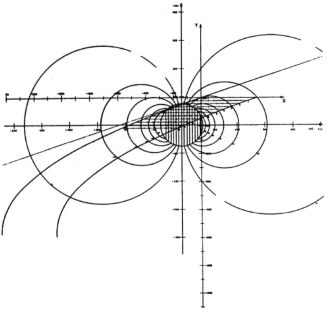

Fig. 14 A calculated locus of train at $V_1 = 15\,\text{m/s}$, $A = 75\,\text{m}$, $V\text{max} = 80\,\text{m/s}$

I-13

5 A calculated locus of train at V_1 = 25 m/s, A = 75 m, Vmax = 80 m/s

parison of simulation results.

est adaptable.

table.

hly adaptable.

tly adaptable.

w related.

elationship.

Vmax (m/s)		60			70			80		
V_1 (m/s)		15	25	35	15	25	35	15	25	35
A (m)	25	△	△	△	△	△	△	△	△	△
	50	△	△	△	△	△	△	△	△	△
	75	△	△	Ⓒ	△	Ⓓ	Ⓔ	Ⓒ	Ⓐ	Ⓑ
	100	△	Ⓓ	Ⓔ	Ⓓ	Ⓓ	Ⓓ	Ⓔ	Ⓑ	Ⓓ

I-14

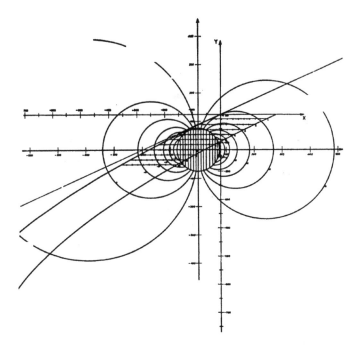

Fig. 16 A calculated locus of train at V_1 = 35 m/s, A = 75 m, Vmax = 70 m/s

I-15

EXTREME WIND SPEEDS AT 129 STATIONS

IN THE CONTIGUOUS UNITED STATES

Michael J. Changery

National Climatic Center

Asheville, North Carolina

and

Emil Simiu and James J. Filliben

National Bureau of Standards

Washington, D.C.

ABSTRACT

The purpose of this report is to present information on recorded and predi
speeds at 129 airport stations in the contiguous United States at which reliabl
available over a number of consecutive years. This information is provided to
documentation from which appropriate decisions can be made on values of design
be specified in building codes and standards, and on special projects. Include
report are: recorded wind speeds and anemometer elevations; predicted wind spe
probability distributions of the largest values; estimates of the sampling erro
in the predicted wind speeds; a description of the statistical procedure used in
sis of the data; and a discussion of the results of the analysis.

KEYWORDS: Building (codes); probability distribution functions; statistical an
storms; structural engineering; wind pressure; wind speeds.

·1· INTRODUCTION

The purpose of this report is to present information on extreme wind speeds at 129 airport stations in the contiguous United States at which reliable wind records are available over a number of consecutive years.

This information consists of:

1.1 Extreme yearly wind speeds, and the corresponding wind directions, recorded at each of the 129 stations. These data were obtained by the National Climatic Center from the original records. Thus, reading errors of original records and errors of transcription that have been determined to be present in Local Climatological Data (LCD) monthly and annual summaries have been eliminated. The vast majority of the originally recorded data consisted of fastest-mile speeds. These have been listed without modification in the report. However, at a few stations, some of the recorded data consisted of fastest observed one-minute speeds. These have been transformed into fastest-mile speeds using a relation given in Section 2.1. It is these fastest-mile speeds that have been listed in the report in lieu of the orginally recorded fastest observed one-minute data. The stations and dates at which fastest-minute speeds were originally recorded are listed in Section 2.1.

1.2 Anemometer elevations at which the largest yearly wind speeds were recorded.

1.3 Largest yearly wind speeds reduced to an elevation of 10 m above ground (corrected speeds). These were obtained by using an expression given in Section 2.4.

1.4 Results of the statistical analysis of the corrected wind speed data. These results include:

* for each of the 129 sets of data, the predicted wind speeds corresponding to various return periods, based on the assumption that the Type I probability distribution of the largest values is a valid description of the extreme wind speeds

* for those sets of data that are best fit by a Type II probability distribution of the largest values, the predicted wind speeds corresponding to various return periods, based on that distribution

I-17

* estimates of the lower bound of the standard deviation of the errors inherent
  in the predicted speeds
* estimates (obtained by the method of moments) of the standard deviation of the
  errors inherent in the predicted speeds

Extreme wind speed predictions have been included for mean recurrence intervals of u
1,000,000 years. However, in the writers' opinion, physical considerations suggest that
dictions corresponding to mean recurrence intervals beyond a few hundred years should be
regarded with caution.

A brief description of the procedure used in the analysis of the data is presented in
Section 3. Section 3 includes a summary, and Sections 3.2 and 3.3 a discussion of the
results of the statistical analysis. A sample of the information described under items 1
through 4 above is included in Section 4 of the report.

It is noted that at a number of stations the extreme yearly wind speed data may not
provide a reliable basis for predicting extreme speeds. The results of the statistical
analysis for these stations should therefore be regarded with caution. Stations for which
such caution is in order are listed in Appendix I.

2. WIND SPEED DATA

2.1 Fastest Observed One-Minute Speeds

It was indicated in Section 1 that the vast majority of the original data used in thi
report consisted of fastest-mile wind speeds, i.e., speeds averaged over a time interval (
seconds) $t = 3600/v_f$, and v_f = fastest-mile wind speed in miles per hour. However, at the
following stations the original recorded maximum annual wind speed data consisted of faste
observed one-minute speeds during the periods indicated below:

Atlanta, Georgia	(1961 through 1963)
Indianapolis, Indiana	(1962-1963)
Boston, Massachusetts	(1954 through 1958)
Lansing, Michigan	(1955 through 1958)
Sault Ste Marie, Michigan	(1956 through 1965)

According to Reference 1, studies of the relationship between fastest observed one-
minute to fastest-mile wind speeds undertaken at four weather stations "showed the mean
regression between the two types of observation to be

I-18

$$v_f = 9.55 + 0.999 \, v_m \qquad\qquad (1)$$

where v_f = fastest-mile speed in miles per hour and v_m = fastest-minute speed in the same hour as the fastest-mile, in miles per hour. Since the slope is very near unity and the mean difference very near 10, it has been assumed for some time that adding 10 mph to the fastest-minute would give an approximation to the fastest-mile." It is this relation which - in the absence of other information - has been used in this report.

While the writers are not certain that Eq. 1 provides a correct relation between v_f and v_m, they note that it results in estimates of v_f that are conservative from a structural safety point of view.

2.2 Roughness Conditions at Airport Stations

In an attempt to ensure that the terrain roughness conditions are uniform among all the sets of data being analyzed, only airport stations have been considered herein. In principle, it may be assumed that at such stations open exposure conditions prevail. Nevertheless, the mere fact that wind speed measurements are taken at an airport station does not necessarily ensure that the wind climatological conditions reflected by these measurements are identical, from the standpoint of the terrain exposure, to those prevailing at a different airport. For example, it is noted in Reference 2 that the estimated 50-year wind at Chicago Midway Airport is about 15 mph less than at the Chicago O'Hare airport. The probable reason for this difference is that the terrain around the Chicago Midway Airport is relatively heavily built-up. Similar considerations might explain to some extent the difference between the estimated 50-year winds at the Washington National Airport and the Baltimore-Washington International Airport, which are estimated in this report to be 66 mph and 75 mph, respectively. Thus, in interpreting airport data for the purpose of developing wind maps, it is appropriate to take into account the possibility that, at the airport of concern, the terrain exposure conditions might differ somewhat from those defined as "open" (e.g., in Reference 3).

2.3 Variation of Wind Speed with Height Above Ground

To ensure the micrometeorological homogeneity of the data at any given station it is necessary to reduce all the wind speeds recorded at that station to a common elevation. The elevation chosen for this purpose is 10 m above ground.

The mean wind profile near the ground in homogeneous terrain is given by the well-known logarithmic law, which may be written in the form:

$$U(z) = \frac{\ln \frac{z}{z_0}}{\ln \frac{10}{z_0}} U(10) \tag{2}$$

where z = height above ground and z_0 = roughness length, both expressed in meters. In open terrain, z_0 may vary from, say, 0.03 m to 0.10 m. In this report the reduction of the data to an elevation of 10 m is based on the assumption z_0 = 0.05 m. It can be verified that the errors inherent in the assumption z_0 = 0.05 m -- when in fact the values z_0 = 0.03 m or z_0 = 0.10 m were correct -- are small (of the order of 1% or 2%).

An approximation to Eq. 2 is given by the power law

$$U(z) = (\frac{z}{10})^{\sigma} U(10) \tag{3}$$

where, for open terrain conditions, it is generally assumed σ = 1/7 (3). It is noted that Eq. 2, and therefore its approximate equivalent given by Eq. 3, is valid for mean wind speeds averaged over a relatively long time interval, e.g., one hour. The question thus arises of expressing the variation with height of the fastest-mile wind speed, which is averaged over a relatively short time (30 to 90s or so).

To obtain an approximate expression for the fastest-mile wind profile, note that it may be assumed, approximately,

$$\frac{U_{pk} - U_{fm}}{U_{pk} - U} \approx \frac{1}{2} \tag{4}$$

where U_{pk} = peak wind speed, U_{fm} = fastest-mile speed, and U = hourly mean speed (see, e.g., Reference 4, p. 62). The expression for U_{pk} can, in open terrain, be written as

$$U_{pk}(z) = U(z) + 3\,\overline{u'^2}^{1/2} \tag{5}$$

where $\overline{u'^2}^{1/2}$ = r.m.s of longitudinal velocity fluctuations, and

$$\overline{u'^2}^{1/2} \approx \frac{U(10)}{\ln \dfrac{10}{z_o}} \qquad\qquad (6)$$

where z_o is expressed in meters (see Reference 4, pp. 45 and 54).

It can be verified by using Equations 2, 4, 5, and 6 that, within the anemometer elevation range of interest in this report, it is possible to write approximately

$$\frac{U_{fm}(10)}{U_{fm}(z)} \approx \frac{U(10)}{U(z)} \quad \frac{(1 + z-10 \; 0.02)}{10}$$

where z is expressed in meters. The errors inherent in Equation 7 are of the order of -1 to 3%, the higher errors being on the conservative side (i.e., yielding slightly higher fastest-mile values at 10 m above ground than would be obtained by a more "exact" expression). Eq. 7 has been employed to obtain the corrected speeds at 10 m ground in this report.

3. STATISTICAL ANALYSIS

3.1 Objective of Statistical Procedure

Probabilistic considerations, as well as available empirical evidence suggest that the asymptotic probability distributions of the largest values with unlimited upper tail are an appropriate model for the behavior of the largest yearly wind speed. There are two such distributions, known as the Type I and Type II distributions of the largest values, whose cumulative distributions functions, $F_I(v)$ and $F_{II}(v)$, respectively, are of the form

$$F_I(v) = \exp\left[-\exp\left(-\frac{v-\mu}{\sigma}\right)\right]; -\infty < v < \infty \;;$$

$$-\infty < \mu < \infty; \; 0 < \sigma < \infty$$

and

$$F_{II}(v) = \exp\left[-\left(\frac{v-\mu}{\sigma}\right)^{-\gamma}\right]; \; \mu < v < \infty \;; \qquad\qquad (8)$$

$$-\infty < \mu < \infty; \; 0 < \sigma < \infty; \; \gamma > 0$$

in which μ, σ, and γ are location, scale, and tail length parameters, respectively. Actually, the Type I distribution may be shown to be a Type II distribution with $\gamma = \infty$ (see Reference 4, p. 422); however, it is convenient to refer to it separately.

The data were analyzed using -- with minor modifications -- a computer program listed in Reference 5. For convenience, the main features of the procedure used in the analysis of the data are summarized in this section.

I-21

this stage is the so-called maximum probability plot correlation coefficient criterion. The
probability plot correlation coefficient is defined as

$$r_D = \text{Corr}(X,M) = \frac{\Sigma(X_i - X)[M_i(D) - \overline{M(D)}]}{\{\Sigma(X_i - \overline{X})^2 \, \Sigma[M_i(D) - \overline{M(D)}]^2\}^{1/2}} \qquad (9)$$

in which $\overline{X} = \Sigma X_i/n$; $\overline{M(D)} = \Sigma M_i(D)/n$; n = sample size; and D = probability distribution
tested. The quantities X_i are obtained by a rearrangement of the data set: X_1 is the
smallest, X_2 is the second smallest; and X_i the ith smallest of the observations in the set.
The quantities $M_i(D)$ are obtained as follows. Given a random variable X with probability
distribution D and given an integer sample size n, it is possible from probabilistic
considerations to derive mathematically the distributions of the smallest, second smallest,
and generally the ith smallest values of X in a sample of size n. There are various quantities that can be utilized to measure the location of the distribution of the ith smallest
value X_i (e.g., the mean, the median, or the mode). It is convenient to use the median as
a measure of location in Eq. 9 -- these medians of the distribution of the ith smallest value
being denoted by $M_i(D)$.

If the data set was generated by the distribution D, then aside from a location and
scale factor, X_i will be a approximately equal to $M_i(D)$ for all i, and so the plot of X_i
versus $M_i(D)$ [referred to as probability plot] will be approximately linear. This linearity
will, in turn, result in a near unity value in r_D. Thus, the better the fit of the distribution, D, to the data, the closer r_D will be to unity.

The procedure just described makes use of 46 extreme value Type II distributions defined
by various values of γ from 1-25 in steps of 1, from 25-50 in steps of 5, from 50-100 in
steps of 10, from 100-500 in steps of 50, from 500-1,000 in steps of 250, and $\gamma = \infty$. For
any given data set, 46 probability plot correlation coefficients are computed corresponding
to these distributions, and the distribution with the maximum probability plot correlation
coefficient is chosen as the one which best fits the data. The final result from this first
stage is a value, γ_{opt}, of γ corresponding to the estimated best fitting distribution.

The second stage in the procedure consists of estimating the location and scale parameters, μ and σ, respectively, in Eqs. 7 and 8 for the observed data set and for the determined optimal value, γ_{opt}, as determined in stage 1. Estimates of the location and scale follow directly from the basic probability plot approach. If a least-squares line is fit to the probability plot corresponding to γ_{opt}, then the computed intercept and slope of the fitted line serve as estimates for the unknown location and scale parameters, μ and σ. In terms of the X_i and $M_i(D)$, these estimated location and scale values, μ and σ, are as follows:

$$\hat{\sigma} = \frac{\Sigma(X_i - \overline{X})[M_i(D) - \overline{M(D)}]}{\Sigma[M_i(D) - M(D)]^2} \tag{10}$$

$$\hat{\mu} = \overline{X} - \hat{\sigma}\,\overline{M(D)} \tag{11}$$

The third and final stage in the procedure determines the predicted wind speed, v_N, for various intervals N of interest. The estimate for v_N is

$$v_N = \hat{\mu} + \hat{\sigma}G_{X\gamma_{opt}} \quad (1 - \frac{1}{N}) \tag{12}$$

in which γ_{opt} = the optimal value of γ (as determined in stage 1); $\hat{\mu}$ and $\hat{\sigma}$ are the estimates of the location and scale parameters, μ and σ in Eqs. 7 and 8 (as determined in stage 2); and $G_{X\gamma_{opt}}$ (p) = the percentage point function of the best fitting extreme value distribution. If $\gamma_{opt} \neq \infty$ (i.e., if a member of the extreme value type II family provides the best fit), then

$$G_{X\gamma_{opt}} (p) = (-\ln p)^{-1/\gamma} \tag{13}$$

If $\gamma_{opt} = \infty$ (i.e., if the extreme value type 1 distribution provides the best fit), then

$$G_{X\gamma_{opt}} (p) = -\ln(-\ln p) \tag{14}$$

In effect, the procedure described in this section is an automated equivalent of probability paper plotting in which 46 types of probability paper, corresponding to 46 extreme value distributions, would be used and in which fitting would be carried out on the basis of the least-squares method, rather than by eye.

3.2 Estimation of Sampling Errors

As indicated in Section 1, the computer output of Section 4 includes estimates of the standard deviation of the sampling errors, i.e., errors that are a consequence of the limited size of the data sample from which the Type 1 distribution parameters are estimated. Two such estimates were used. One estimate is based on the method of moments and has the following expression given by Gumbel in Reference 6 (PP. 10, 174, and 228):

$$SD(\hat{v}N) = [\frac{\pi^2}{6} + 1.1396(\underline{y - 0.5772})\frac{\pi}{\sqrt{6}} + 1.1(y - 0.5772)^2]^{1/2} \frac{\hat{a}}{n} \qquad (15)$$

in which $SD(\hat{v}_N)$ = the (estimated) standard deviation of the sampling error in the estimation of the N-year wind

$$y = -\ln [-\ln (1 - \frac{1}{N})] \qquad (16)$$

$\hat{a}$ = the estimated value of the scale parameter; and n = the sample size.

A lower bound for the estimated sampling error is given by the following expression:

$$SD_{CR}(\hat{v}_N) = (0.60739y^2 + 0.514y + 1.10866)^{1/2} \frac{\hat{a}}{n} \qquad (17)$$

where the notations are the same as in Equation 15. Equation 17 is commonly referred to as the Cramer-Rao lower bound [7].

3.3 Summary of Results

The results of the analyis are summarized in Table 1, in which the following notations are used:

n = sample size

$\overline{X}$ = sample mean

s = sample standard deviation

v_{max} = sample maximum

γ_{opt} = value of optimal tail length parameter (see Section 3.1)

$\hat{v}_n$ = estimated extreme wind corresponding to a n-year return period, based on Type I distribution

$ppcc$ = probability plot correlation coefficient (see Section 3.1) for Type I distribution

I-24

$\hat{v}_{50}$ = estimated 50-year wind speed

$SD(\hat{v}_{50})$ = estimated standard deviation of sampling error for 50-year wind speed

3.4 Type I Versus Type II Distribution

Of the 129 stations listed in Table 1, 15 stations [marked with the superscript (c) in Table 1 and listed in Appendix 1] have been noted to have largest yearly speed records that may not provide a reliable basis for predicting extreme winds. The remaining 114 stations may be divided into three categories characterized by the value of the optimal tail length parameter γ_{opt}, as shown in Table 2.

Table 2 Classification of Stations According to Value of γ_{opt}

Category	Range of γ_{opt}	Number of Stations	Percentage
I	$13 \leq \gamma_{opt} < \infty$	89	78%
II	$7 \leq \gamma_{opt} < 13$	11	10%
III	$2 \leq \gamma_{opt} <$	14	12%

The sample size for the stations of Table 2 varies between n = 10 and n = 45.

It is noted that the percentage of Table 2 are in qualitative agreement with those found from the analysis reported in Reference 8, in which all sample sizes were n = 37. This tends to confirm the hypothesis advanced in Reference 8 to the effect that, for stations in well-behaved wind climates, the best fit of a Type II (rather than Type I) distribution to a set of extreme wind data might be attributed to a sampling error in the estimation of the tail length parameter. This hypothesis does not exclude the possibility that stations exist for which a Type II distribution might provide an appropriate description of the wind climate; however, according to the results of both Reference 8 and Table 2, the number of such stations, if they exist, is very likely to be small. Thus, it appears justified to assume, as in Reference 8, that the Type I distribution of the largest values provides in general a better description of the wind climate than Type II distributions with small values of the tail length parameter (say, $2 \leq \gamma \leq 12$).

3.5 Largest Wind Speed in a Sample of Size N and the N-Year Wind

It is shown in Reference 9 (see also Reference 4, p. 423) that, if a variate X has a Type I distribution, the mode of the largest value in a sample of n values of X is very nearly equal to the value of the variate corresponding to the mean return period n (recall

I-25

that the mode of a variate X is the value of that variable most 1
given trial). It can be seen from Table 2 that, for most sets fo
ratio $v_{max}/\hat{v}_n$ is indeed close to unity.

4. WIND SPEED AND DIRECTION DATA AND COMPUTER OUTPUT

BISMARCK, N.D. (1940-1977)

THE SAMPLE NUMBER OF OBSERVATIONS =

THE SAMPLE MEAN

THE SAMPLE STANDARD DEVIATION =

THE SAMPLE MINIMUM =

THE SAMPLE MAXIMUM

DATE	ANEMOMETER ELEVATION (FT)	FASTEST MILE WIND SPEED AND DIRECTION (RECORDED AT ANEMOMETER ELEVATION)
06/02/40	43.	57. NW
03/15/41	43.	65. NW
07/07/42	43.	66. NW
07/12/43	43.	72. W
08/08/44	43.	72. S
05/06/45	43.	61. NW
.	.	.
.	.	.
12/08/73	20.	45. NW
05/01/74	20.	45. W
01/11/75	20.	54. NW
02/02/76	20.	47. NW
05/28/77	20.	45. SW

RETURN PERIOD (IN YEARS)	PREDICTED EXTREME WIND BASED ON EXTREME VALUE TYPE I DISTRIBUTION	ESTIMATED STANDARD DEVIATION SAMPLING ERROR CRAMER-RAO	ESTIMATED STANDARD DEVIATION SAMPLING METHOD OF MOMENT
2.0	57.45	.83	.83
3.0	59.84	1.01	1.05
4.0	61.36	1.16	1.24
5.0	62.49	1.27	1.40
.	.	.	.
10.0	65.83	1.63	1.89
.	.	.	.
50.0	73.17	2.48	3.05
.	.	.	.
100.0	76.27	2.85	3.55
.	.	.	.
1000.0	86.53	4.09	5.23
.	.	.	.
10000.0	96.77	5.34	6.92
100000.0	107.01	6.60	8.61
1000000.0	117.25	7.86	10.31

REFERENCES

[1] H. C. S. Thom, "Prediction of Design and Operating Velocities for Large Steerable Radio Antennas," Large Steerable Radio Antennas – Climatological and Aerodynamic Considerations, Annuals of the New York Academy of Sciences, Vol. 116, Art. 1, pp. 90-100, New York, N.Y., 1964.

[2] H. C. S. Thom, "Engineering Climatology of Design Winds, with Special Reference to the Chicago Areas," Proceedings, Symposium on Wind Effects on High-Rise Buildings, Northwestern University, Evanston, IL., March 1970.

[3] American National Standard Building Requirements for Minimum Design Loads in Buildings and Other Structures A58.1, American National Standards Institute, New York, N.Y., 1972.

[4] E. Simiu and R. H. Scanlan, Wind Effects on Structures: An Introduction to Wind Engineering, Wiley-Interscience, New York, N.Y., 1978.

[5] E. Simiu and J. J. Filliben, Statistical Analysis of Extreme Winds, NBS Technical Note No. 868, National Bureau of Standards, Washington, D.C., June 1975.

[6] E. J. Gumbel, Statistics of Extremes, Columbia University Press, New York, N.Y., 1958.

[7] F. Downton, "Linear Estimates of Parameters in the Extreme Value Distribution," Technometrics, Vol. 8, No. 1, February 1966, pp. 3-17.

[8] E. Simiu, J. Bietry and J. J. Filliben, "Sampling Errors in Estimation of Extreme Wind Speeds," Journal of the Structural Division, ASCE, March 1978, pp. 491-501.

[9] E. Simiu and B. R. Ellingwood, "Code Calibration of Extreme Wind Return Periods," Journal of the Structural Division, ASCE, March 1977, pp. 725-729.

Table 1 Summary of Resul

		n	$\bar{x}$	$s/\bar{x}$	v_{max}	$v_{max}/\hat{v}_n$
1.	Birmingham, Alabama	34	46.6	0.139	62.3	1.00
2.	Montgomery, Alabama	28	45.3	0.185	76.7	1.20
3.	Prescott, Arizona	17	52.2	0.169	66.0	0.96
4.	Tucson, Arizona	30	51.4	0.167	77.7	1.09
5.	Yuma, Arizona	29	48.9	0.157	65.1	0.98
6.	Fort Smith, Arkansas	26	46.6	0.150	60.7	0.98
7.	Little Rock, Arkansas	35	46.7	0.206	72.2	1.03
8.	Fresno, California	37	34.4	0.140	46.5	1.00
9.	Red Bluff, California	33	52.1	0.141	67.3	0.97
10.	Sacramento, California	29	46.0	0.223	67.8	0.97
11.	San Diego, California	38	34.5	0.130	46.6	1.02
12.	Denver, Colorado	27	49.2	0.096	62.3	1.02
13.	Grand Junction, Colorado	31	52.7	0.102	69.9	1.07
14.	Pueblo, Colorado	37	62.8	0.118	79.2	0.98
15.	Hartford, Connecticut	38	45.1	0.151	66.8	1.08
16.	Washington, D.C.	33	48.3	0.135	66.3	1.04
17.	Jacksonville, Florida[c]	28	48.6	0.206	74.4	1.04
18.	Key West, Florida[c]	19	51.0	0.337	89.5	1.06
19.	Tampa, Florida[c]	10	49.6	0.163	65.1	1.05
20.	Atlanta, Georgia	42	47.4	0.195	75.5	1.06
21.	Macon, Georgia	28	45.0	0.169	59.7	0.96
22.	Savannah, Georgia[c]	32	47.6	0.202	79.3	1.13
23.	Boise, Idaho	38	47.8	0.111	61.9	1.01
24.	Pocatello, Idaho	39	53.3	0.128	71.6	1.02
25.	Chicago, Illinois	35	47.0	0.102	58.6	1.00
26.	Moline, Illinois	34	54.8	0.141	72.1	0.98
27.	Peoria, Illinois	35	52.0	0.134	70.2	1.02
28.	Springfield, Illinois	30	54.2	0.111	70.6	1.04
29.	Evansville, Indiana	37	46.7	0.130	61.3	1.00
30.	Fort Wayne, Indiana	36	53.0	0.125	69.0	1.00
31.	Indianapolis, Indiana	34	55.4	0.200	93.0	1.12
32.	Burlington, Iowa	23	56.0	0.164	71.9	0.95
33.	Des Moines, Iowa	27	57.7	0.147	79.9	1.04
34.	Sioux City, Iowa	36	57.9	0.157	88.1	1.10
35.	Concordia, Kansas	16	57.6	0.160	73.7	0.97
36.	Dodge City, Kansas	35	60.6	0.099	71.5	0.95
37.	Topeka, Kansas	28	54.5	0.150	78.8	1.08
38.	Wichita, Kansas	37	58.1	0.146	89.5	1.13
39.	Louisville, Kentucky	32	49.3	0.136	65.7	1.00
40.	Shreveport, Louisiana	11	44.6	0.121	53.4	1.00
41.	Portland, Maine	37	48.3	0.179	72.8	1.04
42.	Baltimore, Maryland	29	53.9	0.123	71.2	0.99
43.	Boston, Massachusetts[c]	42	56.3	0.172	81.4	1.05
44.	Nantucket, Massachusetts[c]	23	56.7	0.141	71.3	0.97
45.	Detroit, Michigan	44	48.9	0.140	67.6	1.01
46.	Grand Rapids, Michigan	27	48.3	0.209	66.8	0.94
47.	Lansing, Michigan	29	53.0	0.125	67.0	0.98
48.	Sault Ste Marie, Michigan	37	48.4	0.159	67.0	0.99
49.	Duluth, Minnesota	28	50.9	0.151	69.6	1.01
50.	Minneapolis, Minnesota	40	49.2	0.185	81.6	1.14
51.	Jackson, Mississippi	29	45.9	0.155	64.4	1.03
52.	Columbia, Missouri	28	50.2	0.129	62.6	0.97
53.	Kansas City, Missouri	44	50.5	0.155	75.2	1.06
54.	St. Louis, Missouri	19	47.4	0.156	65.7	1.06
55.	Springfield, Missouri	37	50.1	0.148	71.2	1.04
56.	Billings, Montana	39	59.4	0.135	84.2	1.06
57.	Great Falls, Montana	34	59.0	0.110	74.2	1.00
58.	Havre, Montana	17	58.0	0.159	77.7	1.03
59.	Helena, Montana	38	55.2	0.118	71.2	1.14
60.	Missoula, Montana	33	48.3	0.122	70.9	1.14
61.	North Platte, Nebraska	29	62.0	0.108	74.4	0.96
62.	Omaha, Nebraska	42	55.0	0.195	104.0	1.28
63.	Valentine, Nebraska	22	60.6	0.142	74.1	0.95
64.	Ely, Nevada	39	52.9	0.117	70.1	1.02
65.	Las Vegas, Nevada	13	54.7	0.128	70.1	1.05
66.	Reno, Nevada	36	56.5	0.141	76.6	1.00
67.	Winnemucca, Nevada	28	50.2	0.142	62.6	0.95
68.	Concord, New Hampshire	37	42.9	0.195	68.5	1.08
69.	Albuquerque, New Mexico	45	57.2	0.136	84.8	1.09
70.	Roswell, New Mexico	31	58.2	0.153	81.6	1.03
71.	Albany, New York	40	47.9	0.140	68.5	1.06
72.	Binghamton, New York	27	49.2	0.130	63.8	1.00
73.	Buffalo, New York	34	53.9	0.132	78.6	1.11
74.	New York, New York[c]	31	50.3	0.143	61.4	0.93
75.	Rochester, New York	37	53.5	0.097	65.4	0.99
76.	Syracuse, New York	37	50.3	0.121	67.2	1.03
77.	Cape Hatteras, N. Carolina[c]	45	58.0	0.214	103.0	1.14
78.	Charlotte, N. Carolina	27	44.7	0.168	64.6	1.05

Table 1   Summary of Results (Continued)

79.	Greensboro, N. Carolina	48	42.3	0.180	66.8	1.07	6	.98239	63	4
80.	Wilmington, N. Carolina(c)	26	49.9	0.218	84.3	1.14	4	.97311	80	7
81.	Bismarck, North Dakota	38	58.3	0.096	68.9	0.96	-	.96635	73	3
82.	Fargo, North Dakota	36	59.4	0.185	100.5	1.17	5	.97527	89	6
83.	Williston, North Dakota	16	56.5	0.117	69.3	1.00	-	.98686	75	6
84.	Cleveland, Ohio	35	52.7	0.125	68.5	1.00	-	.97810	70	4
85.	Columbus, Ohio	26	49.4	0.133	61.3	0.96	-	.97303	67	4
86.	Dayton, Ohio	35	53.6	0.142	72.0	1.00	-	.98306	74	4
87.	Toledo, Ohio	35	50.8	0.177	82.2	1.13	11	.98360	75	5
88.	Oklahoma City, Oklahoma	26	54.0	0.110	69.3	1.03	30	.99422	71	4
89.	Tulsa, Oklahoma	35	47.9	0.145	68.3	1.05	150	.98115	67	4
90.	Portland, Oregon	28	52.6	0.196	87.9	1.16	4	.96907	81	7
91.	Roseburg, Oregon	12	35.6	0.169	51.1	1.14	2	.95775	53	6
92.	Harrisburg, Pennsylvania	39	45.7	0.164	64.4	1.00	-	.98849	66	4
93.	Philadelphia, Pennsylvania(c)	23	49.5	0.115	62.4	1.00	150	.99008	66	4
94.	Pittsburgh, Pennsylvania	18	48.4	0.120	59.6	1.00	-	.98221	65	5
95.	Scranton, Pennsylvania	23	44.6	0.107	54.2	0.99	-	.98344	58	3
96.	Block Island, Rhode Island(c)	31	61.4	0.142	86.2	1.06	7	.97591	85	5
97.	Greenville, South Carolina	36	48.5	0.226	71.9	0.95	-	.98512	78	6
98.	Huron, South Dakota	39	61.4	0.132	78.8	0.96	-	.97919	83	4
99.	Rapid City, South Dakota	36	61.0	0.087	70.5	0.96	-	.92675	75	3
100.	Chattanooga, Tennessee	35	47.8	0.218	73.9	1.04	12	.98779	76	6
101.	Knoxville, Tennessee	33	48.8	0.141	65.9	1.01	-	.98045	68	4
102.	Memphis, Tennessee	21	45.4	0.137	60.7	1.04	10	.97829	63	5
103.	Nashville, Tennessee	34	46.8	0.171	70.2	1.06	8	.98665	69	5
104.	Abilene, Texas	34	54.7	0.192	99.9	1.27	3	.93065	82	6
105.	Amarillo, Texas	34	61.0	0.117	80.7	1.03	-	.97987	80	4
106.	Austin, Texas	35	45.1	0.122	58.0	1.00	-	.97715	60	3
107.	Brownsville, Texas(c)	35	43.7	0.185	66.1	1.04	20	.99440	66	5
108.	Corpus Christi, Texas(c)	34	54.5	0.288	127.8	1.45	2	.87186	92	9
109.	Dallas, Texas	32	49.1	0.132	66.8	1.03	30	.99392	67	4
110.	El Paso, Texas	32	55.4	0.087	66.7	1.00	-	.97912	69	3
111.	Port Arthur, Texas(c)	25	53.1	0.181	81.0	1.09	11	.99098	80	6
112.	San Antonio, Texas	36	47.0	0.183	79.5	1.14	3	.96353	70	5
113.	Salt Lake City, Utah	36	50.6	0.142	69.0	1.00	-	.99384	70	4
114.	Burlington, Vermont	34	45.7	0.160	66.5	1.05	14	.98801	66	4
115.	Lynchburg, Virginia	34	40.9	0.149	53.4	0.96	-	.95527	57	4
116.	Norfolk, Virginia(c)	20	48.9	0.182	68.9	1.03	-	.99284	74	7
117.	Richmond, Virginia	27	42.2	0.152	61.3	1.08	35	.98752	60	4
118.	North Head, Washington	41	71.5	0.141	104.4	1.09	3	.94329	98	5
119.	Quillayute, Washington	11	36.5	0.085	41.9	1.01	-	.95120	45	3
120.	Seattle, Washington	10	41.9	0.080	49.3	1.00	-	.91433	51	4
121.	Spokane, Washington	37	47.8	0.133	64.6	1.01	-	.97885	65	4
122.	Tatoosh Island, Washington	54	66.0	0.106	85.6	.99	-	.98811	85	3
123.	Green Bay, Wisconsin	29	56.6	0.212	103.0	1.24	4	.94101	88	8
124.	Madison, Wisconsin	31	55.7	0.190	80.2	1.00	45	.98207	85	6
125.	Milwaukee, Wisconsin	37	53.7	0.121	67.9	0.97	-	.98670	71	4
126.	Cheyenne, Wyoming	42	60.5	0.093	72.6	0.98	-	.96510	75	3
127.	Lander, Wyoming	32	61.2	0.160	80.4	0.96	-	.97563	88	6
128.	Sheridan, Wyoming	37	61.5	0.116	82.0	1.04	-	.97227	81	4
129.	Elkins, West Virginia	10	51.1	0.160	68.5	1.08	13	.98313	75	11

(a) Corresponding to a Type I distribution
(b) Estimated by method of moments
(c) At this station the largest yearly wind speed data may not provide a reliable
     basis for predicting extreme winds - see Appendix 1

APPENDIX 1

STATIONS AT WHICH THE LARGEST YEARLY WIND SPEED DATA MAY NOT
PROVIDE A RELIABLE BASIS FOR PREDICTING EXTREME WINDS

As indicated in some detail, for example, in Reference 4 (p. 84), in a h
region most of the speeds in a series of the largest annual winds are conside
the extreme speeds associated with hurricanes. It may then be argued that in
regions the series of the largest annual speeds may not, in certain cases, pro
statistical information on winds of interest to the structural designer.

For this reason, caution is in order in using the results of the statisti
for the following stations at which hurricane winds may occur:

> Jacksonville, Florida
>
> Key West, Florida
>
> Tampa, Florida
>
> Savannah, Georgia
>
> Boston, Massachusetts
>
> New York, New York
>
> Cape Hatteras, North Carolina
>
> Wilmington, North Carolina
>
> Philadelphia, Pennsylvania
>
> Block Island, Rhode Island
>
> Brownsville, Texas
>
> Corpus Christi, Texas
>
> Port Arthur, Texas
>
> Norfolk, Virginia

ANALYSIS OF HIGH WIND OBSERVATIONS FROM VERY TALL TOWERS

Keikichi Naito

Isao Tabata

Noboru Banno

Katsumi Takahashi

Meteorological Research Institute

Tokyo, Japan

ABSTRACT

This paper describes the analyzed results of high winds, such as typhoons, observed in Japan and its vicinity from tall towers which vary in height from 200 to 400 meters. The results show that the vertical profile of the average wind velocity is well simulated by the power law: the obtained values of power are 0.12, 0.23, 0.46 and so on, and depend upon the local topography upwind. When the upwind roughness of the ground surface, or the unobstructed "clearness" which represents the reciprocal character of upwind roughness, are introduced to model the characteristics of the strong wind, the power decreases with larger degree of clearness, and vice versa. The power is expected to be around 0.12 or 0.15 in extreme clearness. This seems to agree with what is called the seventh-root formula obtained in wind tunnel experimentation. The power is considered usually to be less than 1.0, but will be over 1.0 in severely obstructed situations. The turbulent intensity decreases generally with the increase of height. It shows the following height dependency: for example it is $0.52 \, Z^{-0.31}$ for good clearness, and $1.02 \, Z^{-0.42}$ and $2.14 \, Z^{-0.54}$ for intermediate, or worse clearness, respectively, where Z represents the height in meters. These results show that the profile of turbulent intensity depends upon the effect of the local terrain. The gust factor shows characteristics somewhat similar to those of turbulent intensity. The energy spectrum of turbulence is simulated quite well by the $-5/3$ power law, but the local topography affects height variation in the distribution of small spectral peaks.

KEYWORDS: High wind analysis; gust winds; tall towers; wind turbulence spectrum.

1. INTRODUCTION

There have been only a limited number of high wind observations taken in Japan an
vicinity by the use of very tall towers. Under these circumstances, therefore, observ.
of typhoon 17 (T17/70) on the Iwojima Island in 1970, typhoon 2 (T02/70) on the Okinaw.
Island in 1970 and the severe gust at Kawaguchi in spring 1970 are very valuable for w
analysis. These tower locations ate denoted I, O, and K, respectively, for simplicity
location I, a Loran tower height is 410 meters high, whereas at location O and at loca
respectively, broadcasting tower (165 m) and the NHK broadcasting tower (313 m) were u:
Each tower is equipped with aerovanes at several heights, where observations were cont
ously undertaken for about one and a half years. Since aerovanes were attached to the
side only at the one direction of azimuth, the effective observations were confined to
strong winds mentioned above. Figure 1 shows the observation locations.

On the campus of the Meteorological Research Institute at the Tsukuba Scientific
a meteorological observation tower height of 213 meters has been installed and meteoro.
instruments and data analysis facilities have been gradually installed. Winds have be
observed for one and a half years. During this period, such a strong wind as a typhoo
not been observed. This circumstance results in a wind over 10 m/s at the top of the
being specified as a strong wind for the data taken.

This paper describes the analyzed results based on these strong winds. For vert
profile of wind speed, data of ten-minute averages were used, and the power law obtai
simulation was examined. The characteristics of turbulence, turbulent intensity, gus
tor and spectrum are investigated.

2. VERTICAL PROFILE OF STRONG WIND

2.1 Typhoons and a Severe Spring-Gust [1]

For the vertical profile of strong wind, an investigation will be made concernin
following power law,

$$u \propto Z^P$$

where u and Z represent wind velocity and height, respectively. Figure 2 illustrates
examples of the vertical profile. They are well simulated by the power law and the s

holds in all other cases. The obtained value of the power p has no large variation at the same location, but the difference of the power p is seen clearly among those at the different locations. Table 1 shows the average power at each locations.

For a relatively weak wind, p should depend upon the atmospheric stability or the vertical gradient of air temperature. For a strong wind, however, the stability would be almost neutral and then p should be affected largely by a topographical difference.

Figure 3 shows the topography of the three locations. The tower at the location 0 stands on an isolated and small hill, where a plateau and a long stretch of hills rich in trees are to the wind direction. For the power p at the location 0, the height from the plateau surface is taken as Z. When the height from the base of the tower is adopted as Z, a significant kink appears in the vertical distribution. However, for the use of the height above mentioned, this kink disappears and hence fits the power law very well as a whole. K is located in the Kanto Plains with a small topographic undulation, a grass field, and a small residential area to the wind direction.

At the location I, there are no obstructions to the wind, whereas at the location 0, there are many obstructions. The condition of location K is somewhere between these two. These topographical conditions yield the following results: At the least rough surface (I-location), the value of p is smallest, whereas at the roughest surface (0-location), it is largest. In the intermediate surface (K-location), it is between these values. These results will be examined in detail in the following sections.

2.2 Strong Wind at the Tsukuba Science City

This section describes the analyzed results of strong wind over 10 m/s at the top of the tower for a year from October 1977 to September 1978. The ten minute averages of wind speed were obtained hourly. It is noted as illustrated in Figure 4 that the power law applies well. Table 2 shows the monthly statistics of p. It is rather insufficient to have reliable statistics in July due to a small number of strong wind samples. The standard deviation is not always small, but it is easily seen that the value of p is small in May, June and August and that it is large in October and November. This change of p should be examined.

The change of p is considered usually due to the change of the atmospheric stability. Unfortunately it has become clear there is a problem in the accuracy of the measurement of

air temperature at the meteorological observation tower. At present an effort has been paid
to improve the accuracy, but the analyzed result of the atmospheric stability is not yet
ready to show. However, in the case of a strong wind, vertical variation of air temperature
is not an influential factor as explained in the previous section, whereas local topography
upwind may be an important factor.

The Tsukuba Scientific City Area is generally a flat terrain, but there are many blocks
of buildings of research laboratories and pine groves which obstruct the air flow near the
ground. These circumstances yield the results that the wind profile obtained at the 200
meter tower might be affected by near areas with obstructions of over 20 meters in height.
A direction where such obstructions exist is termed in this paper an "obstructed azimuth"
for convenience; an unobstructed approach is called a "clear azimuth."

All the horizontal direction is divided into the obstructed azimuth, O, and the clear
azimuth, C, at the center of the meteorological observation tower (see Figure 5). In this
figure the distance of 2 km from the center is undertaken. In the obstructing area, buildings of over 35 meters or even 30 meters in height are very few, thereby the distance of
2 km is almost sufficient. In fact the area within 31 km from the center does not show any
significant difference from that within 2 km in this case. It should be noted further that
obstructions over short distances influences the wind quite a bit.

There is an interesting result found in comparison of the azimuthal characteristics
shown in Figure 5 with the windrose obtained each month. In May, June and August when the
value of p is small, the wind comes from the clear azimuth (C-azimuth) very frequently and
in the other months the wind does from the obstructed azimuth (O-azimuth) frequently. This
fact is illustrated in Figure 6 showing the windroses in June and October, and corresponds
to the correlation between the upwind topography and the power p as mentioned in the previous section. In addition it should be noted that both the topography and the power in May,
June and August resemble those of the case at the location K.

So far this correlation has been examined rather qualitatively, but it is desired to
investigate more or less quantitatively. Therefore the term "clearness" is introduced in
regard to upwind topography. Clearness is defined as the percentage frequency of wind
coming from the C-azimuth and is obtained by comparing the windrose with the azimuthal
characteristics as shown in Figure 5. Clearness seems to be reciprocal to surface roughness.

At present clearness is evaluated, very roughly, to around 85 percent in June and 28 percent in October. In Figure 7, the dots show the monthly clearness C and power p. As a whole, it is not clear in this figure that there is a good correlation between the clearness C and the value of p, but in the region of C over 60 percent, p decreases with the increase of C, as shown in solid line. For the value of C over 85 percent an extrapolation is made in the figure with a broken line.

It is very interesting to note that, at the C of 100 percent, p is in the region ranging from 0.12 to 0.15, as indicated by the broken line. As mentioned in the previous section, at the location I the local topography is extremely open to the wind and p is 0.12. This gives good agreement with the extrapolation of C at 100 percent.

Furthermore there is a well known result which is called the seventh-root formula obtained by a wind tunnel experiment: the p value of 1/7(= 0.142) occurs in the generally accepted power law profile for the tubulent boundary layer of a flat plate in a wind tunnel [2]. Obviously this experiment corresponds to the case at the C of 100 percent and verifies the extrapolation. With these considerations, the trend of solid and broken lines would be an understandable and acceptable fact.

As a matter of fact the definition of clearness in this section would be beyond satisfaction. It should be defined by taking into account the distance between observation point and obstruction position, the height of obstruction, the size of obstructing area and so on. Thus the value of C under 60 percent in Figure 7 would be under-estimated if compared with that obtained by the use of the satisfactory definition of clearness. For this reason, the position of dots with C under 60 percent should be shifted to the right considerably.

Another thing to note with the aid of the standard deviation in Table 2: the value of p is possible in some cases to be over 1.0 even in such flat terrain as the Tsukuba area. This should be pointed out here, since the value of p is usually considered to be less than 1.0. [2]

3. TURBULENT CHARACTERISTICS

The fluctuation of wind speed was observed by the instantaneous measurement of the aerovane propeller rotation as mentioned in Section 2.1. In estimating the turbulent characteristics, ten minute sampling was adopted in the case of a significant trend in wind, whereas in the rest of the cases thirty minute sampling was adopted. In the observation described in the Section 2.2, supersonic anemometers are used for the turbulence. Howe the observation system has not been completed yet for strong wind and analysed results not ready to show in this paper.

3.1 Turbulent Intensity

Turbulent intensity is defined as follows:

$$\text{Turbulent intensity} = \frac{\text{standard deviation of wind speed fluctuation } \sigma_u}{\text{average wind speed } \bar{u}}$$

The values of this turbulent intensity are averaged at each observation height and plotted against the height as is shown in Figure 8. The turbulent intensity decreases the height of observation. It shows surprisingly linear dependency with height in this figure. Hence the turbulent intensity in a vertical profile is possibly simulated by t following power law:

$$\frac{\sigma_u}{\bar{u}} = r z^{-q} \tag{2}$$

where r and q are constant.

Table 3 shows empirical equations obtained by the use of equation (2) and Figure 8 The values of r and q increase with the order of locations I, K and 0, as indicated in table. These values are clearly affected by the surface roughness or the upwind clearn as seen at the vertical profile of the average wind speed. The power law obtained expe mentally should be examined in details for various terrains.

There is another point to be noted in Figure 8. The height variation at the locat shows an inversion at above 200 meter height. This phenomena was noticed only at the location I, but there is not reason to cause this inversion in the profile of average w and the surface roughness. It might be caused by the internal waves in the atmosphere, it is not verified at all at present.

3.2 Gust Factor

Gust factor is defined by the following expression:

$$\text{Gust factor} = \frac{\text{maximum wind speed } u_m}{\text{average wind speed } \bar{u}}$$

Analysed results show that all the characteristics of gust factor are similar to those of turbulent intensity. Generally the gust factor is large at low altitudes and small at high altitudes. Moreover there appears quite the same inversion at the location I as that of the turbulent intensity. The obtained values of the gust factor disperse very much, but it could be seen in Figure 9 that the gust factor tends to decrease with the increase of average wind speed.

3.3 Energy Spectrum of Turbulence

Figure 10 shows one of the examples of the calculated spectra. The trend of this example is quite general. In the range of higher frequency especially more than 0.02 Hz, the so-called -5/3 power law holds. However, a further examination in detail shows spectral fine structures. In Figure 11 frequencies of spectral peaks over 2dB high locally are plotted at the observation heights. The spectral fine structures in this figure were obtained by the use of observations arbitrarily picked up. In this figure, in the case of the I location, the distribution of the spectral peaks remains rather unchanged at all the altitudes. In the case of the K location, the range of the peaks seems to be somewhat dispersed at the low height. At the location O, it is clear that the range of the peaks is much dispersed at the lower altitudes. These facts seem to show the effect of the surface roughness as seen in the vertical profile.

4. CONCLUSION

Strong winds, such as typhoons, were analyzed based on the observations by the use of tall towers ranging from some 200 meters to 400 meters. The following was concluded for the average wind speed.

1) In the vertical distribution, the power law fits very well.

2) The powers are 0.12, 0.23, 0.46 and so on, which depend upon the upwind terrains. In the case of small roughness in the surface or open clearness in the upwind, the power is small and vice versa.

3) The minimum value of the power seems to agree with what is called the seventh root formula empirically obtained by a wind tunnel experiment. In the worse case of clearness, the power is possible to be over 1.0, beyond our expectation.

The following was concluded in the turbulence characteristics.

4) The turbulent intensity decreases generally with the increase of height by showing the following height dependency: for example, it is $0.5 \ Z^{-0.31}$, $1.02 \ Z^{-0.42}$, and $2.14 \ Z^{-0.54}$ in the case of good, intermediate and worse clearness, respectively, where Z is the height in meters.

5) Turbulent intensity distribution, as described in (4), is clearly affected by the terrain. Moreover, at high altitudes unusual inversion can be present in some cases.

6) Gust factors show a similar trend to turbulent intensity, except that the former disperse more than the latter.

7) The energy spectrum is expressed by the -5/3 power law, but fine spectral peaks are affected by the terrain.

ACKNOWLEDGMENTS

The authors wish to express their thanks to Dr. Tsutomu Takashima, Meteorlogical Research Institute, for his many useful suggestions.

REFERENCES

[1] Naito, K., Takahashi, K., Kinase, R., Toyama, N., and Ito, S., 1973, Strong Winds Observed from Very Tall Towers, Chokoso Tower Shiryo, Japan Meteorological Agency, No. 5, pp. 20-32 (In Japanese).

[2] Sutton, O. G., 1953, Micrometeorlogy, McGraw Hill, New York, p. 238.

Iwojima (I)	Kawaguchi(K)	Okinawa(O)
$Z^{0.12}$	$Z^{0.23}$	$Z^{0.46}$

Table I   Power law obtained over the different terrains.

Table 2.   Statistics of the exponent p in the power law
Obtained in Tsukuba Science City

Month	Oct.,'77	Nov.,'77	Dec.,'77	Jan.,'78	Feb.,'78	Mar.,'78	Apr.,'78	May,'78	Jun.,'78	(Jul.,'78)	Aug.,'78	Sep.,'78
Sample Number	44	70	92	132	141	77	119	735	185	(15)	129	35
p^{*1}	0.450	0.482	0.396	0.404	0.399	0.391	0.372	0.290	0.258	(0.418)	0.271	0.369
$S.D.p^{*2}$	0.087	0.258	0.159	0.242	0.201	0.183	0.225	0.283	0.182	(0.170)	0.128	0.417

＊1   Average value of p

＊2   Standard deviation of p

Iwojima (I)	Kawaguchi(K)	Okinawa(O)
$0.52\,Z^{-0.31}$	$1.02\,Z^{-0.42}$	$2.14\,Z^{-0.54}$

Table 3   Power law for turbulent intensity.

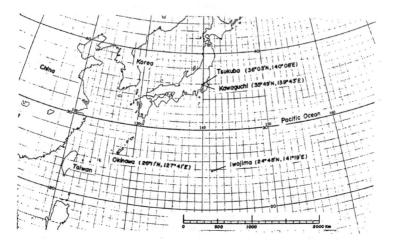

Fig. 1 Locations of the observation points.

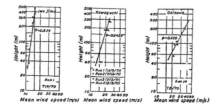

Fig.2 Vertical profiles of the mean speed of strong wind

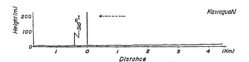

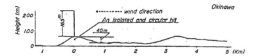

Fig.3 Topographical conditions and observed wind directions
around the towers.

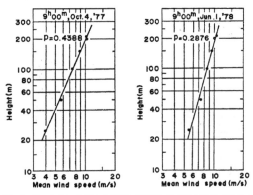

Fig.4 Examples of the vertical profile of the mean wind speed
larger than 10m/s at the tower top in Tsukuba Science City.

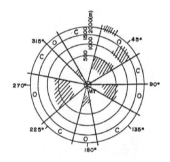

MT : Meteorological observation tower.
 C : Clear azimuth viewed from MT.
 O : Obstructed azimuth viewed from MT.
Shaded area : Blocks of buildings and pine groves.

Fig. 5   Azimuthal pattern of the clearness, C,
and obstruction, O.

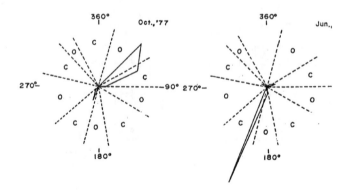

Fig. 6 Examples of monthly windrose and azimuthal pattern
of the clearness and obstruction.

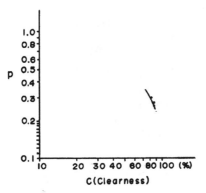

Fig.7 Correlation between the clearness C
and the power P.

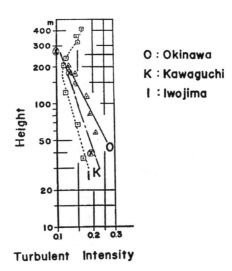

O : Okinawa

K : Kawaguchi

I : Iwojima

Fig. 8 Height variation of turbulent intensity

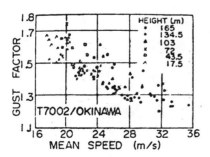

Fig. 9 Correlation between the gust factor and the mean wind speed.

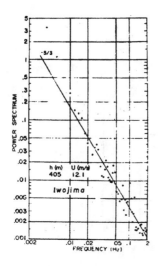

Fig. 10 An example of calculated energy spectrum of turbulence.

I-44

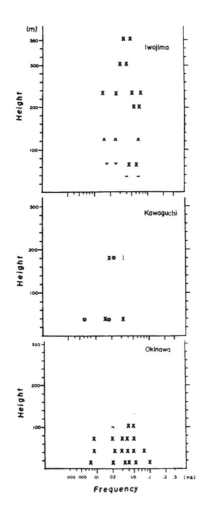

Fig. 11 Local peaks in the spectra of wind
speed fluctuations at various heights.

I-45

A PROPOSAL FOR A NEW PARAMETER IN ASSESSING SEISMIC DISASTER

Michio Otsuka

International Institute of Seismology and Earthquake Engineering

Building Research Institute

ABSTRACT

Out of the recent progress in modern seismology, especially in the rapidly increasing body of knowledge about the nature of what the fault generating process looks like, a prospect of practical earthquake prediction in the near future seems reasonable.

Although a general methodology is not at hand for complete forecasting, long-term estimating of earthquake occurrences in limited areas based on past seismicity studies has already been put to practical use in many reported instances. Though this technique is still insufficient for prediction, since it lacks temporal accuracy about the occurrence of the earthquake, it should prove invaluable and be reflected in disaster reduction strategy. What is needed still is the methodology to associate the expected fault geometry to the plausible disasters.

The purpose of this paper is, then, to search for a technique to evaluate the influence of the impending earthquake based only on a geometrical estimate of a 'should be' fault and without any dynamic information.

Through repeated trials, the author has arrived at the conclusion that the intensity of the seismic vibration is closely related to the solid angle extended from the observation site toward the periphery of the fault.

KEYWORDS: Earthquake prediction; fault dynamics; seismic disaster parameters.

There are at least two categories among modern seismological research items which seem to be promising for future earthquake engineering. One among them is the study of the focal process of an earthquake and the other is earthquake prediction.

a) Focal Process

Integration of seismograms to depict what sort of dynamic process has caused that event in relation to the tectonic system of the focal area has come into popular practice among seismologists. This technique can easily be extended to guess the pattern of a seismogram which should be recorded at any arbitrary site. This feature seems to be of great value since it is expected to produce invaluable data for disaster reduction strategy. However, a lack of knowledge to enable the prediction of the detailed focal process in the present stage causes a bottleneck for the practicality of this approach.

b) Earthquake Prediction

Earthquake Prediction is divided into two categories depending on the duration in time after issuance of warning until the occurrence of the earthquake: long-term, and short-term prediction. It goes without saying that short-term prediction is much more effective for reducing loss of human lives and property. From the standpoint of building construction with longer lifespan, however, long-term prediction is useful as well and should be taken into account also. The complicating circumstance is that for long-term prediction, no detailed information is available. What is needed for the technique to make the best use of long-term prediction, therefore, is that it reflects the meager knowledge about the impending earthquake to disaster reduction strategy.

It may be premature to discuss the successful use of this information since a considerable amount of inaccuracy cannot be avoided now, but it is the firm belief of the author that every possible effort should be made to minimize the possibility of earthquake disaster by making the maximum use of even the small amount of knowledge now on hand.

2. AN EXAMPLE OF LONG TERM PREDICTION -- SEISMICITY STUDY OF KURIL-HOKKAIDO REGION

Figure 1 shows the chain of the focal domains of the events in recent years compiled by Utsu (1972 a, b). Utsu remarked that focal domains of each earthquake are strictly bounded and do not superpose each other. He also pointed out that domain C looks likely to experience an occurrence of a fairly large earthquake since this area has not experienced any since 1894.

Shimazaki (1972) associated crustal deformation observed at the eastern part of Hokkaido to the compression of continental lithosphere subjected to the stress of the Pacific Plate. As was anticipated, an earthquake with magnitude 7.4 occurred on June 7, 1973, off Nemuro Penninsula situated at the eastern tip of Hokkaido.

This is an example of successful long-term prediction of earthquakes based only on gaps of seismicity and of crustal deformation. The technique exemplified herein is one regarded to be the most reliable among various methods proposed toward the prediction of earthquakes through many successful reports (Sykes (1976), Borovik, et al. (1971), Ohtake, et al. (1977)).

It must be noted, however, that information obtained from this kind of study is not about the detailed mechanism of the focal process, but about the vague location of the focal region.

3. PARAMETER M AND Δ

The trial is customarily made to associate the seismic damage with the magnitude M of that earthquake and with the distance Δ from the epicenter. Various formulas are successfully proposed in this regard proving that this technique is promising for application to seismic disaster assessment. However, it should be noted that each of the two parameters has its own inadequateness judging from seismological view point.

a) Magnitude M

Since magnitude M of the earthquake is basically determined based on the maximum amplitude of seismograms, it appears to be the most effective measure to assess a seismic disaster. Unfortunately, however, it has been shown that the magnitude scale is accompanied by an inherent skew to represent the real potentiality of seismic ground motion originating from the frequency characteristics of seismographs. It seems fatal, moreover, that tendency of skew is more pronounced in the region of "large earthquakes" to which earthquake engineering should mainly be concerned.

b) Epicentral Distance Δ

Epicentral distance Δ is defined as the distance between the epicenter and the site. Association of seismic disaster with epicentral distance is only varied when a major part of seismic energy is radiated from the hypocenter. According to the modern understanding of seismology, hypocenter is just a point from which the fracture started and the major part of seismic energy is radiated from a rather wide region around the fault.

In addition, parameters M and Δ involve some physical confusion. Use of these two parameters for assessing the seismic disaster does not suit the type of long-term prediction of earthquakes described above.

4. NEW PARAMETER

Seismic disaster or intensity of earthquake is regarded as a complicated function of motion. It is hardly possible to express it with a limited number of parameters, but as a primary step to the study, a parameter is developed for expressing the acceleration in place of M and Δ.

For that purpose we start from a simple a priori assumption connecting seismic peak acceleration and source geometry.

1) Point Source

As shown in Figure 2, a simple dynamic situation is considered that a single force F is exerted on a point P. It is assumed that seismic peak acceleration observed at O is expressed by

$$\alpha = K \frac{F}{r} \tag{1}$$

where r is the distance between P and O, and k is a constant.

2) Dipole Source

As is shown in Figure 3, when dipole force is exerted on P instead of a single force, the peak acceleration at O is expressed as

$$\alpha = kF\left(\frac{1}{r - \frac{\ell}{2}\cos\theta} - \frac{1}{r + \frac{\ell}{2}\cos\theta}\right) = k\frac{F\ell}{r^2}\cos\theta \tag{2}$$

expanding the Eq. 1. The quantity Fl is the dipole moment and θ represents the azimuth of Vector PO in relation to the dipole reference.

3) Dipole Shell Source

Let us consider a group of dipoles lined up so as to construct a shell with positive and negative surfaces as is shown in Figure 4. The force exerted per unit area of the surface is designated as F' and surface element as dS. Peak acceleration can be obtained as Eq. 3 by integrating Eq. 2.

$$\alpha = \int k \frac{F'\ell}{r^2} \cos\theta \, dS \tag{3}$$

where integration ranges over the entire surface.

While, using the relation $r^2 \, d\omega = \cos\theta \, dS$ (dω is solid angle extended by dS), Eq. 3 can be written

$$\alpha = \int kF'l \, d\omega = kF'l \, \Omega \tag{4}$$

where Ω represents the total solid angle extended from 0 toward the entire surface.

The strain ϵ that ground material exerts is given by

$$\vdots = \frac{D}{\ell}$$

where D denotes the dislocation due to the formation of the fault. Denoting the rigidity of the material around the focal area as μ, F' can be expressed as

$$F' = \mu \epsilon = \mu \frac{D}{\ell} \tag{5}$$

merging Eq. 5 with Eq. 4, we get

$$\alpha \quad k\mu \, D\Omega \tag{6}$$

Eq. 6 signifies that if μ and D are constant for an earthquake, α is proportional to Ω Taking logarithm of Eq. 6, we get

$$\log \alpha = C + \log \Omega \tag{7}$$

Eq. 7 shows that when plots are made of peak accelerations α against the solid angle Ω toward the fault plane, both on logarithmic scale, they should be aligned on a straight line with gradient of unity.

5. MIYAGIKEN-OKI EARTHQUAKE

Figure 5 shows schematically the fault geometry of Miyagiken-Oki Earthquake worked out by Seno et al. (1978) which occurred on June 12, 1978. Reported peak accelerations for this earthquake are plotted against the solid angle extended to the fault plane from the site on log-log sheet in Figure 6. It is observed in Figure 6 that plots are aligned on a single line with a gradient of unity as expected which shows that our a priori postulation is not physically absurd.

6. VOLCANIC FRONT

Grouping of the plots in Figure 6 into two parts is made by fitting a line to show the approximate trend. Solid and open circles show the sites which exhibited peak accelerations more and less than average, respectively. Replotting the solid and open circles to the original sites given in Figure 7, it is remarkable that the distribution of plots show an extremely interesting feature from a geophysical view point. That is, the position of the border between the two symbols looks very similar to the so-called volcanic front line, across which seismic wave dissipation characteristics are believed to be markedly different because of difference in heat flow.

For reference sake, the same grouping is applied to the traditional technique using epicentral distance, and plots are made in Figure 8. Note that no physically interesting and systematic interpretations are available.

7. CONCLUSION

It should be admitted that several clear drawbacks do exist for the technique proposed here. For example, information on the minute structure of fault generating dynamics is entirely too thin, and a dipole source is not appropriate from modern seismological understanding. These are merely two points, among others, which need further refinement.

REFERENCES

[1] Borovik, N., Mansharina, L., and Treskov, A., On the Possibility of Strong Earthquakes
     in Pribaykalia in the Future, Izv. Acad. Sci. USSR, Phys. Solid Earth, English Trans-
     lation, (1) 13-16 (1971).

[2] Ohtake, M., Matsumoto, T., Latham, G. V., Temporal Changes in Seismicity Preceding
     Some Shallow Earthquakes in Mexico and Central America, Bulletin International Insti-
     tute of Seismology and Earthquake Engineering 15, 105-123 (1977).

[3] Shimazaki, K., Focal Mechanism of a Shallow Shock at the Northwestern Boundary of the
     Pacific Plate: Extentional Feature of the Oceanic Lithosphere and Compressional
     Feature of the Continental Lithosphere, Phys. Earth Planet. Inter. 6, 397-404 (1972).

[4] Sykes, L., Aftershock Zones of Great Earthquakes, Seismicity Gaps, and Earthquake
     Prediction for Alaska and Aleutians, U. Geophys. Res. 76, 8021-8041 (1971).

[5] Utsu, T., Large Earthquake Near Hokkaido and the Expectancy of the Occurrence of a
     Large Earthquake Off Nemuro, Rep. Cook. Comm. Earthquake Prediction, 7, 7-13 (1972 a).

[6] Utsu, T., Aftershocks and Earthquake Statistics, TV, J. Fac. Sci. Ser. 7, 4: 1-42
     (1972).

Fig. 1 Seismicity in the Kuril-Hokkaido
region

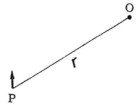

Fig. 2 Point source

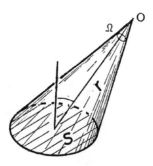

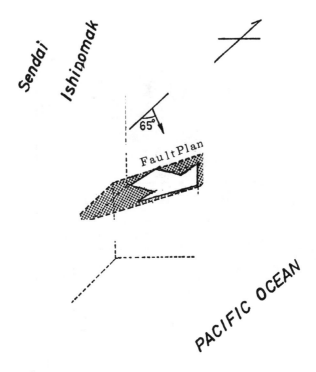

Fig. 5 Schematic representation of the Miyagiken Oki earthquake fault

II-9

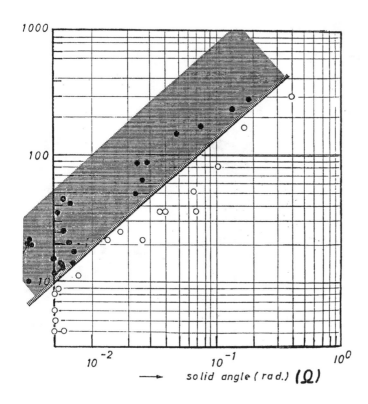

Fig. 6 Peak acceleration against the solid angle

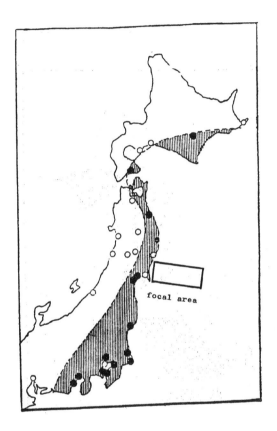

focal area

Fig. 7 Distribution of magnitude of peak acceleration
(grouped by α-Ω relation)

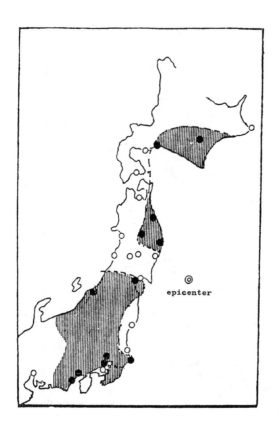

epicenter

Fig. 8 Distribution of magnitude of peak acceleration
(grouped by α-Δ relation)

SIMULATION OF EARTHQUAKE GROUND MOTION AND

ITS APPLICATION TO DYNAMIC RESPONSE ANALYSIS

Tetsuo Kubo

Makoto Watabe

Structural Engineering Department

Building Research Institute

Ministry of Construction

ABSTRACT

Through the relation of Fourier transformation, two types of stochastic modeling of earthquake ground motion are introduced. Twenty samples of synthetic motion, in each case, are generated by use of these two modelings simulating a certain recorded motion. Using these motions, characteristics of synthetic motions such as the cumulative energy distribution (the integration of square acceleration), the maximum elastic response and the maximum elasto-plastic response are evaluated. From a statistical viewpoint, the results are compared with those obtained from the recorded earthquake motion in an attempt to make use of synthetic motions for an engineering application.

KEYWORDS: Dynamic response analysis; ground motion; modeling; synthetic earthquake motion; waveforms.

INTRODUCTION

For aseismic design of a structural system, there would be two alternatives to d
an earthquake input to the system. One is to use a waveform of motion, and the other
use a design spectral curve such as that in the literature [1]. When using the forme
dure, the following two manners can be used to determine a waveform for structural de

The one is to determine a waveform either by use of real strong motion waveforms
those recorded at El Centro, California and at Hichinohe Harbour, Japan with consider
upon such as subsoil conditions, or by use of real waveforms of intermediate-size ear
recorded at the prospective construction site. It has been recognized that earthquak
motions, even recorded at the same site location, have often yielded quite dissimilar
acteristics to dynamic response properties. Hence, there would arise a question "Cou
specific motion of a past earthquake be representative of a future ground motion at a
site location?".

The other is to determine a waveform by use of a stochastic process having approp
properties in a statistical sense. Recently by a seismological approach, time trace o
earthquake motion can be calculated determining the source mechanism and physical prop
of wave propagation path. However, considering complexity and irregularity in the ini
of seismic wave and its transmission paths, stochastic modeling of earthquake motion i
realistic and feasible approach for an engineering application.

GENERATION OF SYNTHETIC EARTHQUAKE MOTION PROCESS

Under certain mathematical conditions, a time function corresponds uniquely to it
Fourier transform through the relations

$$F(\omega) = \int_{-\infty}^{\infty} f(t) \, e^{-j\omega t} \, dt \tag{1}$$

$$F(t) = \frac{1}{2\pi} \int_{-\infty}^{\infty} F(\omega) \, e^{j\omega t} \, d\omega \tag{2}$$

where f(t) and F(ω) represent a time function and its corresponding Fourier transform,
respectively, and j denotes the imaginary unit, i.e. $j^2 = -1$. Fourier transform in Eq
be expressed by a pair of an amplitude function and a phase function through

II-14

$$F(\omega) = A(\omega) \exp\left[-j\phi(\omega)\right] \tag{3}$$

where

$$A(\omega) = |F(\omega)| \tag{4}$$

$$\phi(\omega) = -\tan^{-1}\left[\text{Im}[F(\omega)]/[\text{Re}[F(\omega)]]\right] \tag{5}$$

are Fourier amplitude spectral function and Fourier phase spectral function, respectively. Eq. 1 through 5 indicate that time function $f(t)$ and a pair of functions $A(\omega)$ and $\phi(\omega)$ relate uniquely with each other through the relation of Fourier pair.

Provided that $f(t)$ is a real time function such as an earthquake accelerogram studied in this paper, Eq. 2 can be written in discrete form by

$$f(t) = \sum_{n=0}^{N/2} a_n \cos(\omega_n t - \phi_n) \tag{6}$$

where

$$a_n = A_n \Delta\omega/2\pi \quad (n = 0 \text{ and } N/2), \quad a_n = A_n \Delta\omega/\pi \quad \text{(otherwise)}$$
$$A_n = A(\omega_n), \quad \omega_n = n\Delta\omega = n\, 2\pi/T_D, \quad \text{and} \quad \phi_n = \phi(\omega_n) \tag{7}$$

Let Fourier amplitude and Fourier phase spectral functions obtained from recorded earthquake accelerogram $f_o(t)$ be denoted by $A_o(\omega)$ and $\phi_o(\omega)$ respectively. The following two stochastic modelings of synthetic earthquake motion are introduced.

The first, called the Modeling I hereafter, is to determine Fourier phase spectrum by analysis of a real earthquake motion and to determine Fourier amplitudes by uniformly distributed random numbers, i.e.

$$g_i(t) = \sum_{n=0}^{N/2} a_{in} \cos(\omega_n t - \phi_{on}) \quad (i = 1, 2, \dots) \tag{8}$$

in which i denotes the i-th individual member from the ensemble and a_{in} are random numbers distributed uniformly over the range of 0 and 1.

The second, the Modeling II, is to establish Fourier amplitude spectrum from a real earthquake motion giving random phase angles distributed uniformly in the interval of 0 and 2π to Fourier phase spectrum; thus

$$g_i(t) = \sum_{n=0}^{N/2} a_n \cos(\omega_n t - \phi_{in}) \quad (i = 1,2, \ldots)$$

in which ϕ_{in} represent random phase angles uniformly distributed over the range
and 2π.

SYNTHETIC EARTHQUAKE MOTIONS

Modeling I Since both the mean value and the auto-correlation function are
functions of time [2], the process reveals nonstationarity.

The amplitudes in Eq. 8 are given by random numbers in the range of 0 and 1
paring the synthetic motions with a recorded motion, one should make an appropri
tion upon amplitudes of the generated waveform. Many intensity scales have been
estimate a seismic intensity. Among them, the scale introduced by Arias [3] is
this investigation. Determining the intensity scale of a synthetic motion equal
real motion, the waveform of synthetic earthquake motion $f_i(t)$ is given through

$$f_i(t) = c_i \, g_i(t)$$

where c_i is a scaling factor obtained from

$$c_i = \sqrt{\int_0^{T_D} f_0^2(t) \, dt \Big/ \int_0^{T_D} g_i^2(t) \, dt}$$

in which $f_0(t)$ represents the waveform of a recorded earthquake motion.

Modeling II Both the mean value and the auto-correlation function are time
Average over time of any member from the ensemble equals its corresponding avera
ensemble [2]. It leads to the conclusion that, in addition to being stationary,
is ergodic.

It is easily shown that the intensity of process $g_i(t)$ in Eq. 9 is essentia
that of recorded motion $f_0(t)$. Since the process has been concluded stationary,
multiplied by an appropriate deterministic time function to reveal adequate rise
decaying of intensity of waveform during shaking. Let the energy of waveform at
defined through

II-16

$$E_i(t) = \int_0^t f_i^2(s) \, ds \qquad (12)$$

and suppose synthetic earthquake waveform $f_i(t)$ is expressed by the product of process $g_i(t)$ and deterministic time function $\zeta(t)$,

$$f_i(t) = \zeta(t) \, g_i(t) \qquad (13)$$

Assuming that the energy of a synthetic earthquake waveform at time t would be expected to equal the corresponding energy of the real earthquake waveform, deterministic intensity function $\zeta(t)$ is given as follows

$$\zeta(t) = \sqrt{T_D \frac{d}{dt} E_0(t)/E_0(T_D)} \qquad (14)$$

or in a normalized form,

$$\zeta(\tau) = \sqrt{\frac{d}{dt} e_0(\tau)} \qquad (15)$$

in which τ and e_0 denote normalized time variable t/T_D and normalized energy of waveform $E_0(t)/E_0(T_D)$, respectively.

In this investigation, considering conditions upon a form of function $\zeta(t)$, an intensity function is determined through the following manner. In Figure 1a, the curve of a function for energy of a waveform is shown. Divide axes x and y into m and n pieces, respectively, and obtain the intersecting points with the curve. Connecting these points in succession by straight lines, approximate the curve with the pieces of straight line. Consequently in this study, the deterministic intensity function is obtained by a step function as shown in Figure 1b.

Earthquake Waveform   The real waveform, the East-West component at Hachinohe Harbour recorded during the Tokachi-oki earthquake of May 16, 1968, is selected. Figure 2 presents the time trace of the real accelerogram. Twenty samples of synthetic motion, in each case, are produced by use of these two modelings simulating the recorded waveform. In Figures 3 and 4, time traces of synthetic earthquake motion are shown. It can be recognized that plots of motions from the Modeling I reveal quite similar features with one another, while they show higher fluctuation than the plot of the real earthquake motion.

CUMULATIVE ENERGY DISTRIBUTION

Energy distribution of waveform associated with time for a synthetic motion is examined by terms of the so-called "normalized cumulative energy function,"

$$e_i(\tau) = E_i(t)/E_i(T_D) \tag{16}$$

where τ and E_i denote normalized time variable t/T_D and the energy of waveform defined previously by Eq. 12, respectively. For motions from the Modeling II, the energy function is expected to coincide with that obtained from the real motion with certain amount of fluctuation. Hence, examination is carried out only for motions from the Modeling I.

Cumulative energy functions both for synthetic motions and for the real motion are presented in Figure 5, in which x and y axes represent normalized time variable τ and normalized cumulative energy distribution e_i, respectively. Six curves in the figure show the cumulative energy functions for cases described as follows; the thick solid curve for that of the recorded motion, the thin solid curve for the mean value of those obtained from the twenty samples of synthetic motion, two dot-dashed curves for the mean value plus or minus ten times the standard deviation of those obtained from the synthetic motions, and two dashed curves for the maximum and the minimum of those obtained from the synthetic motions.

The results indicate that energy distribution functions for waveforms from the Modeling I reveal quite similar plots with one another. Consequently, forms of envelope curve which specifies intensity varying with time for synthetic motions are similar with one another. It can be observed that the general form of envelope plots for synthetic motion is similar to that obtained from a real earthquake motion.

MAXIMUM ELASTIC RESPONSE SPECTRA

Figure 6 and 7 show the results of maximum elastic response spectra for motions of the selected component from both the Modeling I and the Modeling II, respectively. Axes x and y represent the natural period of an oscillating system and the maximum pseudo velocity response associated with a fraction of the critical damping of 0.05 in the system. In these figures, the thick solid curve, the thin solid curve and two dashed curves represent the results of the maximum response subjected to the recorded motion, of the mean value of maximum responses subjected to the synthetic motions, and the maximum and the minimum of maximum responses subjected to the synthetic motions, respectively. The shaded zone shows the range where responses fall in value within the interval of the mean value plus or minus one standard deviation taken across the responses obtained from the twenty samples of synthetic motion.

Modeling I Similarity between the maximum velocity response spectrum and Fourier amplitude spectrum suggests that the plot of the thin solid curve will be flat. The results in Figure 6 reveal a fairly flat form. The magnitude of responses in the intermediate range of periods is smaller for the synthetic motions than that for the real motion. For motions from the Modeling I, one cannot observe the dominant period of motion included which is one of the characteristic features of earthquake motion in engineering sense. Therefore, the Modeling I is concluded not adequate for engineering practice.

Modeling II In Figure 7, the thin solid curve coincides with the thick solid curve with good agreement over the entire range of period. It is an interesting result that the responses subjected to a recorded motion usually fall in the shaded zone representing the interval of the mean value plus or minus one standard deviation obtained across the synthetic motions. The twenty responses subjected to the synthetic motions associated with a certain natural period of an oscillating system have a tendency to fall in value around the mean value obtained from these twenty responses. Though distribution of responses cannot be concluded Gaussian by x^2 examination with confidential level of 50 percent, numerical analysis yields the result that the maximum response subjected to a synthetic motion is not greater than the mean value plus one standard deviation with probability of 85 percent, while in case of Gaussian random distribution the corresponding probability is 84.13 percent.

MAXIMUM ELASTO-PLASTIC RESPONSE SPECTRA

The responses of a linear system subjected to a stochastic process can be theoretically determined by variations of statistical quantities associated with the input process. For a on-linear system, in general cases, one is forced to conduct numerical analyses. In this case, input excitation is essentially expressed explicitly as a waveform of motion. For synthetic motions from the Modeling II, maximum elasto-plastic response spectra are evaluated and are compared with those obtained from a real motion.

Presented in Figure 8 is a schematic rule for the Degrading Tri-Linear hysteresis model [ ] employed in this investigation which is one of representative models for a reinforced

gal, respectively, and the ratio of the second stiffness to the initial one is 1/4.
Multiplying amplitudes of waveform by scaling factors 1.640 and 2.187 so as resulting peak
accelerations of the real motion to be 300 gal and 400 gal, and assuming a fraction of the
critical damping of 0.05, the maximum displacement responses are obtained both for synthetic
motions and for the recorded motion. Using logarithmic scale diagrams, the results are shown
in Figures 9a and 9b. The legends in these figures such as the thick solid curve, the thin
solid curve, the dashed curves and the shaded zone are identically similar to those in
figures showing the results of elastic response spectra.

Observation upon these figures leads to the evidence that the response subjected to a
recorded motion and the mean value of responses subjected to the synthetic motions coincide
well with each other over the entire range of natural periods. It should be noticed that
responses obtained from a real motion usually fall in the shaded zone. On an average, in
sixteen or seventeen cases out of twenty, the responses subjected to synthetic motions do
not exceed in value greater than the mean value plus one standard deviation obtained from
the responses subjected to the synthetic motions.

CONCLUDING STATEMENT

Through the correspondence of a time function with its Fourier transform, two types
of stochastic modeling of synthetic earthquake motion are introduced. Using this modeling,
synthetic earthquake motions are produced which simulate a certain selected recorded motion.
In an attempt to use these synthetic motions for engineering practice, characteristics such
as the cumulative energy distribution, the maximum elastic response and the maximum elasto-
plastic response are obtained, and are compared with those obtained from the recorded motion.
Dynamic properties of synthetic motions are summarized in the following:

1) Motions from the Modeling I have nonstationarity in intensity with time, and reveal
quite similar cumulative energy distributions with one another. Having quite similar forms
with one another, the general form of envelope functions from synthetic motions coincide
well with that from the recorded motion. In the maximum elastic response analysis, certain
dominant periods included in the recorded motion cannot be synthesized by the Modeling I.

2) For motions from the Modeling II, the mean value of responses both in elastic analysis and in elasto-plastic analysis coincides with the response obtained from the recorded
II-20

motion with good agreement over the entire range of periods of an oscillating system. In both analyses, the maximum response subjected to an individual member from synthetic motion falls in the value not greater than the mean value plus one standard deviation obtained from analysis across the synthetic motions with a probability of higher than 80 percent.

Although motions from the Modeling I have been concluded to be inadequate for an engineering application, they reveal quite a specific feature in intensity with time. It is necessary that a mathematical model of synthetic earthquake motion be developed that considers the phase angles of waveforms for extensive application of synthetic motions to engineering practice.

ACKNOWLEDGMENT

The authors wish to express their sincere thanks to Mr. Norio Suzuki, Graduate Student, University of Tokyo, Tokyo for his great contribution to this paper. For other components of real earthquake accelerograms, similar analysis and observation has been carried out [2,5]. The first-named author is indebted to Professors H. Umemura and H. Aoyama, University of Tokyo, Tokyo for their continuous guidance and encouragement through the study.

REFERENCES

[1] Newmark, N. N., and Rosenblueth, E., "Fundamental of Earthquake Engineering," Prentice-Hall, pp. 229, 1971.

[2] Kubo, T., and Suzuki, N., "Simulation of Earthquake Ground Motion and Its Application to Response Analysis," Trans. Archi. Inst. Japan, No. 275, pp. 33-43, January 1979 (in Japanese).

[3] Arias, A., "A Measure of Earthquake Intensity," in "Seismic Design for Nuclear Power Plants," ed. R. J. Hansen, MIT Press, pp. 438-483, 1970.

[4] Fukada, Y., "On the Load-Deflection Characteristics of R/C Structure," Proceedings 40th Kanto District Symposium, Archi. Inst. Japan, pp. 121-124 (1969) (in Japanese).

[5] Kubo, T., Suzuki, N., Aoyama, H., and Umemura, H., "Simulation of Synthetic Earthquake Motions for Response Analyses," Proceedings 49th Kanto District Symposium, Archi. Inst. Japan, pp. 197-204 (1978) (in Japanese).

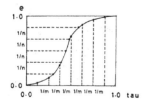

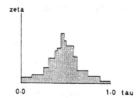

Fig. 1 Deterministic intensity function.

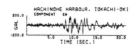

Fig. 2 Time trace of the recorded motion,
the EW component, Hachinohe Harbour.

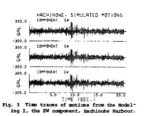

Fig. 3 Time traces of motions from the Modeling I, the EW component, Hachinohe Harbour.

Fig. 4 Time traces of motions from the Modeling II, the EW component, Hachinohe Harbour.

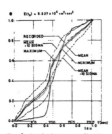

Fig. 5 Cumulative energy distribution for the EW component, Hachinohe Harbour.

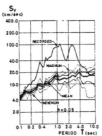

Fig. 6 Elastic response spectra for both the recorded and synthetic motions from the Modelling I, the EW component, Hachinohe Harbour.

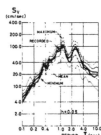

Fig. 7 Elastic response spectra for both the recorded and synthetic motions from the Modelling II, the EW component, Hachinohe Harbour.

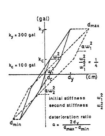

Fig. 8 Hysteresis rule of the Degrading Tri-Linear model.

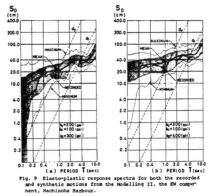

Fig. 9 Elasto-plastic response spectra for both the recorded and synthetic motions from the Modelling II, the EW component, Hachinohe Harbour.

COMPARISON OF VERTICAL COMPONENTS OF STRONG-MOTION

ACCELEROGRAMS FOR WESTERN UNITED STATES AND JAPAN

Tatsuo Uwabe

Port and Harbour Research Institute

Ministry of Transport

ABSTRACT

Characteristics of the vertical component of 187 strong-motion accelerograms recorded in the Western United States were studied. According to the analysis of the ratio of maximum vertical acceleration to maximum horizontal acceleration, the average ratio of vertical peaks to horizontal peaks is 0.48. Significant correlations between maximum horizontal acceleration, epicentral distance, and earthquake magnitude could not be found. The vertical maximum acceleration occurs near the occurrence time of the horizontal maximum acceleration. But, the coincidence of vertical and horizontal peaks is very rare. At the time when the safety factor against sliding and overturning for gravity-type structures is smallest, the ratio of the vertical acceleration to the resultant horizontal acceleration is less than one-third. A comparison was made of the ratios of vertical peaks to horizontal peaks for Western United States and for 574 strong-motion accelerograms from Japan. The result of this comparison is as follows: The ratio of vertical peaks to horizontal peaks in the Western United States is larger than that in the Japanese recordings. The difference between the accelerograms of the two countries is believed to be caused by dissimilar instruments, dissimilar installation conditions, and differences in focal depth and site conditions. This difference should be kept in mind when the digitized accelerograms of both countries are used together.

KEYWORDS: Earthquake engineering; earthquake resistant structures; earthquakes; ground motion.

INTRODUCTION

Strong-motion accelerograms consist of three perpendicular components of ground acceleration: vertical and two horizontal. It is generally recognized that vertical acceleration time histories usually have smaller maximum amplitudes and higher frequency components as compared with horizontal acceleration time histories. Hence, the effects of vertical accelerations on stability of civil engineering structures and buildings are often relatively unimportant. However, case studies of earthquake damage records indicate that the vertical motion may have severe influence on the stability of a soil embankment or piled blocks. [1] Some accelerograms whose vertical components were larger than their horizontal components have been recorded. It is, therefore, important to investigate the characteristics of vertical components of accelerograms recorded during past earthquakes.

In order-to collect more reliable data for the seismic design, engineering features of vertical motions including peak amplitude, ratios of vertical peaks to horizontal peaks and predominant frequency were investigated in the study by analysis of 187 strong-motion accelerograms recorded in the western United States. Difference in occurrence time between the vertical and horizontal peaks also were analyzed.

As the number of accelerograms is not sufficient, accelerograms recorded in foreign countries are used. But there are differences between accelerograms of two countries. Therefore, it is necessary to compare accelerograms of two countries before these data are used together. From this viewpoint, a comparison was made of the ratios of vertical peaks to horizontal peaks for the western United States and for 574 strong-motion accelerograms from Japan.

STRONG-MOTION DATA

The strong-motion accelerograms used in this study were derived from the "Catalogue of Strong-Motion Earthquake Records, Volume I, Western United States, 1933-1971." [2] These accelerograms have been extracted from the Volume II tapes [3] which contain corrected accelerograms uniformly processed by the California Institute of Technology. [4] The accelerograms have been band-pass filtered to retain only those components between 0.07 Hz (in case of 70- and 35-mm film records, the cutoff frequency of 0.125 Hz [5]) and 25 Hz.

These data were obtained from 49 earthquakes whose magnitude range from 3.3 t
data contain 187 records recorded on the ground or in the basements of buildings.
records, 90 percent correspond to the magnitude range of 5.0 to 6.9. Eighty-one p
the data are from shallow focal depth (less than 20 km) earthquakes. Fifty-nine pe
the data were recorded on "alluvium" sites, 24 percent of the "intermediate" sites,
percent on the "hard rock" sites. [6] Sixty-five percent of the records are for ep
distances less than 50 km, 18 percent between 50 and 100 km, and 17 percent are mor
100 km.

Six different types of strong-motion accelerograms that have a variety of natu
periods, sensitivities, and recording media are described in Table 1. [7,8] Some o
are or have been used in the United States, and some are used in Japan.

CHARACTER OF VERTICAL ACCELERATION

Ratio of Maximum Vertical Acceleration to Maximum Horizontal Acceleration

Figure 1 presents the maximum vertical accelerations (A_{vmax}) of western United
strong-motion data plotted versus the maximum horizontal accelerations (A_{hmax}) on a
grid. Circles denote the San Fernando (1971) earthquake record, and dots indicate o
strong-motion data for the period in 1931-1971. In this figure, ratios of the verti
to the horizontal peaks (A_{vmax}/A_{hmax}) are also indicated by straight lines. The ave
the ratios is 0.48, and the standard deviation is 0.19.

The maximum horizontal acceleration in Figure 1 means the larger one of the two
tal components, but the real horizontal acceleration at every moment is the resultan
two components, and the maximum resultant acceleration in the horizontal direction i
larger than the largest horizontal acceleration in a single component. Therefore, t
A_{vmax}/A_{hmax} would become smaller, if the maximum horizontal resultant acceleration w
instead of the maximum horizontal acceleration.

Factors Affecting Ratio of Maximum Vertical to Horizontal Acceleration

According to Figure 1, some of the data show large ratios of A_{vmax}/A_{hmax}. If i
determined what factors tend to cause large ratios, this may be useful in the develo
site-specific ground motion parameters for earthquake-resistant design. The influen

various factors including horizontal acceleration, epicentral distance, earthquake magnitude, subsoil condition, and focal mechanism on the ratio A_{vmax}/A_{hmax} were examined on an empirical basis.

The average vertical to horizontal acceleration ratio for records whose horizontal peaks are less than 50 gal is 0.51. When horizontal peak ranges 50 to 100 gal, the average ratio is 0.49. When the horizontal peak is more than 100 gal, the average ratio is 0.46. From the trend, it can be seen that the average ratio decreases slightly as the maximum horizontal acceleration increases. Significant correlations between the ratio of vertical to horizontal peak and epicentral distance, earthquake magnitude, subsoil condition or focal mechanism could not be found.

Frequency Characteristics of Vertical Motion

It is generally recognized that the vertical motion contains more relatively high frequency components. The predominant frequency is taken as the frequency at which the peak amplitude of the Fourier Spectrum occurs. The source of the Fourier Spectra used in this study is Volume IV of the series "Analysis of Strong-Motion Earthquake Accelerograms" published by the Earthquake Engineering Research Laboratory of the California Institute of Technology. [9] The predominant frequencies of the vertical accelerations were plotted against those of the horizontal accelerations in Figure 2. The predominant frequencies of the vertical accelerations are less than 12 Hz and those for horizontal accelerations are nearly all less than 7 Hz. The predominant frequencies of the vertical acceleration are not always larger than that of the horizontal acceleration, but a few predominant frequencies of vertical accelerations show relative high values.

Figure 3 presents the predominant frequencies of vertical acceleration plotted against epicentral distance. Small dots show the data recorded on the "alluvium" sites, circles denote data on "hard rock" sites, and large dots show "intermediate" sites. The result is that the upper bound of the predominant frequency decreases with the increase of the epicentral distance.

Difference In Occurrence Times Between Vertical Peaks and Horizontal Peaks

There are few strong-motion accelerograms whose maximum vertical and horizontal component occur at the same time. Therefore, discussion based on the maximum values of accelerograms treat the extreme cases. This is on the safe side from a design viewpoint. But, for

II-27

rational earthquake-resistant design, it is desirable to know the ratio of vertical t
izontal component at the time when the structures become most unstable during earthqu

Figure 4 presents the frequency distribution of the difference in occurrence tim
(units; seconds) between vertical and horizontal motion. The positive sign denotes t
the maximum vertical accelerations occur after the maximum horizontal accelerations,
negative sign means opposite. This figure shows that although the vertical peaks occ
the occurrence time of the horizontal peaks, the coincidence of vertical and horizont
maximum acceleration is very rare.

From the viewpoint of the seismic design, it is desirable to know the ratio of v
to horizontal components at the time when the structures become most unstable during t
earthquakes. Consider the stability of a rigid block on a solid level plane analyzed
seismic coefficient method. A resultant force due to gravity and seismic force and it
direction are expressed by the following equations.

$$R = mg \sqrt{(1 \pm k_v)^2 + k_h{}^2} \tag{1}$$

$$K = \tan\theta = \frac{k_h}{1 \pm k_v} \tag{2}$$

where

R = instantaneous resultant force acting on the block

m = mass of the block

g = acceleration of gravity

k_v = instantaneous seismic coefficient in the vertical direction

k_h = instantaneous seismic coefficient in the horizontal direction

K = instantaneous resultant seismic coefficient

θ = angle between directions of resultant force and gravity

The block becomes more unstable against sliding and overturning with increase of the
resultant seismic coefficient which tends to maximize as k_h increases and as k_v become:
negative. From the digitized records, the resultant accelerations (A_{rh}) of the two co:
nents in the horizontal direction are calculated and the resultant seismic coefficient:
each instant are obtained from Eq. 2. Figure 5 presents, for the 187 western United S:
records, the vertical acceleration (A_v) and the horizontal acceleration (A_{rh}) at the t:

II-28

when the resultant seismic coefficient reaches a maximum. Usually the maximum resultant seismic coefficient occurs at the time when the horizontal acceleration is almost maximum, but the corresponding vertical acceleration at that time is not large. Figure 5 shows that the ratios A_v/A_{rh} are always almost less than one-third at this time.

COMPARISON WITH JAPANESE STRONG-MOTION DATA

Figure 6 presents comparison between vertical and horizontal peaks of strong-motion data in Japan. [10] The data contain 574 records which have been recorded on the ground. These data, recorded by SMAC-B2 strong-motion accelerographs, are uncorrected. The average of the ratio A_{vmax}/A_{hmax} is 0.33 and the standard deviation is 0.18. Comparison of Figures 1 and 6 indicate that the average ratio from Japanese data is smaller than that from U.S. data, and that the ratios of Japanese data decrease more clearly in accordance with increase of horizontal peak accelerations than that of U.S. data. The differences between Figures 1 and 6 might be associated with the differences in the installation conditions, dissimilar instruments, and differences in focal depth and subsoil conditions. Each of these possible causes is discussed below.

Installation condition. All Japanese records in Figure 6 were recorded on the ground, but most of the United States data are from accelerographs installed in the basements of buildings. It was pointed out by Housner [11] that the motion in the basement floors of the buildings is almost the same as the motion on the ground without large coupling between the ground and the structures. Also, in the recent paper by Boore, Oliver, Page, and Joyner [12], it is pointed out that the peak horizontal acceleration is smaller on the average at the base of large structures than at the base of small structures. In order to determine whether there is any difference between the A_{vmax}/A_{hmax} ratios for data obtained on the ground and on the basement floor in Japanese data, the relevant strong-motion data recorded in Japan are presented in Figures 7 and 8. Figure 7 shows the records in 1978 of the Izu Oshima-Kinkai earthquake and Figure 8 presents the records in 1978 of the Miyagi-Ken-Oki earthquake. [13] Circles denote the records on the ground, and dots show the records on the basement floors. In case of the 1978 Izu Oshima-Kinkai earthquake, the average ratio of records on the ground is 0.33, and that of records on the basement floors is 0.45. In the event of the 1978 Miyagi-Ken-Oki earthquake, the mean ratio of data on the ground is 0.37, and that of data on the basement floors is 0.43. Though the data are not sufficient, it seems that the A_{vmax}/A_{hmax} ratio of the records on the ground is smaller than that of the record on the basement floors.

Type of instruments. The Japanese data in Figure 6 were recorded by SMAC-B2 accelerographs. This SMAC-B2 type of instrument has different frequency response characteristics than the accelerographs used in the United States. Table 1 shows the characteristics of the accelerographs in the United States and SMAC-B2 type of instruments. [7,8] The frequency response characteristics of SMAC-B2 accelerographs shown in Figure 9 indicate that this type of instrument has diminished response at frequencies above 2 Hz. For example, the amplification associated with the frequency of 10 Hz is one-third. On the other hand, the data in the United States are corrected in the frequency range of 0.07 (or 0.125) to 25 Hz. This fact is important, because peak ground accelerations are often associated with high frequency components of ground motion. A correction program for SMAC-B2 records was reported recently by Iai, Kurata, Tsuchida, and Hayashi [14] but has not been used on the data examined here, because it is not the main object of this report to investigate. To further examine the effects of different frequency characteristics, the United States accelerograms were converted to SMAC-B2 type accelerograms, using the filter in Figure 9 and fast Fourier Transform (FFT) method. The filter is

$$H(f) = \frac{1}{1 - (\frac{f}{f_n})^2 + 2h_n (\frac{f}{f_n}) i} \qquad (3)$$

where

$H(f)$ = filter

f = frequency (Hz)

f_n = natural frequency ($=\cdot> 0.14$ Hz)

h_n = damping ($= 1.0$)

The accelerograms transformed to frequency domain by FFT were filtered through Eq. 3, and were transformed to time domain. Figure 10 presents peaks for the converted accelerograms plotted against those for original United States accelerograms. Dots show the maximum horizontal acceleration and circles show the maximum vertical acceleration. The vertical peaks in the U.S. records are attenuated more by filtering through Eq. 3 than are the horizontal

peaks. The post-filtering maximum vertical accelerations are plotted against the maximum
horizontal accelerations in Figure 11. The average ratio is 0.43, and the standard deviation
is 0.17. The comparison between Figure 1 and 11 indicates that the ratios A_{vmax}/A_{hmax}
decrease slightly as a result of filtering through Eq. 3. In the Japanese data in Figure 6,
there is a trend of decrease A_{vmax}/A_{hmax} as A_{hmax} increases. The average ratio for post-
filtering records in Figure 11 is 0.48, when the horizontal peak is less than 50 gal. When
the horizontal peak ranges from 50 to 100 gal, the average ratio is 0.42. When the horizontal peak is more than 100 gal, the average ratio is 0.39. The trend of decrease A_{vmax}/A_{hmax}
with increase of A_{hmax} in Figure 11 is clearer than in Figure 1.

    Focal depth. The focal depths of the records in the western United States, as described
in the section on strong-motion data, are shallow; usually less than 20 km. On the other
hand, the focal depths of earthquakes in Japan are deep, usually more than 20 km. Figure 12
shows A_{vmax}/A_{hmax} plotted versus the focal depth. Circles show the records for Japan, and
dots show filtered acceleration data for the western United States. The line in this figure
connects the average ratio of each focal depth. The trend is that the ratio decreases as
focal depth increases.

    Subsoil condition. Figure 13 presents the horizontal and vertical peak acceleration
data plotted versus epicentral distance. The left part of Figure 13 is the horizontal acceleration, and the right part is the vertical acceleration. Dots show the strong-motion data
recorded on the alluvium sites in the San Fernando earthquake, and circles show the data
obtained in Japanese ports from the earthquakes whose magnitude range from 6.4 to 6.7.
Figure 13 indicates that the horizontal accelerations recorded in Japanese ports whose subsoils consist of very soft layers are larger than those in the San Fernando earthquake, and
that the difference between the vertical accelerations of both data are relatively small.
As the vertical peaks are almost the same, the larger horizontal peaks will come lower
ratios of the vertical to horizontal peaks. This difference of subsoil condition between
the two countries might be one of the reasons why the A_{vmax}/A_{hmax} ratio of the data in
Japan is smaller than that of the records in the western United States.

CONCLUSIONS

    From the analysis of vertical components of strong-motion accelerograms in the western
United States, the following conclusions were drawn:

a) The average ratio of the maximum vertical acceleration to horizontal acceleration is about one-half.

b) Significant correlations between this ratio and maximum horizontal acceleration, epicentral distance, or earthquake magnitude could not be found.

c) The predominant frequency of the vertical component is less than 12 Hz. Significant relationships could not be found between the predominant frequencies of vertical and horizontal components.

d) The predominant frequency of the vertical component decreases with increasing distance.

e) The vertical maximum acceleration occurs near the occurrence time of the horizontal maximum acceleration. But, the exact coincidence of vertical and horizontal peaks is very rare.

f) With respect to sliding and overturning for gravity-type structures, at the time when the angle of the resultant force vector with the vertical is largest, the ratio of the vertical acceleration to the resultant horizontal acceleration is almost always less than one-third.

The result of a comparison of ratios of vertical peaks to horizontal peaks for western United States and Japan is as follows:

a) The ratio of vertical peaks to horizontal peaks in accelerograms for the western United States is larger than that in the Japanese recordings.

b) The difference between the data from the two countries is consistent with variations to be expected because of dissimilar instruments, dissimilar installation conditions, and differences in focal depth and site conditions.

ACKNOWLEDGMENT

This report was prepared during the author's visiting research at the U.S. Army Engineer Waterways Experiment Station (WES) sponsored by the Office, Chief of Engineers (OCE). It is part of ongoing work at WES in Civil Works Investigation Studies 31246 entitled "Dynamic Stresses and Permanent Deformations of Earth Structures in Response to Seismic Loadings," which is monitored for OCE by Mr. Ralph W. Beene. His visiting research at WES was recognized as part of the research personnel exchange program of U.S.-Japan Panel on Wind and Seismic Effects pursuant to the resolution of the tenth meeting of the panel.

The author wishes to express his thanks to Dr. Paul F. Hadala, Chief of the Earthquake Engineering and Geophysics Division (EE&GD), Dr. Ellis L. Krinitzsky, Chief, Engineering Geology Research Facility, and Mr. Frank K. Chang and Dr. Arley G. Franklin of EE&CD who reviewed the manuscript and offered many useful suggestions. He also expresses his appreciation to Dr. William F. Marcuson, III, of EE&CD for his efforts on this visiting research in WES.

REFERENCES

[1] Fujino, Y., Sasaki, Y., and Hakuno, M., "Slip of a Friction-Controlled Mass Excited by Earthquake Motions," Bulletin of Earthquake Research Institute, University of Tokyo, 53 (1978).

[2] Chang, F. K., Miscellaneous Paper S-73-1, Report 9, U.S. Army Engineer Waterways Experiment Station, CE, Vicksburg, Mississippi (1978).

[3] Trifunac, M. D., and Lee, V. E., "Routine Computer Processing of Strong-Motion Accelerograms," EERL 73-03, California Institute of Technology (1973).

[4] Hudson, D. E., Brady, A. G., Trifunac, M. D., and Vjayaraghavan, A., "Strong-Motion Earthquake Accelerograms, Corrected Accelerograms, and Integrated Ground Velocity and Displacement Curves, Vol. II, Part A," EERL 71-50, California Institute of Technology (1971).

[5] Trifunac, M. D., Brady, A. G., and Hudson, D. E., "Strong-Motion Earthquake Accelerograms, Corrected Accelerograms, and Integrated Ground Velocity and Displacement Curves, Vol. II, Part G," EERL 73-52, California Institute of Technology (1973).

[6] Trifunac, M. D., and Brady, A. G., "On the Correlation of Seismic Intensity Scales with the Peaks of Recorded Strong Ground Motion," Bulletin of the Seismological Society of America, 65 (1975).

[7] Halverson, H. T., "The SMA-1 Strong-Motion Accelerograph," Fourth Symposium on Earthquake Engineering, Roorkee, India (1970).

[8] Hudson, D. E., "Ground-Motion Measurement," Earthquake Engineering (Prentice-Hall, Inc., New Jersey, 1970), pp. 107-125.

[9] Hudson, D. E., Trifunac, M. D., Udwadia, F. E., Vjayaraghavan, A., and Brady, A. G., "Strong-Motion Earthquake Accelerograms, Fourier Amplitude Spectra Volume IV, Part A-Y," Earthquake Engineering Research Laboratory, California Institute of Technology.

[10] Uwabe, T., Noda, S ,. Kurata, E., and Hayashi, S., "Characteristics of Vertical Component of Strong-Motion Accelerograms," Ninth Joint Meeting of U.S.-Japan Panel on Wind and Seismic Effects, UJNR (1977).

[11] Housner, G. W., "Interaction of Building and Ground During an Earthquake," Bulletin of the Seismological Society of America, 47, 3 (1957).

[12] Boore, D. M., Oliver, III, A. A., Page, R. A., Joyner, W. B., "Estimation of Ground Motion Parameters," U.S. Geological Survey, 78-509 (1978).

[13] Strong-Motion Earthquake Observation Council, "January 14, 1978 Izu Oshima-Kinkai Earthquake," No. 13, "June 12, 1978 Miyagi-Ken-Oki Earthquake" No. 14, National Research Center for Disaster Prevention, Science and Technology Agency (1978).

[14] Iai, S., Kurata, E., Tsuchida, H., and Hayashi, S., "Integration of Stron
Accelerograms," Tenth Joint Meeting, U.S.-Japan Panel on Wind and Seismic
UJNR (1978).

Table 1 Characteristics of Strong-motion Accelerographs

Characteristics	USCGS Standard	Teledyne AR-240	Teledyne RPT-250	Teledyne RMT-280	Kinemetrics SMA-1	New Zealand MO2	Akashi SMAC-B2
Period, sec	0.043-0.085	0.055-0.065	0.05	0.05	0.4	0.03	0.14
Sensitivity mm/0.1 g	5.5-19.7	5.0-7.5	1.9	$\pm$ 200 cps FM deviation/$\pm$ 1 g	1.9	1.5 Horizontal	6.5
Recording range, g's	0.01-1.0	0.01-1.0	0.01-1.0	0.01-1.0	0.01-1.0	0.01-1.0	0.006-0.5
Damping % Critical	60	55-65	60	60	60	60	100
Damping Mechanism	Magnetic	Electromagnetic	Electromagnetic	Electromagnetic	Magnetic	Oil Paddle	Air Piston
Recording Speed, cm/sec	1	2	1	3-3/4 in./sec (9.53 cm/sec)	1	1.5	2
Recording Medium	Photo Paper	Photo Paper	70 mm (Type II Perforated)	1/4-in. Magnetic Tape	70 mm film	35 mm film (Unperforated)	Waxed Paper
Recording Drive	DC motor	DC motor	DC motor	DC motor	AC motor	DC motor	Hand-Wound Spring
Recording Duration	1.25 min	7 sec after last strong motion	5 sec after last strong motion	7 sec after last strong motion	25 min	47 (70) sec	3 min
Time Marking	2/sec	2/sec at $\pm$ 1%	2 sec at $\pm$ 2%	2/sec at $\pm$ 2%	2/sec $\pm$ 1%	Trace interrupt 5 and 50 cps 0.1%	1/sec
Size, in. (cm)	13x20x45 (33x51x114)	14x16x16 (36x41x41)	8-3/4x10-1/2x 19-1/2 (22x27x 50)	9x15x19 (23x38x48)	8x8x14 (20x10x36)	7x7x17 (18x18x43)	15x21x21 (54x54x37)
Weight, lb (kg)	135 (61)	60 (27)	30 (14)	42 (19)	25 (11)	20 (9)	220 (100)

Figure 1. Maximum Vertical Acceleration Versus Maximum Horizontal Acceleration

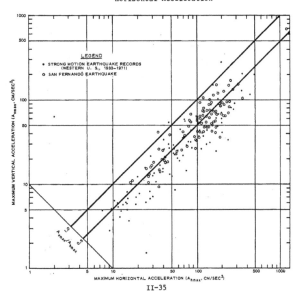

Figure 2. Predominant Frequency of Vertical Acceleration and Horizontal Acceleration

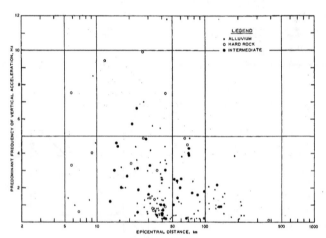

Figure 3. Predominant Frequency of Vertical Acceleration Versus Epicentral Distance

II-36

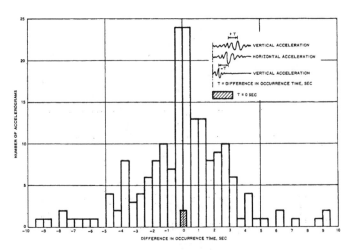

Figure 4. Frequency Distribution of Difference in Occurrence Time Between Vertical and Horizontal Peaks

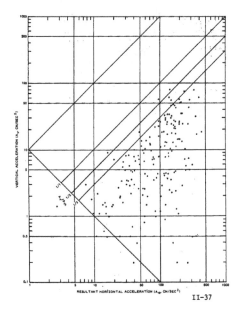

Figure 5. Vertical Acceleration
Versus Resultant Horizontal Accel-
eration at the Time When Resultant
Seismic Coefficient is Maximum

II-37

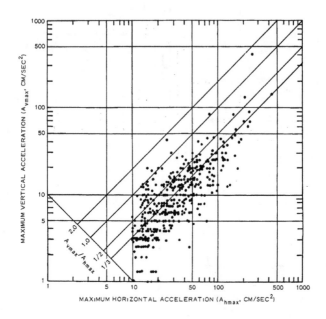

Figure 6. Maximum Vertical Acceleration Versus Maximum Horizontal
Acceleration for Japanese Recordings

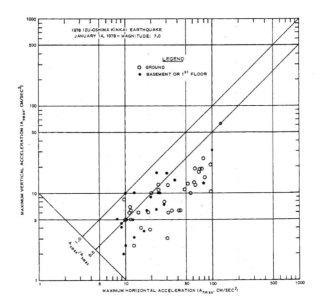

Figure 7. Maximum Vertical Acceleration Versus Maximum Horizontal
Acceleration for Japanese Recordings

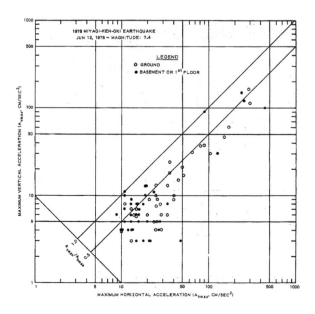

Figure 8. Maximum Vertical Acceleration Versus Maximum Horizontal
Acceleration for 1978 Miyagi-Ken-Oki Earthquake

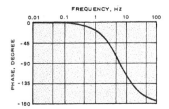
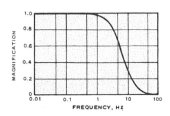

Figure 9.   Characteristics of SMAC-B2 Accelerograph

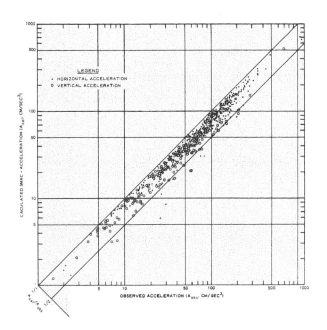

Figure 10.   Calculated SMAC Acceleration Versus Observed Acceleration

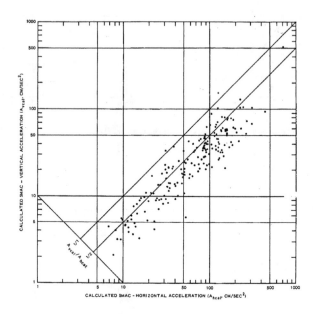

Figure 11.   Calculated SMAC Vertical Acceleration Versus Calculated
SMAC Horizontal Acceleration

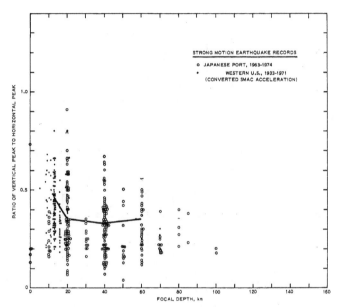

Figure 12. Ratio of Vertical Peak to Horizontal Peak Versus Focal Depth

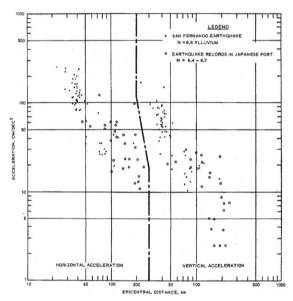

Figure 13. Maximum Acceleration Versus Epicentral Distance for Western United States and Japan

EARTHQUAKE OBSERVATION SYSTEMS

IN AND AROUND STRUCTURES IN JAPAN

Kiyoshi Nakano

Yoshikazu Kitagawa

Building Research Institute

Ministry of Construction

ABSTRACT

Ever since the observation system for underground earthquake motions was undertaken in 1934, efforts have been made to obtain exact knowledge of the characteristics of ground motions, the dynamic interaction behavior between building and subsoil, and the earthquake motions input to structures.

The number of observation sites has increased year by year. There are now more than 200 in number, and much valuable data has been accumulated. In this paper the state-of-the-art of underground earthquake observation in Japan is reviewed.

KEYWORDS: Earthquake observation systems; simulation models; spectral analysis; transfer function; underground amplitudes.

1. INTRODUCTION

In Japan, there is an extensive history of underground earthquake observation. In 1934, earthquake observations were performed from the bottom of wells in the center of Tokyo by means of mechanical type accelerographs (Ishimoto's type) to make comparisons with observations on the ground surface. In the 1950's, the advantages of earthquake observations at the bottom of bore-holes were emphasized, and a bore-hole seismometer of the electromagnetic type was successfully developed.

Around 1965, with the advent of high-rise buildings, long-span bridges, and nuclear power facilities, a remarkable increase in research and investigation has taken place with special reference to earthquake observation projects. The observation of earthquake motions in and around a structure began about the same time to yield data from the analyses of soil-structure interaction problems. Instrumentation has improved year by year, underground observation projects have increased in number, and there are now more than 200 in Japan.

2. THE PRESENT STATE OF EARTHQUAKE OBSERVATION

In order to make inquiry about the state-of-the-art on the observation of underground earthquake motions, the authors obtained information from universities, institutes, construction companies, the Port and Harbor Research Institute, the Center of the Disaster Prevention Agency, electric power companies, the P.W.R.I., and other organizations. The earthquake observation systems of the organizations mentioned above are classified into two types; type A is a set of observations of a structure and its surrounding subsoil, while type B is a set of observations of the subsoil only.

The locations and instrumentation of some observation stations obtained from the inquiry are listed in Table 1 (A) and (B). [1] In Table 1 (A), S means an underground observation point separate from a construction. These observation systems are also reported in the publications of some organizations such as the Architectural Institute of Japan.

2.1 Purpose of Earthquake Observation

The purposes for performing underground observations in the field of earthquake engineering are many and vary according to the interest of the individual organizations which carry out each project. A review of various projects is still in progress, but they may be summarized now as follows:

1) to investigate the characteristics of earthquake motions,

2) to investigate changes in depth of wave shape, and maximum acceleration in relation to ground conditions,

3) to investigate input motions to structures and their spectral characteristics,

4) to investigate the interaction effects between ground motions and the responses of super-structures, piles and underground structures,

5) to investigate the characteristics of earthquake motion within rock or rock-material which may be regarded as the so-called base.

6) to investigate the propagation characteristics of earthquake motions (amplitude and phase) along a line by installing a horizontal array of seismometers on the ground surface or underground.

2.2 Earthquake Observation System

Object of Observation -- Earthquake observations within soft soil deposits are quite large in numbers. This is due to the majority of earthquake observation projects being planned in association with structures in urban areas which are developed on areas of soft soil deposits. As seen in Table 1 (A) and (B), the earthquake observation systems are classified into two types. The systems in type A amount to approximately half the total number, indicating that the soil-structure interaction problem is one of great interest. At the same time the underground seismometers are usually installed along double vertical lines; one just below the structure and the other apart from the structure by a distance which is greater by approximately twice the structure's width, which may be considered enough to eliminate influences from the presence of the structure. The systems in type B are installed in order to investigate the characteristics of ground motions and of underground structures and are performed together with the measurements of micro-tremors and the seismic prospecting test in the usual manner.

A percentage breakdown of the number of type A earthquake observations which have been recorded is as follows: 51% - university, apartment, office, and hotel buildings; 11% - electric power facilities; 10% - bridges and banks; 10% - dams; 9% - models (reactors and other rigid structures); 9% - tanks, towers, and miscellaneous.

Location of Underground Seismometer -- The deepest seismometer in type A and B is usually installed at the upper boundary of the base layer (rock or rock-like formation) with

N-value of the penetration test more than 50 or shear wave velocity more than 400-800 m/sec. Table 2 shows the statistics of the number of seismometers in the projects of Types (A) and (B), and Table 3 shows the statistics of the greatest depth of observation. As for the distribution of the depths of underground observation, the embedded depths of seismometers range as follows: at ground surface (included rock), 5%; at a depth between ground surface and 20 m below the surface, 50%; at a depth of between 20 m and 60 m, 30%; at a depth of between 60 m and 100 m, 7%; at a depth deeper than 100 m, 3%; where the maximum depth is about 3500 m.

System of Seismometer -- In general, a system of electromagnetic seismometers consists of pickups, amplifier and recorder. Pickups of servo type (32%) and moving coil type (62%) are usually employed. The former is advantageous in general since the uniform frequency and phase characteristics can be maintained from 0 up to approximately 300 Hz.

Amplifier -- Amplifiers such as a logarithmic amplifier, auto-gain-control amplifier, binary-gain amplifier, and so on are tried and used in order to enlarge the dynamic range of records. The auto-gain amplifier is the one most frequently used.

Recorder -- The recorder of analog magnetic-tape type is replacing the conventional optical type. The delay device utilizing integrated circuit memory is becoming more popular year by year. In some delay devices, the digitized record which is stored in integrated circuit memory not only actuates the earthquake observation system, but also automatically establishes adequate gain of the amplifier after the actuation. The delay time is usually in the range of 3 to 5 seconds.

For any of the observation systems a crystal clock and a simple voice recorder such as those used in public radio broadcasting are employed as time-mark devices to make time-interconnections between the results of observations from the different recorders of an earthquake motion.

Consequently, the uniform frequency characteristics of the overall system is usually maintained from 0.3 up to 30 Hz with long span stability of instruments, and the measurements of acceleration are overwhelmingly large in number. Table 4 shows the statistics of objective quantities to be aimed at for measurement. It should be noted that great importance is attached to acceleration as the basis of earthquake engineering. However simultaneous observation of acceleration, velocity and displacement have tended to increase in number recently.

3. EXAMPLE OF OBSERVATION

Among the observation systems listed in Table 1 (A) and (B), the B.R.I., P.W.R.I. of the Ministry of Construction, and the University have carried out observation at several places, for example at numbers 6, 48, 49 and 42 in type A, and at number 16 in type B. The details of the earthquake observations of these five places are described in the following paragraphs.

3.1 The University Building (No. 6) in Type A [2]

The subsoil below the building is Kanto loam layer, alternative layers of fine and sandy loam, alternative layers of fine sand and silty or clayey sand, and a gravel layer, in that order from the surface. The water-level is 8.8 m below the ground surface. The soil profile and penetration test results (N-values), and the result of the seismic prospecting test are shown in Figure 1.

The building is a six-story reinforced concrete structure with a one basement level. It is 15.8 m x 84 m in plan, as shown in Figure 2. The west part of the building was first constructed during 1962-1963 with three additions by 1970. Each construction stage is indicated in Figure 2. The main lateral resisting elements of this building are the walled-type reinforced concrete frames, which have a walled girder (spandrel) and a walled column, in the longitudinal direction, and reinforced concrete shear walls in the transverse direction. In both directions there are slightly reinforced concrete and concrete block partition walls that are considered to be non-structural members. The building is supported by individual footings without piles at 5 m below the level.

The fundamental natural period of the building determined by forced vibration tests is 0.29 sec in the longitudinal direction, and 0.32 sec in the transverse direction. The rocking vibration is predominant in the transverse direction.

Electromagnetic seismometers were used to measure the earthquake motion under, around, and inside the building. The locations of the seismometers are shown in Figure 3. Seismometers A, B, C, D and E are placed along the vertical line from the top of the building to the soil layer 42 m below the surface. Seismometers F, G and H are arranged along the vertical line also, and are 60 m south of the above group in a horizontal direction. Seismometer K is arranged at a depth of 82 m below ground level. Each underground seismometer was

installed at the bottom of a bore-hole, which was refilled after the installation so that
seismic waves would not be affected by the bore-hole. Examples of the acceleration records
are shown in Figure 4.

3.2 The Test Model Building (No. 48) in Type A [3]

The subsoil is composed of loam, clay soil and fine sand. The soil profile, the penetration test results (N-values), and the result of the seismic prospecting test are shown in
Figure 5.

The foundation of the test model building is reinforced concrete with a mat foundation.
The super-structure of the model building consists of a concrete slab of 1.0 m which is
supported by four steel pipe columns. The model building is 2.0 m high, and 4.0 m in plan,
as shown in Figure 6.

The fundamental natural frequency of the soil-foundation-building system obtained by the
vibration test is 3.41 Hz. The vibrator was set on the foundation during the test. It
should be noted that vibration due to the relative displacement is predominant.

The location of the seismometers on the model and in the soil is shown in Figure 7. The
pendulum of the seismometers is heavily damped (h ÷ 10) and the output voltage which has a
natural frequency of 3.5 Hz is proportional to the acceleration. Typical examples of waves
in the soil-foundation building are shown in Figure 8.

3.3 The Apartment House (No. 49) in Type A [4]

The subsoil is composed of soft alluvial silt and clay deposits, the depth of which is
about 25 meters. The soil profile, the number of blows in the standard penetration tests
(N-values), and the velocity of the S waves are shown in Figure 9.

The building is an eleven-story apartment house made of steel reinforced concrete, as
shown in Figure 10. It is supported on reinforced concrete bored piles. This type of construction is called the Benoto method. The main structural components of the building are a
walled frame in a transverse direction and a rigid frame in a longitudinal direction.

The forced vibration tests confirm that the fundamental natural frequencies of the
structure are 2.2 Hz in a transverse direction, and 3.6 Hz in a longitudinal direction,
respectively.

Servo-mechanism type accelerometers are installed at two places. The accelerometers
are set in the building pile and soil at one place, and in the soil at the other place. The

four accelerometers, one placed at the top of the building and three in the soil, are connected to the double integrator in order to obtain the record of displacement. The earth
pressure gauges and pore water pressure gauges of a differential transformer type are
installed both under and beside the foundation of the building. The location of the instruments is shown in Figure 11. The frequency characteristics of the accelerometer, of which
the undamped natural frequency is 5.0 Hz, is almost flat (more than 95% relative sensitivity)
in the overall range from 0.3 to 30 Hz. Examples of acceleration records are shown in
Figure 12.

3.4 The Spherical Tank (No. 42) in Type A [5]

The soil profile and the S-wave travel-time curve measured at the boring hole for
installation of the underground seismometers are shown in Figure 13.

The tank is a spherical shell supported by 12 steel pipes (700 x 9) with a pair of
steel-bar bracings (2 x 800) between them, and the heights to the crown and the circumference
and the diameter of the shell are 20.4 m, 11.5 m and 17.9 m, respectively, as shown in Figure
14.

The electromagnetic accelerometers (Tuss type, f = 5 Hz, h = 0.6) are installed at 5
observation points along two vertical lines, as shown in Figure 14. An example of the acceleration records is shown in Figure 15.

3.5 The Ukishima Park (No. 16) in Type B [6]

The earthquake observations at Ukishima Park represent one of the four underground
observation projects carried out at four stations around Tokyo Bay. The subsoil is classified into several layers as follows; reclaimed soil, silt, clay and sand. The soil profiles
and the penetration test results are shown in Figure 16 together with the location of the
seismometers.

Four seismometers have been installed, one on the ground surface, one at a depth of 27 m
below the ground surface, one at a depth of 67 m, and one at a depth of 127 m, respectively.
Each accelerometer, whose pendulum is heavily damped (h ÷ 10) with a natural frequency of
4.0 Hz, has three components; one in the vertical direction, and two in the horizontal
direction. An example of the records triggered is shown in Figure 17.

At some sites mentioned above, the digitized earthquake observation data have been published by the B.R.I. and the P.W.R.I. [7] of the Ministry of Construction, Japanese Government.

On the other hand the analyses are widely made on the records observed for moderate earthquake motions in accordance with the purpose of observation. The most representative examples of analyses are as follows:

Distribution of Underground Amplitude -- The amplification of layered subsoil, in which the depth to the layered boundary is changed, is one of the basic research purposes, and is investigated with the quantities of the maximum acceleration, the root mean square amplitude, and the amplitude after band-limited filtering. The maximum amplitudes observed on two vertical lines, i.e., building-subsoil line and soil line, are generally normalized by those at the ground surface or at the base layer, and the average amplitude ratio of maximum accelerations with standard deviations is investigated.

Spectral Analysis -- Fourier spectra, power spectra and response spectra are usually used with a smoothing technique such as the window types of Hanning, Perzen and so on. The spectral techniques of the maximum entropy method and auto-regressive method in the information theory also have come into use recently. In addition, the running spectra technique is frequently used in order to examine time variations of spectra.

Transfer function -- The transfer function is calculated as the spectral ratio between two spectra of different records. Consequently, a considerable amount of variation is seen in the observed spectral ratios. The fluctuation increase is seen in the spectral ratios between the top and bottom of the building, between the base of the building and the ground surface, and between the ground surface and the deeper soil layers. The characteristics of phase difference between the ground surface and the deeper soil layers have begun to be analyzed recently.

Simulation Models -- The response analysis is generally carried out by means of one dimensional wave propagation theory, lumped mass model and Finite Element model. Through comparisons between the observed and the computed values, the validity of the models is examined. The average equivalent damping coefficient and shear strain of soil deposits are also estimated with various treatments of models.

CONCLUDING REMARKS

The present state of current earthquake observation projects in Japan have been reviewed in this paper. The number of observation sites is increasing year by year, and much valuable data have been accumulated. In relation to the increase of data, the disclosure of the data obtained at any observation site, for cooperation and coordination between earthquake observation projects would be expected, and the free use of this data would be fruitful in the progress of the research. However, more systematic, more orderly planned, and better equipped measurement systems are required. Hereafter possible technical improvements should be extended widely in underground observations for moderate and strong earthquake motions as follows: stability and accuracy of instruments; wider range measurements in amplitude and frequencies; more intentional choice of the site; proper overall array arrangement of instruments; automatization of data processing and so on.

ACKNOWLEDGMENT

In preparing this paper, the authors would like to express their thanks to the researchers who responded with information about earthquake observation system through the mail survey. The authors also wish to express their thanks to Dr. T. Tanaka, of the Earthquake Research Institute for many valuable discussions, and to Mr. T. Kajima and Miss M. Makishima of B.R.I. for providing the illustrations.

REFERENCES

[1] Kitagawa, Y., "Observation System for Underground Earthquake Motions in Japan," Building Research Institute Research Paper No. 73, Building Research Institute, 1977.

[2] Osawa, Y., Kitagawa, Y. and Ishida, K., "Response Analysis of Earthquake Motions Observed in and around a Reinforced Concrete Building including Building Subsoil System," Fifth W.C.E.E., 1973.

[3] Osawa Y., Kitagawa, Y. and Irie, Y., "Evaluation of Various Parameters on Response Analysis of Earthquake Motions Including Soil Building Systems," Sixth W.C.E.E., 1977.

[4] Ohta, T., Niwa, M., Ando, H. and Ujiyama, M., "Study on Seismic Responses of Soft Alluvial Subsoil Layers by Simulation Analysis," Sixth W.C.E.E., 1977.

[5] Minami, T., Osada, K. and Osawa, Y., "Earthquake Observation on a Spherical L.P.G. Tank and the Surrounding Ground," Proceedings, of Fourth J.E.E.S., 1975.

[6] Iwasaki, T., Wakabayashi, S. and Tatsuoka, F., "Characteristics of Underground Seismic Motions at Four Sites around Tokyo Bay," Eighth UJNR, 1976.

[7] "Report on Digitized Earthquake Accelerograms in a Soil-Structure S
     1976.

[8] "Strong-Motion Earthquake Records Obtained at Undergrounds," P.W.R.

Table 1 (A) Earthquake Observation Project in Type A.

No.	Type of Building	Type of Construction, Number of Stories, Type of Foundation	Organization	Opening Date	Observation Points	Objective Quantity
1	Dam	Gravity l=462m h=145m -	Electric Power Development Co.	1959	EL +399m,+444m,+486m ⑤ GL (Rock)	ACC. Vel.
2	Dam	Rockfill l=258m h=105m -	Electric Power Development Co.	1961	Dam Top ① GL (Rock)	ACC.
3	Dam	Rockfill l=202m h=115m -	Electric Power Development Co.	1961	Dam Top ① GL (Rock)	ACC. Vel.
4	Reactor & Footing	RC, S - -	Architectural Inst. of Japan	1963	RF, BF ③ GL,GL-11m,-17m	ACC.
5	Dam	Rockfill l=355m h=128m	Electric Power Development Co.	1964	EL +449m,+485m,+524m ⑤ GL (Rock)	ACC.
6	University Building	RC 6F individual	Univ. of Tokyo	1964	RF,4F,1F,GL-23m,-42m ⑧ GL,GL-23m,-41m, GL-5m,-82m	ACC.
7	High school Building	RC 2F Pile (l=13m)	Kyoto Univ.	1965	RF,2F,1F ⑥ GL,GL-1m,-10m,-20m	Vel.
8	Rigid Structure Model No.1	RC 1F mat	Architectural Inst. of Japan, Kajima Corp.	1965	RF,1F,B1F ⑧ GL,GL-2.2m,-4.4m,-14.4m,-19.7m	ACC.
9	Rigid Structure Model No.2	RC 2F mat	Architectural Inst. of Japan, Kajima Corp.	1965	RF,2F,1F, ⑤ GL,GL-20m	ACC.
10	University Building	RC 5F+B1F pile (l=19m)	Tokyo Inst. of Tech.	-	B1F ⑤ GL-10m,-40m,-50m	ACC.
11	Tower	S 81m	Ministry of Const.	1967	GL,+81m,GL ⑤ GL	ACC.
12	Rigid Structure Model No.3	RC 2F mat	Architectural Inst. of Japan, Shimizu const.	1967	RF,1F,B1F ⑤ GL-3m,-14m,-20m	ACC.
13	Office Building	S 36F +B3F mat	Kajima Corp.	1968	RF,23F,13F,B2F (SMAC) B2F,GL-80m	ACC.
14	Apartment Building	SRC 15F steel pile	Waseda Univ.	1968	RF (SMAC),GL-4m,-44m	ACC.
15	Model Bldg. with PC pannel	RC 1F H steel pile (l=10m)	Nippon Steel Corp., Univ. of Tokyo	1968	RF,1F,GL-5m ③ GL-5m,-15m	ACC.
16	Model Building	RC 3F mat	Japan Electric Association	1968	RF,3F,2F,1F ③ GL.	ACC.
17	Dam	Fill l=1593m h=38m	Electric Power Development Co.	1968	EL+1200m, +1225m,+1240m	ACC. Vel.
18	University Building	SRC 9F pile (l=12m)	Tohoku Univ.	1969	RF,7F,5F,1F	ACC.
19	Reactor Building	RC 5F+B1F mat	Tokyo Electric Power Co.	1970	RF,5F,Footing,GL-14m,-24m,-50m ③ GL-1m	ACC.
20	Model Building	S 5F individual	Kyoto Univ.	1970	RF,4F,1F ② GL-3m,-12m	ACC.
21	Office Building	S 40F+B3F mat	Kajima Corp.	1970	40F,30F,20F,10F,B2F (SMAC) ⑥ GL-20m,-90m	ACC.
22	University Building	RC 5F+B1F	Nihon Univ.	1970	B1F ③ GL-3m,-15m,-30m	ACC.
23	Apartment Building	S 11F steel H pile (l=12m)	Chiba Univ., Nippon Steel Corp.	1970	11F,6F,1F,GL-12m,-5m ⑥ GL-1m,-13m,-30m	ACC.
24	Reactor Building	RC 5F +B1F mat	Tokyo Electric Power Co.	1970	RF,5F,7F,RF ⑤ EL-4m,-1m,-4m	ACC.

No.	Building type	Structure	Owner/Corp.	Year	Measurements	Objective Quantity
25	Apartment Building	B, 7F, PC pile (1×12m)	Univ. of Tokyo, Taisei Corp.	1971	RF,1F,GL-4m,-12m,-24m ⓖ GL,GL-4m,-12m,-24m	Objective Quantity
26	Apartment Building	RC, 5F, continuous	Ministry of Const.	1971	RF,4F,BF,GL-2.5m,GL-10.2m,GL-30m ⑧ GL-2.5m,-10.2m GL-2.5m	ACC. Dis.
27	Electric Power Facilities	RC, -, steel pile	Chugoku Electric Power Co.	1971	Turbine frame GL+8.1m,-3.8m steel pile GL-15.1m,-23.35m ⑧ GL-1.0m,-23.5m	ACC.
28	Office Building	RC, 10F, steel pile	Central Research Inst. of Electric Power Industry	1971	RF,6F,4F,2F,B1F,B3F GL-30m(tip of steel pile) ⓖ GL,GL-200m	ACC.
29	Office Building	S, 3F, individual	Nippon Telegraph & Telephone Public Co.	1971	RF,3F,1F (SMAC) GL-5m,-15m	ACC.
30	Apartment Building	SRC, 14F, PC pile	Ministry of Const.	1972	14F,8F,Basepoint ⑧ GL-9m,-16m,-26m,-60m	ACC.
31	High school building	RC, -, RC pile(1×32m)	Architectural Inst. of Japan	1972	RF,1F,GL-32m	ACC. Vel. Earth Pres
32	Experiment Building	RC, 5F+B2F, mat	Shimizu Const.	1972	RF,1F,B2F,GL-20m	ACC.
33	Apartment Building	S, 11F, earth drill pile (1×11m)	Shimizu Const.	1972	11F,1F,GL-10m	ACC.
34	Apartment Building	SRC, 10F, pier	Kajima Corp.	1973	RF,BF,GL-10m,-20m ⑧ GL-2m,-20m	ACC.
35	Steam Power Station	RC, 5F, steel pipe pile (1×20m)	Chugoku Electric Power Co., Ohbayashi-gumi Ltd.	1973	FL-38m,-55m,-23.75m ⑧ GL-73m,+70m	ACC.
36	Reactor Building	RC, 5F+B1F, mat	Chugoku Electric Power Co. Kajima Corp.	1973	RF,5F,1F,BF,GL-35m,-70m ⑧ GL+0.5m	ACC.
37	Office Building with Tower	Building SRC,15F+B5F Tower S, h=140m mat	Tokyo Electric Power Co. Kajima Corp.	1973	Tower GL +200m,+107m, Building 15F, B5 (SMAC)	ACC. Vel.
38	Tank	S, 25m, steel pipe pile (1×38m)	Shimizu Const.	1973	Top,Footing,GL-2m,-18m,-39m ⑧ GL-2m,-18m,-38m,-71m	ACC.
39	Office Building	S, 30F, mat	Ohbayashi-gumi Ltd.	1973	RF,15F,1F,B2F,GL-17m	ACC.
40	Office Building	SRC, 12F+B2F, mat	Kumagai-gumi General Contract Co.	1973	12F,6F,B2F ⑧ GL-19.6m	ACC. Dis.
41	Apartment Building	RC, 18+B1F, pier (1×20m)	Kajima Corp.	1974	RF,9F,B1F,GL-24m(tip of pier) ⑧ GL-1m,GL-27m	ACC.
42	Tank	S, 20m, steel pipe pile (1×31m)	Univ. of Tokyo	1974	Top, Footing ⑧ GL,GL-17.4m,-43.4m	ACC.
43	Chimney	S, 230m, steel pipe pile (1×40m)	Tokyo Electric power Co., Mitsubishi Heavy Industries Co.	1974	GL+222m,+112m,+0.5m GL-10m,-31m,-38m	ACC. Dis.
44	Plant tower	S, 50m,30m, steel pipe pile (1×45m)	Mitsui Petrochemical Co., Taisei Corp.	1974	T49,T30,P1,P2, GL-10m,-45m,-85m	ACC. Dis.
45	Foundation of Shaking Table	S, 1F, mat	The National Research Cente. for Disaster Prevention Science & Tech. Agency	1974	BF ⑧ GL,GL-10m,GL-40m	ACC.
46	High school building	RC, 3F, PC pile	Mitsui Const.	1974	RF,1F ⑧ GL,-1m,-7.5m,-66m	ACC.
47	Laboratory building	SRC, 7F+B3F, continuous	Central Research Inst. of Electric Power Industry	1975	PF,7F,6F,4F,3F,1F,B1F,B3F	ACC.
48	Main Building	S, 1F, mat	Ministry of Const.	1975	BF, Footing ⑧ GL-0.5m×2,-13m,-41m	ACC.
49	Apartment Building	SRC, 11F, pier (1×26m)	Ministry of Const., Technology Center for National Land Development, Kajima Corp.	1975	11F,1F,GL-1½m (pier GL-13m,-26m) GL-2m,-15m ⓖ GL-2m,-5.5m,-9m,-13m,-30m,-100m	ACC. Dis.

Table 1 (B) Earthquake Observation Project in Type B.

No.	Location	Organization	Opening Date	Observation Points	Objective Quantity
1	Kushiro, Hokkaido	Hokkaido Univ.	1965	GL, GL-20m	ACC.
2	Hamamatsucho, Tokyo	Kajima Corp.	1966	GL, GL-20m, -40m	ACC.
3	Tomakomai, Hokkaido	Hokkaido Univ.	1966	GL, GL-30m. -50m	ACC.
4	Shinjuku, Tokyo	Takenaka Co.	1967	GL, GL-1.3m, -13m, -27m, -81.6m	ACC.
5	Sapporo, Hokkaido	Hokkaido Univ.	1967	GL, GL-60m	ACC.
6	Saijo, Ehime	Shikoku Electric Power Co.	1968	GL, GL-10m, -20m	ACC.
7	Yokohama, Kanagawa	Takenaka Co.	1969	GL-4m, -18m, -31m	ACC.
8	Ohtemachi, Tokyo	Waseda Univ.	1969	GL, GL-26m, -86m	ACC.
9	Abeno, Ohsaka	Hukui Univ.	1970	GL, GL-10m, -30m	ACC.
10	Kiyose, Tokyo	Ohbayashi-gumi Ltd.	1970	GL, GL-10m, -20m, -180m	ACC. Dis.
11	Kiolcho, Tokyo	Taisei Corp.	1970	GL-2m, -33m, -83m	ACC. Dis.
12	Kobe, Hyogo	Nikken Sekkei Ltd.	1970	GL-2.2m, -54.5m, -83.9m	ACC.
13	Nishiuwa, Ehime	Power Co.	1970	EL-10m, -10m, -30m	ACC.
14	Shinjuku, Tokyo	Kajima Corp.	1970	GL, GL-24.5m	ACC.
15	Yatsu, Chiba	Ministry of Const.	1970	GL, GL-70m, -110m	ACC.
16	Kuwasaki, Kanagawa	Ministry of Const.	1970	GL, GL-27m, -67m, -127m	ACC.
17	Kannonzaki, Kanagawa	Ministry of Const.	1971	GL, GL-80m, -120m	ACC.
18	Yokohama, Kanagawa	Fujita Corp.	1971	GL-1.5m, -13.2m, -60.2m	ACC.
19	Tateyama, Chiba	The National Research Center for Disaster Prevention Science & Technology Agency	1971	GL.	Vel.
20	Maruyamacho, Chiba	The National Research Center for Disaster Prevention Science & Technology Agency	1971	GL-30m	Vel.
21	Chishimacho, Ohsaka	Fukui Univ.	1971	GL, GL-20m, -33m, -60m	ACC.
22	Morinomiya, Ohsaka	Fukui Univ.	1972	GL, GL-22m, -60m	ACC.
23	Minato, Tokyo	Ohbayashi-gumi Ltd.	1972	GL, GL-10m, -26.5m	ACC.
24	Yokohama, Kanagawa	Kajima Corp.	1972	GL, GL-24m	ACC.

II-57

					Objective Quantity
25		Japanese National Railways	1972	GL,GL-15.5m,-31.5m	
26	Kawaguchi, Saitama	Ministry of Const.	1972	GL-7m,-14m,-21m,-30m,-43m,-60m	ACC.
27	Kitasuna, Tokyo	Ministry of Const.	1972	GL,GL-8m,-20m,-30m,-48m,-60m	ACC.
28	Wakayama	Toanenryo Industrial Co., Taisei Corp.	1972	GL), GL-31m, -57m	ACC. Vel.
29	Chiyoda, Tokyo	Shimizu Const.	1972	GL-1m, -25m, -50m	ACC.
30	Tomiyamacho, Chiba	The National Research Center for Disaster Prevention Science & Technology Agency	1972	GL-50m	Vel.
31	Tomiyamacho, Chiba	The National Research Center for Disaster Prevention Science & Technology Agency	1972	GL-50m	Vel.
32	Amagasaki, Hyogo	Fukui Univ., Ohsaka Inst. of Technology	1973	GL, GL-15m, -30m, -60m	ACC.
33	Toyosu, Tokyo	Shimizu Const.	1973	GL-2m, -18m, -38m, -17m	ACC.
34	Kouto, Tokyo	Ministry of Transport	1973	GL, GL-20m, -50m, -90m	ACC.
35	Funabashi, Chiba	Ministry of Transport	1973	GL, GL-20m, -50m, -90m	ACC.
36	Nagoya, Aichi	Ministry of Transport	1973	GL, GL-20m, -50m, -90m	ACC.
37	Taisho, Obsaka	Ministry of Transport	1973	GL, GL-20m, -50m, -90m	ACC.
38	Urayasu, Chiba	Muto Institute of Structural Mechanics	1973	GL, GL-10m, -41.4m	ACC. Dis.
39	Narashino, Chiba	Takenaka Co.	1973	GL-2.5m	ACC.
40	Iwatsuki, Saitama	The National Research Center for Disaster Prevention Science & Technology Agency	1976	GL-3510m GL-80m	ACC. Vel.
41	Ohsaka, Obsaka	Fukui Univ. Ohsaka Inst. of Technology	1974	GL, GL-60m	ACC.
42	Hachiohji, Tokyo	Waseda Univ.	1974	GL, GL-14m, -25m	ACC.
43	Yokohama, Kanagawa	Ministry of Const.	1974	GL-3m, -13m, -38m, -160m	ACC.
44	Tokyo International Airport, Tokyo	Ministry of Transport	1974	GL 6, -50m, -67m	ACC.
45	Hachinohe, Aomori	Hachinohe City, Nippon Steel Corp.	1975	GL, GL-2m	ACC.
46	Kawasaki, Kanagawa	The National Research Center for Disaster Prevention science & Technology Agency	1975	GL-600m	Vel.
47	Fukui, Fukui	Fukui Univ.	1976	GL, GL-18m, -80m, -175m (Rock)	ACC.
48	Asuwayama, Fukui	Fukui Univ.	1976	GL	ACC.

Number of Seismomenters	Percentage (%)
≦ 2	23
3 ~ 4	45
5 ~ 6	20
7 ~ 8	7
≧ 9	5

Table 2   Statistics of Number of Seismometers

Depth (m)	Percentage (%)
≦ 20	57
20 ~ 40	17
40 ~ 60	11
60 ~ 80	3
80 ~100	4
100 ~200	2
≧200	1
Outcrop Rock	5

Table 3   Statistics of Largest Depth of Seismometers

Objective Quantity	Percentage (%)
Acceleration (cm/sec^2)	87
Velocity (cm/sec)	8
Displacement (cm)	5

Table 4   Statistics of Objective Quantity for Measurement

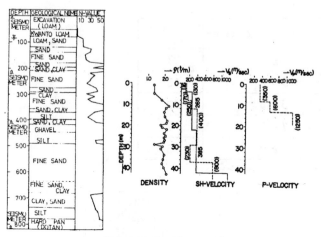

Fig. 1   Soil Profile and Underground Structure

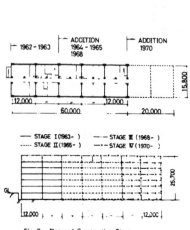

Fig. 2   Plan and Construction Stage

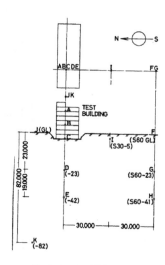

Fig. 3   Location of Seismometers

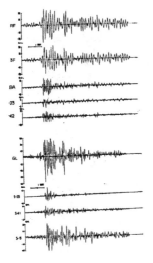

Fig. 4   Example of Acceleration Records

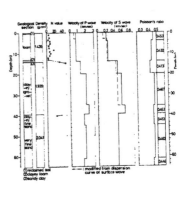

Fig. 5   Soil Profile and Underground Structure

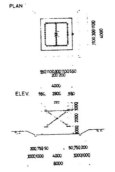

Fig. 6   Plan and Elevation

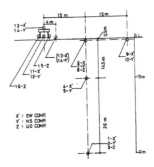

Fig. 7   Location of Seismometers

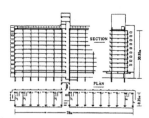

Fig. 8 Example of Acceleration Records

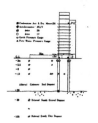

Fig. 9 Soil Profile and Underground Structure

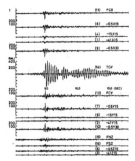

Fig. 10 Plan and Section

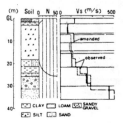

Fig. 11 Location of Seismometers

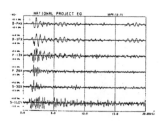

Fig. 12 Example of Acceleration Records

II-62

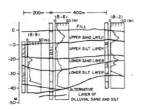

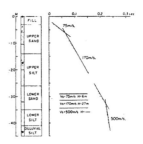

Fig. 13 Soil Profile and S-Wave Travel Curve

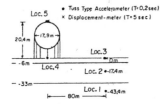

Fig. 14 Location of Seismometers

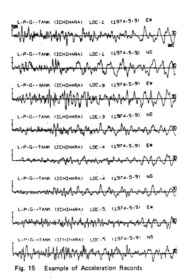

Fig. 15 Example of Acceleration Records

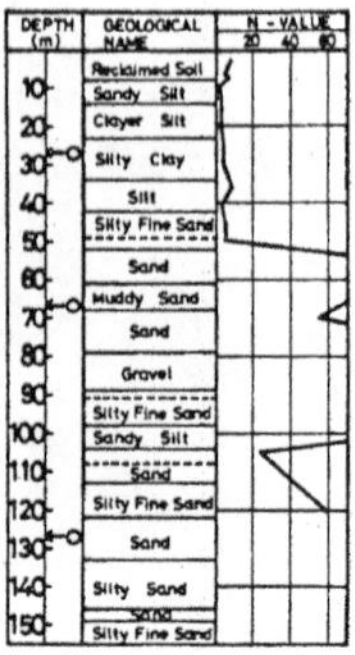

DEPTH (m)	GEOLOGICAL NAME	N-VALUE 20 40 60
	Reclaimed Soil	
10	Sandy Silt	
20	Clayer Silt	
30	Silty Clay	
40	Silt	
50	Silty Fine Sand	
60	Sand	
70	Muddy Sand	
80	Sand	
90	Gravel	
100	Silty Fine Sand	
110	Sandy Silt	
120	Sand	
	Silty Fine Sand	
130	Sand	
140	Silty Sand	
150	Sand	
	Silty Fine Sand	

Fig. 16   Soil Profile and
Location of Seismometers

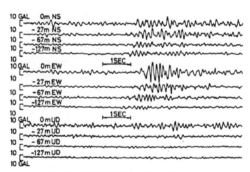

10 GAL   Om NS
10   − 27m NS
10   − 67m NS
10   −127m NS
10
10 GAL   Om EW   1SEC
10   − 27m EW
10   − 67m EW
10   −127m EW
10
10 GAL   Om UD   1SEC
10   − 27m UD
10   − 67m UD
10   −127m UD
10 GAL

Fig. 17   Example of Acceleration Records

SINGLE-STORY RESIDENTIAL MASONRY CONSTRUCTION

IN UNIFORM BUILDING CODE SEISMIC ZONE 2

G. Robert Fuller

U.S. Department of Housing and Urban Development

Washington, D.C.

ABSTRACT

The U.S. Department of Housing and Urban Development's (HUD) Minimum Property Standards (MPS) require that all masonry construction in Seismic Zone 2 have partial reinforcement in accordance with the Uniform Building Code (UBC). Since there is a lack of behavioral data on the resistance of single-story masonry houses to earthquake forces, the housing industry has objected to the added cost of providing this reinforcement. HUD therefore contracted with the University of California at Berkeley to determine the behavior of masonry structures under seismic loads, by testing full scale specimens on the shaking table. This paper is a report of the related results of material, timber roof connections and shaking table tests. HUD also engaged the Applied Technology Council (ATC) of Palo Alto to develop design and construction criteria for adoption in the HUD-MPS. Upon completion of that contract in early 1980, a separate report will be presented to the UJNR Panel.

KEYWORDS: Minimum property standards; partially reinforced masonry; residential; roof diaphragms; seismic resistance; single-story; shaking table tests.

INTRODUCTION

"Partial reinforcement" is required for all residential masonry constructio[n]
Zone 2, if compliance with HUD's Minimum Property Standards (MPS) [1]* is to be [r]
The Uniform Building Code (UBC) [2] is cited in the MPS as being the acceptable [c]
There is, however, a lack of data on the resistance of single-story masonry hous[e]
to earthquakes. Justification of partial reinforcement was therefore questioned
the housing design and construction industry, particularly in the western states
Arizona, Utah, and Colorado, where masonry is the predominate form of constructi[c]

Consequently, HUD contracted with the Earthquake Engineering Research Cente[r]
the University of California at Berkeley to test the seismic behavior of full-sca
story residential masonry buildings on the shaking table. This paper therefore [p]
findings from these tests, plus cyclic tests conducted on timber roof to masonry
connections.

As previously stated, HUD's MPS requires compliance with UBC; masonry constr
Seismic Zone 2 must be partially reinforced. This MPS requirement applies to all
produced under HUD programs or covered by HUD mortgage insurance.

The Minimum Property Standards are considered as minimum requirements for th
construction of adequate housing. Where local codes or standards permit lower re
than the MPS, the latter shall apply. Local Acceptable Standards (LAS) to modify
however, may be issued by HUD Field Offices, with concurrence by HUD Headquarters

In August 1974, the HUD office in Phoenix, Arizona issued an LAS to cover th
construction of masonry buildings. Specifically, unreinforced masonry was not pe
masonry walls of single-story residences located in Seismic Zone 2 would require
following amounts of vertical reinforcement: [3]

1) No. 4 bars at all building corners, at wall ends, at all door and window
and in wall sections 2'-0" (0.6 m) wide or less. Openings larger than 12'0" (3.7
require special analysis.

2) No. 4 bars spaced at not over 12'0" (3.7 m) o.c.

3) All reinforcement to be matched with dowels embedded in foundation walls

---

* Numbers in brackets indicate references.

The promulgation of this standard was questioned by representatives of local industry, on the grounds that compliance with the requirements would lead to increased costs and unnecessarily high factors of safety. To address this problem, the research program was organized to yield the following information: [3]

1) Dynamic behavior of simple masonry structures subjected to simulated earthquake motions;

2) Reinforcement requirements, if any, for adequate transverse and in-plane resistance of typical masonry house construction for level of seismic activity associated with Seismic Zone 2.

3) Adequacy of typical roof-masonry wall connection details to withstand earthquake forces.

OBJECTIVE

The overall objective of this investigation was to subject typical masonry houses to carefully controlled simulated earthquakes and to observe their structural behavior. Partially reinforced and unreinforced wall panels, fabricated of hollow-core concrete masonry units and hollow clay bricks, were tested. [3]

Four house specimens, similar to that shown in Figure No. 1, were tested, each was 16' x 16' (4.9 m x 4.9 m) in plan and had 8' - 8" (2.6 m) high walls. Walls were connected at the top by a timber truss roof diaphragm with trusses arranged parallel or perpendicular to the horizontal motion of the shaking table. To account for the reduction of mass resulting from the scaled-down plan dimensions, concrete slabs were bolted to the roof. [4]

Each structure was tested with a series of horizontal base motions of increasing intensity. The three simulated earthquakes utilized in the program were the 1952 Taft, 1949 El Centro, and 1971 Pacoima Dam accelerograms. All houses were designed so that simultaneous transverse and in-plane response of both reinforced and unreinforced panels could be measured. [5]

Another objective of the research program was to analyze the behavior of typical timber roof connections used in single-story masonry residential construction when subjected to cyclic loads. To assess the adequacy of these connections in transferring roof inertia loads, five basic types of roof-to-wall connections (nineteen models) were subjected to displacement controlled load tests, both in-plane (along the wall) and out-of-plane. The five types of connections contained loadbearing and nonloadbearing wall sections and either gabled truss or flat roof construction. By examining the mode of failure and code allowable loads on bolted and nailed connections, the margin of safety inherent in current code requirements could be determined. [6]

Although inadequate seismic resistance of poorly designed or constructed masonry buildings has been given heavy emphasis in post-earthquake survey reports, no experimental studies had been reported. The overall purpose of the study was to evaluate the earthquake resistance of typical masonry dwelling construction in less seismically active regions of the U.S. Once this was accomplished, reinforcement could be determined. Therefore, during the project's planning phase, an important aspect was to model structures which would be simple in concept and yet contain the most significant components, such as wall panels with and without openings, and with wall-to-footing and roof-to-wall connections.

MATERIAL PROPERTIES

Since the final objective of the research was to arrive at reinforcement requirements, the number of parameters had to be limited. Therefore, the types of masonry units, mortar, grout and reinforcement had to be kept to a minimum. In addition, all materials had to be commercially available and typical of those commonly used in Arizona. [3]

Concrete Block Units

House specimens No. 1, 2 & 4 were constructed of standard two-core hollow block with nominal dimensions of 6" x 4" x 16" (15 cm x 10 cm x 41 cm). Two different shapes of block were used; corner return and standard blocks in House No. 1, and only standard units in House No. 2 and 4. Dimensions and properties of typical concrete masonry units are shown in Table No. 1 (House No. 4) [7] and masonry strengths are shown in Table No. 2. [5] Typical units are shown in Figure 2. [3]

Hollow Clay Brick Units

House No. 3 was constructed of 4" x 6" x 12" (10 cm x 15 cm x 30 cm) standard hollow clay bricks. Two shipments of brick were received at different times; Shipment No. 2 had

38% voids (Figure 3). Five bricks were selected at random and were tested for strength. The average strength of bricks in Shipment No. 1 was 4970 psi (34268 kN/m^2) and in Shipment No. 2, 4060 (27994 kN/m^2). (Table No. 2).

Compression tests were conducted on two- and ten-unit high prisms and diagonal compression tests were conducted on 36" (91.4 cm) square panels. The average strength of two-unit prisms was 1220 psi (8412 kN/m^2) for Lot 1 and 2280 psi (15721 kN/m^2) for Lot 2. The corresponding values for ten-unit prisms were 2120 psi (14617 kN/m^2) and 1710 psi (11790 kN/m^2), respectively. Critical tensile strength was determined to be 47.2 psi (325 kN/m^2) and 81.5 psi (562 kN/m^2) for the two shipments. [8]

Mortar

Type S mortar, as specified by UBC, was used in all houses. Volume proportions were 1: 1/2: 4 1/2 for cement: lime: sand. For each batch, two mortar samples in 2" x 4" (5.1 cm x 10.2 cm) cylinders were taken and air cured in the laboratory under the same conditions as masonry specimens. The resultant, average, 29-day mortar strengths are shown in Table No. 2. [5]

Grout

A fine sand and cement grout was used in all partially reinforced wall panels. One part of Type I cement and three parts of top sand were mixed with water to achieve the desired consistency. Control specimens were cast in typical concrete block cells and stored in the laboratory under the same conditions as wall panels. Compressive strengths were determined by cutting the filled cells into 2" (5.1 cm) cubes and testing on a standard testing machine. A minimum of 2,000 psi (13,790 kN/m^2) strength was specified and exceeded by all samples as shown in Table No. 2. [5]

Reinforcement

Partial reinforcement was provided in Houses No. 1, 2 and 3 by medium grade steel #4 bars. House No. 4 was tested twice with no reinforcement; then 2 - #3 bars were placed in each wall. In House No. 1, average yield (fy) and ultimate strengths were 54,000 psi (372,330 kN/m^2) and 80,000 psi (551,600 kN/m^2), respectively. The corresponding values for

III-5

No attempt was made to obtain additional properties of steel, since stress levels were not proved significant in determining structural response. [3]

Joint reinforcement of 9 gage, truss type steel wire was provided in the top two bed joints of each wall and corner unit of Houses No. 1 and 4. No joint reinforcement was provided in Houses No. 2 and 3.

SHAKING TABLE TESTS

Test Specimens

The first test structure was considered exploratory to observe behavior of masonry buildings under realistic earthquake loads. Limitations of types of masonry components and size of the shaking table resulted in design and fabrication of the structure shown in Figure No. 4. Four 8' (2.4 m) long wall panels and four corner L-shaped units were used. Two wall panels and two corner units were reinforced with #4 bars and all reinforcement was lapped 20" (51 cm) with dowels embedded in the footings. The footings were then attached to the shaking table to prevent horizontal translation and rotation.

At the beginning of the testing program, only translation of the walls was prevented. During later stages, however, in-plane walls were bolted down to prevent observed uplift. Footings under all four corners were also prestressed against the shaking table.

In House No. 1 the roof was anchored by three symmetrically placed bolts to the masonry wall panels and corner units. Overall structural integrity was provided by the roof diaphragm, which consisted of trusses and plywood sheathing.

To account for scaling down of the overall building plan and reduction in weight, six concrete slabs were bolted to the plywood sheathing. A prototype total roof load of 20 psf (964 N/m^2) was assumed, resulting in a total roof weight of 15,000 lbs (6,804 kg). [3]

Compared to current construction practice, the geometry of the first test specimen was over simplified. It was therefore decided, after consultation with professional engineers and home developers in Arizona, to revise the second test house configuration.

House No. 2 consisted of four independent wall panels with either door or window openings. In-plane walls were 16' (4.9 m) in length and transverse walls were 14' (4.3 m) long. Two walls were unreinforced; the other two walls had a #4 bar in each end cell.

III-6

Double steel angles were used for lintels above doors and windows, and large openings had steel beams with the top flange trimmed and a steel plate welded to the bottom flange.

Corner units were omitted to prevent interaction between in-plane and transverse walls, and wall elements were designed to represent center parts of longer wall panels in a typical dwelling. [3]

In order to compare clay brick to concrete block wall construction, House No. 3 was constructed almost identically in House No. 2. Some minor adjustments were made to the dimensions to account for the difference in length of the two material units. [7]

Another modification was to introduce continuous bracing into the roof system to minimize the effect of flexible roof trusses when transverse to the direction of motion. This reduced, but did not eliminate, rocking of the roof trusses which apparently absorbed part of the energy from the concrete slabs.

House No. 4 was different from all previous test specimens. A rectangular plan was used and separate 4'-0" (1.2 m) wide wall panels were constructed. Initially no reinforcement was provided. Then, after two tests were conducted, two #3 bars were placed and grouted in each of the four wall panels. There were two objectives considered: [7]

1) To establish a lower bound of critical base acceleration for a completely reinforced house.

2) To determine the influence of large out-of-plane displacements on relatively slender piers having different reinforcement details.

Shaking Table Test Procedures

Each of the first two houses was subjected to a series of single horizontal component base motions. These motions varied in intensity from just barely perceptible (peak base acceleration less than 0.05 g) to extremely severe (base acceleration over 0.5 g). Three different types of earthquakes were simulated within this range of intensities:

1) The N-S component of the 1940 El Centro,

2) the S69E component of the 1952 Kern County (Taft), and

3) the S74W component of the 1971 San Fernando earthquake recorded at Pacoima Dam. [3]

Response of the structures was recorded by an array of displacement and acceleration measuring devices. Displacements were measured by direct current transducers or potentiometers and accelerations by strain gage or force-balance accelerometers. Displacements of

the structure relative to the shaking table were measured from a rigid frame inside the
structure. This rigid frame was well braced and extremely stiff, with natural frequencies
considerably higher than significant response frequencies of the test structure.

Each house was tested under a wide spectrum of conditions. In addition to the variations of applied base motions, roof truss orientations as well as base fixity of in-plane
wall footings were changed. [3] After initial testing, cracked masonry elements were also
repaired with a fiberglass based surface bonding mortar, and then retested.

The basic philosophy was to subject the structures to a wide range of earthquake
motions. By measuring and visually observing the structure's behavior, it was possible to
identify initiation of significant structural distress and to compare simulated seismic
effects with what could be expected in a Zone 2 area. [8]

All tests conducted on the first two houses had only a horizontal component. Then,
after initial failure and subsequent repair of the unreinforced panels, four runs on the
third house had vertical input motions.

This third structure was subjected to simulated earthquakes over a period of two weeks
Characteristic features and intensities of applied motion were varied, and the roof structure was rotated 90° for several of the tests. After initial failure, a surface bonding
cement was used to repair the walls. The parameters of the first 17 tests were identical
to those of House No. 2, and the results were very similar. But then in Run 18 (Pacoima/
300) and Run 19 (Pacoima/400), transverse unreinforced walls failed. These same walls had
survived similar testing in House No. 2. After these walls were repaired, vertical table
motions of 50% to 60% of horizontal intensity were introduced. In general, this caused a
10% - 15% increase in recorded amplitudes. [8]

For comparable peak accelerations, the Pacoima record seemed to be more with a particular base motion and with the roof trusses oriented either parallel or transverse to the
table motion. A total of 29 test runs were conducted on the specimen in a two-week period,
roughly an equal number for each roof orientation. Evaluation of test results indicated th
following: [3]

1) Response of test specimens to base motions was complex, and affected by roof orien
tation, in-plane wall base fixity, and cracking of unreinforced walls.

2) Overall response limits were dependent upon roof orientation and vibratory characteristics. In-plane response was governed by inertial effects transmitted from the roof. Out-of-plane response depended upon constraints and vertical load provided by the roof.

3) Both out-of-plane walls were capable of significant transverse displacement in the "cantilever" mode without structural distress. Inasmuch as transverse walls, because of their much smaller rigidity, resisted little of the roof inertial load, their performance was governed by the amount of transverse moment capacity.

4) Nominal amounts of vertical reinforcement prevented signficant damage.

5) Surface bonding material used to repair cracked unreinforced walls appeared to be effective. Following repair, test specimens had to be subjected to increased base motions before cracking reoccurred. [3]

House No. 2

This test specimen was also tested over a two-week period. The roof was first oriented with the trusses in line with the table motion; about midway in the test sequence the roof system was rotated 90°. In all, 32 tests were run.

Because of different geometry of in-plane walls, House No. 2 was stiffer than House No. 1. Therefore, smaller displacements were recorded for similar base motions. Other signficiant observations were: [3]

1) Transverse walls displaced consistently greater than in-plane walls. Relative displacement of the roof was greater than out-of-plane wall displacement. The roof diaphragm was very flexible when trusses were perpendicular to the table motion.

2) Because of torsional aspects, inherent in the design response of the structure, simultaneous twisting and bending were observed in the two panels perpendicular to table motion. However, during both roof orientations, transverse displacements did not cause substantial cracking.

3) The unreinforced in-plane panel failed during Test 32 (Pacoima motion with 0.519 g peak acceleration), when it was not in a loadbearing configuration. This indicates that the roof orientation had a large influence on the measured response of the walls.

4) Horizontal cracking at the midheight of the unreinforced transverse wall occurred; caused by lateral inertial forces. [3]

The following observations were noted during the testing of the subject specimen: [8]

1) The Pacoima record appeared to be the most critical for the same peak acceleration. Failure occurred in the out-of-plane, unreinforced wall at 0.386 g. Both unreinforced panels underwent significant structural damage at 0.480 g, but the house did not collapse.

2) In addition to sliding relative to shear walls, the roof deformed unevenly and overall torsional motion took place. The roof structure did not appear to be very rigid.

3) Structural response appeared mostly at frequencies of between 2 Hz and 7 Hz. For the rigid structure, the table motion contributed greatly. Acceleration peaks also shaped the overall response.

4) Based on the results of these tests, 0.35 g peak base acceleration was approximately where strong, nonlinear behavior and structural distress began. There seemed to be no fundamental difference between the seismic response of concrete block and clay brick.

5) Introduction of vertical motion increased displacements by 10% to 15% for similar accelerations.

6) Reinforcement consistently made a difference between structural integrity and failure. However, the reinforcement was stressed minimally, while the grouted cavities withstood the forces.

7) Repairing cracked masonry elements with surface bonding restored some of the flexural strength of panels. However, surface bonding was only partially successful where panels were subjected to significant in-plane shear or shear combined with flexure. [8]

House No. 4

The initial tests require careful examination, since these were conducted on totally unreinforced structures. The structure was subjected to a base motion similar to the El Centro accelerogram, with both horizontal and vertical peak accelerations.

Some cracking occurred in all the panels, but the structure did not approach failure. Transverse panels and piers fared better than in-plane walls. These transverse panels, however, did eventually fail as a consequence of shear, since they couldn't accommodate large diaphragm displacements. [7]

The Taft signal with horizontal and vertical peaks of 0.287 g and 0.219 g was the second strong base motion used. Crack patterns indicated that both in-plane walls developed lateral load resistance largely from friction. Transverse walls were able to withstand significant displacements and unreinforced piers deflected or arched at about midheight. Rigid attachment of the roof structure to these elements apparently improved their behavior.

Final tests were conducted after wall panels had cracked and after reinforced cells were grouted. Reduction of displacement amplitudes indicated an overall strengthening and stiffening of the structure.

Other important observations from the tests are as follows:

1) For in-plane walls, a horizontal acceleration of below 1/3 g did not cause failure. Signs of failure were observed, however, at this value or at a slightly higher intensity. The critical level of excitation appeared to be slightly higher for transverse walls which were unreinforced.

2) Transverse walls accommodated large displacements because of transverse flexibility of plywood roof diaphragm. In-plane walls, however, tended to develop uplift modes in responses to overturning moments.

3) Doweling of vertical reinforcement to foundations did not appear to be required for transverse panel strength. Dowels, however, had a signficant effect on in-plane walls when accelerations exceeded 0.45 g. [7]

Shaking Table Tests - Conclusions

A large number of simulated earthquake tests were conducted on four masonry houses containing both unreinforced and reinforced components. Although the test structures had a similar configuration, roof truss orientations and base motions at various intensities were different. An overall evaluation of the shaking table tests yields the following: [5]

1) Strong base motions of peak accelerations greater than 0.3 g caused significant cracking in unreinforced elements. Partial vertical reinforcement provided a definite beneficial effect, for base motions up to 0.6 g, by inhibiting cracking and by limiting permanent displacements. No structural failure was recorded for partially reinforced walls.

2) Unreinforced wall transverse strength was closely related to bond strength between mortar and masonry units. Cracked transverse walls experienced large out-of-plane

displacements under base motions on the order of 0.3 g. Because of arching and support restraint, however, walls did not become unstable. This stability of unreinforced masonry walls needs further study.

3) Two #4 bars at 14'-0" (4.3 m) spacing in transverse walls were adequate to resist all force levels up to 0.6 g peak acceleration.

4) The strength of in-plane unreinforced walls was governed by overturning; but nominal shear stresses only reached a peak of about 30 psi (207 kN/m^2). Therefore, the dowelled vertical reinforcement at each of the 8'-9" (2.4 m) wind piers in House No. 1 and 16'-0" (4.9 m) long walls in Houses No. 2 and 3, was sufficient to resist overturning and prevent failure.

5) Detailed evaluation of test data is still underway, and specific reinforcement requirements for masonry houses in Seismic Zone 2 have not yet been determined. The amount of reinforcement is dependent upon the safety factor considered appropriate for residential construction. Low safety factors would not require reinforcement, whereas safety factors over 3 may require at least partial reinforcement. A more detailed analysis of the test results, including the safety factor effects, will be required. [5]

CYCLIC LOAD TESTS OF TIMBER ROOF CONNECTIONS

Timber Roof Connection Specimens

This investigation was conducted to determine strength and cyclic behavior of typical roof connections used for masonry dwellings constructed in less seismically active areas of the U.S. Five types of roof connections were cyclically loaded by dynamic actuators. Specimens were subjected to either in-plane (parallel to wall) or out-of-plane (transverse to wall) horizontal loads. Both loadbearing and non-loadbearing models of gabled truss and flat roof construction were tested. [6]

The study clearly distinguished between in-plane behavior of connections with inertial effects transmitted from the roof diaphragm oriented parallel to the wall, and transverse behavior when the force was applied normal to the wall. In addition, bolt size was considered a variable. Each model had an 8'-0" (2.4 m) long masonry wall panel with bolts,

4'-0" (1.2 m) apart, embedded in the wall. Wall heights were varied and both hollow concrete block and hollow clay brick, were used. Wall panels were reinforced with 2 - 4# bars at each end.

The first two specimens represented a portion of bearing wall with four truss joists, at 2'-0" (61 cm) on center, nailed to a single 2" x 6" (5.1 cm x 15.2 cm) top plate. Two truss joists were also secured to the top plate by metal framing anchors (see Figure No. 5).

The third connection specimen represented a gable end of a roof system with trusses parallel to the wall. "Truss-ties", at 6'-0" (1.8 m) on center, were used to connect the bottom chord to the wall segment. In addition, the bottom chord was attached to the top plate through solid wood blocking at 4'-0" (1.2 m) on center (see Figure No. 6).

Specimens four and five represented flat roof construction with ledger boards bolted to the wall face. The nominal thickness of the ledger was 3" (7.6 cm) for one test and 2" (5.1 cm) for the other. Both groups were subjected to horizontal forces transverse to the wall. Some tests were run with an added vertical roof load of 180 plf (2.65 kN/m) simulated by placing weights above the ledger. [6]

Timber Roof Connection Test Conclusions

On the basis of measured strength and observed behavior of tested connections, the following was concluded:

1) Cyclic shear forces caused local deformations resulting in simultaneous slip and pull-out of nailed timber connections. Therefore, safety factors based on code provisions had to be substantially reduced.

2) The truss rafter to top plate and masonry wall connections tested in the first two groups of specimens adequately resisted Seismic Zone 2 forces.

3) The size of bolts in the first three groups of tests did not affect connection strength. Minimum 1/2" (12.7 mm) diameter bolts, with embedment lengths equal to or greater than UBC code requirements, are recommended.

4) Further testing of the ledger connection with horizontal forces parallel to the wall is recommended. Anchorage devices would not be effective initially. Therefore, the parallel forces would be resisted solely by anchor bolts.

5) Response characteristics measured in the fourth group of tests (flat connections) were not significantly affected by either the use of hollow clay the addition of the gravity roof loads.

CONCLUSION

This research at the University of California at Berkeley, conducted und was just completed in August 1979. Results of the shaking table and cyclic 1 tests will be published by HUD's Office of Policy Development and Research. final reports will be available from the National Technical Information Servi Springfield, VA 22161, USA.

Films were also taken of each of the shaking table tests and, after edit available for viewing at future UJNR Panel meetings. A slide presentation wi pared, to be used for training HUD field personnel.

Planning of the experimental research program was coordinated, under a s contract, with the Applied Technology Council (ATC-5) Advisory Panel. ATC wi design and construction recommendations for single-family masonry dwelling co U.S. Seismic Zone 2 areas. The final criteria is intended to be incorporated mum Property Standards (MPS) or in the MPS Manual of Acceptable Practice (MAP

REFERENCES

[1] U.S. Department of Housing and Urban Development, "Minimum Property Sta Two-Family Dwellings," HUD 4900.1, 1973 Edition, Washington, D.C.

[2] International Conference of Building Officials (ICBO), "Uniform Buildin 1973 Edition, Whittier, CA.

[3] P. Gulkan, R. L. Mayes, and R. W. Clough, "Experimental Study of Reinfo Single-Story Masonry Houses - Vol. 1," (Draft) (1979).

[4] P. Gulkan, R. L. Mayes, and R. W. Clough, "Response of Single-Story Bri to Simulated Earthquakes," Draft for Fifth International Brick Masonry October 1979 (Brick Institute of America, McLean, Virginia, 1979).

[5] P. Gulkan and R. L. Mayes, "Experimental Investigation of Reinforcement for Simple Masonry Structures in Moderately Seismic Areas of U.S.," Dra U.S. National Conference on Earthquake Engineering, Stanford University (Earthquake Engineering Research Institute, Berkeley, CA, 1979).

[6] P. Gulkan, R. L. Mayes, and R. W. Clough, "Strength of Timber Roof Conn Subjected to Cyclic Loads," Draft Report No. EERC-79 to HUD, Earthquake Research Center, Berkeley, CA (1979).

[7] Earthquake Engineering Research Center, "Interim Report on Summary of Experimental
     Observations Concerning Behavior of Fourth Masonry House Subjected to Simulated
     Earthquakes," Berkeley, CA (February 1979).

[8] Earthquake Engineering Research Center, "Interim Report on Summary of Experimental
     Observations Concerning Behavior of Third (Brick) Masonry House Subjected to Simulated
     Earthquakes," Berkeley, CA (October 1978).

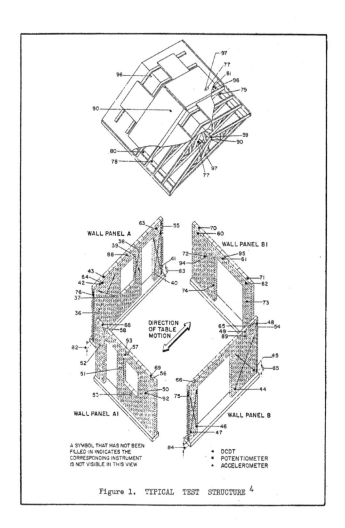

Figure 1. TYPICAL TEST STRUCTURE [4]

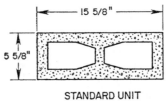

STANDARD UNIT
HOUSES 1 AND 2

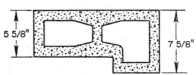

CORNER RETURN UNIT
HOUSE 1

Figure 2. CONCRETE MASONRY UNITS [3]

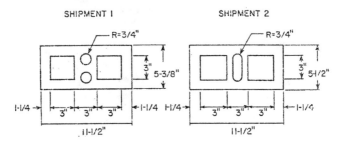

SHIPMENT 1   SHIPMENT 2

HEIGHT 3-5/8"

Figure 3.   BRICK UNITS FOR HOUSE 3 [8]

III-17

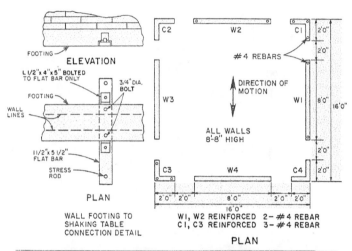

FOOTING

ELEVATION

L1/2"x 4"x 5" BOLTED
TO FLAT BAR ONLY

3/4"DIA.
BOLT

FOOTING

WALL
LINES

1 1/2"x 5 1/2"
FLAT BAR

STRESS
ROD

PLAN

WALL FOOTING TO
SHAKING TABLE
CONNECTION DETAIL

C2   W2   C1   2'0"

2'0"

#4 REBARS

↕ DIRECTION OF
MOTION

W3   W1   8'0"   16'0"

ALL WALLS
8'-8" HIGH

2'0"

C3   W4   C4   2'0"

2'0" 2'0"   8'0"   2'0" 2'0"

16'0"

W1, W2 REINFORCED 2– #4 REBAR
C1, C3 REINFORCED 3– #4 REBAR

PLAN

C2   W3   C3

Figure 4. HOUSE NO. 1 [3]

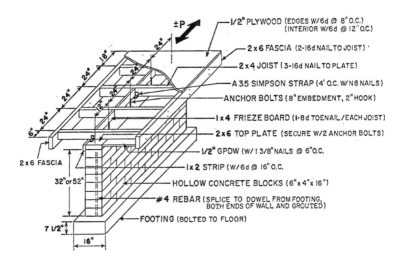

±P

1/2" PLYWOOD (EDGES W/6d @ 8"O.C.)
(INTERIOR W/6d @ 12"O.C.)

2 x 6 FASCIA (2-16d NAIL TO JOIST)

2 x 4 JOIST (3-16d NAIL TO PLATE)

A 35 SIMPSON STRAP (4' O.C. W/N8 NAILS)

ANCHOR BOLTS (8" EMBEDMENT, 2" HOOK)

I x 4 FRIEZE BOARD (I-8d TOENAIL/EACH JOIST)

2 x 6 TOP PLATE (SECURE W/2 ANCHOR BOLTS)

1/2" GPDW (W/ 1 3/8"NAILS @ 6"O.C.

I x 2 STRIP (W/6d @ 16" O.C.

HOLLOW CONCRETE BLOCKS (6"x 4"x 16")

#4 REBAR (SPLICE TO DOWEL FROM FOOTING,
BOTH ENDS OF WALL AND GROUTED)

FOOTING (BOLTED TO FLOOR)

2 x 6 FASCIA

32" or 52"

7 1/2"

16"

Figure 5. TYPICAL TIMBER ROOF TRUSS CONNECTION TEST SPECIMEN [6]

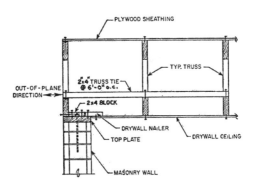

PLYWOOD SHEATHING

TYP. TRUSS

OUT-OF-PLANE
DIRECTION

2 x 4 TRUSS TIE
@ 6'-0" o.c.

2 x 4 BLOCK

DRYWALL NAILER

TOP PLATE

DRYWALL CEILING

MASONRY WALL

Figure 6. GABLE END ROOF CONNECTION TEST SPECIMEN [6]

Table I. Dimensions and Properties of Concrete Masonry Units [7]

Width	Height	Length	Minimum Thickness in		Gross Area	Net Area	Compressive Strength psi		Oven Dry Weight	Concrete Weight
			Face Shell	Web			Net Area	Gross Area		
in	in	in			in²	%			lb	lb/ft³
							(10184 kN/m²)			
5.61	3.63	15.56	1.03	1.06	87.4	53.5	2778	1477	10.84	110.5
(14.2cm)	(9.2cm)	(39.5cm)	(2.6cm)	(2.7cm)	(563.9cm²)		(19154 kN/m²)		(4.92kg)	(1770 kg/m³)

(1) Values given in the table represent the average results from tests or measurements of 5 units.

Table II. Materials for the Test Structures [5]

House No.	Type of Masonry Unit	Masonry Strength psi	Mortar Strength psi	Grout Strength psi	f_y ksi	Reinforced Panels (Fig. 1)
1	6x4x16 in. Conc. block	1,530	2,650	6,350	54.0	W1, W2, C1, C3
2	6x4x16 in. Conc. block	1,983	4,730	3,360	59.3	B, B1
3	6x4x12 in. Clay brick	4,970	2,336	2,536	59.3	B, B1
4	6x4x16 in. Conc. block	2,778	1,944	2,324	54.0	See Note (3)

Notes: (1) Masonry strength is based on net area of units.
(2) Type S mortar of the UBC was specified for all houses.
(3) House 4 was tested tested twice with no reinforcement in any of the four walls before two #3 bars were grouted in each wall.

FIGURE 6   HOUSE NO.3
III-21

FIGURE 6   HOUSE NO. 3 (CONTINUED)
III-22

THE PERFORMANCE OF LAPPED SPLICES IN REINFORCED CONCRETE

UNDER HIGH-LEVEL REPEATED LOADING

Peter Gergely

Cornell University, Ithaca, N.Y.

Fernando Fagundo

University of Puerto Rico, Mayaguez, P.R.

Richard N. White

Cornell University, Ithaca, N.Y.

ABSTRACT

The results of two series of experiments on lapped splices in reinforced concrete is reported; one on half-scale beams and the other on full-scale beams. The effects of repeated loading and transverse reinforcement on splices in constant moment regions have been studied to date.

KEYWORDS: Beams; bond; concrete; design; lapped splices; reinforced concrete; seismic design; splices; testing.

INTRODUCTION

The lapped splice provisions of seismic design codes vary greatly and are generally conservative. Lapped splicing is ordinarily not allowed in regions of a structure where repeated yielding may occur. This restriction most likely originates from research on cycl loading in bond tests where serious degradation may occur, or from observations of splice failures in seismically loaded structures. The effect of transverse steel, which can great enhance splice performance, is not considered explicitly in splice design in codes for eith seismic or nonseismic areas.

Recent research [1] has resulted in expressions for the design of lapped splices in nonseismic areas in which the effects of transverse reinforcement, cover, and bar spacing a explicitly accounted for. The suggested provisions [1] specify a maximum amount of transverse steel (stirrups) above which additional steel is not effective. The limit is given b

$$\frac{A_{tr}}{s} \leq \frac{1500d_b}{f_y} \text{ in.,} \left(\frac{10.5d_{b\ mm}}{f_y} \right) \tag{1}$$

where A_{tr} is the area of transverse steel crossing the expected splitting crack, s is the spacing of stirrups, d_b is the diameter of the main bars, and f_y is the yield strength of t transverse steel.

The research summarized here has studied the effects of high-level repeated flexural loading (0 to Maximum to 0) and the amount and distribution of transverse steel along the splice. Beams tested under reversed loading are not discussed in this paper. The amount o transverse steel used in the beams are expressed in terms of the maximum amount given above

TEST PROGRAM

The half-scale specimens had a span of 6 ft (1.27 m) and a cross section of 6 in. by 10 in. (152 by 254 mm), whereas the corresponding dimensions for the full-scale tests were 21 ft (6.40 m) and 9.5 by 16 in. or 12 by 20 in. (241 by 406 or 305 by 508 mm). All beams were loaded at the third points and the splices were in the constant moment region. The main reinforcement consisted of two #4 (13 mm) bars in the small scale tests and two #8 or #10 (25 or 32 mm) in the large scale tests. Both longitudinal bars were spliced in every beam. Since the only two bars were used, they were in the corners of the stirrups.

III-24

Three small-scale beams were tested under monotonic loading and three similar beams under repeated loading. All splices were 28 d_b long. Four beams had no transverse steel whereas two beams had the maximum effective amount.

Twelve large-scale beams were tested under repeated loading and two under monotonically increasing loading. All beams had web reinforcement ranging from 16% to 210% of the maximum effective amount given by Equation 1. The distribution of stirrups along the splice was variable; three main types of stirrup arrangements were used: (1) one near each end of the splice and one at the center, (2) two stirrups near each splice end, and (3) closely spaced stirrups along the splice. The splice lengths were 30 d_b or 36 d_b.

The concrete strength was nominally 4000 psi (28 MPa).

TEST RESULTS

Only selected preliminary results are discussed here; a forthcoming report will contain a detailed presentation of the test results and a critical evaluation of their significance.

The effect of the amount and distribution of stirrups on splice behavior becomes evident as several pairs of tests are compared. Four half-scale tests are summarized in Table 1. The transverse steel is in terms of the maximum effective amount given in Equation 1. It is seen that this amount was sufficient to allow the beam to reach twice the yield deflection under monotonic loading (beam B), but it was insufficient for cyclic loading (beam D). Beams without transverse steel are obviously unsatisfactory even for monotonic loading; beam A failed at 90% of the expected yield load.

Beam	Load	Steel	ℓ/d_b	Failure	Cycle
A	M	0	28	0.90 P_y	-
B	M	100%	28	2 D_y	1
C	R	0	28	0.85 P_y	
D	R	100%	28	0.90 P_y	25

M = monotonic, R = repeated, P_y = yield load,
D_y = yield deflection measured at mid-span

Table I. Comparison of half-scale beam tests

set of three beams had 1/3, 1, and 2 times the maximum effective amount of transverse steel. The first of these was insufficient to go beyond the yield strength of the beam, whereas the second design permitted inelastic deformation to about twice the yield displacement but for only a few cycles. The beam with about twice the maximum amount of transverse steel did not fail after many cycles at over twice the yield displacement.

The amount and distribution of stirrups has a great effect on the rate of penetration of damage towards the center of the splice. A beam with #10 (32 mm) bars and #3 (9.5 mm) stirrups at 3 in. (76 mm) spacing (giving 1.15 times the maximum effective amount) reached about twice the yield displacement. The yielding of the main bar spread about 12 in. (305 mm) along the splice which was 30 d_b long. Two stirrups at each end yielded. Both the bar strains and the stirrup strains increased continuously, especially at the peak load level, where failure occurred after seven cycles.

The third beam had #4 (12.7 mm) stirrups at 3 in. (76 mm) which provided 2.1 times the maximum effective amount of transverse steel. It did not fail even after many cycles at about twice the yield deflection. The strains in the stirrups tended to equalize as cycling continued beyond bar yield. However, the stirrup strains remained below about half the yield level.

The distribution of stirrups along the splice is more important for cyclic loading than for monotonic loading. Concentration of stirrups near the ends of the splice is beneficial up to the yielding of the bar, but beyond that stage, damage (mainly cracking around the bar) progresses along the splice during cycling and additional stirrups are brought into action. Thus, it is essential to have closely spaced stirrups along the splice. The concrete cover is much less important for repeated high-level loading than for monotonic loading below yield.

The study of the effect of load history generally confirmed previous investigations in bond. Repeated loading up to about 75% of yield had little effect on splice behavior. However, deflections and strains along the bar and in the stirrups increased with cycling above about 80% of yield.

High-level load cylces cause local damage around the splice that affects subsequent low-level behavior. This effect can be seen in Figure 1 where bar and stirrups strains are shown at 40% of the flexural yield load before and after cycles at P_y, 0.8 P_y, and P_y. After a single loading to P_y the bar and stirrup strains increased substantially when reloaded to 0.4 P_y (compare curves 1 and 3). However, 11 further cycles at 0.8 P_y caused little additional damage (cycles 3 and 15). Another loading to P_y caused comparatively more damage because subsequent bar strains were higher as comparison of cycles 15 and 21 shows. This figure illustrates the conclusion that a force level of greater than about 80% of yield is detrimental to splice behavior.

The investigation has shown that the use of about twice the maximum effective amount of stirrups specified in the suggested code revisions by ACI Committee 408 [1] with the stirrups closely spaced along the splice, will permit repeated loading to at least twice the yield deflection. The research is continuing and several additional variables are being studied. Pilot tests indicate that reversed loading or the presence of substantial shear forces both seriously impair the load-carrying capacity of splices unless a large amount of closely space stirrups are used.

ACKNOWLEDGMENTS

This investigation has been supported by the National Science Foundation; however, any opinions, findings, conclusions, or recommendations expressed herein are those of the authors and do not necessarily reflect the views of the NSF.

REFERENCES

[1] ACI Committee 408: "Suggested development, splice, and standard hook provisions for deformed bars in tension," Concrete International, Vol. 1, No. 7, July 1979.

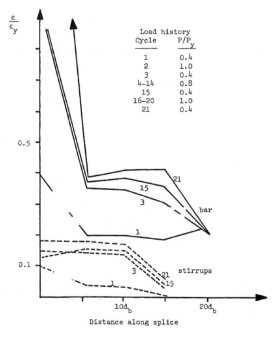

Fig. 1 Strains in main bar and in stirrups at $0.4P_y$ after cycling at higher loads. #10 (32 mm) bars, #3 (9.5 mm) stirrups at 3 in. (76 mm), $\ell_s = 38d_b$.

III-28

DYNAMIC BEHAVIOR OF REINFORCED CONCRETE FRAME STRUCTURES

Keiichi Ohtani

Chikahiro Minowa

Earthquake Engineering Laboratory

National Research Center for Disaster Prevention

Science and Technology Agency

ABSTRACT

This study presents earthquake simulator test of reinforced concrete structures conducted at the Earthquake Engineering Laboratory of the National Research Center for Disaster
Prevention at the Tsukuba New Science City.

Two types of one-story, single-bay reinforced concrete frames in actual size were built
on the shaking table. The two reinforced concrete frames had different slab weights; about
60 tons for the first tested frame and about 130 tons for the second one. Using this large-
scale shaking table, both reinforced concrete frames were subjected to simulated and modified
earthquake ground motions with intensities large enough to cause inelastic behavior and
dynamic property changes to the frames.

Test results of both frames are described, as is the yielding of reinforcing steel bars,
the varying properties of the overall frame responses, natural frequencies, and damping
ratios.

Finally, computer simulations to evaluate inelastic responses and the correlation with
measured performance are described using bi-linear or tri-linear analytical models.

KEYWORDS: Earthquake acceleration-displacement analysis; reinforced concrete structures;
shaking table simulation; structural testing.

1. INTRODUCTION

The nonlinear responses of reinforced concrete building structures have been st
theoretically and experimentally. It may be necessary to verify the analytical corr
with the earthquake simulator test, using the computer simulation methods. However,
ing experimental studies, the test specimens were the element tests or the reduced s
models. The large-scale shaking table of the National Research Center for Disaster
tion has a capacity large enough to excite the actual size reinforced concrete struc
Therefore, the earthquake simulator test of actual-size reinforced concrete structur
planned. One-story, single-bay reinforced concrete frames were adopted for the reas
economy and safety.

Two frames were tested. The test of the first frame was executed in 1977 and t
of the second was in 1978. Both test frames had no walls, so these structures exhib
moment yield patterns. The column section and height of both test frames were ident
In order to obtain different natural frequency, the second frame was given a slab we
about twice as much as the first frame.

In another way, it is a primary interest to verify that our shaking table would
damage to actual-size reinforced concrete structures with the simulated earthquake mc

2. TESTING FACILITIES AND INSTRUMENTATION

The shaking table as used for testing can move in one horizontal direction with
hydraulic driving system. This table is 15 m x 15 m in size, 160 tons in weight. Th
mum displacement and velocity that can be achieved are ± 3 cm and 37 cm/sec. This st
table may be used to subject a specimen weighing up to 500 tons. The natural frequen
the table weight and the oil spring would be considered as 20 Hz approximately. The
tional frequency range would be 0 - 50 Hz.

A total of about 45 transducers were employed to measure the structure responses
tests of the two frames, respectively. The table and the roof acceleration were meas
the accelerometers of strain gage type in three directions. The table and the roof d
ments were measured by IDS type position meters in excitational direction. In order
the position meters, the steel frame was constructed on the shaking table foundation,
shown in Figure 2 and 9. The story displacement is obtained by subtracting the table

displacement from the roof in the use of the electric amplifier. The wire strain gages were arranged on reinforcing steel bars (D19) of each columns, as shown in Figures 4 and 10. In addition to these measurements, the behaviors of the steel frame were observed by accelerometer in order to verify the accuracy of displacement measurements. The test data were recorded with 14 channel data recorders. With 16 mm movie cameras and video systems, the performances of cracks grown in columns were watched and recorded for the tests of both frames.

When the earthquake ground motion is simulated, by the reason of the limitations of the shaking table stroke of 3 cm, the low frequency components of the original earthquake motions may be cut off in order to produce high intensity shakes. Therefore, the table inputs for the earthquake motion tests are obtained as the outputs from the one-degree of freedom systems's responses to the original earthquake motions. The damping constant of this system should be settled as 0.707. The natural frequency of this system would be selected in accordance with the dynamic properties of the test frame. These natural frequencies were decided as 3 Hz for the first frame, 2 Hz for the second frame. The SMAC record of HACHINOHE E-W in Tokachi-Oki earthquake 1968 was adopted as the table input for the earthquake motion tests. The Fourier spectra of original ground motion and measured shaking table motions of both test frames are shown in Figure 1.

3. TEST OF THE FIRST FRAME

The first frame was designed and constructed in accordance with the old A.I.J. Calculation Code for Reinforced Concrete Structure. The general view of the first frame is shown in Photo. 1. The elevation and setting plan of this frame are shown in Figures 2 and 3, respectively. The details of reinforcing steel bar installations and the section of columns and girders are shown in Figure 4.

The total height of this frame was 4.2 m and the column height was 2.8 m. The widths of frame are 8 m in long axis, 6 m in short axis. The long axis of this frame was corresponding to the excitational direction of the shaking table. Each column had a section of 45 cm x 45 cm, and the roof girders in both axes was 60 cm in height, 35 cm in width. The footing of each column was 3 m x 3 m in plan, 80 cm in height. The footings were connected by the tie beams in both axes. The section of tie beam was 80 cm in height, 40 cm in width. The roof slab was 40 cm in thickness, and was strengthened by a beam of 60 cm x 25 cm in long axis. The slab reinforcing steel bars of D10 were installed with 20 cm x 20 cm mesh in both

columns. Each footing was fastened to the shaking table by sixteen 2-inch steel bolts. The total weight of the first frame was about 130 tons, and the weight of the roof was about 60 tons. The dead load axial stress of the columns was considered to be 7.5 kg/cm^2 approximately. The 28 day compressive strength of concrete was 215 kg/cm^2. The tensile strength of the steel bar (D19) was estimated as 5.59 ton/cm^2. The yield point for the steel bars was estimated as 3.78 ton/cm^2.

The test of the first frame was started one month after the removal of slab supports. Ahead of the vibration test, the micro-tremor of this frame was observed and the dominant frequency was estimated about 5.8 Hz.

At first, the free vibration frequency and damping of the frame without cracks was measured by rectangular displacement table motions in amplitude of 1 mm. The frequency of the frame was estimated as 4.76 Hz in this free vibration test. As the second step of the test, the resonance test was carried out. The intensity of sinusoidal excitation was approximately 15 gals. During the excitation of 4.5 Hz sinusoidal motion, near the resonance frequency of this cracked frame, the properties of the frame changed. Therefore, the cracks at columns were enlarged by this test. After this excitation, the properties of the frame were continued to change with every sinusoidal excitation. Finally, the peak frequency and the damping of the resonance curve was determined in 3.5 Hz, 0.02. However, the free vibration frequency after the resonance test was measured 3.96 Hz. Then, the static tensile test was carried out, and the frame spring was estimated as 30 ton/cm by loading the frame laterally at the roof. Adopting this spring and the roof mass, the natural frequency would be calculated to 3.5 Hz.

The first run of the simulated earthquake ground motion was carried out with an intensity of about 0.3 G. The Fourier spectrum for this run is shown in Figure 1. The chart record at this run is shown in Figure 8. The maximum accelerations measured at the table and the roof were about 0.34 G and 0.8 G. The maximum story displacement obtained was approximately 2.9 cm. The strain gages at the foot of the column exhibited the yield of reinforcing steel bars. From the instance of yielding, the frequency of the frame began to decrease. The estimated running gain for this run is shown in Figure 7. Judging from this gain, it is clear that the frame underwent nonlinear behavior and the frame frequency decreased from 3.5 Hz to 2.5 Hz. This frequency change occurred from main shocks in about

5 seconds. The diagram of the acceleration-story displacement relation, which might be considered the approximate relation between story shear and story displacement, is shown in Figure 6. This diagram also described the degrading properties of the frame stiffness. The free vibration frequency and the damping after this run was observed as 3.07 Hz and 0.07. During this run, the cracks of the column grew, but a failure of concrete was not found.

The second run of the simulated earthquake ground motion was carried out with an intensity of about 0.5 G. During this excitation, the cracks were noticed and a few, small broken pieces of concrete fell down. The maximum accelerations were approximately 0.78 G for the table, 1 G for the frames. The maximum displacement was calculated as 4.19 cm. The free vibration frequency and the damping after the second run were 2.25 Hz and 0.08. The results of free vibration for this frame are shown in Figure 5.

4. TEST OF THE SECOND FRAME

This frame was designed and constructed in accordance with A.I.J. Code. The general view of this frame is shown in Photo. 2. The elevation is shown in Figure 9. The details of the reinforcing steel bar installations and sections of columns are shown in Figure 10.

The total height of the second frame was 4.4 m. The height and the section of columns, and the frame widths corresponded to those of the first frame. The section of roof girders in both axes were 100 cm x 45 cm in order to obtain a higher rigidity than the first frame. The roof slab had the dimensions of 11 m in long axis, 9 m in short axis, and 40 cm in thickness. The roof was strengthened by three beams of 100 cm x 35 cm; two in short axis, one in long axis. The footing of each column had the same dimensions in plan as the first frame, and thinner thickness of 60 cm than the case of the first frame. The footing was connected by the tie beams in long axis. There were no anchor steel bars in the columns. The slab reinforcing steel bars of D13 were installed with 20 cm x 20 cm mesh in both sides. Each footing was fastened in the same manner as the first frame. The total weight of the second frame was about 180 tons, the weight of the roof which was considered twice as much as the case of the first frame, was 130 tons approximately. The dead load axial stress in a column was estimated about 15 kg/cm^2. The compressive strength of concrete on the test day was 194 kg/cm^2. The yield point of the steel bar D19 was estimated as 3.35 ton/cm^2, and the tensile strength was 5.03 ton/cm^2.

The test of this frame was carried out in four weeks after the removal of slab supports. The micro-tremor of this frame was observed before the vibration test. The dominant frequency was estimated about 4.8 Hz. During the first measurement of vibration test, the free vibrations were induced in the same way as in the test of the first frame. The dominant frequency was measured 4.0 Hz.

The simulated earthquake ground motion was executed with the full stroke of the shaking table. The Fourier spectrum of the table acceleration is shown in Figure 1. The maximum acceleration and displacement were 0.43 G and 2.5 cm at the table, 0.45 G and 4.9 cm at the frame, respectively. The maximum story displacement was calculated as 4.25 cm. All strain gages on the reinforcing steel bars at the foot and head of columns increased as the steel yielded and went over the measurable range at the first shock of the excitation. Thus, the strain measurement became impossible. From this excitation, the cracks at the top of columns, which are shown in Photo. 3, were caused. At the foot of the columns, pieces of the concrete broke off.

Using the table and the frame acceleration records, the estimated running gain was obtained with a frame time of 2 sec and a shift time of 0.5 sec, which is shown in Figure 11. By the running gain for this excitation, it was obvious that the second frame underwent non-linear behavior. The frame frequency decreased from 4 Hz to 1.5 Hz during the simulated earthquake motion test. The diagram of acceleration-story displacement relation which is shown in Figure 12, also expressed the degrading properties of the frame stiffness.

5. COMPUTER SIMULATION AND ITS CORRELATION WITH EARTHQUAKE MOTION TEST

The computer simulations were carried out for the first run test of the first frame and for the second frame.

Judging from the diagram of the acceleration-story displacement relation, as shown in Figure 6, the restoring force characteristics in the first frame are seen to possess a bi-linear relation and story displacement. Moreover, the stiffness degradation would be found in the estimated running gain and the diagram of acceleration-story displacement. Thus, for the computer simulation, the bi-linear peak oriented mechanism, which is depicted in Figure 13, was adopted. The time history correlation between computer simulations and test results are shown in Figure 17. The computed relation between acceleration and story displacement is shown in Figure 14.

Judging from Figure 12, the second frame would be considered to have the tri-linear restoring force characteristics. Therefore, the tri-linear peak oriented mechanism was adopted for the second frame. The computation model is depicted in Figure 15. The XE3 was the cross point of the skelton curve and the straight line which passed the origin parallel to the second gradient of the skelton curve. The time history correlations between the computer simulation and the test results are shown in Figure 18. The computed diagram is shown in Figure 16.

Both computer simulations produced close correlations to the test results in the acceleration time histories.

Photo. 1 General View of the First Frame

Photo. 2 General View of the Second Frame

Photo. 3 Damage of the top of column (2nd Frame)

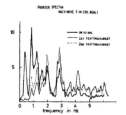

FOURIER SPECTRA
NACHINOHE E-W(199.8GAL)

ORIGINAL
1st TEST(MEASURED)
2nd TEST(MEASURED)

frequency in Hz

Fig. 1 Fourier Spectra
for Simulated Earthquake
Ground Motion

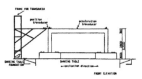

Fig. 2 Elevation of the
First Frame

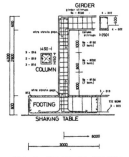

Fig. 3 Setting Plan
of the First Frame

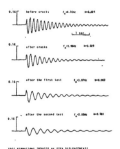

FREE VIBRATIONS INDUCED BY STEP DISPLACEMENTS

Fig. 5 Records of the
Free Vibration Tests

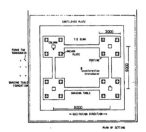

Fig. 4 Structural Details
of the First Frame

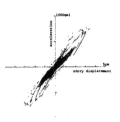

Fig. 6 Relation between
Acceleration and Story
Displacement

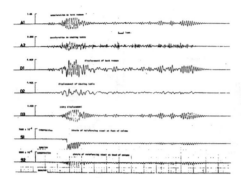

Fig. 8   Chart Record of the 1st Run (1st Frame)

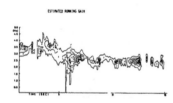

Fig. 7   Estimated Running Gain
(1st Frame)

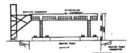

Fig. 9   Elevation of the 2nd Frame

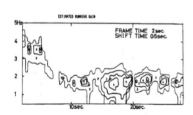

Fig. 11   Estimated Running Gain
(2nd Frame)

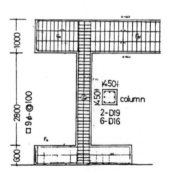

Fig. 10   Structural Details
of the 2nd Frame

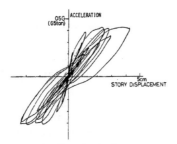

Fig. 12 Relation Between
Acceleration and Story
Displacement

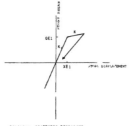

BILINEAR, STIFFNESS DEGRADING

Fig. 13 Computational
Model for the 1st Frame

Fig. 14 Computed Diagram
for the 1st Frame

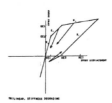

TRILINEAR, STIFFNESS DEGRADING

Fig. 15 Computational
Model for the 2nd Frame

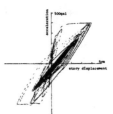

Fig. 16 Compured Diagram
for the 2nd Frame

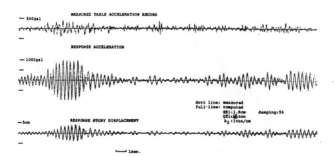

Fig. 17 Time History Correlations for the First Frame

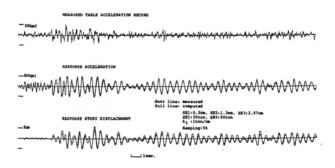

Fig. 18 Time History Correlations for the Second Frame

SEISMIC RESPONSE OF PRECAST CONCRETE WALLS

James M. Becker

Carlos Llorente

Peter Mueller

Massachusetts Institute of Technology

Cambridge, Massachusetts

ABSTRACT

Large panel precast concrete structures have been widely utilized in major seismic regions throughout the world. The seismic behavior of such structures is strongly dependent upon the characteristics of both the horizontal and vertical connections. The limiting behavior of precast systems, however, is basically dependent upon the horizontal connection. The influence of horizontal connections can be studied in terms of the behavior of a simple wall--a vertical stack of panels having only horizontal connections.

This paper reports on research into the seismic behavior of simple precast concrete walls. The research was carried out through the development of computer-based modeling techniques capable of including the typical behavioral characteristics associated with horizontal joints. The model assumes that all nonlinear, inelastic behavior is concentrated in the connection regions and that the precast panels remain linear elastic. This assumption allows the precast panels to be modeled as statically condensed 'super-elements' and the connection regions as interface elements. The above modeling technique allows for nonlinear-inelastic seismic analysis that is capable of handling both rocking type motions throughout the height of the structure and slippage due to shear in the plane of the connection.

A series of parametric studies are presented to illustrate the potential influence of rocking and slip on precast walls using both regular reinforcement and post-tensioning. These studies demonstrate the period elongation associated with the non-liner elastic rocking phenomenon. Shear slip is found to occur only when friction coefficients are extremely low or when the normal forces across the connections are low. This latter case occurs only in low buildings or in the upper floors of tall buildings.

The paper concludes with a brief discussion of the design imp
Particular attention is paid to the problems stemming from the for
ated with rocking and shear slip.

KEYWORDS: Dynamic analysis; friction; post-tensioning; precast co

shear walls.

INTRODUCTION

Large panel precast concrete structures have been widely utilized in major seismic regions throughout the world. This form of construction raises many unusual problems in terms of aseismic design because of the connection regions inherent in precast construction. These connection regions are one of the main sources of the nonlinear and inelastic behavior associated with the seismic response of these structures. At the same time, the likely concentration of nonlinear and inelastic behavior in the connection region facilitates the nonlinear modeling of these structures. This paper reports on research that uses a nonlinear modeling approach to study the influence of the horizontal connection on the seismic response of precast concrete walls.

The paper begins with the description of precast configurations typical of American construction practice. An isolated or simple wall is abstracted from these typical precast configurations and the behavioral basis for its nonlinear modeling is discussed. A finite element model, using contact type elements for connections and linear-elastic substructures for wall panels, is presented. The seismic response predicted by this model is presented in the form of a series of parametric studies. The results of these studies are discussed in light of their aseismic design implications.

SEISMIC BEHAVIOR OF SIMPLE PRECAST WALLS

The cross-wall structure illustrated in Figure 1 is typical of many of the precast systems currently being used by the American precast concrete industry. The system is typified by the use of one-way precast prestressed floor planks spanning between the load-bearing cross walls. As a basis for understanding the seismic behavior of these systems, they can be idealized as being constructed of three basic structural elements (see Figure 1). The primary lateral load-carrying element is a simple wall, typified by an isolated load-bearing cross wall. The simple wall is constructed of a vertical stack of solid panels and therefore has only horizontal connections. The other lateral load-carrying element is a composite or coupled wall. This element consists of simple walls joined together by vertical connections or lintels. The third element is the floor system acting as a series of horizontal

diaphragms and tying together the various wall elements. It is important to note that the
horizontal connections found in both simple and composite walls also act, in general, as
connections in the precast floor systems.

Precast floor systems often have significantly reduced in-plane stiffness [2,4,10,19]
when compared to cast-in-place systems. This reduced stiffness and the difficulty in achieving adequate tensile continuity between floor elements poses problems in both dynamic analysis and design of large panel structures [19,24]. However, in order to concentrate the
ensuing discussion on the modeling and behavior of the simple wall, it will be assumed
that the floors as well as the foundations behave in a rigid fashion.

As is seen in Figure 1, the simple wall is often found as a primary lateral load-
carrying element in cross-wall systems. However, the interest in the behavior of simple
walls extends beyond this explicit use, for it serves as a bound for the behavior of composite walls (see Figure 2) and also allows for the detailed investigation of the possible
influence of the horizontal connection on seismic response.

Because of normal precast construction procedures and the effects of shrinkage and
creep, it is reasonable to consider horizontal connections to act as precracked planes. Thus
it is to be expected that crack opening will be concentrated in the region of the horizontal
connections. In this sense, the simple wall may be thought of as a precracked, reinforced or
prestressed concrete beam with known crack locations. Crack opening and closing along a
horizontal connection is usually referred to as rocking. However this term must be clarified
because the motion is not that associated with the rocking of rigid bodies; rather a smooth,
continuous opening and closing occurs in which plan sections no longer remain plane in the
regions immediately adjacent to the connection (Figure 3).

In the modeling of large panel precast concrete structures it is typical to consider
that solid panels remain linear elastic [15,17,20]. This assumption is reasonable in light
of the expected role of these elements in the structure's seismic response. Typical wall
panels can be assumed to behave in an elastic-brittle fashion in regards to in-plane normal
forces and whatever nonlinear-inelastic behavior that may exist can be effectively lumped
into the behavior of the connection regions.

The assumption of elastic-brittle behavior is based upon the general lack of confinement
of panel concrete in areas of maximum expected compressive forces. The typical single or
double layer of wall reinforcement (usually not tied together) cannot be expected to provide

III-44

confinement for the concrete. However, the weakness to be expected in the connection regions
may protect the walls from compressive failure through their inability to transfer the necessary forces. This makes it possible to assume that the nonlinear-inelastic behavior is concentrated in the connection region and that the panels remain linear elastic.

Vertical reinforcement across the horizontal connection plane is generally kept to a
minimum, especially when compared to cast-in-place construction. Many techniques have been
developed for facilitating the necessary connection of vertical reinforcement (grouted voids,
splice sleeves, welded details, use of vertical connection regions, post-tensioning, etc.).
Few if any allow for the economic development of the amount of vertical steel normally
expected of load-bearing cast-in-place shear walls. While this lack of vertical reinforcement may seem to contradict normal aseismic design practice it may have two potential benefits. First, as suggested by Abrams and Sozen [1], it may provide a softer wall, in which
the seismic forces will be less than those associated with a stiffer, more highly reinforced
wall, and second the compressive force that can be developed across the potentially vulnerable horizontal connection is directly related to the amount of tensile steel in the wall.

The limited experimental data [8,9,12] on the compressive strength of the platform type
horizontal connections typical of American systems (Figure 4) indicates significantly reduced
strength when compared to the wall panel (in general less than 50-60%) and reduced stiffness
(50% or less of the wall's elastic modulus). To the author's knowledge, no experimental data
has been published on the cyclic behavior of horizontal joints under compression and/or shear
in a range that is meaningful to the ultimate seismic behavior of these joints.

The transfer of shear along the horizontal connection may be accomplished by the
following mechanisms: (a) coulomb friction, μN, (b) shear friction, associated with transverse reinforcement across the crack, $\mu A_s f_y$, and (c) by mechanical means such as keying and
metal details. The first two mechanisms may be accompanied by significant shear displacement
(slip) and may exhibit ductile behavior [7,13,14,25,26]. The latter mechanism requires
smaller shear displacements and may exhibit ductile or brittle behavior, depending on the
detail used.

It is typical to assume that the ultimate shear strength of a connection, considering
only friction transfer mechanisms, is the product of the friction coefficient, μ, and the
sum of the normal force, N, and the steel yield force, $A_s f_y$. However, it should be noted
that the latter contribution ($\mu A_s f_y$) to the ultimate strength in shear can only be developed

when there is relative slip in the cracked plane. Thus the shear resistance is initially given by the coulomb friction component (μN), which requires no relative slip to be active.

In general, the lower bound for the friction coefficient is considered to be around 0.7 [3,22,25]. However, under the unique conditions of smooth panel edges and grouted or dry packed interfaces, along with cyclic loading, a lower value may be expected [14]. Values as low as 0.4 [5] and even 0.2 [7] have been suggested.

Regardless of the particular shear transfer mechanism, global shear slip along the entire length of the connection may be possible (Figure 3). It should be noted that such global slip represents an unconfined yield mechanism when only coulomb friction is involved. The actual displacement is only controlled by reversals of acceleration or by the stiffness of secondary elements if present.

Modeling of Behavior

While the connections in large precast panel buildings pose difficult design problems, the concentration of nonlinear and inelastic behavior around connections and penetrations facilitates the modeling of these structures. Based upon the discussion in the previous section, the following modeling assumptions were made in order to study the seismic response of the simple wall:

- Precast panels remain linear elastic--this assumes that the panels are either solid or that they only have minor penetrations and that they are adequately reinforced.
- All nonlinear and inelastic behavior occurs in the connection regions.
- The floor system acts as a series of rigid diaphragms.
- The foundation is rigid.

The response of the simple wall was modeled using finite element techniques [3,6,15,18, 20]. The finite element idealization used is illustrated in Figure 5. Due to the assumptio of linear elastic wall panels, the degrees of freedom of a wall panel, which are not needed to interconnect it to other components, can be eliminated through static condensation. This leads to the repetitious use of 'super elements' or substructures (Figure 5).

Connection Modeling

There are two main sources of nonlinear and inelastic behavior in horizontal connections: opening of the connection and the associated shifting of the neutral axis and slippage along the connection plane (Figure 6). Both the connection panel interfaces along which opening and slip most likely occur, and the connection material between the interfaces are modeled by an interface or contact element [21] of infinitesimal height (Figure 7). Thus the properties both the connection panel interface and the connection material are condensed together to provide the constitutive relationship for the contact element. For simplicity the connection material was assumed to be linear elastic in compression and shear (with reduced modulus). Across the connection panel interface zero tensile strength (vertical reinforcement is modeled separately) and a rigid-plastic coulomb friction shear transfer mechanism were assumed. The combined stress-deformation relationship of the contact element is thus nonlinear-elastic in compression and tension and elasto-plastic in shear (Figure 8).

Physical slippage in the connection-panel interface occurs when the shear stress exceeds the shear strength. The shear strength is given by the coefficient of friction times the compressive stress across the connection (Figure 8), thus coupling the shear behavior to the behavior normal to the connection. This implies that the level at which slip occurs at a specific point will not remain constant, but will vary with the level of overturning moments. This problem was handled by assuming that the slip level remained constant within a given time step.

The deformation displacement relation for the contact or interface element is as follows (Figure 6):

$$\left\{ \begin{matrix} \Delta v \\ \Delta u \end{matrix} \right\} = BU$$

where Δv and Δu are the axial and shear deformations of the connection respectively, and

$$B = \frac{1}{2} \begin{bmatrix} 0 & (1+r) & 0 & (1-r) & 0 & -(1-r) & 0 & -(1+r) \\ (1-r) & 0 & (1-r) & 0 & -(1-r) & 0 & -(1+r) & 0 \end{bmatrix}$$

$$u^T = [u_1 \ v_1 \ u_2 \ v_2 \ u_3 \ v_3 \ u_4 \ v_4]$$

$$r = 2s/L.$$

The restoring nodal forces $F_r(t)$ for a typical element are evaluated as follows

$$F_r(t) = \frac{1}{2} bL \int_{-1}^{+1} B^T \left\{ \begin{matrix} \sigma(t) \\ \tau(t) \end{matrix} \right\} dr \qquad \text{III-47}$$

$$\sigma(t) = \sigma(t - \Delta t) + \Delta\sigma$$

$$\tau(t) = \tau(t - \Delta t) + \Delta\tau$$

where $\sigma(t)$ = axial stress at time t

   $\tau(t)$ = shear stress at time t

   $\Delta\sigma$ = increment in axial stress during time interval Δt

   $\Delta\tau$ = increment in shear stress during time interval Δt

   b = thickness of connection

   L = length of interface element

The nodal restoring forces $F_r(t)$ were evaluated numerically using two integration points per element. Four and eight interface elements corresponding to 8 and 16 integration points, respectively, were used along a connection, and the smaller number was found to provide sufficient accuracy. The vertical reinforcing steel was modeled using elasto-plastic truss elements between panel edges. In the case of ungrouted post-tensioning bars a truss element connects the top of the structure with the base. In this case the bar is only allowed to yield in tension; however, yielding of such a bar seldom, if ever, occurs. Reinforcing and post-tensioning steel was assumed to provide only vertical continuity; its contribution to shear transfer was not considered.

Time Step Integration

The dynamic analysis was carried out using an integration scheme based on the central difference method. This method is computationally very efficient within each time step; in conjunction with diagonal mass and damping matrices, there is no need for solving a system of equations. A global stiffness matrix does not need to be assembled and all operations can be performed at the element level. Unfortunately, the method is only conditionally stable, that is for a time step

$$\Delta t \leq \Delta t_{cr} = \frac{T_n}{\pi}$$

where T_n is the smallest period of the system. Because precast panel buildings are fairly rigid structures, the time step required for numerical stability can be very small. In the parametric studies reported in the next section, a time step in the order of 1/1200 = 0.000833 sec was required. Mass distribution is also a potential problem, since, to avoid infinite frequencies, masses must be allocated to all degrees of freedom.

SEISMIC RESPONSE

Using the analytic model previously described, a series of parametric studies have been carried out to examine the influence of horizontal connections on the seismic response of simple walls. Specifically, the effects of such behavioral aspects as rocking and shear slip on the magnitude of the seismic force levels and deformations are examined. Studies were carried out on both five-and ten-story walls, with different reinforcement patterns for providing vertical continuity and for various earthquake records.

The simple wall was constructed out of panels 7.32 meters wide by 2.44 meters high and 20 cm thick. (Figure 9) A connection region with a height of 30.5 cm was assumed. The wall panels had a composite modulus of 25.2×10^6 kPa ($\nu = 0.15$) and the connection modulus was one half that of the wall panels. Shear slip was assumed to have a secondary stiffness of one hundredth that of the connection's primary shear stiffness. The lumped floor mass was 68.3 tons (51.7 tons at roof level) (see Figure 10) and the gravity load for each floor level was 670.4 kN (507.1 kN at the roof level).

Six different cases of vertical continuity across the horizontal connection were considered. The first is a linear elastic case, which serves as a reference for response comparisons. For this baseline case the ten-story structure had a fundamental period of 0.62 seconds, while the five-story structure had a fundamental period of 0.17 seconds. Two cases are post-tensioned identically by ungrouted bars, but have different friction coefficients ($\mu = 0.2$ and 0.4). In both the 5- and 10-story walls a uniform prestress of 1400 kPa was applied. The three remaining cases have vertical continuity provided by Grade 40 reinforcement, either evenly distributed along each connection level (0.25% and 1.00%), or concentrated at the ends of the panel (0.5%). All three reinforced cases have the same friction coefficient ($\mu = 0.4$).

The majority of the parametric runs were made using an artificial earthquake generated to match the Newmark-Blume-Kapur response spectra for 2% damping. To provide a comparison for these results, several analyses were made using the El Centro (N.S. Component, 1940) and Taft (Kern County N21E, 1952) records. The response spectra of all three earthquakes for 5%

critical damping (which was used in the analyses) are given in Figure 11. Table 1 presents a summary of the various parametric studies upon which the following discussion of seismic response will be based.

Figure 12 presents three different cases that are intended to illustrate the basic characteristics of both the rocking and the shear slip phenomena. All three cases refer to the ten-story structure being subjected to the artificial earthquake with a peak acceleration of 30% g. The horizontal connection was modeled with eight contact elements (16 integration points) for these three cases. Case 1 is the linear elastic reference case. The only difference between the present wall and a cantilever beam is the reduced stiffness of the connection. In this case the maximum base shear was 3954 kN. As was to be expected, the axial strain and shear stress distribution corresponds to normal beam theory. The axial strains and shear stresses are given for the time of maximum overturning moment.

In Case 2 tension across the connection is solely transferred through ungrouted post-tensioning bars. The shear friction coefficient is 0.4. The maximum base shear has dropped to a value of 2476 kN. This decrease in maximum base shear is attributable to an effective elongation in the fundamental period of the wall and an ensuing shift in the response spectra. This elongation of the fundamental period is easily observed by comparing the time histories of Case 1 and Case 2. The elongation of the fundamental period is caused by a rocking motion. This rocking motion is evident from the axial strain distribution presented in Figure 12 for the time of the maximum crack length. In the tension region the axial strain is obtained by dividing the crack opening by the height of the connection. As can be easily seen, plane sections no longer remain plane in the region of the connection. The shear stress distribution shows effects of local shear slip relieving potential stress concentrations at the tip of the crack [11]. Note that in the region immediately adjacent to the crack tip the shear stress is directly proportional to the compressive axial strain.

Case 3 is the same as Case 2 with the exception that the shear friction coefficient is now 0.2. In this case global shear slip is occurring at the time of maximum crack opening. Global shear slip has two significant effects. First shear slip provides both a source of energy dissipation and a force isolation mechanism. This can be observed in the time history of Case 3. The higher frequency components observed in Case 2 are missing and the maximum base shear is reduced to 1940 kN (the product friction coefficient times dead load plus post-tensioning is 1787 kN). Second the shear stress distribution is now basically

III-50

proportional to the compressive stress distribution and becomes a triangular stress block in accordance with the simplified assumption of linear elastic behavior in compression. The fact that the peak compressive and shear stresses coincide and are concentrated near the connection ends, when global slip occurs, enhances the changes of shear compression type failure in the connection region.

It is interesting to note that in all three cases the maximum compressive axial strains are almost identical. The major difference is in the concurrent shear associated with these compressive strains. Also note the magnitude of the shift in the location of the neutral axis between the first case and the latter two cases. Figure 12 has illustrated the potential basic differences between precast and cast-in-place walls. In the following sections the influence of horizontal joint behavior on the seismic response is discussed in greater detail.

The Effect of Rocking Motion on Response

As was mentioned above, when a precast wall undergoes a pure rocking motion (i.e., no global shear slip) it exhibits a softening behavior with little or no energy dissipation. (It is assumed that the energy dissipated by local shear slip is negligible). As long as the connection material properties remain nonlinear-elastic, rocking is a nonlinear-elastic phenomenon which lends only to an apparent elongation of the wall's fundamental period.

Results typical of rocking motion are summarized in Table 1(a) and in Figure 13. The results in Figure 13 have been normalized with respect to the linear elastic response for 25% peak acceleration. As can be observed, there is a decrease in the maximum force levels attained. This decrease in forces can be explained by noting that any elongation in the fundamental period will cause a downward shift in the acceleration spectrum (Figure 11), leading to lower forces. The opposite is true for deformations, for which an increase in period will only cause increased deformations relative to the elastic response. This trend can also be observed in the results for both the El Centro and Taft earthquakes (see Figure 11 and Table 1(d). It must be noted, however, that for different earthquakes or for stiffer buildings the period elongation associated with rocking could lead to the opposite results, that is an increase in force levels.

A major concern in the use of ungrouted post-tensioning systems in aseismic design is the possible yielding of the post-tensioning bars. Figure 14 shows the time history for

III-51

the post-tensioning bar closest to the walls edge in the 30% g rocking case (note: this is the same as Case 2 in the earlier discussion). Even for this extreme case, where the non-linear-elastic material assumption for the horizontal connection is of questionable validity, the post-tensioning bar is not stressed close to its yielding level.

The Effect of Shear Slip on Response

The magnitude of the friction coefficient, together with the compressive resultant due to gravity loads, post-tensioning (if present), and the stressing of mild reinforcement (if present), determine whether global shear slip will occur. The effect of global slip is to limit seismic force levels and to dissipate energy as has been previously noted [5,20]. To illustrate this global slip and shear limiting phenomenon, Figure 15 presents selected response time histories for both the post-tensioned ($\mu = 0.2$), and the mild reinforcement (1.0%) cases. Note the uniform peaks of the shear force time history of level 4 in the mild reinforcement case and how each peak is associated with some level of global slip. This same isolation phenomenon can be observed, to a smaller extent, at the base of the post-tensioned case. As can be seen in Figure 16, the cases with the lower friction coefficient ($\mu = 0.2$) experience global slip along all connection levels and consequently have the lower force levels. The post-tensioned cases with the higher friction coefficient ($\mu = 0.4$) undergo pure rocking motion only with localized slip occuring; thus their force levels are the highest. It is interesting to note, in Figure 16c, that the peak global slips, for the $\mu = 0.2$ post-tensioned cases, are located from the third through the fifth levels. This can be attributed to a combination of higher mode contributions and a decreasing shear strength with height.

The shear strength of the three mildly reinforced cases for both 5- and 10-story walls lies between the shear strength of the two post-tensioned cases in the lower levels, and below in the upper levels. As a result, these three cases experienced global slip in the upper levels and only localized slip in the lower levels (see Figure 16 a & b). The resulting base shears and moments lie between the two post-tensioned cases. The magnitude of global slip indicates that mild reinforcement will be engaged in developing a shear friction mechanism and that this should be accounted for in future modeling. The yielding of the mild reinforcement also contributed to a diminished force level due to some limited energy dissipation.

To illustrate the effect of rocking and global shear slip on the overall response, the time histories for both roof deflection and overturning moment are presented in Figure 17 for the 10 story post-tensioned ($\mu = 0.2$) case subjected to the El Centro earthquake (25% g). In addition to these time histories, the figure shows two load-deformation (base overturning moment-roof deflection) plots for selected segments of the time histories. The load deformation plot for the response between seconds 1 and 2.1 shows the possibility for hysteretic damping through global shear slip. This is illustrated by the full hysteresis loop. It must be noted, however, that this form of inelastic deformation represents an unconfined yield mechanism and could lead to problems not dissimilar to those of soft story systems. In particular, degradation of a horizontal joint directly threatens the overall stability. The second load-deformation plot shows the response between seconds 4 and 5. This plot demonstrates the nonlinear elastic nature of pure rocking motion. During this period no global shear slip occurs, and thus the load-deformation plot exhibits a closed hysteresis loop, in which no energy is dissipated.

Effect of Rocking and Slip on Force Distribution

Rocking motion and global shear slip have a significant effect on the distribution of forces along the horizontal connection and within the precast wall panels. As can be seen in Figure 16 a and c, cracked lengths on the order of 70% of the total connection width were obtained for the 10-story wall subjected to earthquakes with a peak acceleration of 25% g. When coulomb friction is assumed, the transfer of shear occurs only in the unopened portion of the connection in direct relationship with the normal stresses.

The relationship between shear and normal stresses is illustrated in Figure 18. It presents a series of time histories for the 10-story post-tensioned ($\mu = 0.2$) case subjected to the El Centro earthquake (25% g). This particular case was run using eight contact elements (16 integration points) for the horizontal connection. At the bottom of Figure 18 the base shear and cumulative base shear slip time histories corresponding to those in Figure 17 are presented. Above the shear stress time histories for the two extreme integration points (pts. 1 and 16) and one integration point near the center (pt. 8) are shown. Superimposed on the shear time histories are the envelopes of shear strength (i.e., shear slip level $\mu\sigma(t)$). When global slip occurs, the shear strength envelopes and the shear stresses coincide. This condition occurs only three times in the ten seconds of the analysis. Figure 19 presents the

III-53

shear stress – shear deformation hysteresis loops for the same integration points. Because the axial stress varies and is not necessarily in phase with the shear stress, the hysteresis loops in Figure 19 do not show the simple elastoplastic behavior associated with constant axial stress levels.

The concentration of the peak shear and axial stresses near the connection ends leads to a thrusting action that can create severe biaxial stress conditions in the connection as well as in the panel corners. Of particular concern is the upper corner of the panel, in which significant tensile stresses are found, indicating the need for care in the design of panel reinforcement. This thrusting condition is illustrated in Figure 20.

Figure 20 presents the flow of principal stresses in the bottom two panels for the post-tensioned case discussed above (10-story wall, El Centro, 0.25 g, PT, $\mu = 0.2$). The shown deformed configuration occurs one cycle after the wall has undergone the maximum global slip in the opposite direction. The stresses shown by dashed lines are not expected to occur in the actual structure. These stresses can be attributed in part to the distribution of mass in the model. Upon opening, the corner of the wall panel should be essentially stress free; however, because there are lumped masses and consequent inertia forces in these wall panel corner nodes, significant stresses are present.

CONCLUSIONS

This paper has examined the role of the horizontal connection in the seismic response of precast concrete panel buildings. The study of a simple precast wall allowed for the isolation of the effects of nonlinear and inelastic horizontal connection behavior.

The seismic response of a simple precast wall is found to be mainly governed by a nonlinear-elastic rocking phenomenon. The opening of the horizontal connection, associated with rocking, leads to a progressive softening of the structure with increasing excitation level. Thus, rocking leads to an elongation of the apparent fundamental period of the wall, with a consequent increase or decrease in the seismic response, depending upon the nature of the earthquake.

It was found that global slippage between two panels would occur only in limited situations. These situations included extremely low coefficients of friction, the upper levels of non-post-tensioned walls and shorter walls. It was concluded that global slippage could not normally be counted upon for seismic force isolation. It should be noted that in all of the reported work the possible effects of vertical acceleration have not been considered. Such effects may have a significant influence when considering behavior dependent upon gravity loads.

Rocking coupled with local and global shear slip leads to significant force concentrations in both the corners of panels and the ends of connections. It is felt that such force concentrations, unless specifically designed for, will lead to a progressive deterioration of the panel corners and connection ends in a significant earthquake. The concern for these force concentrations, along with a hesitancy to use a unconfined yield mechanism to limit forces, suggests caution in relying on beneficial effects of global shear slip mechanisms in the design of simple walls. It is recommended that, unless specific design provisions are made, the force levels in horizontal connections should be kept significantly below ultimate capacity for design earthquakes.

The work presented in this paper is based on the limited experimental data available about the behavior of panelized buildings. This data calls for simple and therefore also questionable assumptions with respect to material behavior (non-degrading shear-slip relationship, etc.). Accordingly, the computer studies presented are limited in their scope. However, it is felt that the basic behavioral aspects presented are correct and point to the major problems in the aseismic design of panelized buildings. While the presence of weak horizontal joints is a drawback relative to cast-in-place shear walls, this does not need to be true for vertical connections. Studies indicate [16,23] that the presence of vertical connections in panelized buildings may even improve their seismic performance. It is obvious that, for such construction to achieve wide acceptance in seismic regions, extensive experimentation is required along with continuing theoretical studies.

ACKNOWLEDGMENTS

This research on the Seismic Resistance of Precast Concrete Panel Buildings is being conducted at the Massachusetts Institute of Technology and has been sponsored by the National Science Foundation under Grant ENV 75-03778 and Grant PFR-7818742. In addition, the authors would like to acknowledge the assistance of Joseph Burns, Maria Kittredge, and Sally Brunner in the preparation of this paper.

REFERENCES

[1] Abrams, D. P., and Sozen, M. A., "Experimental Study of Frame-Wall Interaction in Reinforced Concrete Structures Subjected to Strong Earthquake Motions," Civil Engineering Studies, Structural Research Series No. 460, University of Illinois, Urbana.

[2] Anicic, D., "Experimental Determination of the Stiffness of the Floor Diaphragam Loaded by Horizontal Forces," unpublished, personal communication, October, 1977.

[3] Becker, J. M., and Llorente, C., "Seismic Design of Precast Concrete Panel Buildings," Proceedings of a Workshop on Earthquake-Resistant Reinforced Concrete Building Construction, Vol. 3, University of California, Berkeley, July, 1977.

[4] Bernander, K. G., and Liefendahl, T., "Shear Force Transmission Through Joints in Precast Hollow Floors Subjected to Bending in the Plane of the Floor," Nordisk Betong, No. 6, 1976, pp. 23-26 (in Swedish).

[5] Brankov, G., and Sachanski, "Response of Large Panel Buildings for Earthquake Excitation in Nonelastic Range," Proceedings of the Sixth World Conference on Earthquake Engineering, New Delhi, India, 1977.

[6] Frank, R. A., "Dynamic Modeling of Large Precast Panel Buildings Using Finite Elements with Substructuring," MIT Publication No. R76-36, Department of Civil Engineering, Massachusetts Institute of Technology, Cambridge, Massachusetts, August, 1976.

[7] Hanson, N. W., "Seismic Test of Horizontal Joints," Design and Construction of Large Panel Concrete Structures, Supplemental Report C, prepared for U.S. Department of Housing and Urban Development, Portland Cement Association, January, 1979.

[8] Iyengar, S., and Harris, H. G., "Behavior of Horizontal Joints in Large Panel Precast Concrete Buildings," Report No. M78-1, Structural Models Laboratory, Department of Civil Engineering, Drexel University, Philadelphia, Pennsylvania, February, 1978.

[9] Johal, L. S., and Hanson, N. W., "Horizontal Joint Tests," Design and Construction of Large Panel Concrete Structures, Supplemental Report B, prepared for U.S. Department of Housing and Urban Development, Portland Cement Association, November, 1978

[10] Lewicki, B., Building with Large Prefabricates, 1st ed., Elsevier Publications Co., London, 1968.

[11] Llorente, C., Becker, J. M., and Roesset, J. M., "The Effect of Non-Linear-Inelastic Connection Behavior of Precast Panelized Shear Walls," paper presented at the ACI Convention, Toronto, Canada, April, 1978, submitted for inclusion in Session Volume on Mathematical Modeling of Reinforced Concrete Structures (ACI 442).

[12] Lugez, J., and Zaravcki, A., "Influence of Horizontal Joints on the Resistance of Prefabricated Panel Elements of Bearing Wall," Cahiers du Centre Scientifique et Technique du Batiment, No. 103, Paris, October, 1969.

[13] Mattock, A. H., "Shear Transfer Under Monotonic Loading, Across on Interface Between Concretes Cast at Different Times," Final Report, Part 1, Department of Civil Engineering, University of Washington, Seattle, Washington, September, 1976.

[14] Mattock, A. H., "Shear Transfer Under Cyclically Reversing Loading, Across an Interface Between Concretes Cast at Different Times," Final Report, Part 2, Department of Civil Engineering, University of Washington, Seattle, Washington, June, 1977.

[15] Mehlhorn, G., and Schwing, H., "Tragverhalten von aus Fertigteilen Zusammengesetzten Scheiben," Forschungsberichte aus dem Institute für Massivbau der Technischen Hochschule, No. 33, Darmstadt, 1976.

[16] Mueller, P., and Becker, J. M., "Seismic Characteristics of Composite Precast Walls,"
      Proceedings of Third Canadian Conference on Earthquake Engineering, Montreal, Canada,
      June, 1979.

[17] Petersson, H., "Analysis of Building Structures," Chalmers University of Technology,
      Department of Building Construction, Gothenburg, 1973.

[18] Petersson, H., and Bäcklund, J., "SERFEM: A computer program for in-plane analysis of
      plates by the finite element method," Chalmers University of Technology, Department of
      Building Construction, Gothenburg, 1974.

[19] Polyakov, S., Design of Earthquake-Resistant Structures, 1st ed., MIR Publishers,
      Moscow, 1974.

[20] Powell, G., and Schricker, V., "Ductility Demands of Joints in Large Panel Structures,"
      ASCE Fall Convention, San Francisco, October 1977, Preprint 3022.

[21] Schafer, H., "A Contribution to the Solution of Contract Problems with Aid of Bond
      Elements," Computer Methods in Applied Mechanics and Engineering, Vol. 5, 1975.

[22] Suenaga, Y., "Study of Rationalization of Horizontal Dry Joints in Box-Frame-Type
      Precast RC Structures," Bulletin of Faculty of Engineering, Yokohama National
      University, Yokohama, Japan, March 1976.

[23] Thiel, V., "The Role of Vertical Connections in the Seismic Response of Precast
      Composite Walls," thesis presented to the Massachusetts Institute of Technology, at
      Cambridge, Massachusetts, in 1979, in partial fulfillment of the requirements for the
      degree of M.S. in Civil Engineering.

[24] Unemori, A. L., Roesset, J. M., and Becker, J. M., "Effect of In-Plane Floor Slab
      Flexibility or the Response of Shear Wall Buildings," paper presented at the ACI
      Convention, Houston, Texas, October/November, 1978, submitted for inclusion in Session
      Volume on Mathematical Modeling of Reinforced Concrete Structures, (ACI 442).

[25] Verbic, B., "Behavior of Joint Elements of a Building Constructed of Large Panels by
      the System 'Vranica' Under Quasistatic Cyclic Loading," Institute za Materijale i
      Konstruckcije, Sarajevo, April, 1977.

[26] White, R. N., and Gergely, P., "Shear Transfer in Thick Walled Reinforced Concrete
      Structures Under Seismic Loading," Final Report, NSF Grant No. ATA73-03178, Department
      of Structural Engineering, Cornell University, Ithaca, New York, May, 1978.

TABLE 1 SUMMARY OF PARAMETRIC STUDIES

Case	Shear (kN)	Moment (kN-m)	Deflection (cm)	Cracked %	Length Level	Global (mm)	Slip Level
10%g peak accel.	1118	16275	2.97	29.0	0	-	-
15%g peak accel.	1462	21494	4.34	50.0	0	-	-
20%g peak accel.	1707	25614	6.83	64.0	0	-	-
25%g peak accel.	2323	32736	6.96	70.0	0	-	-
30%g peak accel.	2470	36200	8.60	78.0	0	-	-

(a) Rocking Study, 10 Story Wall, PT, μ = 0.4, Artificial Earthquake

Case	Shear (kN)	Moment (kN-m)	Deflection (cm)	Cracked %	Length Level	Global (mm)	Slip Level
Linear Elastic	3295	47000	5.50	-	-	-	-
PT, μ = 0.2	1894	29950	6.06	65.0	0	2.9	3
PT, μ = 0.4	2323	32736	6.96	70.0	0	-	-
R/C, .25%, μ = 0.4	1915	28146	6.08	71.1	0	4.7	9
R/C, 1.0%, μ = 0.4	2052	31991	5.86	62.8	0	8.1	9
R/C, .5%, μ = 0.4	1994	30511	5.85	70.8	0	5.2	9

(b) Reinforcement Study, 10 Story Wall, Artificial Earthquake, 25%g

Case	Shear (kN)	Moment (kN-m)	Deflection (cm)	Cracked %	Length Level	Global (mm)	Slip Level
Linear Elastic	1533	12598	0.54	-	-	-	-
PT, μ = 0.2	1126	9422	0.57	13.9	0	0.9	1
PT, μ = 0.4	1467	11751	0.53	29.0	0	-	-
R/C, .25%, μ = 0.4	1211	10490	0.88	38.1	1	3.4	4
R/C, 1.0%, μ = 0.4	1185	9590	0.81	24.3	0	3.4	4
R/C, 0.5%, μ = 0.4	1242	9778	0.83	31.6	0	3.4	4

(c) Reinforcement Study, 5 Story Wall, Artificial Earthquake, 25%g

Case	Shear (kN)	Moment (kN-m)	Deflection (cm)	Cracked %	Length Level	Global (mm)	Slip Level
Artificial μ = .2	1894	29950	6.06	65.0	0	2.9	3
Artificial μ = .4	2323	32736	6.96	70.0	0	-	-
El Centro μ = .2	1860	29035	6.90	70.2	0	6.7	5
El Centro μ = .4	2640	36000	8.44	72.0	0	-	-
Taft μ = .2	1790	29700	6.28	68.0	0	2.6	4
Taft μ = 0.4	2012	31640	7.45	75.2	0	-	-

(d) Earthquake Study, 10 Story Wall, Post-Tensioned, 25%g

III-58

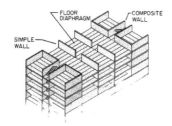

FIGURE 1   BASIC STRUCTURAL ELEMENTS
OF LARGE PANEL PRECAST
CONCRETE BUILDINGS

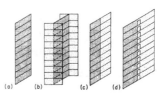

FIGURE 2   TYPICAL PRECAST WALL
CONFIGURATIONS (Shading
denotes a simple wall)

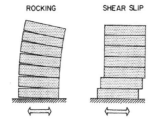

FIGURE 3   ROCKING AND SHEAR SLIP
MECHANISMS

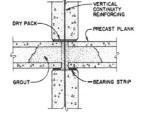

FIGURE 4   TYPICAL PLATFORM
CONNECTION

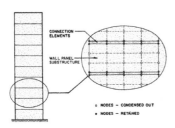

FIGURE 5   FINITE ELEMENT MODEL

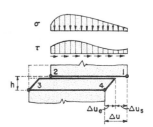

FIGURE 6   CONNECTION ELEMENT

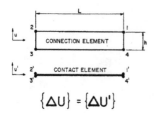

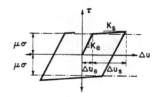

$$\{\Delta U\} = \{\Delta U'\}$$

FIGURE 7   CONTACT ELEMENT

FIGURE 8   SHEAR STRESS-SHEAR DEFORMA-
TION AND SLIP RELATIONSHIP

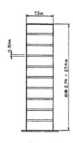

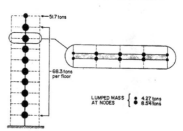

FIGURE 9   10-STORY SIMPLE WALL
(wall thickness = 0.20m)

FIGURE 10   MASS DISTRIBUTION IN SIMPLE WALL

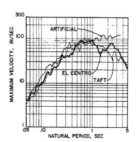

FIGURE 11   RESPONSE SPECTRA, 5% DAMPING,
1.0g PEAK ACCELERATION

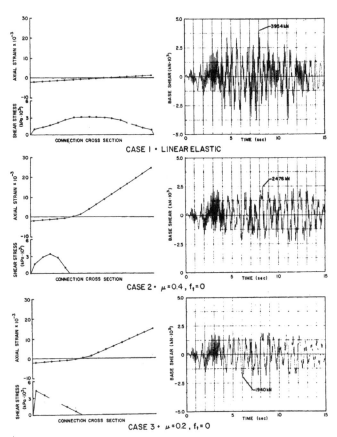

FIGURE 12 COMPARISON OF AXIAL STRAIN AND SHEAR STRESS AT TIME OF MAXIMUM
BASE SHEAR, AND BASE SHEAR TIME HISTORIES, 10 STORY, P.T.,
ARTIFICIAL 0.30g.

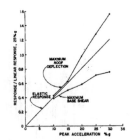

FIGURE 13 EFFECT OF ROCKING IN A
SIMPLE WALL, 10 STORY
P.T., μ = 0.4, ARTIFI-
GIAL EQ.

REINFORCEMENT−1.0%, μ=0.4

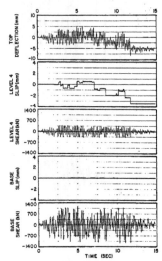

FIGURE 15 SELECTED TIME HISTORIES, 5-S

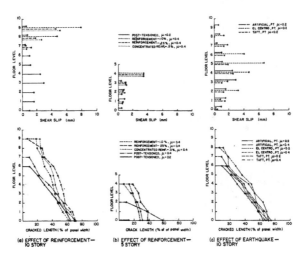

(a) EFFECT OF REINFORCEMENT— 10 STORY

(b) EFFECT OF REINFORCEMENT— 5 STORY

(c) EFFECT OF EARTHQUAKE— 10 STORY

FIGURE 16   EFFECT OF REINFORCEMENT AND EARTHQUAKE ON SHEAR SLIP AND CRACK LENGTH

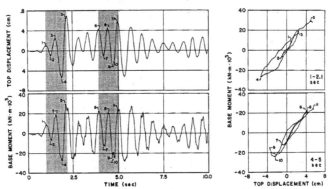

FIGURE 17   BASE MOMENT-TOP DISPLACEMENT RELATIONSHIP, 10 STORY, P.T. $\mu = 0.2$, ARTIFICIAL 0.25g

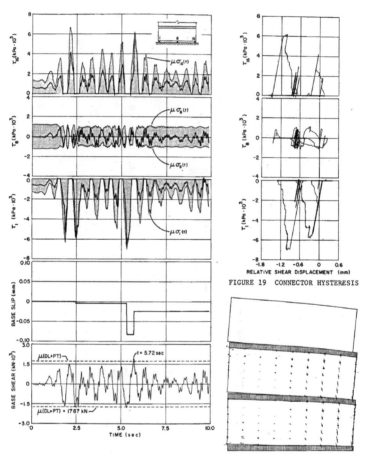

FIGURE 18   SELECTED TIME HISTORIES AT BASE
OF 10 STORY WALL, P.T. μ = 0.2,
ARTIFICIAL 0.25g

FIGURE 19   CONNECTOR HYSTERESIS

FIGURE 20   STRESS DISTRIBUTION
AT BOTTOM, AT t = 5.72 sec
(Arrows indicate tension)

III-64

SEISMIC RESPONSE ANALYSIS OF THE ITAJIMA BRIDGE

THROUGH USE OF STRONG MOTION ACCELERATION RECORDS

Masamitsu Ohashi

Toshio Iwasaki

Kazuhiko Kawashima

Public Works Research Institute

Ministry of Construction

ABSTRACT

In analyzing the seismic behavior of highway bridges constructed on soft soil deposits, it is important to take into account soil-structure interaction effects. In this paper, the seismic response of a bridge pier-foundation is analyzed from earthquake acceleration records taken simultaneously from the pier crest and the ground surface near the bridge. Four motions were used in the analysis, i.e., two were induced by two earthquakes with magnitudes of 7.5 and 6.6, and two by their aftershocks. In the former two earthquakes, the maximum accelerations were 186 and 441 gals on the ground surface, and 306 and 213 gals on the pier top. Analyses of frequency characteristics of the motions showed that the predominant frequencies of the pier-foundation were always almost identical to the fundamental natural frequency of the subsoil. Analytical models were formulated to calculate the seismic response of the pier-foundation assuming the subsoil and pier-foundation to be a shear column model with an equivalent linear shear modulus and an elastically supported beam on the subsoil, respectively. Bedrock motions were computed from the measured ground surface motions and then applied to the bedrock of the analytical model. The seismic responses of the pier-foundation were thus calculated and compared with the measured records and produced a good agreement.

KEYWORDS: Bridge-pier foundations; bridge seismology; earthquake frequency characteristics; ground surface accelerations; foundation structure response.

INTRODUCTION

In the past, numerous highway bridges have suffered extensive damage due to strong motion earthquakes. Seismic damage to bridges consisting of simply supported girders or trusses rested on massive piers and abutments were commonly caused by foundation failures resulting from excessive ground deformations and/or loss of stability and bearing capacity of the foundation soils. As a direct result, the substructures often tilted, settled, or sometimes overturned, and these large displacements of the supports caused relative shifting of the superstructures, induced failures of the bearing supports, and even caused dislodging of the spans from their supports.

It has been well recognized from these evidences that the influences of surrounding sub-surface ground are very important for the seismic responses of foundations deeply embedded in the ground, and considerable interests were concentrated on the soil-structure interaction effects of such structures through model experiments and theoretical analyses. However very limited researches have been undertaken in studying seismic responses of actual foundations during high intensity seismic excitations since little data have been available for such purposes. This investigation presents the results of observation of seismic motions on a bridge pier and on the ground surface nearby with high accelerations which were measured during four earthquakes, and also correlates the pier motions measured with the pier motions analyzed using the ground surface acceleration records.

STRUCTURAL AND SITE CONDITIONS OF THE BRIDGE

The Itajima Bridge studied is a five-span simply-supported plate girder bridge as shown in Figure 1. The strong-motion acceleration observations have been conducted since 1966 at a crest of one of the piers and on the free field ground surface located about 400 m apart from the pier. Two SMAC-B2 type accelerographs are set at both sites to measure accelerations in the longitudinal, transverse and vertical directions of the bridge axis.

Ground surveys were performed at both sites and information on soil profile, N-value of standard penetration test, and shear wave velocities were obtained. Figure 2 shows the soil profiles and N-values at both sites. It is recognized from the results that the ground conditions are essentially the same between the two sites, i.e., the soil profiles consist of upper soft alluvium loam to fine sand formations with the averaged N-value of approximately

III-66

7 and lower stiff diluvium gravel formations with the averaged N-values of 30 or more. The shear wave velocities of the upper and lower formatiohs were estimated to be approximately 130 and 480 m/sec, respectively. Figure 3 shows the geological structures around the Itajima Bridge. The stiff diluvial gravel formations are recognized to be underlaid by mesozoic cretaceous formation. According to Figures 2 and 3 it can be recognized that the ground conditions are continuous and almost identical between the two sites. The gravel formations were assumed to be the bedrock at the sites in calculating seismic responses of the ground and foundation in the following paragraph.

ANALYSES OF STRONG-MOTION ACCELERATION RECORDS

Strong-Motion Acceleration Records

Four simultaneous strong-motion acceleration records have been obtained at the Itajima Bridge as summarized in Table 1 which were induced by four earthquakes, i.e., main and after shocks of both the Hyuganada Earthquake of April 1, 1968 and the Bungosuido Earthquake of August 6, 1968, which are designated herein as A, B, C and D Earthquakes, respectively. Figures 4, 5, 6 and 7 show the acceleration records thus obtained in the longitudinal and transverse directions of the bridge axis. In the D-Earthquake, accelerations on the pier could not be recorded, unfortunately, and only the maximum value of acceleration was obtained. It should be noted here that although the seismic response accelerations developed at the pier crest were very high, superstructures and foundations of the Itajima Bridge suffered no structural damages through any of these four earthquakes.

Figure 8 represents amplifications of maximum accelerations between the ground surface and the crest of the pier. It can be recognized from this result that the amplifications of maximum accelerations are very much different between the Hyuganada Earthquake (A- and B-Earthquakes) and the Bungosuido Earthquake (C- and D-Earthquakes), i.e., the amplification factors are in the range of 1.1 - 1.7 in the case of A- and B-Earthquakes, whereas they are in the range of 0.4 - 0.6 in the C- and D-Earthquakes. The frequency characteristics of the motions were than investigated. Predominant frequencies of the records are summarized in Table 2, based on the power spectra presented in Figure 9.

Characteristics of Ground Motions

Based on Figure 9 and Table 2, it can be understood that the ground motions induced during the A-Earthquake have the predominant frequencies of 1.5 Hz and 4.5 Hz in the longitudinal motion, and 1.3 Hz and 4.0 Hz in the transverse. motion. The frequencies of 1.5 Hz and 1.3 Hz are the most predominant ones whereas the frequencies of 4.5 Hz and 4.0 Hz are the secondary predominant ones. The ratio between the most predominant frequency f_1 and secondly predominant frequency f_2 are approximately 3.0 in both longitudinal and transverse motions. The geological situation at the site consists of two formations with significantly different stiffnesses, i.e., alluvium loam ∿ silt formation and diluvium gravel formation. In such case that subsoils with shear wave velocity V_s and thickness H are rested horizontally on the stiff formation, the natural period T_n of the subsoils can be estimated by Eq. 1 for the n-th mode of shear vibration.

$$T_n = \frac{1}{2n-1} \cdot \frac{4H}{V_s} \tag{1}$$

The ratio between T_1 and T_2 (T_1/T_2) are always evaluated as 3.0 in such case, which well explains the characteristics of ground motions developed during the A-Earthquake. It can be understood therefore that the predominant frequencies observed in the ground motions would represent the fundamental and second natural frequencies of the subsoils. Since there are not any predominant frequencies beyond 1.3 ∿ 1.5 Hz and 4.0 ∿ 4.5 Hz in the ground motions, it is considered that the incident motions in the bedrock themselves have these frequencies as their predominant frequencies.

According to the ground condition presented in Figure 2, the fundamental natural period T_1 of the subsurface ground at the free field observatory can be estimated by Eq. 2 taking the diluvium gravel formation as bedrock.

$$T_1 = \sum_i \frac{4Hi}{V_{si}} = 4 \left(\frac{4}{107} + \frac{4}{125} + \frac{5}{145} + \frac{3.5}{125} \right) = 0.53 \text{ sec} \tag{2}$$

Consequently the fundamental natural frequency f_1 of subsurface ground is estimated as 1.9 Hz which is considered to be the natural frequency during low amplitude vibration. Assuming that the difference of fundamental natural frequency f_1 between 1.9 Hz derived by Eq. 2 and the values shown in Table 2 depends on the strain dependence of soil modulus, the ratio of shear wave velocity which would be developed during the A-Earthquake V_{si} and that during the low amplitude vibration V_{so} would approximately be estimated as follows.

$$\frac{V_{si}}{V_{so}} = \frac{1.3 \sim 1.5}{1.9} = 0.68 \sim 0.79 \tag{3}$$

Consequently the ratio of soil stiffness during the A-Earthquake G_1 and that corresponding to low amplitude strain level G_o can be estimated as follows.

$$\frac{G_1}{G_2} \propto (\frac{V_{si}}{V_{so}})^2 = 0.46 \sim 0.62 \tag{4}$$

According to laboratory experiments concerning the strain dependence of soil stiffnesses, such a reduction of soil stiffness is reasonable.

The explanation of ground motion characteristics for the A-Earthquake can also be applicable for the ground motions of the B- and C-Earthquakes, i.e., the ground motions of the B-Earthquake have predominant frequencies of 1.9 Hz and 5.1 Hz in the longitudinal motion, and 1.5 Hz and 4.9 Hz in the transverse motion, which shows that the ratios between two predominant frequencies are approximately 3.0 in both directions. The ground motions of the C-Earthquake have predominant frequencies of 1.4 Hz, 3.7 Hz, 4.2 Hz and 4.8 ∼ 5.2 Hz in both the longitudinal and transverse motions. Although there are many apparent predominant frequencies, it would be considered that the frequencies of 1.4 Hz and 4.3 Hz represents the fundamental and second mode frequencies, respectively, which again shows that the ratio between two predominant frequencies are approximately 3. The frequencies of 3.7 Hz and 4.8 ∼ 5.2 Hz are considered to be the predominant frequencies of the incident motions in the bedrock.

Characteristics of Pier Motions

It is understood from the power spectra of Figure 9 that in the A-Earthquake the most predominant frequencies of the pier motions are 1.5 Hz in the longitudinal direction and 1.3 Hz in the transverse direction, which are approximately equal to those of ground motions. Also all of the four acceleration records during the A-Earthquake are narrow banded waves with a single frequency predominant. It is also understood from Table 2 that in the B-Earthquake the predominant frequencies of the pier motion are approximately 1.6 Hz in the

III-69

transverse direction, which is very close to the predominant frequency of the ground motion of 1.5 Hz. In the longitudinal direction, however, the pier motion has several predominant frequencies. The most predominant one is approximately 1.8 Hz which is again very close to the predominant frequency of the ground motion of 1.9 Hz. As is the case of A-Earthquake, the acceleration records on the ground in both directions and the acceleration record on the pier top in the transverse direction are narrow banded waves with a single frequency predominant. On the other hand, the records during the C-Earthquake are somehwat different from those for the A- and B-Earthquakes described as above. The most predominant frequencies of the pier motions are approximately 1.4 Hz in both the longitudinal and transverse directions which are different from the most predominant frequencies of the ground motions, i.e., approximately 3.7 Hz in both directions. It should be noted here, however, that the ground motions have a predominant frequency close to the 1.4 Hz although they are not the most predominant.

It can be recognized from these results that the acceleration records of pier always contain the motions in the range of 1.3 - 1.8 Hz as the most predominant ones in both the longitudinal and transverse directions, which is considered to correspond to the lowest natural frequency of the subsurface ground. The frequency response functions were then computed between the pier motions and the ground motions as shown in Figure 10. It can be recognized from Figure 10 that amplifications of the frequency response functions are rather small for the range of frequency of 4 Hz or more because of low pass filter effects of the foundation. The amplifications vary almost evenly between 0.5 Hz and 4 Hz although the apparent natural frequency of the foundation is not evaluated.

It would be deduced from these considerations that the most predominant frequency that is always contained in the motions of the pier is significantly influenced by the lowest natural frequency of the subsurface ground so that the pier vibrates in accordance with the motion of the subsurface ground nearby.

Nonsteady Vibrations of Pier and Ground

In order to investigate the time dependence of predominant frequencies of the pier and ground motions, running power spectra were computed. One of the results is displayed in Figure 11 for the A-Earthquake. Based on the running power spectra changes of predominant frequencies of the motions in time domain were estimated as shown in Figure 12. For the

A- and C-Earthquakes, it can be observed that the frequencies are rather low at the main parts of the motion where a large amplitude of vibration occurred. For instance, in the case of the transverse motion of the A-Earthquake, the predominant frequency is approximately 2 Hz around 4 seconds in time. The predominant frequency then decreased to 1.3 Hz around 12 – 18 seconds, and again recovered to 1.7 Hz around 30 seconds. It should be noted here that the predominant frequency developed at the beginning and ending of the motions are very close to that estimated by Eq. 2 by employing the shear wave velocity of subsurface ground at low strain amplitude.

It should also be noted here that the time dependence of the predominant frequencies described as above are almost identically developed in both the pier and the ground motions. Such characteristics of the time dependence of motions suggest that the nonsteady responses of the foundation were not derived from degradation of the soil stiffness just around the foundation but were caused by the nonsteady responses of subsurface ground nearby.

SEISMIC ANALYSIS PROCEDURE OF FOUNDATION

A discrete analytical model as shown in Figure 13 was formulated to calculate earthquake responses of the pier foundation. The equation of motions of the system can be written as

$$(\underline{M}_p + \underline{M}_e)\,\ddot{\underline{u}}_p + \underline{C}_p\dot{\underline{u}}_p + \underline{K}_p\dot{\underline{u}}_p + \underline{C}_e(\dot{\underline{u}}_p - \dot{\underline{u}}_g) + \underline{K}_e(\underline{u}_p - \underline{u}_g) = \underline{0} \qquad (5)$$

where,

$\underline{M}_p$ = mass matrix of the foundation

$\underline{M}_e$ = mass matrix of surrounding soils

$\underline{C}_p$ = damping matrix of the foundation

$\underline{K}_p$ = stiffness matrix of the foundation

$\underline{C}_e$ = damping matrix expressing radiational dampings

$\underline{K}_e$ = stiffness matrix expressing springs between foundation and surrounding soils

$\underline{u}_p,\ \dot{\underline{u}}_p,\ \ddot{\underline{u}}_p$ = absolute displacement, velocity and acceleration vectors of foundation

$\underline{u}_g,\ \dot{\underline{u}}_g$ = absolute displacement and velocity vectors of subsurface ground

in which the subsurface ground motions of $\underline{u}_g$ and $\underline{u}_g$ are assumed to be specified. Denoting as

$$\underline{M}\,\underline{\ddot{u}}_p + \underline{C}\,\underline{\dot{u}}_p + \underline{K}\,\underline{u}_p \;=\; \underline{C}_e\,\underline{\dot{u}}_g + \underline{K}_e\,\underline{u}_g \tag{7}$$

The vector $\underline{u}_p$ can be conveniently decomposed into a quasi-static displacement vector $\underline{u}_{ps}$ and a dynamic displacement vector $\underline{u}_{pd}$, i.e.,

$$\underline{u}_p \;=\; \underline{u}_{pd} + \underline{u}_{ps} \tag{8}$$

By definition of quasi-static displacement in the form as

$$\underline{K}\,\underline{u}_{ps} - \underline{K}_e\underline{u}_g \;=\; \underline{0} \tag{9}$$

$\underline{u}_{ps}$ can be written as

$$\underline{u}_{ps} \;=\; \underline{K}^{-1}\underline{K}_e\underline{u}_g \;=\; \underline{K}_s\underline{u}_g \tag{10}$$

Substitution of Eqs. 8 and 9 into Eq. 7 gives

$$\underline{M}\,\underline{\ddot{u}}_{pd} + \underline{C}\,\underline{\dot{u}}_{pd} + \underline{K}\,\underline{u}_{pd} \;=\; \underline{M}\,\underline{K}_s\ddot{u}_g + (\underline{C}_e - \underline{C}\,\underline{K}_s)\,\underline{\dot{u}}_g \tag{11}$$

Usually the damping term on the right hand side of Eq. 8 is less significant when compared with the inertia terms so that it can be dropped from the equation without introducing significant errors. Then Eq. 11 can be written as

$$\underline{M}\,\underline{\ddot{u}}_{pd} + \underline{C}\,\underline{\dot{u}}_{pd} + \underline{K}\,\underline{u}_{pd} \;=\; \underline{M}\,\underline{K}_s\underline{\dot{u}}_g \tag{12}$$

Eq. 12 can be solved by modal-superposition procedures provided that the damping matrix on the left hand side of the equation is assumed to be triangularized in the same manner as the mass and stiffness matrices in the form of the critical damping ratio.

The pier motions were computed by using the analytical procedure described in the pre-ceeding paragraph based on the measured ground motions for the A- , B- , and C-Earthquakes, and they were compared with the measured motions.

The bedrock motions were computed from the measured ground surface motions by the de-convolution procedure taking account of the strain dependence of the shear moduli and hysteretic damping ratio of the subsoils as shown in Figure 14. The subsoils and foundation were idealized by a one-dimensional shear column model with equivalent linear soil properties and one-dimensional elastic beam supported elastically by the surrounding subsoils, respec-tively. The weight of a girder supported by the pier was idealized as an additional mass lumped at the crest of pier. The lowest natural frequencies of the pier and surrounding subsoils thus estimated are shown in Table 3.

The response accelerations of pier were then calculated based on Eq. 12 by applying the bedrock motions at the bottom of the shear column model of subsoils. The comparative plots of both the theoretical and measured accelerations at the crest of pier are shown in Figures 15, 16 and 17 for the A- , B- , and C-Earthquakes, respectively. The damping ratios assumed in the analyses are shown in Table 3. It is recognized from the results that fairly good agreements are obtained for the motions in the A- and B-Earthquakes. However, the correla-tion for the motion in the C-Earthquake is appreciably less satisfactory and further precise investigations are needed to clarify the frequency characteristics of the foundation.

CONCLUSIONS

Based on the results presented, the following conclusions may be deduced:

1) Seismic responses of the deeply embedded foundation are significantly influenced by the effects of surrounding subsurface soils. The most predominant frequencies which are always contained in the motions of the pier crest are prescribed by the lowest natural fre-quency of the subsurface ground so that the pier foundation responses in accordance with the motion of the subsurface ground nearby.

2) The nonsteady responses developed on the pier foundation are almost idenitical with those observed on the ground surface, which suggest that the nonsteady responses of the pier

foundation are not derived from degradation of the soil stiffness just around the foundation but are primarily caused by nonsteady responses of the subsurface ground nearby.

3) Seismic response accelerations of the foundation can be calculated with fairly good accuracy by the analytical procedure presented herein from the free-field ground accelerations measured near the foundation for earthquakes which induce ground accelerations at the bridge site with the most predominant frequencies lower than the fundamental natural frequency of the foundation.

REFERENCES

[1] Iwasaki, T., "Earthquake-Resistant Design of Bridges in Japan," Bulletin of Public Works Research Institute, Volume 29, Public Works Research Institute, Ministry of Construction, May 1973.

[2] Kuribayashi, E. and Iwasaki, T., "Dynamic Properties of Highway Bridges," Fifth World Conference on Earthquake Engineering, Roma, December 1972.

[3] Iwasaki, T., Wakabayashi, S., Kawashima, K. and Takagi, Y., "Strong-Motion Acceleration Records from Public Works in Japan (No. 3)," Technical Note of the Public Works Research Institute, No. 34, Public Works Research Institute, Ministry of Construction, 1978.

[4] Yamahara, H., "Ground Motions During Earthquakes and the Input Loss of Earthquake Power to an Excitation of Buildings," Soils and Foundations, Vol. 10, No. 2, 1970.

[5] Duke, C. M., et al., "Strong Motions and Site Conditions -- Hollywood," Bulletin of the Seismological Society of America, Vol. 60, No. 4, 1970.

[6] Matsushima, Y., "Structure-Soil Interaction for Finite Element Hysteretically Damped Systems, Journal of Architectural Institute of Japan, No. 199, 1972.

[7] Iwasaki, T. and Kawashima, K., "Seismic Analysis of Highway Bridge Utilizing Strong-Motion Acceleration Records," Proceedings of Second International Conference on Microzonation, San Francisco, California, USA, 1978.

Table 1   Strong Motion Acceleration Records at The Itajima Bridge

Earthquake No.	Earthquake	Date	Richter Magnitude	Epicentral Distance[km]	Maximum Acceleration (Gal)			
					Pier Motion		Ground Surface Motion	
					Longitudinal	Transverse	Longitudinal	Transverse
1	The Hyuganada Earthquake	1968.4.1.09:42	7.5	101	209	306	169	186
2	The Hyuganada Earthquake (Aftershock)	1968.4.1.16:13	6.3	99	38	66	34	42
3	The Bungosuido Earthquake	1968.8.6.01:17	6.6	11	199	231	441	353
4	The Bungosuido Earthquake (Aftershock)	1968.8.6.13:21	5.3	19	100	63	227	172

Table 2   Predominant Frequencies of Strong Motion Acceleration Records
Measured on Pier Crest and Ground Nearby [Hz]

Earthquake No.	Earthquake	Longitudinal		Transverse	
		Ground	Pier Crest	Ground	Pier Crest
1	The Hyuganada Earthquake	1.5	1.5	1.3	1.3
2	The Hyuganada Earthquake (Aftershock)	1.9	1.8,2.0,2.3,3.7	1.5	1.6
3	The Bungosuido Earthquake	3.7	1.4	3.7,4.4	1.4
4	The Bungosuido Earthquake (Aftershock)	4.2	-	4.4	-

Table 3   Lowest Natural Frequencies and Damping Ratio Assumed
in Seismic Analyses

Earthquake	Surrounding Ground		Caisson-Pier Foundation		Hystretic Damping Ratio of Subsoils(%)	Radiational Damping Ratio(%)
	1st	2nd	1st	2nd		
The Hyganada Earthquake (A-Earthquake)	1.6	4.9	2.8	7.8	10	20
The Hyuganada Earthquake (After Shock) (B-Earthquake)	1.9	5.7	3.1	8.3	5	20
The Bungosuido Earthquake (C-Earthquake)	1.6	4.7	2.7	7.7	8 - 12	20

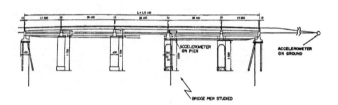

Fig. 1 General View of The Itajima Bridge

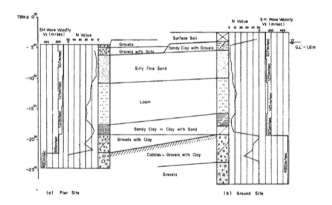

Fig. 2 Soil Profile at the Pier Site and Ground Observatory Site

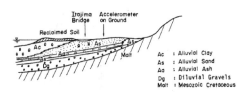

Fig. 3 Geological Condition Arround The Itajima Bridge

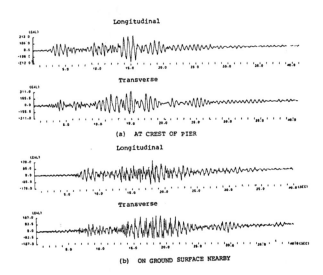

(a) AT CREST OF PIER

(b) ON GROUND SURFACE NEARBY

Fig. 4 Acceleration Records At the Itajima
Bridge During the Hyuganada Earthquake
of April 1, 1968 (A-Earthquake)

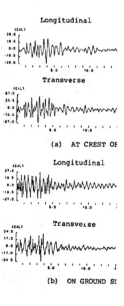

Fig. 5

Acceleration Records at the Itajima
Bridge During An Aftershock of the
Hyuganada Earthquake of April 1, 1968
(B-Earthquake)

(a) AT CREST OF

(b) ON GROUND SI

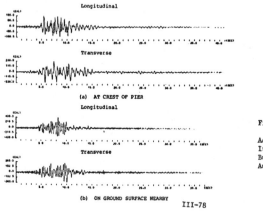

(a) AT CREST OF PIER

(b) ON GROUND SURFACE NEARBY

Fig. 6

Acceleratio
Itajima Bri
Bungosuido
August 6, 1

III-78

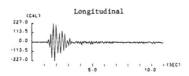

Longitudinal

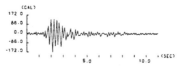

Transverse

(a) ON GROUND SURFACE NEARBY

Fig. 7 Acceleration Records At The Itajima Bridge During
An Aftershock of the Bungosuido Earthquake of
August 6, 1968 (D-Earthquake)

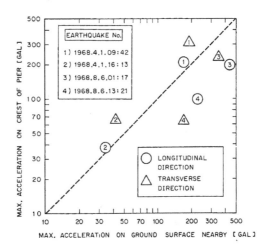

MAX. ACCELERATION ON GROUND SURFACE NEARBY [GAL]

Fig. 8 Amplifications of Maximum Acceleration Between
The Pier Crest and Ground Surface Nearby

III-79

Fig. 9   Power Spectra of Acceleration Records at the Itajim

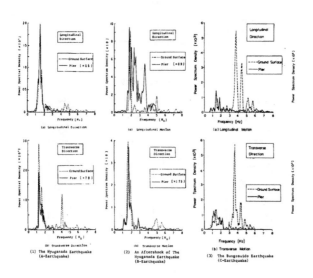

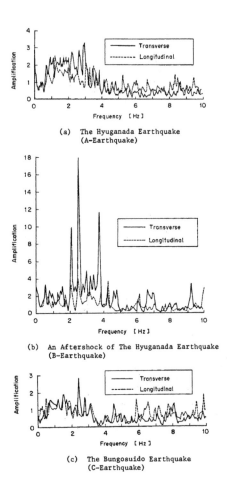

(a) The Hyuganada Earthquake
(A-Earthquake)

(b) An Aftershock of The Hyuganada Earthquake
(B-Earthquake)

(c) The Bungosuido Earthquake
(C-Earthquake)

Fig. 10 Frequency Response Functions Determined by Measured Strong Motions
Between Pier Crest and Ground Surface

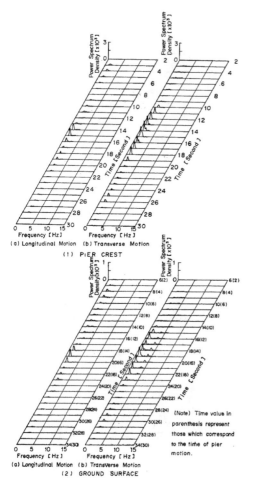

(a) Longitudinal Motion   (b) Transverse Motion
(1) PIER CREST

(Note) Time value in parenthesis represent those which correspond to the time of pier motion.

(a) Longitudinal Motion   (b) Transverse Motion
(2) GROUND SURFACE

Fig. 11 Running Power Spectra of Strong Motions During
The Hyuganada Earthquake (A-Earthquake)

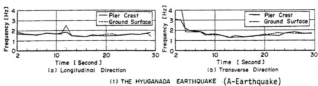

(1) THE HYUGANADA EARTHQUAKE (A-Earthquake)

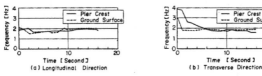

(2) AN AFTERSHOCK OF THE HYUGANADA EARTHQUAKE (B-Earthquake)

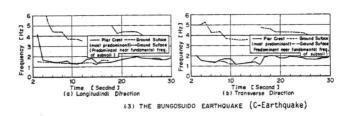

(3) THE BUNGOSUIDO EARTHQUAKE (C-Earthquake)

Fig. 12 Time Dependence of Predominant Frequencies of
Pier and Ground Motions

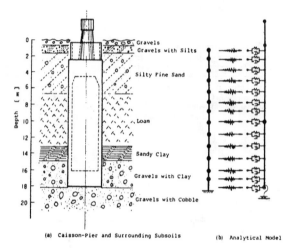

(a) Caisson-Pier and Surrounding Subsoils   (b) Analytical Model

Fig. 13 Analytical Model of Subsoils and Foundation

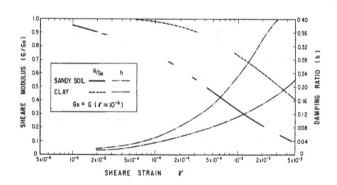

Fig. 14 Sheare Strain Dependence of Shear Modulus and
Hysteretic Damping Ratio Assumed In Seismic
Analyses

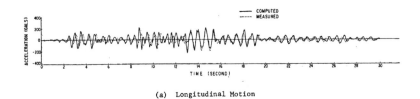

(a) Longitudinal Motion

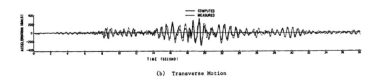

(b) Transverse Motion

Fig. 15 Correlation of Seismic Response Acceleration At the Pier Crest
(The Hyuganada Earthquake; A-Earthquake)

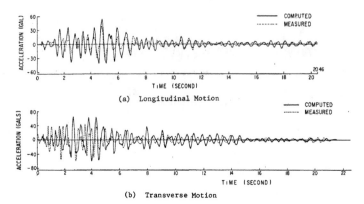

(a) Longitudinal Motion

(b) Transverse Motion

Fig. 16   Correlation of Seismic Response Acceleration At The Pier Crest
(An Aftershock of The Hyuganada Earthquake; B-Earthquake)

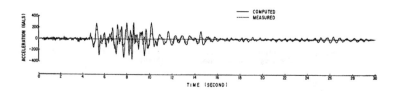

Fig. 17   Correlation of Seismic Response Acceleration At The Pier Crest
(The Bungosuido Earthquake; C-Earthquake)

III-86

FIELD AND LABORATORY DETERMINATION OF SOIL MODULI

William F. Marcuson III

Joseph R. Curro, Jr.

U.S. Army Engineer Waterways Experiment Station

Vicksburg, Mississippi

ABSTRACT

A field geophysical investigation was performed to determine the shear and Young's moduli as a function of depth for a site in southern Ohio. This investigation included crosshole, downhole, and surface refraction investigation techniques. A supplementary set of laboratory resonant column tests was performed with the Drnevich resonant column device. Laboratory undisturbed specimens were excited in both the longitudinal and torsional modes to obtain both Young's and shear moduli as a function of strain. Both laboratory and field data are presented, compared, and discussed.

An idealized soil profile for boring 821-UD is presented. This soil profile includes recommended design values of moduli and damping. At this boring location, bedrock lies approximately 36 ft (11 m) below the ground surface. The 36 ft (11 m) thick soil deposit has been subdivided into four layers whose shear moduli range from 3.4×10^3 to 86×10^3 psi (23.4×10^3 to 592×10^3 kPa). The moduli increase as a function of depth. The Young's moduli for this soil profile range from 10×10^3 to 225×10^3 psi (59×10^3 to 1551×10^3 kPa) and also increase with depth. The internal damping of the soil deposit was found to range from 3 to 5 percent for low dynamic strain amplitudes and is constant with depth. The bedrock has a shear modulus of 125×10^3 (861×10^3 kPa) and a Young's modulus of 370×10^3 psi (2551×10^3 kPa).

KEYWORDS: Field testing; laboratory testing; dynamic properties; shear modulus; damping; wave velocities; geophysical; resonant column test.

INTRODUCTION

In order to evaluate the dynamic response of soil or a structure founded on soil, knowledge of stress-strain properties of the foundation material is required. When using current computer analysis techniques, the initial elastic moduli (very low strains, generally 10^{-4} cm/cm or less) are required along with the variation in moduli as a function of strain.

These initial moduli (shear and Young's) can be determined in situ using various geophysical techniques. Or, resonant column tests can be conducted in the laboratory on undisturbed samples to obtain moduli as functions of strain level.

Since the moduli discussed above would be needed for dynamic foundation design of four proposed buildings and a test loop structure [13], a test program was formulated that consisted of field and laboratory investigations to provide the required moduli. In addition, Poisson's ratios were determined from the investigations to aid in the dynamic design.

Therefore, the primary purpose of this project was to determine values of shear and Young's moduli for the foundation materials underlying the five proposed structures (designated as X-340-1 through X-340-4 and X-774), located as shown in Figure 1. Additionally, building X-340-4 was arbitrarily selected as an example case to illustrate the procedure used for determining dynamic design values.

The purpose of this paper is to compare dynamic soil properties determined in situ and in the laboratory and to discuss the rationale used in obtaining design values.

SITE DESCRIPTION

The construction site is located in the vicinity of Portsmouth, Ohio, near the confluence of the Ohio and Scioto Rivers. Overburden soils at the site were primarily Pleistocene river- and lake-bed deposits. Figure 1 is a geologic map of the site [13]. The map outlines the aerial distribution of the glacial lake-bed deposits and early alluvial deposits. The lake-bed deposits, consisting of fat, thinly laminated clays, are largely confined to the old Scioto River channel.

The area outlined as lake deposits in Figure 1 shows the occurrence of fat, laminated clay either at the surface or at some depth beneath the surface. Lean clays, silts (alluvium), and a fill material which was placed during the period 1953 to 1955, lie above the fat clay in various thicknesses over a large portion of the outlined area. Early river alluvial silty clay and silt occupy that portion of the valley fill outside the fat clay boundaries.

Up to 60 ft (18 m) (averaging 35 to 40 ft (10.5 to 12.0 m)) of early river-and lake-bed deposits in the old valley still remain at the site. These deposits are mostly fat, laminated clays, lean clays, and fine silts, with a 2 to 10 ft (0.6 to 3 m) thick layer of weathered clayey gravel at the base. A 1 to 10 ft (0.3 to 3 m) thick layer of residual soil and colluvium mantles a large part of the valley. Rock, consisting of shale and sandstone, underlies the soil [13].

FIELD INVESTIGATION.

The field investigation was comprised of two phases: (a) drilling and sampling and (b) geophysical investigation.

Drilling and Sampling

During this phase of the field investigation, 27 holes were drilled at the site for in situ seismic test use and to obtain undisturbed samples for laboratory testing. The holes were drilled in sets of three with one set near each end of the four proposed building locations and one set at the test loop location. The three holes were oriented to form an L-type configuration with the distance from the boring at the vertex of the L to the boring at the end of each leg being approximately 20 ft (6 m). Undisturbed samples were obtained from the boring at the vertex of the L for each set with a 5 in. (12.7 cm) Hvorslev fixed piston sampler. The remaining holes were drilled with a Waterways Experiment Station (WES) modified fishtail bit with upward baffles. After the holes had been drilled to the desired depth (2 to 5 ft (0.6 to 2 m) into rock), 3-in. (7.6 cm) ID PVC (polyvinyl chloride) pipe was installed in the boreholes, and the annular space between the casing and walls of the borings was grouted with a mixture of cement, bentonite, and water that, after setting up, had the approximate consistency of soil.

Geophysical Investigation

The geophysical investigation conducted at the site consisted of nine series of in situ seismic tests. Each series was comprised of crosshole, downhole, surface refraction seismic, and surface vibratory tests.

Crosshole Test Procedure

The crosshole tests discussed herein were conducted using a portable 24-channel refraction seismograph. A recorder displayed the data while operating at a paper speed of approximately 35 ips (89 cm/sec). Timing lines were displayed on the oscillogram at 10 sec intervals. The test procedure to obtain shear wave (S-wave) data was a modified technique that used the same vibratory source and control package as that described by Ballard [1]. The primary difference in the two procedures was concerned with the data acquisition package, i.e., no signal enhancement was used. Rather, the vibratory source was swept through a frequency range while monitoring the output of the two geophones placed at the same elevation in the receiver boreholes (the vibrator was producing vertical oscillations, thereby transmitting a vertically polarized shear wave). When an acceptable response was received at a specific frequency, this frequency was then interrupted by a tone burst generator to send a specific number of cycles of that frequency to the receiver units. The source geophone was displayed simultaneously along with both receiver geophones on the oscillograph. In this manner, the origin of the source pulse could easily be recognized, and the time difference that pulse and the signal arrival at each of the two receivers could be determined.

The tests to obtain S-wave data were conducted by first placing the source and receivers at the same elevation near the top of the boreholes, then pulsing the source unit several times recording both transmitted and received signals. After a satisfactory record had been obtained, the units were then repositioned 5 ft (1.5 m) deeper. The procedure above was repeated at this and each succeeding 5 ft (1.5 m) depth until reaching the bottom of the boring. Since these holes were relatively shallow, their verticality could be checked by visual observation using either flashlight or mirror sources. In all cases, the boreholes were vertical within $\pm$ 3 in. ($\pm$ 7.6 cm), thereby eliminating the need for a formal deviation survey.

The compression wave (P-wave) data were obtained in a similar manner, except that the vibrator and its associated instrumentation were not used. Instead, exploding bridgewire (EBW's) were used at each 5-ft (1.5-m) increment as the P-wave source.

Data obtained from the crosshole tests were the times required for P- and S-waves to propagate from the source to a point of detection. These times were then divided into the distance between source and receiver geophones to provide apparent velocities. If a nearby higher velocity layer exists, the wave will refract and travel along that layer, thus traveling along a faster path than the direct distance path. A computer program [2] for crosshole seismic interpretation, based on Snell's law of refraction, is used to determine true velocities by accounting for zones of high velocity contrast.

Downhole Test Procedure

This test, designed to provide data for determination of vertically oriented P- and S-wave velocities, was conducted using the same seismograph and recorder as those used for the crosshole tests.

The receiver geophone was placed at a depth of 5 ft (1.5 m) in a borehole, and a vertical hammer blow to a steel plate positioned approximately 1 ft (0.3 m) from the mouth of the boring on the ground surface was used as the P-wave source. A geophone adjacent to the hammer impact point provided zero time. S-wave tests were then conducted by placing a large wooden plank on the ground surface and striking the plank at each end thereby reversing polarity of the horizontally polarized S-wave which facilitates identification. The procedures for obtaining P- and S-wave data were then repeated with the receiver geophone at 5 ft (1.5 m) deeper increments until the bottom of the borehole was reached.

Surface Refraction Seismic Test Procedures

Conventional surface refraction seismic tests were conducted at the site using the 24-channel unit. Refraction lines run at the site were 240 ft (73 m) in length. Geophones were spaced at 10 ft (3 m) intervals with the first geophone positioned 5 ft (1.5 m) from an explosive source. Both forward and reverse transverses were made at the site so that true velocities, in addition to apparent velocities, could be determined and depths to refracting interfaces computed.

Surface Vibratory Test Procedure

In order to provide a necessary input parameter (near-surface S-wave velocity) to the crosshole computer program, surface vibratory tests [5] were conducted near each crosshole set. Rayleigh (R) waves may be generated using a controlled energy source such as a vibrator. In this case, the electromagnetic vibrator used to conduct the crosshole tests was also used as the R-wave source. The vibrator was operated at several frequencies, and the R-wave was monitored by the geophones used for the surface refraction seismic tests, which were placed in a line at intervals of 1 ft (0.3 m). In this manner, the velocity of the R-wave can be determined from a plot of R-wave phase arrival time versus distance for each specific frequency. Then, a corresponding wavelength can be calculated by dividing the frequency into the velocity. Wave velocities thus derived are considered to be average values for an effective depth of one half the wavelength [10]. S-waves, for practical purposes, can be considered to have the same velocity as R-waves.

Since the vibrator used at this site was of limited force output (50 lb (222 N)), the depth of investigation was limited to approximately 5 ft (1.5 m). These data could then be compared with those obtained for the shallow crosshole test position.

Elastic Moduli and Poisson's Ratio Determination

The P- and S-wave velocities determined from the above-mentioned test procedures should be converted to elastic parameters such as shear modulus G, Young's modulus E, and Poisson's ration v for use in foundation design. This conversion can be accomplished using theory of wave propagation in elastic media [8] and the total mass density ρ of the medium.

FIELD TEST RESULTS

The results determined from the in situ seismic tests conducted at the proposed location of building X-340-4 are presented and discussed below. It appeared feasible to the authors to select one location as being "typical" and to discuss the test results obtained there.

Proposed Building X-340-4 Location

Two series of in situ seismic tests were conducted at this location, one near each end of the proposed building location as shown in Figure 2.

Crosshole Tests

These tests consisted of two crosshole sets. Borings 821, 822, and 823 comprised one set near the south end of the proposed building, whereas the second set, near the north end of the building, was made up of borings 824, 825, and 826 as shown in Figure 2.

Borings 821 and 824 were used as the seismic source boreholes for their respective sets, and borings 822, 823, 825, and 826 were employed as receiver holes for their respective sets. The seismic source and receiver (geophone) placements used in the performance of the crosshole test are shown in Figure 3. True P- and S-wave velocities determined from the crosshole data are presented alongside the receiver and source locations. Note that only one P- and S-wave velocity is shown for each test elevation although true P- and S-wave velocities were determined from the seismic source borehole to each of the receiver boreholes for each crosshole set. In comparing the velocities from the source to each receiver, the spread was so narrow (< 10%) that it was thought advantageous to average the two P-wave velocities and likewise, the two S-wave velocities.

Downhole Tests

Two downhole tests located as shown in Figure 2 were conducted, one using boring 821 and the other using boring 824. The results of the downhole tests conducted in borings 821 and 824 are presented in Figures 4 and 5, respectively. As shown, average and incremental P- and S-wave velocities were determined from the downhole data.

Surface Refraction Seismic Tests

These tests consisted of four traverses (two lines), the orientation and location of which are shown in the test layout, Figure 2. The time versus distance plots for these traverses, designated as S-15 through S-18, are shown in Figures 6 and 7. These figures also present apparent velocities, true velocities, and associated depths to refracting layers.

Surface Vibratory Tests

One surface vibratory test was conducted at boring 821 and another at boring 824. The results of these tests yielded average R-wave velocities of 375 and 235 fps (114 and 72 mps) for the near-surface material (< 5 ft (< 1.5 m) depth) at borings 821 and 824, respectively. These velocities were used as true S-wave velocity inputs for the near surface to the crosshole computer program [2].

Data Interpretation

P- and S-Wave Velocity Zoning

The P- and S-wave velocity results determined from the crosshole, downhole, surface refraction seismic, and surface vibratory tests at this building location were analyzed and interpreted to produce the P- and S-wave velocity profiles shown in Figures 8 and 9, respectively. The P- and S-wave zones will not necessarily align with one another. This is primarily due to a large increase in P-wave velocity as percent saturation of the soil reaches 99-100 percent, whereas S-wave velocity is not affected.

Referring to Figure 8, five velocity zones were delineated from the P-wave data. The velocities presented for the zones were usually weighted averages based on data quality. A 1400 fps (425 mps) velocity was established for the near-surface zone, which averaged about 7 ft (2.1 m) in thickness. The second velocity zone averaged 4900 fps (1500 mps) and was about 20 ft (6 m) thick. This zone was underlain by a lower velocity layer (2800 fps (850 mps)), which ranged from 5 to 13 ft (1.5 to 4 m) in thickness and is best delineated by the uphole seismic data. The next two velocity zones (7100 and 9400 fps (2165 and 2865 mps)) could be collapsed into one. However, since the top of the 9400 fps (2865 mps) zone coincides with the top of shale, it was believed advantageous for analysis purposes to define top of rock and to keep the zones separate. The 7100 fps (2165 mps) zone interpreted from the crosshole and downhole tests in borings 824 through 826 is probably indicative of a silt, sand, and gravel mixture cemented to form the conglomerate noted in the laboratory classification of samples from boring 824 in this depth range.

Referring to Figure 9, five S-wave velocity zones were interpreted from the S-wave data. A 350 fps (105 mps) velocity was established for the near-surface zone and averaged about 7 ft (2.1 m) in thickness. The underlying zone had a 500 fps (150 mps) average and a thickness of about 15 ft (4.6 m). The third zone averaged 700 fps (215 mps) and was approximately 12 ft (3.7 m) thick. The two deeper zones (1750 and 2000 fps (535 and 610 mps)) could probably have been combined; however, for the same reason given for the P-wave zoning, they were

III-94

not. It will be noted that the value of wet unit weight applicable for each zone is also presented in Figure 9. These values were obtained from the undisturbed samples used in the resonant column tests and from laboratory tests conducted on samples from borings near the in situ seismic test locations [13].

Shear and Young's Moduli and Poisson's Ratio (G, E, ν)

The interpretation of the data was continued by establishing zones from the velocity profiles and determining values of (G), (E), and (ν) for these zones. This was accomplished by overlaying the P- and S-wave velocity profiles (Figures 8 and 9). This overlay produced seven zones. Then, values of G, E, and ν were computed for each zone using the P- and S-wave velocities and wet unit weights that fell into each zone. After analyzing G, E, and ν values for the various zones, it was evident that the controlling factor was the S-wave velocity zoning because the P-wave velocity variations had little effect on E and ν values and of course, no effect on G values. Therefore, the seven newly established zones could be reduced to the five original zones shown on the S-wave profile (Figure 9). The final field interpretation for this building location showing the five zones with their associated G, E, and ν values is presented in Figure 10.

LABORATORY INVESTIGATION

Undisturbed samples taken in the field were transported by truck to the WES. Once in the WES laboratory, the samples were opened and classified. The samples were evaluated to see which were representative of field conditions according to the boring logs and general knowledge of the site [13] and which samples could be tested in the laboratory based on lack of evidence of disturbance. Resonant column tests were performed on the representative samples found to be suitable for testing, i.e. those which did not contain particles of gravel or roots.

Resonant Column Test Equipment

The equipment used in this series of resonant column tests is known as the Drnevich torsional resonant column apparatus. Resonant column apparatus normally excite the specimen only torsionally; however, this equipment differs in that it also includes an electromagnetic oscillator for longitudinal vibration of the specimen. A more detailed description of this equipment is given elsewhere [4].

Test Procedures

All tests were performed on a 2.8 in. (7.1 cm) diam by 7 in. (17.8 cm) long undisturbed specimens. These specimens were isotropically consolidated so as to correspond to the in situ mean principal effective stress or octahedral stress, $\overline{\sigma}_{oct}$. The in situ mean principal effective stress was computed assuming normally consolidate conditions [13] and a coefficient of earth pressure at rest of 0.5.

After primary consolidation was complete, the drainage valve was closed and the B valve was checked and recorded [3]. The specimen was subjected to sinusoidal excitation in both the longitudinal and torsional modes. Longitudinal and torsional vibrations were not conducted simultaneously. The frequency of vibration was varied until resonance was obtained and data were recorded such that the modulus and strain amplitude values could be computed. In order to obtain the variation of modulus as a function of strain amplitude, the driving force was increased and the general test procedure was repeated.

The procedure above is a brief description of that used by the WES. A more detailed testing procedure, along with a complete discussion of the analysis and calculations required to obtain shear and Young's moduli from the raw data, is given elsewhere [4].

Resonant Column Test Results

Undisturbed samples from various depths in borings 800 UD, 803 UD, 806 UD, 809 UD, 812 UD, 815 UD, 818 UD, and 821 UD were tested in the resonant column device [3]. A total of 37 specimens were tested.

Reference 3 contains plots of modulus and damping versus dynamic strain amplitudes for all the resonant column tests. In order to familiarize the reader with the format of the resonant column data and its interpretation, the test results obtained from boring 821 UD samples will be discussed. The location of boring 821 UD is shown in Figure 2.

Sample 2 was taken at a depth of 2.7 to 4.2 ft (0.8 to 1.3 m) and was considered to represent the surficial layer to a depth of 14 ft (4.3 m). This material was classified in the laboratory as a CH clay and tested in the resonant column device at a mean effective

III-96

principle stress of 1.7 psi (11.7 kPa). Figure 11 presents the results in terms of dynamic modulus versus dynamic 0-peak strain amplitude for the resonant column tests conducted on this specimen.

Figure 12 presents the damping ratio as a function of dynamic strain amplitude for the resonant column tests conducted on sample 2. The damping ratio varies from approximately 5 to 10 percent over the strain range of 10^{-6} to 10^{-3} in./in. (rad/rad) (cm/cm). In general, the damping determined in the torsional mode is approximately two thirds of that determined in the longitudinal mode. As expected, damping generally increases with increasing strain [6,7,9,12].

Sample 9 was taken at a depth of 20 to 21.8 ft (6.1 to 6.6 m) and was considered to represent a clay layer from a depth of 14 to 22 ft (4.3 to 6.7 m). A resonant column test was performed on the specimen taken from sample 9 which was consolidated to a mean effective principle stress of 6 psi (41.4 kPa). The results of this test are presented in Figures 13 and 14.

Sample 11 was taken at a depth of 24.5 to 26.6 ft (7.5 to 8.1 m) and was considered to be typical of the 5 ft (1.5 m) thick layer between a depth of 22 and 27 ft (6.7 and 8.2 m). This material was a silt and was tested in the resonant column apparatus at a mean effective principle stress of 9.5 psi (65.5 kPa). Shear and Young's moduli and damping ratios as a function of dynamic strain amplitudes are presented in Figures 15 and 16.

Sample 13 taken at a depth of 29.2 to 30.5 ft (8.9 to 9.3 m) was considered to be representative of the soil layer between a depth of 27 and 31 ft (8.3 and 9.4 m). This sample was classified a CL clay and tested in the resonant column device under a mean principal effective stress of 10.75 psi (74 kPa). The dynamic modulus and damping ratio as a function of dynamic strain amplitude are presented in Figures 17 and 18, respectively.

Below a depth of 31 ft (9.4 m) the soil is extremely gravelly. Samples obtained were disturbed to such an extent as to be considered unsatisfactory for testing. Rock was found at a depth of about 36 ft (11 m).

DISCUSSION OF LABORATORY TEST RESULTS

The resonant column data obtained from specimens trimmed from samples taken in boring 821 UD have been presented. A soil profile will be developed for this boring in terms of shear modulus, Young's modulus, and internal damping ratio.

III-97

Shear Modulus

Figure 19 summarizes the shear modulus data obtained from testing four specimens from samples of boring 821 UD. Also shown in Figure 19 is a soil profile depicting th layers that were observed in the boring. Note that the near-surface clay layer is 14 (4.3 m) thick and has a shear modulus at low strain levels of about 3.5×10^3 psi (24 kPa) (10^3 kPa = 1 MPa). As the shear strain is increased to a dynamic strain level of 10^{-3} rad/rad, the shear modulus decreases to about 1.7×10^3 psi (11.7 MPa). Immediat below this 14 ft (4.3 m) thick surficial layer is an 8 ft (2.4 m) thick layer of clay. From the resonant column test results, it appears that this material has a shear modul about 2.8×10^3 psi (19.3 MPa) at very low strains. This modulus decreases to about 1 10^3 (6.9 MPa) as the strain level increases to 10^{-3} rad/rad. Below a depth of 22 ft (there exists a 4 ft (1.2 m) thick silt layer. This material exhibits a shear modulus 17 x 10^3 psi (117 MPa) at low strain levels. As the strain increases to a dynamic str amplitude of about 10^{-4} rad/rad, dynamic shear modulus decreases to a value of 5.6 x 1 (38.6 MPa). Below the 4 ft (1.2 m) thick silt layer there exists a 5 ft (1.5 m) thick clay layer between a depth of 26 and 31 ft (7.9 and 9.4 m). This layer exhibits a she modulus of about 5.6×10^3 psi (38.6 MPa) at low strain levels. This modulus decrease about 2.8×10^3 psi (19.3 MPa) as the strain level becomes greater than 10^{-4} rad/rad.

In reviewing Figure 19 several specific items should be noted. These are:

a) As the strain level increases from 10^{-5} to 10^{-3} rad/rad, the shear modulus is generally reduced by about 50 percent, except in the silt layer at a depth of 22 to 26 (6.7 to 7.9 m). In this layer the shear modulus is reduced to about one third of its original low strain value.

b) One would expect that the modulus would be a function of the mean principal effective stress. Thus, the modulus should increase as a function of depth, other thi being equal. This is not the case in boring 821 UD. The clay layer at a depth of 14 22 ft (4.3 to 6.7 m) is not stiffer than the surficial clay.

c) The 5 ft (1.5 m) thick clay layer immediately below the silt has a modulus th: is significantly lower than the silt layer at a depth of 22 to 26 ft (6.7 to 7.9 m). ' is not unusual but should be noted because it may be important in any future dynamic a or calculations.

Young's Modulus

Figure 20 shows Young's modulus (E) versus dynamic strain amplitude as a function of depth in boring 821 UD. The 14 ft (4.3 m) thick surficial layer has a Young's modulus at low strain levels of about 26×10^3 psi (179 MPa). This value decreases to approximately 15×10^3 psi (103 MPa) when the strain level increases to 5×10^{-5} in./in. (cm/cm). The clay layer between a depth of 14 and 22 ft (4.3 and 6.7 m) is very similar to the surficial layer. This 8 ft (2.4 m) thick clay layer has a Young's modulus of about 26×10^3 psi (179 MPa) at a strain level of 4×10^{-7} in./in. (cm/cm). As the strain increases to a level of approximately 10^{-4} in./in. (cm/cm), the Young's modulus decreases to 9.7×10^3 psi (66.9 MPa). The Young's modulus in the silt layer that lies at a depth of 22 to 26 ft (6.7 to 7.9 m) is approximately 56×10^3 psi (386 MPa) at strain levels below 10^{-6} in./in. (cm/cm). As the strain level increases above 10^{-6} in./in. (cm/cm), this Young's modulus decreases to approximately 20×10^3 psi (138 MPa) when the strain level is about 4×10^{-5} in./in. (cm/cm). The 5-ft- (1.5-m-) thick clay layer located at a depth of 26 to 31 ft (7.9 to 9.4 m) in boring 821 UD has a Young's modulus at low strain levels of about 43×10^3 psi (296 MPa). Again, as the strain increases from a level of around 10^{-6} to 10^{-4} in./in. (cm/cm), the Young's moduli decreases to approximately 16×10^3 psi (110 MPa).

As one looks at this collection of data, it is again apparent that the silt layer at a depth of 22 to 26 ft (6.7 to 7.9 m) is relatively stiff compared with the other layers. The two clay layers above 22 ft (6.7 m) have similar characteristics, with the layer between 14 and 22 ft (4.3 and 6.7 m) being more sensitive to strain than the surficial layer.

Damping Ratio

Figure 21 presents the damping ratio versus dynamic strain amplitude as a function of depth for boring 821 UD. In dynamic analyses and in the calculation of dynamic response of soils, the predominant energy input comes from shear waves. Consequently, shear or torsional damping is the most important consideration, and only shear damping will be discussed here. The longitudinal damping data were sometimes atypical [3] and should be considered suspect (see Figure 14). Close study of Figure 21 reveals that damping is relatively constant as a function of depth for boring 821. For strains less than 10^{-4} rad/rad the damping ratio is on the order of 3 percent. As the strains increase above 10^{-4}, then the damping ratio approaches 12 percent. Based on WES experience, this appears to be reasonable and consistent.

The collection of data presented elsewhere [3] was obtained from 35 resonant column tests. In general, shear and compression moduli decrease as a function of increasing strain [4,6,7] while damping increases.

COMPARISON OF FIELD AND LABORATORY VALUES

Figure 22 shows a boring log for boring 821 UD along with laboratory value s of shear modulus, Young's modulus, and damping as a function of depth at low ($\sim 10^{-6}$) strain levels. The subdivision of the soil profile was based on soil type and laboratory visual classification. These values have been discussed previously.

For comparison purposes, the values that were determined in the field by geophysical techniques are shown immediately to the right of the laboratory values. The reader is reminded that the subdivision of the soil profile is based on field techniques and results from changes in P- and S-wave velocities and is independent of material types or visual classification. Consequently, it is not surprising that the soil profile is not subdivided identically using data obtained in the laboratory and field.

Generally, moduli values obtained in the laboratory are less than in situ determined values. In fact, the values obtained for laboratory tests may be as low as one half the field values [11]. The fact that the data obtained during this investigation compare so poorly is surprising. No consistent trend can be obtained from a comparison of the data presented in Figure 22. For some zones (soil layers) the laboratory values are approximately equal to the field values. For other zones the laboratory values are greater than the field values and vice versa. Because of the scatter in the data, a discussion of the data comparison and the selection of design values are presented below.

Shear and Young's Moduli

Near-Surface Material

In comparing the laboratory and field values, the thickness of the near-surface layer was determined to be 14 ft (4.3 m) in the laboratory. This was based primarily on the boring logs; however, in the field, using velocity data, the near-surface layer was chosen to be only 8 ft (2.4 m) thick. The laboratory-determined shear modulus of 3.5 x 10^3 psi (24 MPa) compares very favorably with the field-determined value of 3.3 x 10^3 psi (22.7 MPa). For practical purposes, these numbers are the same. The Young's modulus computed from the field data was determined to be 9.7 x 10^3 psi (67 MPa). This value seems reasonable; however, it is about one third of the laboratory-determined value, which is 26 x 10^3 psi (180 MPa). This value is judged to be too high. For the top layer, values of shear and Young's moduli selected for design were 3.4 x 10^3 and 10 x 10^3 psi (23.4 and 69 MPa), respectively.

Second Layer

Immediately below the near-surface layer is a second clay layer that ends at 20 to 22 ft (6 to 6.7 m) depending on whether the field seismic data or the laboratory classifications of samples obtained from the boring are accepted. The shear modulus as determined in the laboratory for this layer is 2.8 x 10^3 psi (19.3 MPa). This value is slightly less than half of the shear modulus (6.6 x 10^3 psi) (45 MPa) that was determined in the field. Because the soil is predominantly clay to a depth of 20 ft (6 m), and because the shear modulus should increase with increasing depth, a design shear modulus of about 5 x 10^3 psi (34 MPa) was selected. The values of Young's modulus range from 19.7 to 24.4 x 10^3 psi (136 to 168 MPa) depending on whether one chooses the field or laboratory values. These values are relatively close (within about 20 percent of each other), and this is considered good agreement. A value of 20 x 10^3 psi (138 MPa) has been selected for design purposes.

Third Layer

The third layer selected for design purposes combines the two layers within the depth interval of 22 to 31 ft (6.7 to 9.4 m) in the laboratory-determined values and is assumed to be at a depth of 20 to 30 ft (6 to 9.1 m). If one averages the shear modulus values of 16.7 x 10^3 and 5.56 x 10^3 psi (115 and 38 MPa), one obtains 11.1 x 10^3 psi (76.5 MPa). This value is in close agreement with 12.6 x 10^3 psi (87 MPa) that was obtained in the field, and consequently a value of 12 x 10^3 psi (83 MPa) was assigned for design purposes. In the laboratory, averaging the Young's moduli for the two layers between 22 and 31 ft (6.7 and 9.4 m) yields a value of about 50 x 10^3 psi (344 MPa). This is considerably higher than

Young's modulus of 27.1 x 10^3 psi (256 MPa) as determined by geophysical methods.
engineering judgment, a value of 40 x 10^3 psi (276 MPa) has been selected for desi
poses.

Fourth Layer

From a depth of 30 to 36 ft (9.1 to 11 m), the soil consisted of sandy silts i
gravels and consequently was not suitable for laboratory testing. Shear and young.
of 86 x 10^3 psi (593 and 1550 MPa), respectively, were assumed for design purposes
values are based on the field information.

Foundation Rock

Finally, the shear and Young's moduli of the foundation rock were determined t
125 x 10^3 psi (860 and 2550 MPa), respectively. These values are based solely on t
geophysical data.

Damping Values

If analytical calculations are to be conducted, a damping value of 3 to 5 perc
should be used for low strain calculations. This value is based on resonant column
is valid only to a depth of 31 ft (9.4 m) because the soil below that depth was not
However, because the damping appears to be uniform with depth, extrapolation to the
layer (30 to 36 ft (9.1 to 11 m) in depth) does not seem unreasonable.

SUMMARY

Based on the data and the analysis presented, the soil at boring 821 UD can be
vided and zoned as shown in Figure 22 for the purpose of dynamic analysis of buildi
machine foundations. This figure shows four soil layers above rock. The near-surf
is 8 ft (2.4 m) thick, the second layer is 12 ft (3.6 m) thick, the third layer is
(3 m) thick, and there is a 6 ft (1.8 m) thick layer immediately overlying the rock
is an idealized soil profile that would be suitable for design calculations. The v
shear modulus, Young's modulus, and damping ratio for each of these layers to be us
design are listed in the right hand three columns of Figure 22.

When there exists considerable scatter in the laboratory and field data obtained for a specific soil layer, WES usually places more emphasis on the field data. The reasons for this are:

a) The field investigation usually involves the use of several procedures (such as the crosshole, downhole, surface refraction, and vibratory techniques), which produce redundant data.

b) The laboratory investigation is usually limited to resonant column testing.

c) The data obtained in the laboratory is influenced by such things as sample disturbance, equipment limitations and boundary conditions, etc.

d) Furthermore, the laboratory data is only as valid as the specimen tests is representative of the soil layer. In this regard the field investigation is influenced by a volume of soil measured in terms of cubic meters not cubic centimeters as in the laboratory.

ACKNOWLEDGMENTS

Grateful appreciation is expressed to those colleagues at WES who assisted in the data acquisition phase of this study and to those who offered constructive criticism during the documentation stage. The support of the Energy Research and Development Administration (ERDA) and their permission to published this paper are gratefully acknowledged.

REFERENCES

[1] Ballard, R. F., Jr., "A Method for Crosshole Seismic Testing," Journal Geotechnical Engineering Division, American Society of Civil Engineers, Vol. 102, No. GT12, December 1976, pp. 1261-1273.

[2] Butler, D. K., Skoglund, G. R., and Landers, G. B., "CROSSHOLE: An Interpretive Computer Code for Crosshole Seismic Test Results, Documentation, and Examples," Miscellaneous Paper S-78-8, U.S. Army Engineer Waterways Experiment Station, CE, Vicksburg, Mississippi, July 1978.

[3] Curro, J. R., and Marcuson, W. F., III, "In Situ and Laboratory Determinations of Shear and Young's Moduli for the Portsmouth, Ohio Gasious Diffusion Add-on Site," Miscellaneous Paper S-78-12, U.S. Army Engineer Waterways Experiment Station, CE, Vicksburg, Mississippi, August 1978.

[4] Drnevich, V. P., Hardin, B. O., and Shippy, D. J., "Modulus and Damping of Soils by the Resonant Column Method," Dynamic Geotechnical Testing, American Society for Testing and Materials, STP 654, Philadelphia, Pennsylvania, September 1978, pp. 91-125.

[5] Fry, Z. B., "A Procedure for Determining Elastic Moduli of Soils by Field Vibratory Techniques," Miscellaneous Paper No. 4-577, U.S. Army Engineer Waterways Experiment Station, CE, Vicksburg, Mississippi, June 1963.

[6] Hardin, B. O., and Drnevich, V. P., "Shear Modulus and Damping in Soils: Measurement and Parameter Effects," Journal, Soil Mechanics and Foundations Division, American Society of Civil Engineers, Vol. 98, No. SM6, June 1972, pp. 603-624.

[7] Hardin, B. O., and Drnevich, V. P., "Shear Modulus and Damping in Soils: Design Equations and Curves," Journal, Soil Mechanics and Foundations Division, American Society of Civil Engineers, Vol. 98, No. SM7, July 1972, pp. 667-692.

[8] Kolsky, H., "Stress Waves in Solids," Dover Publications, Inc., New York, N.Y., 1963.

[9] Marcuson, W. F., III, and Wahls, H. E., "Effects of Time on the Damping Ratio of Clays," Dynamic Geotechnical Testing. American Society for Testing and Materials, SPT 654, Philadelphia, Pennsylvania, September 1978, pp. 126-147.

[10] Maxwell, A. A., and Fry, Z. B., "A Procedure for Determining Elastic Moduli of In Situ Soils by Dynamic Techniques," Miscellaneous Paper No. 4-933, U.S. Army Engineer Waterways Experiment Station, CE, Vicksburg, Mississippi, October 1967.

[11] Richart, F. E., Jr., Anderson D. G., and Stokoe, K. H., II, "Predicting In Situ Strain-Dependent Shear Moduli of Soil," Proceedings, Sixth World Conference on Earthquake Engineering, New Delhi, Vol. 6, January 1977, pp. 6-159 to 6-164.

[12] Seed, H. B., and Idriss, I. M., "Soil Moduli and Damping Factors for Dynamic Response Analyses," Report No. EERC 70-10, Earthquake Engineering Research Center, University of California, Berkeley, December 1970.

[13] Taylor, H. M., Jr., et al., "Title I Design Foundation Investigation for Static Loading, Gaseous Diffusion Add-On Plant, Portsmouth, Ohio," Miscellaneous Paper S-77-20, U.S. Army Engineer Waterways Experiment Station, CE, Vicksburg, Mississippi, November 1977.

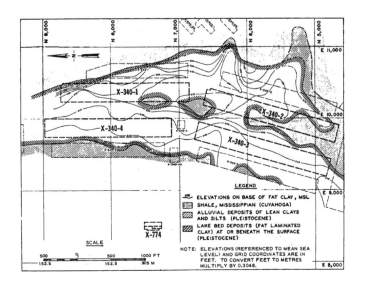

LEGEND

⌐ ELEVATIONS ON BASE OF FAT CLAY, MSL
▦ SHALE, MISSISSIPPIAN (CUYAHOGA)
▨ ALLUVIAL DEPOSITS OF LEAN CLAYS AND SILTS (PLEISTOCENE)
▨ LAKE BED DEPOSITS (FAT LAMINATED CLAY) AT OR BENEATH THE SURFACE (PLEISTOCENE)

NOTE: ELEVATIONS (REFERENCED TO MEAN SEA LEVEL) AND GRID COORDINATES ARE IN FEET. TO CONVERT FEET TO METRES MULTIPLY BY 0.3048.

SCALE

500   0   500   1000 FT
152.5   152.5   305 M

X-774

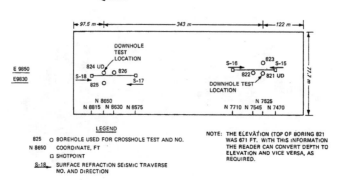

Fig. 2 In Situ Seismic Test Layout, Proposed Building X-340-4 Locatio

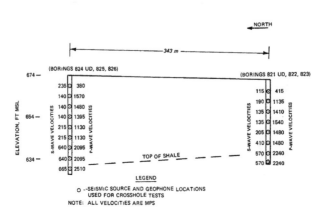

Fig. 3 P- and S-Wave Velocities Determined from Crosshole Tests
Proposed Building X-340-4 Location

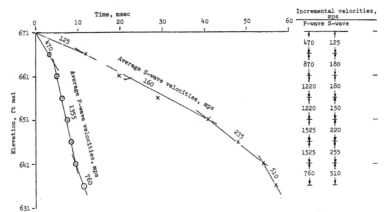

Fig. 4 Arrival Time Versus Depth from Downhole Test, Boring 821,
Proposed Building X-340-4 Location

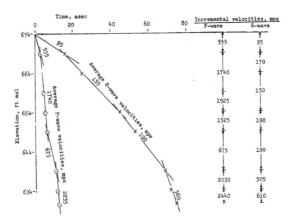

Fig. 5 Arrival Time Versus Depth from Downhole Test, Boring 824,
Proposed Building X-340-4 Location

III-107

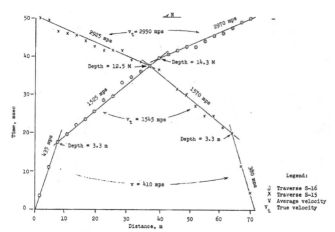

Fig. 6 P-Wave Arrival Time Versus Distance, Proposed Building
X-340-4 Location, S-15 and S-16

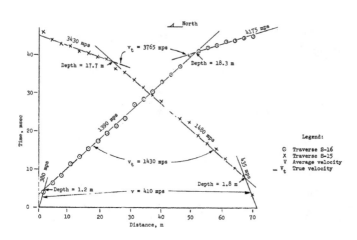

Fig. 7 P-Wave Arrival Time Versus Distance, Proposed Building
X-340-4 Location, S-17 and S-18

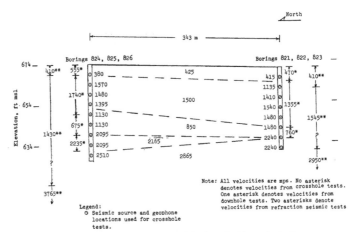

Fig. 8 P-Wave Velocity Profile from In Situ Seismic Tests,
Proposed Building X-340-4 Location

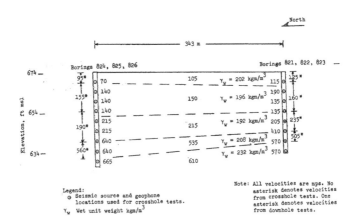

Note: error in γ_w for ex. 202 kgm/m^3 should be 20.2 kN/m^2, etc.

Fig. 9 S-Wave Velocity Profile from In Situ Seismic Tests,
Proposed Building X-340-4 Location

Fig 10. Shear and Young's Moduli Profile from In Situ Seismic Tests,
Proposed Building X-304-4 Location

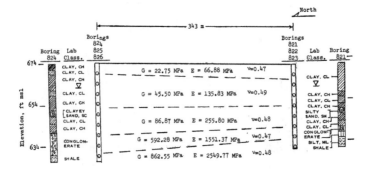

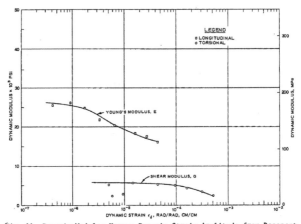

Fig. 11 Dynamic Modulus Versus Dynamic Strain Amplitude from Resonant
Column Tests, Sample: 821 UD No. 2-1, $\bar{\sigma}_{oct}$ = 11.7 kPa,
B = 0.18, Nonsaturated

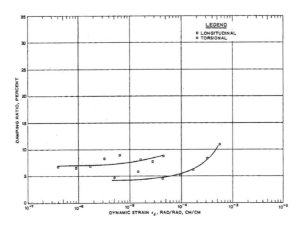

Fig. 12. Damping Ratio Versus Dynamic Strain Amplitude from Resonant
Column Tests, Sample: 821 UD No. 2-1, $\bar{\sigma}_{oct}$ = 11.7 kPa,
B = 0.18, Nonsaturated

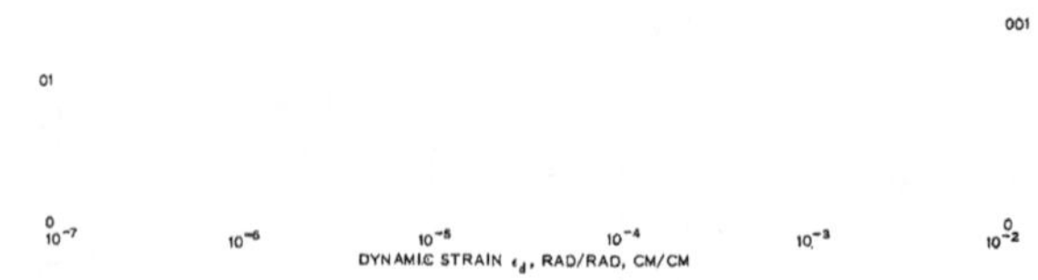

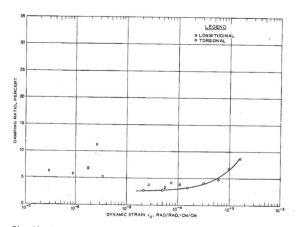

Fig. 13   Dynamic Modulus Versus Dynamic Strain Amplitude from Resonant
          Column Tests, Sample: 821 UD No. 9-1, $\bar{\sigma}_{oct}$ = 41.4 kPa,
          B = 0.68

Fig. 14   Damping Ratio Versus Dynamic Strain Amplitude from Resonant
          Column Tests, Sample: 821 UD No. 9-1, $\bar{\sigma}_{oct}$ = 41.4 kPa,
          B = 0.68

III-112

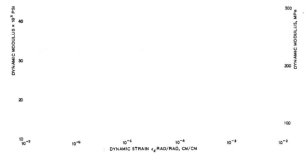

Fig. 15 Dynamic Modulus Versus Dynamic Strain Amplitude from Resonant
Column Tests, Sample: 821 UD No. 11-1, $\bar{\sigma}_{oct}$ = 65.5 kPa,
B = 0.83

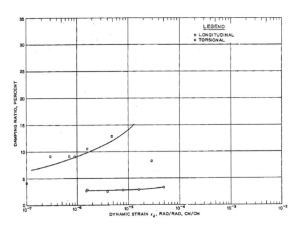

Fig. 16 Damping Ratio Versus Dynamic Strain Amplitude from Resonant
Column Tests, Sample: 821 UD No. 11-1, $\bar{\sigma}_{oct}$ = 65.5 kPa,
B = 0.83

III-113

0
10^{-7} 10^{-6} 10^{-5} 10^{-4} 10^{-3} 10^{-2} 0
DYNAMIC STRAIN ϵ_d, RAD/RAD, CM/CM

Fig. 17 Dynamic Modulus Versus Dynamic Strain Amplitude from Resonant Column Tests, Sample: 821 UD No. 13-1, $\bar{\sigma}_{oct}$ = 74 kPa, B = 0.91

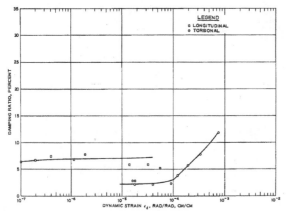

Fig. 18 Damping Ratio Versus Dynamic Strain Amplitude from Resonant Column Tests, Sample: 821 UD No. 13-1, $\bar{\sigma}_{oct}$ = 74 kPa, B = 0.91

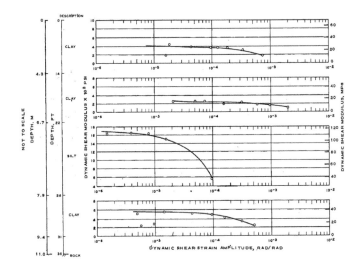

Fig. 20 Dynamic Young's Modulus Versus Dynamic Strain
Amplitude from Resonant Column Tests, Boring 821 UD

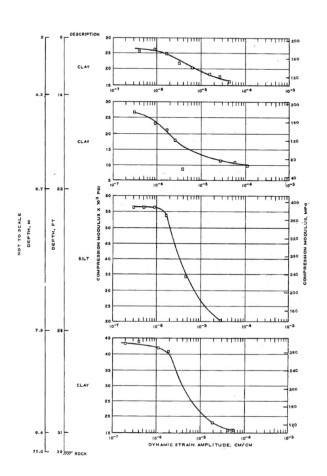

Fig. 21 Damping Ratio Versus Dynamic Strain Amplitude
from Resonant Column Tests, Boring 821 UD

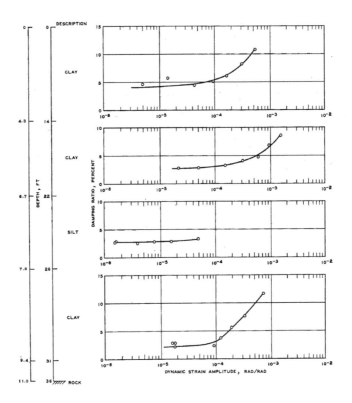

Fig. 22 Field and Laboratory Comparison of Shear an
with Resultant Design Values for Boring 821

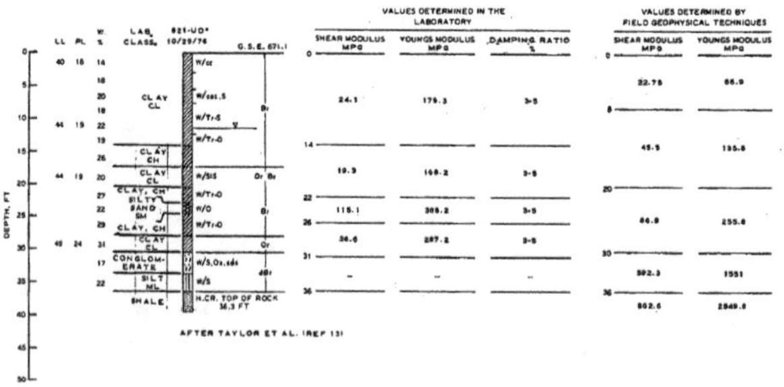

AFTER TAYLOR ET AL. (REF 13)

AN EXPERIMENTAL STUDY ON THE LIQUEFACTION OF

SANDY SOILS IN A COHESIVE SOIL LAYER

Tatsuo Asama

Yukitake Shioi

Public Works Research Institute

Ministry of Construction

ABSTRACT

Experimental and theoretical researches on liquefaction have been made by many research-ers. There is a general understanding that liquefaction occurs when the shear stress reaches a critical value which is determined by the type of soil, its density, its normal stress, and the like. The previous research shows that an essential factor should be stress levels in the soils. But the authors observe that liquefaction chiefly depends upon strain levels of soils. Especially where a sand layer lies on a soft cohesive soil layer, the strain can be amplified by the response of the cohesive soils.

KEYWORDS: Cohesion in soil; liquefaction; sandy soils; strain levels; stress levels.

1. INTRODUCTION

There are a number of research works on the liquefaction of saturated sandy soi
ever, upon examination of the results of researches against the phenomena occurring
structures or soils, they are not always correlated with each other. For example, t
of acceleration and duration and vibration causing liquefaction in the experiments a
observed at the sites of liquefaction. Further, the acceleration is not in correspon
to the damage to the structures in the vicinity. Still further the soils generating
faction are not distinguishable clearly from those generating no liquefaction. In th
Earthquake, liquefaction occurred in the gravel soil along the Kuzuryu River, while i
Niigata Earthquake, slight liquefaction occurred in the dune zone. Further, the phen
of water and sand spouting were seen after the main shock of the earthquake.

Where large scale road construction works are carried out, soils having a high p
bility of liquefaction are distributed extensively as in the estuary of a large river
coastal or reclaimed area. Great is the influence of liquefaction of a sand layer at
time of an earthquake upon the design parameters of a large foundation provided in su
place. It is then necessary to re-examine the captioned phenomenon to clarify its mec
so that adequate measures can be taken.

Before starting the investigation, the process of generation of liquefaction was
theoretically.

Principal elements of an earthquake wave will be the waveform, amplitude, cycle,
duration. These elements represent one or another form of energy respectively. The e
quake wave may be taken as a phenomenon of an energy accumulated in the crust being ti
mitted as a variable dynamic energy in all directions through the destructive phenomer
the seismic center. As such earthquake wave motion comes to the surface, it causes da
structures. The damage occurs generally in the forms of destruction of structures anc
of stability such as overturn, sliding and sinking. The damage due to liquefaction of
soil corresponds to the latter. By liquefaction, the soil loses the bearing force. I
loss of bearing force is due to elevation of the pore water pressure and decrease of t
effective stress of the bearing ground. The cause is the shear strain of the soil. A
from Figure 1, when a shearing strain is brought into the saturated soil by the action
shearing force τ_1, there is generated an excessive pore water pressure σ_0 against the

III-120

decrease in the volume if there is no change in the balance of water. To compensate for the external force P_1, the effective stress σ_1 decreases to σ_1', and so the shear resistance. When the shearing deformation proceeds further to the stage of σ_1' at zero, it represents fluidization or liquefaction of the natural ground. In the case of the foundation of a structure, the shear resistance of the ground decreases with decreasing effective stress so that the bearing force is shortcoming to cause tilting or sinking.

On the other hand, the shearing strain is a kind of work quantity so that it is necessary to look for the source of the corresponding energy. In general, waves are apt to shift from a hard medium to a soft medium but scarcely from a soft medium to a hard medium. Thus, the waves transmitted from the crust are apt to be stored near the surface where relatively soft soil layers are distributed. Then, it is assumed that the energy thus stored gives a shearing strain to the upper saturated sand layers and thus causes liquefaction. From such a point of view, review was made of the soil conditions of the points at which liquefaction occurred in the past earthquakes, and it was found that they had invariably highly strain following soils such as clay or silt distributed thereunder. The case of Fukui Basin is illustrated in Figure 2. Thus, here, the mechanism of liquefaction will be examined according to the foregoing assumption.

2. OBJECTS OF INVESTIGATION

It is intended, in order for reasonable design of the foundation of a structure on a saturated sandy soil, to clarify the mechanism of the phenomenon of liquefaction of the soil at the time of an earthquake and also to examine the load upon, and the properties of strain of, the foundation.

3. METHOD OF INVESTIGATION

In 1977, vibration experiments by means of models were carried out in order to test the validity of the aforementioned assumption.

The experiments were made with the six models shown in Figure 3. Model 1 was designed for a cohesive layer beneath a sand layer. It was prepared with a grouting material and was intended for investigation of the performance upon vibration. Model 2 was intended for investigation of any change in the vibratory performance of said cohesive layer under the influence of a surcharge. Model 3 was intended for investigation of the vibratory

performance of a cohesive layer the upper part of which was restrained by rigid pane.
Model 4 was intended for investigation of the vibratory performance of the cohesive]
having a dry sand layer placed thereover in reference to the case of the saturated sa
layer of Model 5. Model 5 was intended for investigation of the vibratory performanc
a saturated sand layer on the cohesive layer and the process of liquefaction. Model
designed to follow the elevation of pore water pressure and the process of liquefacti
a rubber coat laid over the saturated sand for assumption of an impermeable layer (Ph
and 2).

The materials used in the models and the layout of instruments are shown in Figu
and 5.

For the model of the cohesive layer, plastic materials were used in a thickness
40 cm. They were compounded so as to give a natural frequency of 7 ~ 8 Hz. The sand
that occurring at Takahagi, Ibaraki Prefecture in a proportion of grain sizes adapted
flow. As measuring instruments, 7 accelerometers (A) were installed on a seat and in
cohesive and sand layers as shown in Figure 5. For measurement of the changing soil j
sure, meters (D) were installed on the side of the respective vibration-boxes at the j
corresponding to the cohesive and sand layers. To catch the symptom of liquefaction,
water pressure gauges (P) were installed on those parts of the side wall which corresj
to the sand layer. For measurement of the strain of cohesive and sand layers, a phosj
bronze plate having 8 strain gauges (S) affixed to it was installed at the center of t
respective vibration boxes (Photo 3).

The shaking table was of the specification shown in Table 1.

The inputs were in the forms of irregular waves, sine waves and shock waves, and
were caused to act over the range of 20 gal to 500 gal depending on the respective obj
The irregular wave was of white noise containing frequency components of 5 ~ 40 Hz.]
reproduced by a data recorder and introduced into the control panel of the shaking tat
For the sine waves, frequencies of 5, 10, 15, and 20 Hz were used for Models 1 ~ 3, âf
those of 10 and 20 Hz for Models 4 ~ 6. The shock waves were produced by a method gir
an instantaneous displacement to the shaking table control unit manually, and the subs
free vibrations were measured (Table 2).

Recording was made by a data recorder, pen recorder, or magnetic oscillograph. Analog data recorded on a magnetic tape were converted into actual response waveforms by AD conversion. Sampling was 0.005 seconds.

Waveforms thus obtained were arranged as below.

With respect to the irregular waveforms, there were obtained power spectrums Gx(f) and Gy(f) and cross spectrum Gxy(f) of the waveforms by the accelerometer. Gx(f) is related to the input waveform, and Gy(f) to the ground response waveform. From the relationship between Gx(f) and Gxy(f), the frequency response function H(f) and phase difference δ(f) were calculated according to the formulas

$$H(f) = \frac{Gy(f)}{Gx(f)} \tag{1}$$

$$\delta(f) = \tan^{-1}\left(\frac{I(Gxy(f))}{R(Gxy(f))}\right) \tag{2}$$

where R(Gxy(f)) represents the real portion of Gxy(f), and I(Gxy(f)) the imaginary portion of Gxy(f).

For the sine waveforms, the maximum response values, response magnifications, and the phase differences were obtained from the data of measurement by the accelerometer (A), soil pressure meter (D), pore water pressure gauge (D), and embedded strain gauge (S).

For the shock waveforms, free damping vibrations and damping coefficients were obtained so far as practicable from the free damping waveforms.

4. EXPERIMENT RESULTS

Vibration tests of the models were conducted in the order of irregular waveform, sine waveform and shock waveform. However, for each vibration test, the models or sand were not changed so that the model grounds or, more particularly, sand grounds in the later tests were subject to the compacting and other effects given in the preceding tests.

Responses of the model grounds to the irregular waveforms were arranged in use of the spectrums shown in Figure 6. In Figure 6, the upper diagram shows the power spectrums Gx(f) and Gy(f) of the input and response and the cross spectrum Gxy(f) between them. The middle diagram illustrates the frequency response function H(f), and the lower diagram illustrat

a curve of phase difference (f). Marked o in the diagrams are the response values upon i

of sinusoidal waveforms.

The values of prevailing frequency of vibration at the positions of accelerometers 'i

the respective models as calculated from such response curves were as given in Table 3. .

Model 1, the frequency of vibration of the layer of plastics failed to come to the expect

value of 7 ∿ 8 Hz, and this was found, from the analysis executed later, to be due to the

restraining effect of the side wall.

The process of liquefaction by irregular waveforms is illustrated in Figure 7. In

Model 5, the liquefaction occurred with the irregular waveform at 500 gal and sine wavefor

of 10 Hz at 400 gal. At a higher level of acceleration or with a sine wave of 20 Hz, 400

a phenomenon close to liquefaction appears. In Model 6, liquefaction occurs with an irreg

lar waveform of 200 gal and a sine waveform of 10 Hz, 300 gal.

The values of maximum response magnification of the respective grounds by sine waves

were given as shown in Table 4. The accelerometers A2, A3 and A6 were positioned at the

center of the respective test tanks.

Free vibrations by shock waves tended to damp quickly so that it was difficult to rea

the succeeding waveforms except in Model 2 where a damping frequency of vibration at 6.6 H

and a damping constant of 0.025 was observed.

Subsidence of the upper sand layers in the shaking tests was about 1 cm in Model 1 or

about 5.5 cm in total in Model 5 including 1.9 cm due to liquefaction. In Model 6, it

totaled 3.6 cm including 2.3 cm due to liquefaction.

5. DISCUSSION

The experiments were designed to resolve the problems described in the summary and ha

a character of preliminary examination. The results were not always satisfactory, but we

were able to find some tendencies.

In Model 1, there was observed a good agreement between the irregular waveform and the

sine waveform with respect to the results of vibration of the lower cohesive layer, as sho

in Figure 8. This means that the frequency of vibration obtainable from the function of transfer by the irregular waveform is usable for calculation of the natural frequency. It may be used in the experiments to be conducted hereafter. Marked o in the diagrams are the values of response upon input of sine waveforms.

From the frequency of response at 10.9 Hz in Model 1, the ground constants of the cohesive layer may be estimated as

$$\text{(Propagation speed)} \quad v_s = 4hf_s = 4 \times 40 \times 10.9 = 1744 \text{ cm/sec,}$$
$$\text{(Shear coefficient)} \quad G = v_s^2 \cdot \rho/g = 1744^2 \times 1.0 \times 10^{-3}/980$$
$$= 3.1 \text{ kg/cm}^2, \text{ and}$$
$$\text{(Elastic modulus)} \quad E = 2(1 + v)G = 2(1 + 0.5) \times 3.0 = 9.3 \text{ kg/cm}^2$$

where h = Height of the cohesive layer;

  fs = Frequency of resonance;

  ρ = Density of the cohesive layer;

  g = Gravity; and

  v = Poisson's ratio.

Upon input of these ground constants to a multiple reflection model a resonance appeared at about 11.0 Hz as shown in Figure 8.

In Model 2, water was laid in a depth of 30 cm over the cohesive layer. The results of measurement showed the values of 24.1 Hz and 20.7 Hz respectively at the positions of A2 and A3 as shown in Table 3. In Model 1, they were equal at 10.9 Hz so that it was found that the weight effect would cause greater influence onto the vibration of lower layer. This was the same in the case of sine waveforms and would, therefore, be reliable. But, how to express it in a dynamic model is a problem to be resolved hereafter.

In Model 3, the upper surface of the cohesive layer was restrained by a rigid plate, but its effect was not much and the frequencies of resonance were distributed from 9.6 Hz to 11.1 Hz. In this case, the response to sine waveform agreed well with that to irregular waveform.

Model 4 had a dry sand layer placed on the cohesive layer and could, therefore, be regarded as a composite of Models 2 and 3. It was taken as a reference to Models 5 and 6. In the vibration tests, no liquefaction occurred. Further, as seen from Table 5, there was noted a trend of lower frequency of resonance with higher levels of acceleration. This trend was noted more or less in each of the other models.

Model 5 may be regarded as a saturated Model 4. In the vibration tests, liquefaction occurred with an irregular waveform at 10 Hz, 500 gal. In this case, the strain gauges on

the phosphor-bronze plate showed a corresponding movement. Further, the accele
and A7 exhibited response vibrations at lower response magnification so that it
assumed that there was no complete loss of shear resistance.

The frequencies of vibration prevailing before and after liquefaction are
Table 6. Before and after the liquefaction, the values are more or less decrea
decrease is not substantial. From this, it is considered that although in the
model had the vibrations applied successively the effect of the previous vibrat
respective vibratory performances would not be great.

Further, the level of acceleration at liquefaction was as high as 400 to 5
far greater value than that at the actual liquefaction. Thickness, permeabilit
load, etc. of the sand layer may be considered to be the causative factors.

Model 6 has an impermeable film provided over Model 5. In the vibration w
irregular waveform, liquefaction occurred at an acceleration of 200 gal. This
ferent from the value of 500 gal in the case of Model 5, although both models h
prepared in the same way. In the case of the sine waveform, liquefaction occurr
300 gal. As to the reasons of the liquefaction being caused relatively readily,
the pore water pressure due to the presence of the impermeable film and increase
strain in the sand layer due to reflection of the input waves at the impermeable
considered.

With respect to the prevailing frequencies of vibration before and after th
tion, they changed little; although there was more or less elevation after the 1
and they were similar to those in the case of Model 5.

6. CONCLUSION

The foregoing experiments were conducted as part of the study designed for
of the resistance to earthquake of foundations installed on sand layers over coh
Nevertheless, from the experiments, the following discovered.

1) The saturated sand layer on a cohesive soil is very much subject to the
the vibratory energy accumulated and expanded in the cohesive soil.

2) It is presumed that upon liquefaction of the saturated sand layer, the
energy is converted, for the greater part, into the moment of work.

III-126

3) When the upper soil has a high strength, vibration of the lower cohesive soil is restrained so that the vibration as a whole is reduced.

4) In vibratory analysis of a cohesive soil, the weight of upper soils is by no means negligible. With presence of the weight, there is an elevation of the value of apparent elastic modulus of the lower cohesive soil.

5) The presence of an impermeable layer on the saturated sand layer exerts a great influence upon the liquefaction.

Some other phenomena-deserving study were also noted, and they will be examined through a succeeding series of experiments. As the next stage, it is planned to examine the effects of thickness of the saturated sand layer as well as the cohesive soil, responses of the foundations installed in such soils, etc.

7. ACKNOWLEDGMENTS

The authors are grateful for Mr. Toshio Iwasaki of the Public Works Research Institute, Dr. Fumio Tatsuoka of the Tokyo University and Dr. Eiji Yanagizawa of the Tohoku University for their valuable advice in planning the experiments. Execution of the experiments owes greatly to the assiduous supports of Messrs. Takeaki Tomozawa, Omichi Ukon and Takeshi Onomura, Long Span Bridge Design Center. We should also like to thank Messrs. Ichiro Watanabe and Yoji Yamamoto, Public Works Research Institute, for management of the experiments.

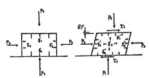

Fig. 1 Effects of shear deformation of soil

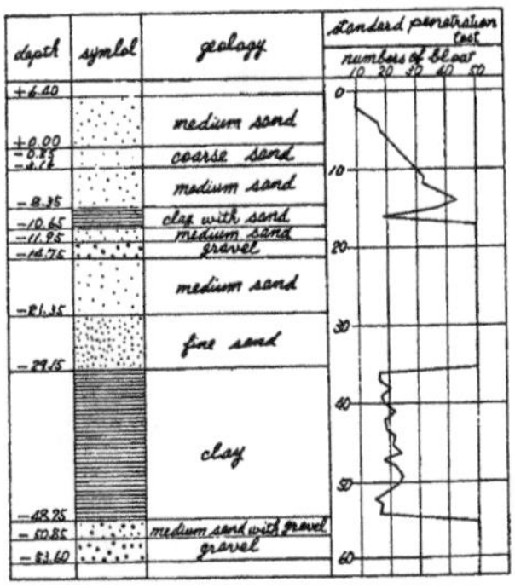

Fig. 2 Typical Column of geology at Fukui

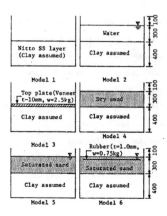

Fig. 3 Kinds of models

Takahagi Sand, Ibaraki
Prefecture

Specific gravity : 2.7
Maximum grain size: 1.0 mm
Minimum grain size: 0.05mm
60% Grain size: 0.6 mm
10% Grain size: 0.3 mm
Uniformity coefficient: 2.0

Fig. 4 Physical properties

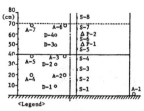

<Legend>

O : Accelerometer
o : Soil pressure meter
△ : Pore water pressure gauge
ı : Embedded strain gauge

Fig. 5 Arrangement of measuring
instruments

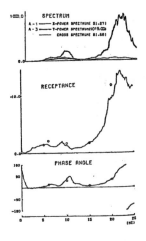

SPECTRUM

A - 1 ── X-POWER SPECTRUM(51.27)
A - 3 ── Y-POWER SPECTRUM(1019.00)
── CROSS SPECTRUM(81.68)

RECEPTANCE

PHASE ANGLE

Fig. 6 Response frequency
characteristics (Model 2)

Fig. 7 Process of
liquefaction
(Irregular waveform)

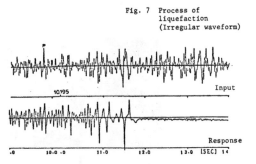

Input

10.195

Response

(Multiple reflection
model)

RECEPTANCE

TRANSFER FUNCTION

PHASE ANGLE

PHASE ANGLE

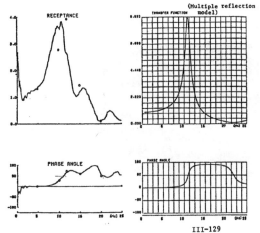

Fig. 8 Response frequency
characteristics
(Model 1)

III-129

Photo 1   Model 4

Photo 2   Model 6 under liquefaction

Photo 3   Model box and measuring device

Table 1 Specifications of shaking table

	Performance
Actuator	Max. exciting force: 25 ton-G(45 kip) or more.
	Max. amplitude: 100mm (4 inches) or more.
	Max. speed: 76 cm/sec(30 inches/sec) or more.
	Frequency range: DC∿100 Hz.
Low frequency oscillator	Oscillating waveform: Sine wave, rectangular wave and triangular wave.
	Oscillating frequency: 0.01 Hz∿1000 Hz or more.
	Output voltage: ±10V (when fixed).
Hydraulic source	Pressure: 210 kg/cm².
	Flow rate: 31 ℓ/min or more to satisfy the foregoing specification.
Shaking table	Linearity: ±0.05%.
	Drift: 0.5%/C (full scale).
	Noise: 0.5% of full scale or less.
	Controllable min. amplitude: ±0.31mm or less.
	Accuracy: 10%.

Table 2 Experimental conditions

Model	Input waveform	Nominal input acceleration	Measurement components Accelerometer	Soil pressure meter	Pore water pressure gauge	Embedded strain gauge	Number of components	Number of records
1	Irregular waveform	100 gal	A-1 A-2 A-3 A-4 A-5	D-1 D-2		S-1 S-2 S-3 S-4	11	3
	Sine waveform	100 gal						4
	Shock waveform	Shaking table						3
2	Irregular waveform	100 gal	A-1 A-2 A-3 A-4 A-5	D-1 D-2		S-1 S-2 S-3 S-4	11	3
	Sine waveform	100 gal						4
	Shock waveform	Shaking table						3
3	Irregular waveform	100 gal	A-1 A-2 A-3 A-4 A-5	D-1 D-2		S-1 S-2 S-3 S-4	11	3
	Sine waveform	100 gal						4
	Shock waveform	Shaking table						3
4	Irregular waveform	20, 50, 100, 150 gal	A-1 A-6 A-2 A-7 A-3 A-4 A-5	D-3 D-4	P-1 P-2	S-1 S-5 S-2 S-6 S-3 S-4	17	3
	Sine waveform	20, 50, 100 gal						6
	Shock waveform	Shaking table						3
5	Irregular waveform	100, 150, 250, 500 gal	A-1 A-6 A-2 A-7 A-3 A-4 A-5	D-3 D-4	P-1 P-2	S-1 S-5 S-2 S-6 S-3 S-4	17	3
	Sine waveform	200, 400, 500 gal						6
	Shock waveform	Shaking table and movable wall impact						5
6	Irregular waveform	100, 200, 300 gal	A-1 A-6 A-2 A-7 A-3 A-4 A-5	D-3 D-4	P-1 P-2	S-1 S-5 S-2 S-6 S-3 S-4	17	3
	Sine waveform	100, 200, 300 gal						6
	Shock waveform	Shaking table and movable wall impact						5

Note) In the tests by sine and shock waveforms in Models 5 and 6, measurement was made after occurrence of liquefaction.

Model No.	1	2	3	4	5	5'	6	6'
Acted acceleration	100	100	100	100	100	250	100	300
A2	10.9	24.1	11.1	21.5	20.0	23.2	18.0	20.5
A3	10.9	20.7	10.3	20.7	18.5	17.3	18.0	17.8
A4	10.9	24.1	9.6	21.1	18.5	17.3	18.0	19.0
A5	10.9	20.7	10.9	20.8	18.3	16.6	18.0	16.8
A6				20.9	18.0	16.3	18.0	17.5
A7				21.1	18.3	17.8	18.0	17.2

Table 3 Prevailing frequenc
of vibrations in th
ground

Note) 5 and 6 for input of 100 gal.
5' for input of 250 gal, and 6' for input of 300 gal.

Table 4 Response magnifications
in ground (at 10 Hz)

Note) 5' having liquefaction at 10 Hz, 400 gal.
6' having liquefaction at 10 Hz, 300 gal.

Input acceleration / Instrument	Prevailing frequency of vibration			
	20 gal	50 gal	100 gal	150 gàl
A-2	22.6 Hz	22.1 Hz	21.5 Hz	21.7 Hz
A-3	22.6	22.9	21.7	19.9
A-4	22.6	22.0	21.1	21.3
A-5	22.6	22.0	20.8	20.0
A-6	22.8	22.1	20.9	20.2
A-7	22.8	22.1	21.1	20.6

Table 5 Input acceleration versus
frequency of resonance

A-2∿A-5 are accelerometers installed in the clay
layer, and A-6 and -7 are those installed in the
sand layer.

Model condition / Instrument	Prevailing frequency of vibration		
	Before liquefaction	Under liquefaction	After liquefaction
A-2		20.8 Hz	
A-3	16.75 Hz	20.0 Hz	16.5 Hz
A-4	18.5 Hz	20.75 Hz	18.0 Hz
A-5	16.75 Hz	19.0 Hz	15.5 Hz
A-6	17.2 Hz		16.25 Hz
A-7	16.75 Hz		16.0 Hz

Table 6 Prevailing frequencies
of vibration before and
after liquefaction

III-132

STRESS-STRAIN BEHAVIOR OF DRY SAND AND NORMALLY CONSOLIDATED

CLAY BY INTER-LABORATORY COOPERATIVE CYCLIC SHEAR TESTS

Hiroshi Oh-oka

Kohjiroh Itoh

Yoshihiro Sugimura

Masaya Hirosawa

Building Research Institute

Ministry of Construction

· ABSTRACT

Inter-laboratory cooperative cyclic shear tests were conducted to obtain fundamental information about stress-strain behavior of dry sand and normally consolidated clay by using various kinds of dynamic shear apparatus which have been developed in Japan.

The tests were carried out in order to examine the characteristics of test apparatus and test procedures under as nearly identical conditions as possible. The results obtained by these tests are compared and discussed.

KEYWORDS: Damping ratios; dynamic soil properties; shear-strain testing of sand and clay; shear modulus; stress-strain soil behaviors; test procedures.

INTRODUCTION

Different soil deposits behave differently in an earthquake. When soil depos: foundations which in turn support structures, it follows that earthquake damage to tures will vary according to the type of soil on which the structure is built. The is important to clarify basic stress-strain behavior of various soils by subjecting cyclic shear stresses, and through analysis of shear modulus and damping ratios.

Up to this time, shear modulus and damping ratios have been measured with seve of apparatus. However, little information has come to the writers' attention concei test data which examined the characteristics of test apparatus and test procedures.

The ad hoc committee, "Dynamic Properties of Soils," headed by Professor Yoshi Yoshimi of the Tokyo Institute of Technology was founded in 1974 to gather fundamen information about stress-strain behavior of sand and clay using several types of api (Yoshimi, et al., 1975, 1976, 1977). Some of these test results are reported here.

CHARACTERISTICS OF THE APPARATUS

The laboratory apparatus for dynamic shear tests can be classified into the fo] three groups.

1) Shear stress is applied to the surface of the specimen; shear stress and st measured directly.

2) Principal stresses are applied periodically on the surface of the specimen; stress and strain are measured directly.

3) The physical quantities such as elastic wave velocity or resonant frequency soil specimens are measured; shear moduli and damping ratios are calculated from the

The First Group

a) Simple shear apparatus

Dynamic simple shear apparatus of the Roscoe type (Roscoe, 1953), which was emp Peacock (1968) and Finn (1971), has scarcely been used in Japan. On the other hand, NGI type devices or the new Kjellman type device (Hara and Kiyota 1977) have been developed for simple cyclic shear tests.

b) Ring torsion apparatus

With the ring torsion apparatus, the ring-shaped specimen is subjected to dynamic shear stresses with inertia torque working on weights which are placed on the specimen. Shear strain is nearly uniform in the specimen, because the height of the specimen is proportional to the radius (Yoshimi and Oh-oka, 1973). With both the simple shear and ring torsion apparatus, cyclic shear stresses can be applied under plane strain condition. However, one has to estimate the effective stresses acting on the side surfaces of the specimen with Jáky's equation, i.e., $K_0 = 1 - \sin\phi'$.

c) Torsional shear device

The advantage of this apparatus lies in being able to clearly control and measure the horizontal effective stresses (Iwasaki, Tatsouka, and Takagi, 1978). However, strain occurs not under plane strain conditions but with a little lateral strain in the strict sense if lateral strain is permitted.

The Second Group

The dynamic triaxial compression apparatus, in which the strain of the specimen is usually symmetric with respect to the vertical axis. The stress condition is extremely clear except at the zone near the upper and lower end of the specimen. It is necessary, however, that shear strain be estimated with axial strain and dynamic Poisson's ratio which is presumed or measured during shear. In the triaxial compression test, the effective mean principal stress at compression is different from that at extension, if the lateral confining pressure is constant. Also the effective mean principal stress changes during shear in spite of its undesirability.

In this respect, special devices have been adapted for the dynamic compression apparatus as follows:

a) An apparatus which is able to apply principal stresses which alternate axial and lateral stress with a phase difference of 180° (Tamaoki 1976). Through this improvement, the effective mean principal stress either becomes constant or the magnitude of the fluctuation becomes smaller.

b) An apparatus with which the dynamic Poisson's ratio can be measured during cyclic shear (Suzuki, Sugimoto, Babasaki, and Kakita, 1976) (Yasuda and Sakai, 1976).

The Third Group

a) Supersonic pulse apparatus

A supersonic pulse is generated at one side of the specimen, and velocities o
and S-wave propagating through the specimen are measured at the other side (Sugimo
and Terada, 1975). Shear modulus and Poisson's ratio are calculated from these ve
Shear strain is estimated to be very small in this test type.

b) Resonant-column apparatus

The specimen is fixed at the bottom, torsional vibration is applied at the to
1975). By changing the frequency of vibration, resonant frequency is found and wi
shear modulus is calculated. Damping ratio is estimated by free vibration at the
frequency by use of the same specimen.

The devices above were developed in the research facilities to which each mem
"Committee on Dynamic Properties of Soils" belong.

TEST CONDITIONS OF INTER-LABORATORY COOPERATIVE CYCLIC SHEAR TESTS_

Special attention was given to the following points:

1) In the series of tests on Toyoura sand, specimens were prepared uniformly
identical skeleton structure of soil specimens as well as the identical dry densiti
cifically, the air-dried sand grains were dropped into the mold from a constant hei
through a flexible tube with a constricted outlet.

2) In the clay tests, artificially consolidated clay samples were prepared for
uniformity, and to minimize sample disturbance during the sampling procedure.

The Tests for Toyoura Sand

1) Common conditions

a) Air-dried Toyoura sand gathered from the same spot was used. The physical
properties of the sand are as follows; specific gravity of 2.64, effective diameter
0.15 mm, uniformity of 1.4, maximum dry density of $1.628g/cm^3$, minimum dry intensit
$1.330g/cm^3$. The photographs of sand grains are shown in Photo 1.

b) The relative density of specimens was about 70%, and the effective mean pr
stress was $1.0 \ kg/cm^2$ during cyclic shear.

2) Special conditions

a) The frequencies used were about 0 Hz (static test), 1 Hz, and 4 Hz; primarily 1 Hz.

b) Inherent apparatus characteristics were various as mentioned before, and as detailed hereafter.

The Tests for Normally Consolidated Clay

1) Common conditions

a) The samples were prepared one at a time with identical marine clay (liquid limit of 102%, plastic limit of 38%) being consolidated one-dimensionally under an overburden pressure of 1.0 kg/cm^2.

b) The degree of saturation of the specimens was about 99%, i.e., specimens were almost totally saturated.

c) The specimens were all cylindrical or disk-shaped.

d) The specimens were reconsolidated for about 24 hours under a mean principal stress of 0.59 kg/cm^2 prior to cylic shear. (The mean water contents after reconsolidation were about 63%.)

e) A confining pressure of 0.59 kg/cm^2 equal to the reconsolidation stress was maintained during shear.

f) The stress wave shape was sinusoidal; the frequency was 1 Hz except in the resonant-column test.

g) During cyclic shear, 100 stress cycles were usually applied to the specimens.

2) Special conditions

a) The unconfined compressive strength on the prepared samples was scattered from 0.7 kg/cm^2 to 1.0 kg/cm^2. Therefore, some difference was produced in the strength characteristics of block samples distributed to each facility.

b) The initial pore water pressure was usually set to zero, but cyclic shear tests in which the initial pore water pressure was not zero were also performed.

c) The recognized differences in the inherent characteristics of the apparatus are as follows:

1. In the resonant-column and the dynamic simple shear tests, stress conditions
during the specimen reconsolidation are equal to K_0-consolidation stresses, which
were applied to the prepared block sample beforehand. While in the dynamic triaxial
compression tests, the specimen has to be consolidated under all-around pressure, in
order to be subjected to fully reversed cycles of shear stresses. Consequently,
consolidation stresses change from those under K_0-conditions to those under all-
around pressure conditions. In such cases, the specimens may be disturbed to some
extent.

2. In the triaxial compression test, maximum shearing stress acts on a plane
inclined to the surface of the specimen by 45°, but not in the resonant-column
and the dynamic simple shear tests. Therefore, for anisotropic soils, shear
test characteristics may reflect the difference of the dynamic properties of
the soils.

3. In the dynamic simple shear apparatus, lateral stresses acting on the side
surface of the specimen cannot be measured. Therefore, it is necessary to
evaluate them with ϕ' obtained from drained shear test or consolidated-
undrained shear test using Jåky's equation, which may allow some error
($\phi = 40.1°$ for sand, $\phi' = 38.5°$ for clay).

4. The methods to obtain equivalent shear modulus and damping ratio differ
according to the apparatus adopted. In the simple shear and the triaxial compression tests, they are directly determined with stress-strain curves. In the
triaxial compression test, however, it is necessary to measure or assume
Poisson's ratio. In the resonant-column test and the supersonic pulse test,
equivalent shear modulus is calculated indirectly in terms of resonant frequency and
S-wave velocity.

5. The strain state in the simple shear test, i.e., the plane strain condition, is different from the axi-symmetrical strain condition in the triaxial
compression test.

6. There are two types of dynamic triaxial compression apparatus. In one type,
lateral total stress is kept constant during cyclic shear; while in the other,
lateral total stress changes periodically with the same amplitude and a 180°
phase difference, comparing with axial total stress.

STRESS STRAIN CURVE AND DYNAMIC POISSON'S RATIO
OF DRY SAND AND ALMOST SATURATED CLAY

Shear stress-strain curves obtained from cyclic shear tests are shown in Figure 1 through Figure 3. Typical data of Toyoura sand obtained from ring torsion shear tests are shown in Figure 1 (a) (b) (c) (d) according to shear strain levels (Hatanaka, 1976). The clay test results in the simple cyclic shear tests and the dynamic triaxial compression tests are also shown in Figure 2 (a) (b) (Hara and Kiyota) and in Figure 3 (a) (b) (Suzuki and Sugimoto), respectively.

It is shown in Figure 1 through Figure 3 that the shapes of the shear stress-strain curves are very similar to each other, irrespective of soil types and test types. However, it can be seen in Figure 1 that for sand, the loop in the second cycle is different from the loop in the tenth and the 100th cycle. On the other hand, for clay, a similar tendency is never seen as shown in Figure 3.

Dynamic Poisson's ratios were measured for dry sand and almost saturated clay. In the supersonic pulse test, P-waves and S-waves were generated independently at the top of specimen, and their velocities were calculated by measuring the time when the waves reached the bottom of the specimen. Dynamic Poisson's ratio ν_d is calculated with the following equation (Sugimoto and Suzuki).

$$\nu_d = \frac{(V_p/V_s)^2 - 2}{2\{(V_p/V_s)^2 - 1\}} \tag{1}$$

where, V_p, V_s are velocities of P-wave and S-wave, respectively. Dynamic shear modulus is also calculated with the following equation:

$$G = \frac{\gamma_t}{g} \cdot V_s^2 \tag{2}$$

where, γ_t is the wet unit weight of specimen, and g is acceleration of gravity. In the triaxial compression test, dynamic Poisson's ratio is estimated by measurement of dynamic lateral strain as well as dynamic axial strain. Lateral strains are estimated by measuring the diameter change of specimen electrically (Suzuki, Sugimoto, Babasaki and Katita, 1976) or measuring the volume change of specimen by measurement of the change of the water height in the cell electrically (Yasuda and Sakai, 1976).

Figure 4 shows the Poisson's ratio of Toyoura sand. It can be seen in Figure 4 (a) t
dynamic Poisson's ratio is almost independent of initial void ratio and effective mean pri
cipàl stress though the data are scattering. It is also recognized in Figure 4 (b) and (c)
that dynamic Poisson's ratio is almost independent of strain level. From the experimental
result mentioned above, dynamic Poisson's ratio may be estimated at about 0.35 for air-drie
Toyoura sand, and this value was used to calculate shear strain from axial strain.

Figure 5 shows the Poisson's ratio of almost saturated clay. The data scatter in the
range from .492 to .518, mainly .500 $\pm$ 0.06. If the clay is saturated thoroughly, Poisson'
ratio becomes about 0.5. However, the value of 0.45 should be used as the dynamic Poisson'
ratio of the clay, because the prepared clay is not saturated in the strict sense and becau
the data obtained with the electric diameter transducer are apt to be somewhat larger than
those obtained with the other methods as shown in Figure 4. The difference of dynamic
Poisson's ratio results in a small difference of shear modulus when the shear modulus G is
evaluated with elastic modulus E and Poisson's ratio ν_d in using the following equation:

$$G = \frac{1}{2(1 + \nu_d)} \cdot E \qquad (3)$$

where

$$E = \frac{\sigma_d}{\varepsilon} \qquad (4)$$

ε is axial strain, and σ_d is deviator stress

EQUIVALENT SHEAR MODULI OF SAND AND CLAY

Typical results of shear moduli obtained from the cyclic shear tests are shown in
Figure 6 through Figure 8. Figure 6 (a) (b) shows the relationship among shear modulus,
shear strain γ (single amplitude) and number of stress cycles N for sand obtained from the
ring torsion shear test and the simple cyclic shear test, respectively. It is seen in both
figures that the shear modulus of sand increases as the shear strain decreases and as the
number of stress cycles increases. Especially, it can be noticed that the shear modulus
increases considerably with the number of stress cycles for the strain level more than

5×10^{-4} strain. This tendency is recognized most clearly from Figure 7 (a) (b), which shows the increase of shear modulus of sand with the number of stress cycles for each strain level.

On the other hand, the results obtained for normally consolidated clay in the simple cyclic shear test and the dynamic triaxial compression test are shown in Figure 8 (a) (b), respectively. It can be seen that for clay, shear modulus decreases slightly with the number of stress cycles, being different from the tendency for sand.

DAMPING RATIOS OF SAND AND CLAY

Typical results of damping ratios of sand and clay are shown in Figure 9 and Figure 10 (a) (b), respectively. Figure 9 shows the data obtained from the ring torsion shear test. It can be seen from this figure that the damping ratio of dry sand decreases considerably as the number of stress cycles increases. It is also recognized that the damping ratio increases as the shear strain increases.

Figure 10 (a) shows the relationship between the damping ratio and the number of stress cycles obtained in the simple cyclic shear test for clay. As shown in this figure, the damping ratio of normally consolidated clay decreases slightly as the number of stress cycles increases.

On the other hand, the damping ratio increases slightly as the number of stress cycles increases in the dynamic triaxial compression test as shown in Figure 10 (b). This tendency is contrary to the result obtained from the simple cyclic shear test. However, in either case it may be found that number of stress cycles has little effect on the damping ratios of clay, which is completely different from the case of dry sand.

COMPARISON OF INTER-LABORATORY TEST RESULTS
ON SHEAR MODULUS AND DAMPING RATIO

In order to compare test results under as similar conditions as possible, appropriate measures were taken in connection with special test conditions. Concerning the special condition (a) mentioned before, the value of the shear modulus G is divided by the unconfined compressive strength q_u, i.e., G/q_u were compared directly, for there were some differences among the block samples distributed to each facility. This step was based upon the assumption that shear modulus was proportional to q_u.

In regard to special condition (b), the data of damping ratio of the specim⟨
non-zero initial pore water pressure were used; those of shear modulus were not ⟨
is because the supplementary test results indicated that the initial pore water ⟨
little effect on damping ratio. However, the shear modulus of the clay with non-
tial pore pressure was smaller than that with initial pore pressure under the sam
confining stress (Suzuki et al., 1976).

The results obtained from the inter-laboratory cooperative cyclic shear test
in Figure 11 through Figure 14 with respect to shear modulus and damping ratio ag
amplitude of shear strain.

The Results of Shear Modulus

It is recognized from Figure 11 and Figure 12 that the results of shear modu
roughly classified into two groups. Especially for sand, the test results of sim
ring torsion, torsion shear, resonant-column and supersonic pulse belong to one g
almost coincides with Shibata's equation (Shibata and Soelarno, 1975). Dynamic t
pression test results belong to the other group, which coincides with the curve p
Seed and Idriss (1970) Shibata's equation was drawn in Figure 11 by assuming the
lus obtained from supersonic pulse test as the initial shear modulus, i.e., shear
very small shear strains.

The difference of the test results of these two groups is caused by the inhe
characteristics of apparatus and test procedures, because the influence of the sp
conditions (a) (b) on the test results were already eliminated.

Cornforth pointed out that the angle of internal friction ϕ of sand obtained
static shear tests under plane strain conditions was larger than that obtained fr
compression tests (Cornforth, 1964). While, according to the inter-laboratory co
test results, shear modulus obtained with the apparatus where shear stress is app
surface of the specimen under nearly plane strain condition, is larger than that
with the triaxial compression apparatus. This result shows a similar tendency of
obtained by Cornforth. Consequently, it may be supposed that the difference of t
results is based on the difference of strain state, i.e., whether axi-symmetrical
system or nearly plane strain system of the test.

Additionally, attention should be paid to the fact that, in the triaxial compression test for clay, close agreement is obtained between shear modulus measured under constant cell pressure and that measured under cell pressure which was changed periodically during shear (Tamaoki, 1976, 1977). It can also be noticed, that the measured values by several triaxial compression apparatus almost agree for both sand and clay.

The Results of Damping Ratio

The results of damping ratio are shown in Figure 13 and Figure 14 plotted versus shear strain. It can be seen from Figure 13 that the damping ratio of sand agrees well, irrespective of the test types, with respect to the values in the tenth cycle. For normally consolidated clay, however, the data are distributed over a wide range as shown in Figure 14 and classification cannot be conducted clearly. It is also recognized that the result obtained from the simple cyclic shear tests differs from that obtained by the resonant-column test, and that there is some differences among the results obtained with the several triaxial compression apparatus. However, the data are distributed within the range presented by Seed and Idriss (1970) as shown in Figure 14.

It should be noticed that for clay, there is no significant difference with respect to shear modulus and damping ratio between the results of the fresh specimen and those of the stage tested specimens mentioned by Silver and Park (1975).

A supplementary test was performed for the clay specimen which was subjected to cyclic stresses of larger amplitude than that intended at the time (Suzuki et al., 1976). The result indicates that shear modulus becomes smaller and damping ratio becomes larger than that of the fresh specimen, respectively at the same shear strain level.

CONCLUSIONS

The following conclusions may be made on the basis of the inter-laboratory cooperative cyclic shear tests on dry sand and normally consolidated clay.

1) The results for Toyoura sand, the relative density of which was about 70% under an effective mean principal stress of 1.0 kg/cm^2, are as follows:

a) Concerning the relationship between shear modulus and shear strain in the tenth cycle, the results were divided into two groups. The first group included the results of simple shear, ring torsion, torsion shear, resonant-column and

III-143

supersonic pulse test, which almost coincided with Shibata's equation. The second
group was composed of the results from dynamic triaxial compression type tests,
which coincided with the curve shown by Seed and Idriss.

b) Concerning the relationship between damping ratio and shear strain in the
tenth cycle, the results agreed well with each other regardless of test types and
procedures.

c) The number of stress cycles had significant effects on shear modulus and
damping ratio of dry sand at large strain levels.

d) Dynamic Poisson's ratio can be measured electrically during cyclic shear
testing.

2) The results for artificial, normally consolidated clay, the water content of which
was about 63% under the effective mean principal stress of 0.59 kg/cm^2, are as·follows:

a) Concerning the relationship between shear modulus and shear strain, the
results were similar to those described for sand. Shear moduli obtained from the
tests of simple shear, ·resonant–column and supersonic pulse tests were larger, than
those obtained from the cyclic triaxial compression tests at the same shear strain
level.

b) Concerning the relationship between damping ratio and shear strain, the
results were scattered in a wide range, being different from those for sand. ·‚The
results were distributed, however, within the range presented by Seed and Idriss.

c) The number of stress cycles had little effect on shear modulus and the
damping ratio of clay.

d) No significant difference was recognized between the results of the fresh
specimen and those of the stage tested specimen.

e) In the triaxial·compression test, close agreement was obtained between
shear modulus measured under constant cell pressure and that measured under cell
pressure which was changed periodically during shear.

ACKNOWLEDGMENTS

The authors are grateful to Professor Yoshiaki Yoshimi of Tokyo Institute of Technology
for his guidance to "The Committee on Dynamic Properties of Soils," and to the members of
the Committee for their cooperation.

The work described in this paper was partially supported by the research funds of the Ministry of Construction.

NOTATION

D_R = relative density

E = elastic modulus

e_o = initial void ration

G = equivalent shear modulus

g = acceleration of gravity

h = equivalent damping ratio

K_o = coefficient of earth pressure at rest

N = number of stress cycles

q_u = unconfined compressive strength

V_p = velocity of P-wave

V_S = velocity of S-wave

γ = shear strain

γ_t = wet unit weight

ε = axial strain

ν_d = dynamic poisson's ratio

ϕ' = angle of internal friction in terms of effective stress

σ_d = deviator stress

σ_{mo}' = effective mean principal stress

τ = shear stress

REFERENCES

[1] Cornforth, D. H., "Some Experiments on the Influence of Strain Conditions on the Strength of Sand," Geotechnique Vol. XIV, No. 2, pp. 143-167, 1964.

[2] Finn, W. D. L., Pickering, D. J. and Bransby, P. L., "Sand Liquefaction in Triaxial and Simple Shear Tests," Journal of the Soil Mechanics and Foundations Division, ASCE, Vol. 97, No. SM4, Proc. Paper 8039, pp. 639-659, 1971.

[3] Hanai, T., "Dynamic Behavior of Soil Obtained in Dynamic Triaxial Test, Reports of Committee on Dynamic Properties of Soils," Building Research Institute, Ministry of Construction (in Japanese), 1976, 1977.

[4] Hara, A. and Kiyota, Y., "Dynamic Behavior of Sand at Very Small to Large Shear Strain," Proceeding of the 11th Annual Meeting of Japanese Society of Soil Mechanics and Foundation Engineering, pp. 331-334 (in Japanese), 1976.

[5] Hara, A. and Kiyota, Y., "Dynamic Shear Tests of Soils for Seismic Analyses," Proceedings of the Ninth International Conference on Soil Mechanics and Foundation Engineering, Tokyo, Vol. 2, pp. 247-250, 1977.

[6] Hatanaka, M., "Dynamic Behavior of Sand Obtained in Ring Torsion Shear Test, Reports of Committee on Dynamic Properties of Soils," Building Research Institute, Ministry of Construction (in Japanese), (1975, 1976).

[7] Iwasaki, R., "Seismic Response Analyses of Ground by Using Data Obtained in Inter-laboratory Cooperative Tests, Reports of Committee on Dynamic Properties of Soils," Building Research Institute, Ministry of Construction (in Japanese), 1977.

[8] Iwasaki, T., Tatsuoka, F. and Takagi, Y., "Shear Moduli of Sands Under Cyclic Torsiona Shear Loading," Soils and Foundations, Vol. 18, No. 1, pp. 39-56, 1978.

[9] Koori, Y., "Shear Moduli and Damping Ratios Measured by Hardin Oscillator," Reports of Committee on Dynamic Properties of Soils, Building Research Institute, Minstry of Construction (in Japanese), 1975, 1976, 1977.

[10] Peacock, W. H. and Seed, H. B., "Sand Liquefaction Under Cyclic Loading – Simple Shear Conditions," Journal of the Soil Mechanics and Foundations Division, ASCE, Vol. 94, No. SM3, Proc. Paper 5957, pp. 689-708, 1968.

[11] Roscoe, K. H., "An Apparatus for the Application of Simple Shear to Soil Samples," Proceedings of the Third International Conference on Soil Mechanics and Foundation Engineering, Vol. 1, pp. 186-191, 1953.

[12] Seed, H. B. and Idriss, I. M., "Shear Moduli and Damping Factors for Dynamic Response Analyses," Earthquake Engineering Research Center Report, No. EERC 70-10, University of California – Berkeley, 1970.

[13] Shibata, T. and Soelarno, D. S., "Stress-Strain Characteristics of Sands Under Cyclic Loading," Proceedings of the Japan Society of Civil Engineers, No. 239, pp. 57-65 (in Japanese), 1975.

[14] Silver, M. L. and Park, T. K., "Testing Procedure Effects on Dynamic Soil Behavior," Journal of the Geotechnical Engineering Division, ASCE, Vol. 101, No. GT10, Proc. Paper 11671, pp. 1061-1083, 1975.

[15] Sugimoto, M., Suzuki, Y. and Terada, K., "Study on Dynamic Properties of Soils – Comparison of Laboratory Test Results with In-Situ Test Results," Proceeding of the Tenth Annual Meeting of Japanese Society of Soil Mechanics and Foundation Engineering, pp. 297-300 (in Japanese), 1975.

[16] Suzuki, Y., Sugimoto, M., Babasaki, R. and Kakita, E., "Measurement of Poisson's Ratio by Means of Dynamic Triaxial Apparatus," Proceeding of the Eleventh Annual Meeting of Japanese Society of Soil Mechanics and Foundation Engineering, pp. 415-418 (in Japanese), 1976.

[17] Tamaoki, K., "Dynamic Behavior of Soil Obtained in Dynamic Triaxial Test with Lateral Stress Changed," Reports of Committee on Dynamic Properties of Soils," Building Research Institute, Ministry of Construction (in Japanese), (1976, 1977).

[18] Yasuda, S. and Sakai, U., "Dynamic Behavior of Soil Obtained in Dynamic Triaxial Test," Proceeding of the Eleventh Annual Meeting of Japanese Society of Soil Mechanics and Foundation Engineering, pp. 423-426 (in Japanese), 1976.

[19] Yoshimi, Y. and Oh-oka, H., "A Ring Torsion Apparatus for Simple Shear Tests," Proceedings of the Eighth International Conference on Soil Mechanics and Foundation Engineering, Moscow, pp. 501-506, 1973.

[20] Yoshimi, Y. et al., "Reports of Committee on Dynamic Properties of Soils," Building Research Institute, Ministry of Construction (in Japanese), 1975, 1976, 1977.

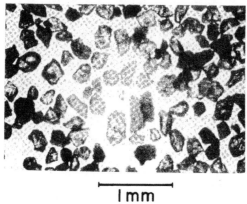

1mm

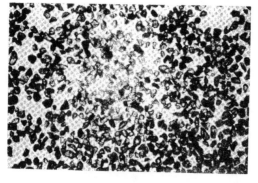

1mm

Photo. 1. Grains of Toyoura sand

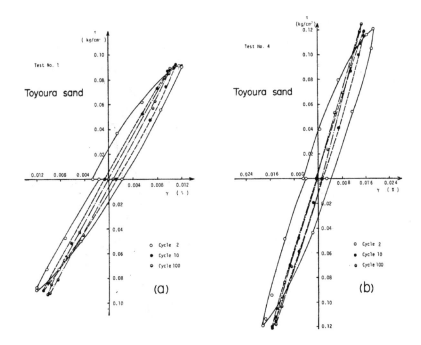

Fig. 1. Stress strain curves obtained from ring torsion shear test
(after Hatanaka)

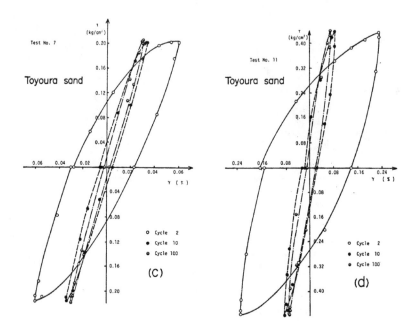

Fig. 1.   Stress strain curves obtained from ring torsion shear test
        (after Hatanaka)

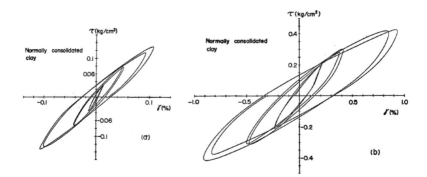

Fig. 2.   Stress strain curves obtained from cyclic simple shear test
(after Hara and Kiyota)

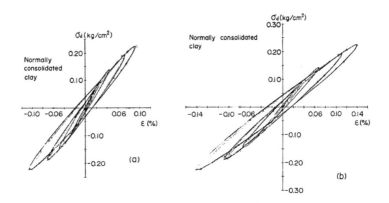

Fig. 3.   Stress strain curves obtained from dynamic triaxial compression
test   (after Suzuki and Sugimoto)

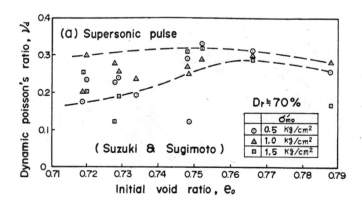

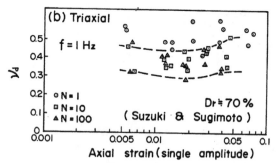

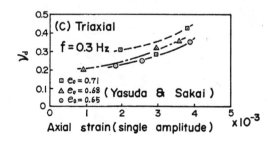

Fig. 4 Dynamic Poisson's ratio of dry sand

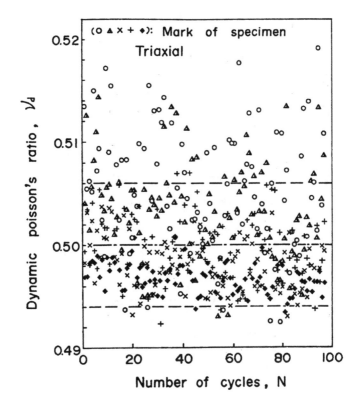

Fig. 5. Dynamic poisson's ratio of saturated clay
(after Suzuki and Sugimoto)

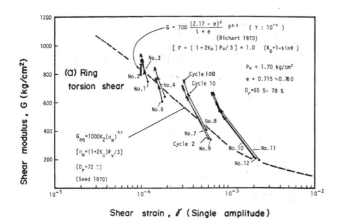

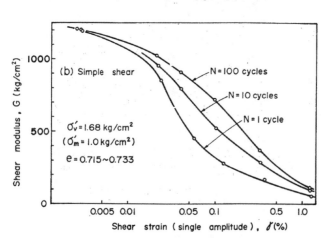

Fig. 6.   Shear modulus versus shear strain relationship
          for dry sand (a)(after Hatanaka), (b)(after Hara
          and Kiyota)

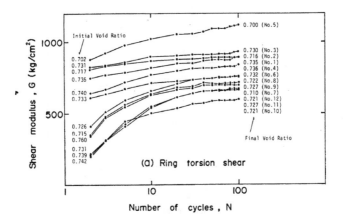

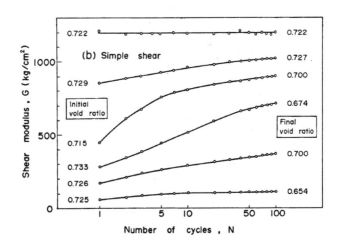

Fig. 7.   Effect of number of stress cycles on shear
modulus for dry sand (a)(after Hatanaka),
(b)(after Hara and Kiyota)

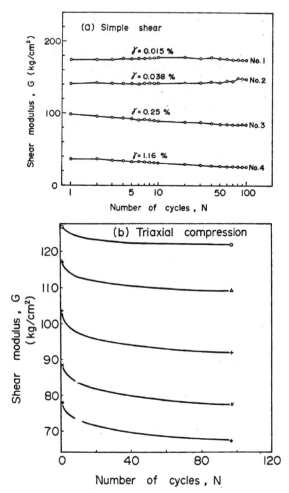

Fig. 8.   Effect of number of stress cycles on shear
modulus for normally consolidated clay
(a)(after Hara and Kiyota), (b)(after Suzuki
and Sugimoto)

ratio

h (%)

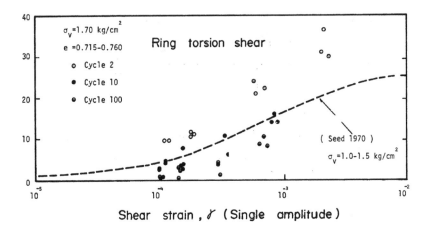

Fig. 9. Damping ratio versus shear strain relationship for dry sand
(after Hatanaka)

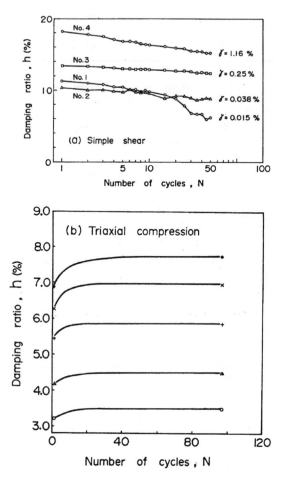

Fig. 10.   Effect of number of stress cycles on damping
ratio for normally consolidated clay
(a)(after Hara and Kiyota), (b)(after Suzuki and
Sugimoto)

Fig. 11.　Shear moduli versus shear strain relationship obtained
from inter-laboratory cooperative test for dry sand

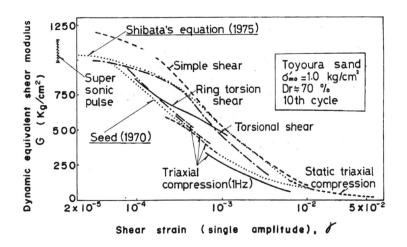

Fig. 12.   Shear moduli versus shear strain relationship obtained
           from inter-laboratory cooperative test for normally
           consolidated clay

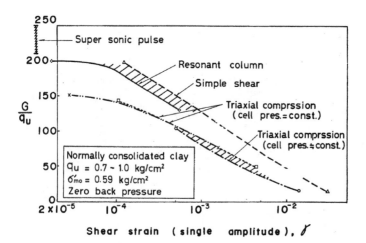

Shear   strain   (single   amplitude), γ

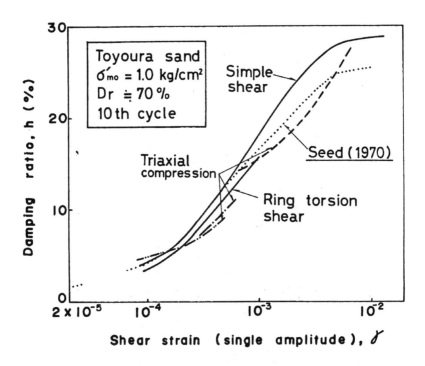

Fig. 13. Damping ratios versus shear strain relationship obtained
from inter-laboratory cooperative test for dry sand

Fig. 14.   Damping ratios versus shear strain relationship obtained
            from inter-laboratory cooperative test for normally
            consolidated clay

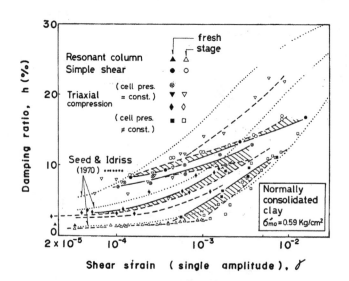

WIND AND STRUCTURE MOTION STUDY FOR PASCO-KENNEWICK BRIDGE

M. C. C. Bampton

Battelle Pacific Northwest Laboratories, Richland, Washington

Harold Bosch, David H. Cheng, and Charles F. Scheffey

Federal Highway Administration

Washington, D.C.

ABSTRACT

A program to study the responses to the natural wind of the 2507 ft (763 m) Pasco-Kennewick Bridge is described using an automatic data collection system. The study consists of three phases: preliminary investigation, data collection and data analysis. The study is sponsored by the Federal Highway Administration and will last two years.

KEYWORDS: Bridges, aerodynamic response, bridges; cable-stayed bridges.

INTRODUCTION

Traditionally wind tunnel tests on scaled section models have been used aerodynamic stability of long-span bridge designs. This procedure is genera provide conservative (safe) results. However, there are still two areas of wind tunnel tests and the interpretation of their results. The first of the differences in turbulence characteristics between the real world and the win effect of these differences on critical wind velocities for the initiation o tural aeroelastic oscillations. The second area of uncertainty is related t energy the structure absorbs from the aerodynamic forces imposed by the wind

To provide prototype information on atmospheric turbulence and the actu. bridge to the turbulence, the Federal Highway Administration is sponsoring a induced motions of the new, cable-stayed bridge between Pasco and Kennewick, specific objective of the study is to determine the responses of an existing span, cable-stayed bridge to the natural wind. This will be accomplished by measurements of the wind structure and bridge motions during a two year peric for this study, which has been awarded to Battelle Pacific Northwest Laborato Richland, Washington, runs through March 1981.

PASCO-KENNEWICK CABLE-STAYED BRIDGE

The Pasco-Kennewick Cable-Stayed Bridge was opened to traffic September currently the longest span, cable-stayed bridge in the United States. The tc the bridge is 2503 ft (763 m) and its center span covers 981 ft (259 m). The bridge is 81 ft (25 m). At the middle of the center span there is a 60 ft (1 clearance between the bridge and the river. The concrete towers from which t hung extend 244 ft (74 m) above the river surface.

Pasco, Kennewick and Richland, collectively referred to as the Tri-Citie on the Columbia River in South-Central Washington State, just north of the Wa border. This part of the State has a semi-arid climate. The prevailing wind ity of the bridge site are either southwesterly, or northeasterly, and have a of 7.9 mph (3.5 m/s). Approximately 4 percent (approximately 342 hrs/yr) of speed exceeds 32 mph (14 m/s). At 50 feet above the ground a gust exceeding 60 mph is expected about once every two years. Higher gusts are expected at lesser frequencies.

In the reach of the river between Pasco and Kennewick, the Columbia flows from the west-northwest to east-southeast. The new, cable-stayed bridge is oriented in the south-north direction. Therefore, the prevailing winds and the vast majority of high winds will approach the bridge at an angle from its upriver side. The upwind fetch in this direction is relatively unobstructed with a surface roughness typical of residential and small commercial buildings. There are no major structures to affect the turbulence.

On the downriver side of the cable-stayed bridge there is an old (completed in 1922) cantilevered-truss bridge which the new bridge is supposed to replace. The two bridges are parallel and the separation between the two bridges is approximately 60 ft (18 m). The effect of the close proximity of these two structures on the wind and the response of the new bridge to it are unknown. There are plans for an eventual demolition of the old bridge, but no date has been set. There are no other structures or topographic features that would cause atypical winds or potentially affect the structural response of the new bridge.

RESEARCH PROGRAM

The research to be conducted by Battelle, under the aegis of the Federal Highway Administration, generally falls into three areas: preliminary investigations, data collection, and data analysis.

Preliminary investigations include: review of previous research on wind induced structural motions with particular emphasis on motions of long-span bridges; study of the reports of wind tunnel tests of a scaled section model of the Pasco-Kennewick Cable-Stayed Bridge; qualitative evaluation of air flow patterns at various locations along the bridge to determine if there is an interaction between the cable-stayed and the old bridge; and evaluation of the adequacy of the instrument system available for use in the measurement or data collection phase of the study. The results of the preliminary investigations will be used to determine specifics of the data collection and analysis phases of the research, such as instrument locations and sampling rates.

The data collection phase of the research includes: physical check-out and calibration of the instrument system provided by the Federal Highway Administration; selection of optimum locations for installation of the instruments; preparation of appropriate mounting brackets and installation of the instruments and collection of wind and bridge motion data. Many of the details of the data collection program are yet to be determined including sampling rates, criteria for initiation of sampling, and the duration of individual sampling periods. These details will initially be determined on the basis of results of the preliminary investigations and will be changed if results of data analysis indicate changes are appropriate.

The primary wind instruments for this study are 3-dimensional propeller anemometers, commonly referred to as Gill anemometers. They provide a nice compromise between the rugged instruments required for high wind speeds on the one hand and the response sensitivity required for turbulence measurements on the other. The three propeller configuration permits full representation of the wind vector and their 3 ft (1 m) distance constant permits resolution of turbulent eddies that are small compared to the major structural components of the bridge.

The primary instruments to be used in measurement of bridge motions are paired, servo-type accelerometers, manufactured by Terra Tech, Inc. The accelerometers will be attached to the deck structures. The accelerometer signals are to be filtered to remove the effects of extraneous high-frequency motions that might be generated by vehicular traffic.

Anemometer and accelerometer signals will be recorded automatically on magnetic tape by a digital recording system using predetermined signal levels to initiate recording. The instrument system will be checked bi-weekly by Battelle staff to determine its operational status.

Data analysis will be conducted in two steps. The initial step will be a screening analysis to determine the general wind and bridge motion characteristics during each measurement period. The second step will be a detailed analysis to determine the specific characteristics of the wind turbulence structure and bridge motions. All data collected during the study will be given a screening analysis; the results of the screening analysis will be used to identify data sets to be examined more closely in detailed analysis.

The screening analysis of the wind data will include: computation of mean wind speeds, directions and vertical angles, and characterization of the variability of the wind by computation of root-mean-square and maximum departures from mean values. Screening analysis of the bridge motion data will include computation of root-mean-square and maximum accelerations.

The detailed analysis will include examination of frequency distributions for the wind velocity components and bridge accelerations, and spectral analysis of the turbulence and motion data. The temporal characteristics of the wind will be determined for each anemometer location. Power spectra will be computed to determine the distribution of wind energy in a frequency domain. Spatial characteristics of the wind will be examined using data from pairs of anemometers. The response of the bridge structure to the wind will be determined through spectral analysis of the accelerometer data. Bending and torsional mode shapes and frequencies will be determined for the bridge. The damping characteristics of the bridge will also be estimated.

The results of the screening and detailed analyses of the wind and bridge motion data will be combined to estimate the critical wind conditions that initiate structural aeroelastic oscillations. These estimates, along with a general discussion of the interpretation of the data, and a description of the procedures by which data collection and analyses are performed, will be presented in the final report at the end of the study. The numercial results of the screening and detailed analyses will be included in the final report as appendices. In addition, the output of the study will include the original data tapes and documented computer programs used in the data analyses.

SUMMARY

In summary, a program to monitor the dynamic responses of the 2507 ft (763 m) Pasco-Kennewick Bridge to the natural wind is described. This research is only a part of an overall field monitoring program nationwide sponsored by the Federal Highway Administration for the purposes of verifying the recent analytical as well as the section model test results, and of increasing the database needed for delineating the characteristics of the natural wind.

CALCULATION OF THE GUST RESPONSE OF LONG-SPAN BRIDGES

Nobuyuki Narita

Hiroshi Sato

Structure Division

Structure and Bridge Department

Public Works Research Institute

Ministry of Construction

ABSTRACT

Discussed in this paper are the random responses of long-span bridges due to the turbulence of natural wind, which are important to the fatigue problem of structural materials and to the serviceability of the bridge for the public. Both the vertical bending and the torsional modes of vibration are considered in performing the calculations, the analytical portion of which is described in a frequency domain and in a time domain as well.

The calculated vibrations are compared with those obtained by field observations conducted at two bridges in order to insure the validity of the method. The calculated results agreed reasonably with the observed ones. It is found that, by setting more rational values for the parameters, gust responses of long-span bridges can be predicted more precisely by the use of the method described.

KEYWORDS: Aerodynamics of bridges; bridge stability; bridge vibration; gust responses; long-span bridges.

INTRODUCTION

It is necessary to insure the aerodynamic stability and the safety of long-span bridges against strong winds right from the early stage of their design. The dynamic phenomena of structures due to wind include self-excited vibrations, vortex-excited vibrations, and buffeting. For the first two phenomena, stability problems can generally be confirmed by means of wind tunnel tests with section models under uniform smooth flow.

Through the tests the critical wind speed for self-excited vibration shall be increased high enough to estimate the structural safety of the bridge, and the vibrational amplitude induced by vortex-shedding shall be decreased small enough to determine the serviceability of the bridge for the public by changing the geometrical configuration of the deck, the hand-rail, and the curb.

The real structures, however, are three dimensional and are exposed to the turbulent natural wind. Then, the relationship between the aerodynamic behaviors of structures observed in the wind tunnel tests and those of real structures under gusty wind should be examined thoroughly. Note a few things have been left unclear till now about the turbulence effects of the aerodynamic behavior of bridges. The followings are qualitatively recognized through the recent boundary layer wind tunnel tests [1, 2, 3,] on the aero-elastic full-models:

1) The critical wind speed of flutter increases in the turbulent flow,

2) The vibrational amplitude of vortex-excitation is suppressed in the turbulent flow, and

3) The intensity of buffeting vibration increases in the turbulent flow.

The effects 1) and 2) show that the dynamic phenomena obtained from the wind tunnel tests in a uniform smooth flow are on the safe side, but the effect 3) requires further investigation of buffeting, whose characteristics cannot be measured correctly in wind tunnel tests under the smooth flow.

Natural wind data and bridge response data at four specific bridges (Kanmon Br., Suchiro Br., Hirado Br., and Suigo Br.) have been collected and processes at the Structure Division of the Public Works Research Institute in order to insure the validity of the wind resistant design applied to the bridge and to establish a more rational wind resistant design method for the buffeting problem. [4] On the other hand some computer programs for calculating the buffeting of long-span bridges have also been developed at the Structure Division,

Works Research Institute. [5] In the first part of this paper the calculation met

vertical bending vibration and the torsional one are described. In the second par

methods are applied to the real long-span bridges in Japan, and the results are co

the observed responses in the last part.

NOTATION

A	Indicial admittance
A_d	Aerodynamic admittance
B_m	Width of bridge model
B_f	Width of full-scale bridge
C_D	Drag coefficient
C_L	Lift coefficient
C_M	Aerodynamic moment coefficient
C_F	$C_L + \alpha C_p$
C	Unsteady aerodynamic coefficient
	ex. $C_{L_\eta}^R$ Lift in phase with vertical bending displacemer
	$C_{M_\phi}^I$ Moment in phase with torsional velocity
D	Drag
f	Frequency
h	Impulse response
I	Polar moment of inertia per unit length of spair
L	Lift
l	Span length of the bridge
M	Aerodynamic moment
m	Mass per unit length of span
p	External force per unit length of span
q	Generalized coordinate
	Order of vibration mode
R	Cross-correlation coefficient of wind
S	Power or cross spectral density
	Time

U	Mean horizontal component of wind speed perpendicular to the bridge axis
	Fluctuating horizontal component of wind speed perpendicular to the bridge axis
w	Fluctuating value of vertical component of wind speed
	Spanwise coordinate of bridge
$1/Z$	Admittance
	Angle of attack
S	Structural damping constant
	Vertical bending displacement
ϕ	Torsional displacement
η_0, ϕ_0	Amplitude of forced vibration during the measurement of unsteady aerodynamic coefficients
μ	Contribution factor of lift to moment
	Air density
ϕ	Vibration mode
ω	Circular frequency
$\omega_{\eta r}$, $\omega_{\phi r}$	Natural circular frequency of r-th order mode
$\wedge$	Expression of Fourier transform
	$\dfrac{d}{dt}$
	Complex conjugate

CALCULATION METHOD

FUNDAMENTAL EQUATIONS

Equations of motion can be expressed as follows when vertical bending vibration and torsional one are considered;

$$\ddot{q}_{\eta r} + 2\zeta_{\eta r}\omega_{\eta r}\dot{q}_{\eta r} + \omega_{\eta r}^2 q_{\eta r} = \frac{\int \phi_{\eta r} p_L dx}{\int m\phi_{\eta r}^2 dx} \tag{1}$$

$$\ddot{q}_{\phi r} + 2\zeta_{\phi r}\omega_{\phi r}\dot{q}_{\eta r} + \omega_{\eta r}^2 q_{\eta r} = \frac{\int \phi_{\phi r} p_L dx}{\int I\phi_{\phi r}^2 dx} \tag{2}$$

IV-9

where $\quad \eta(x, t) = \sum_r q_{\eta r}(t) \phi_{\eta r}(x)$

$\phi(x, t) = \sum_r q_{\phi r}(t) \phi_{\phi r}(x)$

$p_L(x, t) = p_{L\eta}(x, t) + p_{L\phi}(x, t) + p_{LL}(x, t)$

$p_M(x, t) = p_{M\eta}(x, t) + p_{M\phi}(x, t) + p_{MM}(x, t)$

In the above equations, the self-excited forces, $p_{L\eta}$, $p_{L\phi}$, $p_{M\eta}$, and $p_{M\phi}$ are assumed to be

$$p_{ij}(x, t) = \int_{-\infty}^{t} j(x, \tau) \frac{dA_{ij}}{dt}(t - \tau) d\tau$$

where $\quad i = L, M$

$j = \eta, \phi$

$$\frac{dA_{ij}(t)}{dt} = \frac{1}{2\pi} \int_{-\infty}^{\infty} \frac{e^{i\omega t}}{Z_{ij}(\omega)} d\omega$$

$$\frac{1}{Z_{\eta L}(\omega)} = \frac{\frac{1}{2}\rho U^2 (C_{L_\eta}R(\omega) + iC_{L_\eta}I(\omega))}{\eta_o / B_m}$$

$$\frac{1}{Z_{\eta M}(\omega)} = \frac{\frac{1}{2}\rho U^2 B_f (C_{M_\eta}R(\omega) + iC_{M_\eta}I(\omega))}{\eta_o / B_m}$$

$$\frac{1}{Z_{\phi L}(\omega)} = \frac{\frac{1}{2}\rho U^2 B_f (C_{L_\phi}R(\omega) + iC_{L_\phi}I(\omega))}{\phi_o}$$

$$\frac{1}{Z_{\phi M}(\omega)} = \frac{\frac{1}{2}\rho U^2 B_f^2 (C_{M_\phi}R(\omega) + iC_{M_\phi}I(\omega))}{\phi_o}$$

In Eqs. 11 through 14, the unsteady aerodynamic forces [6] can be measured by vibration methods and are assumed to be linear to the amplitude of the motion.

The aerodynamic forces due to the turbulence of natural wind p_{LL} and p_{MM} are a be as follows:

$$P_{LL}(x, t) = \int_{-\infty}^{t} u(x, t) h_U(t - \tau) d\tau + \int_{-\infty}^{t} w(x, t) h_w(t - \tau) d\tau \tag{15}$$

$$P_{MM}(x, t) = \mu B_f P_{LL}(x, t) \quad [7] \tag{16}$$

The impulse responses h_u and h_w can be obtained by inverse transform of aerodynamic admittance Ad. Some of the aerodynamic admittances are given as follows:

1) To relate u to P_{LL}

$$o \; Ad_u(k) = \frac{1}{2}\rho B_f U \frac{dC_F}{d\alpha} \cdot \alpha T(k) \quad [8] \tag{17}$$

where $T(k)$ is Horlock's function [9], and k is $\pi \cdot \dfrac{f \cdot B_f}{U}$

$$o \; |Ad_u(\xi)|^2 = (\rho B_f U C_F)^2 X^2(\xi) \quad [10] \tag{18}$$

where $X^2(\xi) = \left\{ \dfrac{2}{(7\xi)^2}[7\xi-1 + e^{-7\xi}] \right\}^2 \tag{19}$

..... considering the correlation both of chordwise direction and of
vertical direction,

$$X^2(\xi) = \frac{2}{(7\xi)^2}[7\xi-1 + e^{-7\xi}] \tag{20}$$

..... considering the correlation of vertical direction, and $\xi = \dfrac{fB_f}{U}$

In the equations 19 and 20, the decay factor of cross-correlation
coefficient is assumed to be 7.0.

2) To relate w to P_{LL}

$$o \; Ad_w(k) = \frac{1}{2}\rho B_f U \frac{dC_F}{d\alpha} \phi(k) \quad [11] \tag{21}$$

where $\phi(k)$ is Sear's function, and $k = \pi \dfrac{f \cdot B_f}{U}$

The approximations of $\phi(k)$ are as follows:

$$|\phi(k)|^2 = \frac{a + k}{a + (\pi a + 1)k + 2\pi k^2} \, , \, a = 0.1811 \quad [12]$$

$$|\phi(k)|^2 = \frac{1}{1 + 2\pi k}$$

$$o \, |Ad_w(\xi)|^2 = \left(\frac{1}{2}\rho B_f U \frac{dC_F}{d\alpha}\right)^2 X^2(\xi) \quad [10]$$

where $X^2(\xi)$ is the same as equations 19 and 20.

FUNDAMENTAL EQUATIONS IN A MATRIX FORM

When equation 3 - equation 10, equation 15, and equation 16 are equation 1 and equation 2, they can be rewritten as equation 28 and e transform and assuming the following orthogonality.

$$\int_0^1 \phi_{\eta r} \, \phi_{\eta s} dx < \begin{array}{l} = 0(r \neq s) \\ \neq 0(r = s) \end{array}$$

$$\int_0^1 \phi_{\phi r} \, \phi_{\phi s} dx < \begin{array}{l} = 0(r \neq s) \\ = 0(r = s) \end{array}$$

$$\int_0^1 \phi_{\eta r} \, \phi_{\phi s} dx \neq 0$$

$$(-\omega^2 + 2\zeta_{\eta r}\omega_{\eta r}(i\omega) + \omega_{\eta r}^2)\hat{q}_{\eta r}(\omega)$$

$$= \frac{1}{\int_0^1 m\phi_{\eta r}^2 \, dx}\left[\frac{1}{Z_{Ln}(\omega)}\hat{q}_{\eta r}(\omega)\int_0^1 \phi_{\eta r}^2 \, (x)dx + \frac{1}{Z_{L\phi}(\omega)}\hat{f}q_{\phi 1}(\omega)\int_0^1 \right.$$

$$\left. + \int_0^1 \{u\hat{(}x, \, \omega)Ad_u(\omega) + \hat{w}(x, \, \omega)Ad_w(\omega)\}\phi_{\eta r}(x)dx\right]$$

$$(-\omega^2 + 2\zeta_{\phi r}\omega_{\phi r}(i\omega) + \omega_{\phi r}^2\,)\hat{q}_{\phi r}(\omega)$$

$$= \frac{1}{\int_0^1 m\phi_{\phi r}^2\,dx}\left[\frac{1}{Z_{M\eta}(\omega)}\sum_i \hat{q}_{\eta i}(\omega)\int_0^1 \phi_{\eta i}(x)\phi_r(x)dx + \frac{1}{Z_{M\phi}(\omega)}\hat{q}_{\phi r}(\omega)\int_0^1 \phi_{\phi r}^2(x)dx\right.$$

$$\left. + \mu B_f\int_0^1 \{\hat{u}(x,\,\omega)Ad_u(\omega) + \hat{w}(x,\,\omega)Ad_w(\omega)\}\phi_{\phi r}(x)dx\right] \qquad (29)$$

Then, the relation between $\hat{q}_{\eta r}$, $\hat{q}_{\phi r}$ and $\hat{L}_r$, $\hat{M}_r$, are obtained as follows.

$$X_{L_{\eta r}}(\omega)\hat{q}_{\eta r}(\omega) + \sum_i X_{L_{\phi r i}}(\omega)\hat{q}_{\phi i}(\omega) = \hat{L}_r(\omega) \qquad (30)$$

$$X_{M_{\phi r}}(\omega)\hat{q}_{\phi r}(\omega) + \sum_i X_{M_{\eta r i}}(\omega)\hat{q}_{\eta i}(\omega) = \hat{M}_r(\omega) \qquad (31)$$

where

$$X_{L_{\eta r}}(\omega) = (-\omega^2 + 2\zeta_{\eta r}\omega_{\eta r}(i\omega) + \omega_{\eta r}^2\,) - \frac{1}{\int_0^1 m\phi_{\eta r}^2\,dx}\cdot\frac{1}{Z_{L_\eta}(\omega)}\int_0^1 \phi_{\eta r}^2(x)dx \qquad (32)$$

$$X_{L_{\phi r i}}(\omega) = -\frac{1}{\int_0^1 m\phi_{\phi r}^2\,dx}\cdot\frac{1}{Z_{L_\phi}(\omega)}\cdot\int_0^1 \phi_{\phi i}(x)\phi_{\eta r}(x)dx \qquad (33)$$

$$X_{M_{\phi r}}(\omega) = (-\omega^2 + 2\zeta_{\phi r}\omega_{\phi r}(i\omega) + \omega_{\phi r}^2\,) - \frac{1}{\int_0^1 I\phi_{\phi r}^2\,dx}\cdot\frac{1}{Z_{M_\phi}(\omega)}\int_0^1 \phi_{\phi r}^2\,dx \qquad (34)$$

$$X_{M_{\eta r i}}(\omega) = -\frac{1}{\int_0^1 I\phi_{\phi r}^2\,dx}\cdot\frac{1}{Z_{M_\eta}(\omega)}\cdot\int_0^1 \phi_{\eta i}(x)\phi_{\phi r}(x)dx \qquad (35)$$

$$\hat{L}_r(\omega) = \frac{1}{\int_0^1 m\phi_{\eta r}^2\,dx}\cdot\int_0^1 \{\hat{u}(x,\,\omega)Ad_u(\omega) + \hat{w}(x,\,\omega)Ad_w(\omega)\}\phi_{\eta r}(x)dx \qquad (36)$$

$$\hat{M}_r(\omega) = \frac{1}{\int_0^1 I\phi_{\phi r}^2\,dx}\cdot\mu B_f\int_0^1 \{\hat{u}(x,\,\omega)Ad_u(\omega) + \hat{w}(x,\,\omega)Ad_w(\omega)\}\phi_{\phi r}(x)dx \qquad (37)$$

Equations 30 and 31 can be written in a matrix form as follows:

$$
\begin{bmatrix}
X_{L\varphi 1} & & & X_{L\varphi 11} & \cdots\cdots & X_{L\varphi 1n} \\
 & X_{L\varphi 2} & \mathbf{0} & & & \vdots \\
\mathbf{0} & & \ddots & & & \vdots \\
 & & X_{L\varphi k} & X_{L\varphi k1} & \cdots & X_{L\varphi kn} \\
X_{M\varphi 11} & \cdots\cdots & X_{M\varphi 1k} & X_{M\varphi 1} & & \vdots \\
 & & & & X_{M\varphi 2} & \mathbf{0} \\
 & & & \mathbf{0} & & \ddots \\
X_{M\varphi n1} & \cdots\cdots & X_{M\varphi nk} & & & X_{M\varphi n}
\end{bmatrix}
\begin{bmatrix}
\hat{q}_{\varphi 1} \\
\vdots \\
\hat{q}_{\varphi k} \\
\hat{q}_{\varphi 1} \\
\vdots \\
\hat{q}_{\varphi n}
\end{bmatrix}
=
\begin{bmatrix}
\hat{L}_1 \\
\vdots \\
\hat{L}_k \\
\hat{M}_1 \\
\vdots \\
\hat{M}_n
\end{bmatrix}
\qquad (38)
$$

CALCULATION METHOD IN A FREQUENCY DOMAIN

The power spectral density of q_{nr} and $q_{\phi r}$ can be calculated from the inverse matrix in Equation 38, from the complex conjugate of the inverse matrix, and from the power and cross spectral densities of L_r and M_r.

One example of the calculations of cross spectral densities of L_r and M_s is as follow

$$
\begin{aligned}
S_{L_r M_s} &= \lim_{T\to\infty} \frac{\hat{L}_r^* \cdot \hat{M}_s}{T} \\
&= \lim_{T\to\infty} \frac{\mu B_f \int_0^1 \hat{I}(x,\ w)\phi_{\phi r}(x)dx \cdot \int_0^1 \hat{I}(x,\ w)\phi_{\phi r}(x)dx}{T \cdot \int_0^1 m\phi_{nr}^2\ dx \int I\phi_{\phi r}^2\ dx} \\
&= \frac{\mu B_f \int_0^1 \int_0^1 S_L(x,\ y;\ w)\phi_{nr}(x)\phi_{\phi r}(Y)dxdy}{\int_0^1 m\phi_{nr}^2\ dx \cdot \int I\phi_{\phi r}^2\ dx}
\end{aligned}
\qquad (39)
$$

where

$$
\hat{I}(x,\ \omega) = \hat{u}(x,\ \omega)Ad_u(\omega) + \hat{w}(x,\ \omega)Adw(\omega)
\qquad (40)
$$

$$
S_L(x,\ y;\ \omega) = \lim_{T\to\infty} \frac{\hat{I}(x,\ \omega)\ \hat{I}(x,\ \omega)}{T}
\qquad (41)
$$

The cross spectral density $S_L(x, y; \omega)$ can be calculated from the power spectral density of fluctuating lift at reference point $S_{L_0}(\omega)$, using the following assumption [7].

$$S_L(x, y; \omega) = S_{L_0}(\omega)R(x, y; \omega) \tag{42}$$

where

$$S_{L_0}(\omega) = \lim_{T \to \infty} \frac{\hat{L}(x_0^*, \omega)\hat{L}(x_0, \omega)}{T} \tag{43}$$

The power spectral densities of η and ϕ can be calculated from those of $q_{\eta r}$ and $q_{\phi r}$, using Equations 3 and 4. Then the gust responses of the vertical bending mode and the torsional one can be calculated in a frequency domain from the wind data.

CALCULATION METHOD IN A TIME DOMAIN

When the gust responses are calculated without neglecting the effect of coupling different modes to the self-excited forces, it is quite difficult to obtain the impulse responses analytically which relate L_r and M_r to $q_{\eta r}$ and $q_{\phi r}$.

In this paper, the impulse responses are obtained numerically from the inverse transform of the inversion of the matrix in Equation 38.

On the other hand, the aerodynamic forces due to the turbulence of natural wind can be calculated from Equations 15 and 16. $q_{\eta r}$ and $q_{\phi r}$ can be obtained from the convolution of the impulse responses and the aerodynamic force. Then the gust responses of the vertical bending mode and torsional one can be calculated in a time domain from Equations 3 and 4.

SOME EXAMPLES OF CALCULATED GUST RESPONSES

These are some of the assumptions and parameters in the calculation process of the gust responses. Comparisons between the calculated results and observed ones are required in order to insure the validity of the assumptions and to set the proper value for the parameters. This study is just proceeding at Structure Division, Public Works Research Institute. In this paper, some examples of the calculated gust responses are described. The calculation methods in a frequency domain and in a time domain are applied to a cable-stayed girder bridge and to a suspension bridge, respectively.

CALCULATION IN A FREQUENCY DOMAIN

The gust responses of the Suehiro Bridge were calculated in a frequency domain. The Suehiro Bridge, constructed in 1976, is a cable-stayed girder bridge with the span of 110 250 + 110 m.

The gust responses were calculated with regard to the standard deviation at the cent of the main girder by using observed wind data and compared with observed gust responses. Twelve sets of observed data were chosen in which the mean angle of attack ranges from – to 1°, and the mean horizontal wind speed perpendicular to the bridge axis ranges from 7 to 14 m/s and the intensity of turbulence ranges from 0.16 to 0.25. The values of the parameters were determined as in Table 1 by considering aerostatic and aerodynamic charac teristics obtained from wind tunnel testing in a uniform smooth flow. The calculated gus responses are shown in Figures 1 and 2, and they are compared with the observed ones. Fr the results, it is concluded that the gust responses in the vertical bending mode were pr dicted well, but as for torsional mode, the parameters such as the contribution factor of lift to moment should be evaluated more precisely.

CALCULATION IN A TIME DOMAIN

The gust responses of the Kanmon Bridge were calculated in a time domain. The Kanmo Bridge, constructed in 1974, is a suspension bridge with the span of 178 + 712 + 178 m.

The mean horizontal wind speed perpendicular to the bridge axis used in the calculat was about 20 m/s. In order to consider the effect of spanwise correlation, some wind dat series whose correlations to one another are nearly zero were made to act on each segment of the bridge deck. [14] These data series were made from one series by providing a larg lag.

The mean wind speed was calculated for each segment of wind data series, to consider the effect of its increase and decrease. The values of parameters are shown in Table 2. The gust responses were calculated for the five cases in Table 3, and they were compared the observed ones in the same table regarding the standard deviation at the center span. Although more investigations are required concerning the parameters, the calculated resul agree to some extent with the observed ones. The prediction of gust responses in a time domain is expected to become more accurate by evaluating the parameters more precisely.

CONCLUSION

The gust responses of a long-span bridge in the vertical bending mode and in the torsional one can be calculated using the foregoing process, and the more precise evaluation of the described parameters will make the prediction of gust responses more accurate. To further this point of view, the following improvements are required.

1) As for parameters whose more rational values can be obtained in a theoretical way or in an experimental way, use these values.

2) As for the parameters whose more rational values cannot be possibly obtained at present time, get the values to fit the observed responses of the bridge, and use them.

ACKNOWLEDGMENT

The authors are indebted to the Japan Highway Corporation and to the Tokushima Prefecture for obtaining the valuable wind data and the responses at the Kanmon Bridge and the Suehiro Bridge, respectively. The authors also wish thanks to Dr. N. Uehara of the Nikken Consultants, Inc. and to Mr. S. Yuzawa from the Structure Division, Public Works Research Institute, for fine work in programming the calculation and in operating the computer respectively.

REFERENCES

[1] Davenport, A. G., Isymov, N., and Miyata, T., "The Experimental Determination of the Response of Suspension Bridges to Turbulent Wind," Proceedings of Third International Conference on Wind Effects on Buildings and Statistics, 1971.

[2] Melbourne, W. H., "West Gate Bridge Wind Tunnel Tests," August 1973.

[3] Irwin, H. P. A. H., Schuyler, G. D., "Experiments on a Full Aeroelastic Model of Lions' Gate Bridge in Smooth and Turbulent Flow," Laboratory Technical Report, National Technical Report, National Research Council Canada, October 1977.

[4] Okubo, T., Narita, N., Yokoyama, K., "On the Wind Response of the Kanmon Bridge," 7th Joint Meeting, UJNR, Tokyo, May 1975.

[5] Narita, N., Yokoyma, K., "On the Gust Response of Long-span Suspension Bridges," 6th Joint Meeting, UJNR, Washington, D.C., May 1974.

[6] Okubo, T., Narita, N., and Yokoyama, K., "Some Approaches for Improving Wind Stability of Cable-Stayed Girder Bridges, Proceedings of the 4th International Conference on Wind Effects of Buildings and Statistics, 1975.

[7] Konishi, I., Shiraishi, N., and Matsumoto, M., "Studies on Wind Resistant Design of Long-Span Suspension Bridges, Fac. Engineering, University of Kyoto, 1973 (in Japanese).

[8]   Shiraishi, N., and Matsumoto, M., "Studies on Gust Responses of Long-Span Suspension Bridges, Fac. Eng., University of Kyoto, 1976 (in Japanese).

[9]   Horlock, J. H., "Fluctuating Lift Forces on Aerofoils Moving Through Transverse and Chordwise Gusts, Trans. ASME, Journal of Basic Engineering, December 1968.

[10] Daveneport, A. G., "Buffeting of Suspension Bridge by Storm Winds," Proceedings ASCE, Vol. 88, ST3, June 1962.

[11] Konishi, I., Shiraishi, N., and Matsumoto, M., "Fundamental Approach to Aerodynamic Random Response of Structural Sections, The Third Symposium on Wind Effects on Stuctures, 1974 (in Japanese).

[12] Fung, Y. C., "The Theory of Aeroelasticity," J. Wiley, 1955.

[13] Liepmann, H. W., "On the Application of Statistical Concepts to the Buffeting Problem, J. Aeronaut. Sci. 19, 1952.

[14] Konishi, I., Shiraishi, N., and Matsumoto, M., "Studies on Wind Resistant Design of Long-Span Suspension Bridges," Fac. Eng., University of Kyoto, 1975 (in Japanese).

Table 1. Parameters

Aerodynamic admittance; Ad_u; neglected
$\qquad Ad_w$; Eq. (21) with approximations in Eq. (22)

Life curve slope; measured in smooth flow

Decay factor of cross-correlation coefficient; 7

Unsteady aerodynamic coefficient; measured in smooth flow

Structural damping constant; 0.003 (0.02 in logarithmic damping)

Vibration mode; vertical bending; 6 modes torsion; 4 modes

Contribution factor (μ in Eq. (16)); measured in smooth flow
$\qquad (C_M(0)/C_L(0))$

Table 2. Parameters

Aerodynamic admittance; Ad_u; Eq. (17)
$\qquad Ad_w$; Eq. (21) (Both in complex)

Life curve slope; measured in smooth flow

Number of divided segments of the span; 2

Lag of wind data series; 8 sec.

Time length of the divided segments of wind data series; 8 sec.

Unsteady aerodynamic coefficient; measured in smooth flow

Structural damping constant; 0.005 (0.03 in logarithmic damping)

Vibration mode; vertical bending; 5 modes torsion; 2 modes

Contribution factor (μ in Eq. (16)); measured in smooth flow
$\qquad (C_M(0)/C_L(0))$

Table 3.

Start Time of the Data Series	U(m /s)	σ_u(m/s)	σ_w(m/s)	σ_η(cm)	$\sigma_\varphi(\times 10^{-3}\text{rad})$	σ'_η (cm)	$\tilde{\sigma}_\varphi(\times 10^{-3}\text{rad})$
1976, 9, 13 1: 59: 53	18.6	1.2	1.0	1.40	2.48	2.85	1.74
2: 00: 33	19.9	1.4	0.9	2.46	3.56	2.47	3.32
2: 01: 13	20.8	1.1	1.0	2.10	2.93	2.43	2.29
2: 01: 53	19.6	1.4	1.0	3.50	3.19	2.31	2.74
2: 02: 33	22.5	1.7	1.0	3.08	2.53	3.36	3.77

u ; horizontal component of wind speed perpendicular to bridge axis

w ; vertical component of wind speed

U ; mean of u.

σ ; standard deviation (observed)

η ; vertical bending response at the center of the main girder

φ ; torsional response at the center of the main girder

σ' ; standard deviation (calculated)

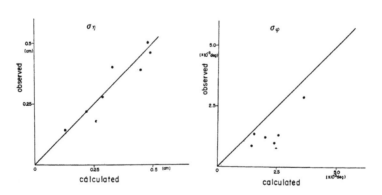

Fig. 1   Fig. 2

DRAFT SEISMIC DESIGN GUIDELINES FOR HIGHWAY BRIDGES

James D. Cooper

Charles F. Scheffey

Federal Highway Administration

Washington, D.C.

Roland L. Sharpe

Ronald L. Mayes

Applied Technology Council

Palo Alto, California

ABSTRACT

Seismic design guidelines for highway bridges in the United States have been under development since 1977. The draft guidelines represent the collective thinking of a distinguished group of academicians, consultants and highway bridge engineers. The guidelines were formulated and based on both the observed performance of bridges during past earthquakes and on recent research conducted in the United States and abroad. They are currently undergoing evaluation by bridge designers and based on their comments are subject to change. The final version of the guidelines will be available in 1981.

KEYWORDS: Bridges; seismic design; design guidelines; bridge.

INTRODUCTION

Significant earthquake engineering research has been initiated by many organizations during the past several years. Structural analyses, response predictions, building codes, and specifications have been refined and updated to provide the engineer with the background and capabilities necessary to design and construct modern buildings to adequately resist the forces developed during periods of strong earthquakes. However, little research had been conducted in the United States prior to the 1971 San Fernando earthquake to insure the satisfactory seismic performance of highway bridges.

Researchers are now focusing their attention on the problem that exist in developing a rational earthquake-resistant design methodology for highway bridges. It is interesting to note that in 1971, 21 countries of the world specified some form of a seismic design coefficient for use in the design of buildings. Of the 21 countries, only four (India, Japan, Turkey, and the United States) had requirements specifically for the design of highway bridges, and it was not until just recently that researchers or practicing engineers in any of the 21 countries took into consideration the dynamic response characteristics of bridge structures in developing their seismic codes and design criteria.

CODE DEVELOPMENT IN UNITED STATES

Until relatively recent times almost all consideration of earthquake forces on structures and relevant code considerations have been concentrated in building construction as opposed to bridge construction.

The 1906 San Francisco earthquake that caused an estimated $50 million of property damage and with subsequent $350 million in fire damage was considered ill fortune, and the city rebuilt in almost identical fashion. It was the 1925 Santa Barbara earthquake with several million dollars in damage that gave impetus to consideration of earthquake design provisions in building codes. In 1927 the simple Newtonian concept that lateral earthquake force is proportional to mass was included in the Uniform Building Code. The Long Beach earthquake of 1933 with at least $50 million in damage emphasized the need for earthquake design; since then changes have continued in the various California building codes.

The first requirment in the United States for the inclusion of seismic loading in the design of highway bridges was in the AASHO, American Association of State Highway Officials

(now American Association of State Highway and Transportation Officials, AASHTO) 1958-59
interim to the 1957 Specifications which were unchanged until the 1975 Interim Specifications. In 1971 the AASHTO Bridge Committee was prepared to present a new earthquake criteria
for adoption. Prior to adoption, the San Fernando earthquake occurred and demonstrated that
the proposed revisions were inadequate and did not seem to apply.

The 12th edition (1977) of the AASHTO "Standard Specifications for Highway Bridges" presented a new approach for designing highway bridges to withstand earthquake forces. Article
1.2.20 of the Specification requires, "In regions where earthquakes may be anticipated,
structures shall be designed to resist earthquake motions by considering the relationship of
the site to active faults, the seismic response of the soils at the site, and the dynamic
response characteristics of the total structure." The specification is generally based on
the 1973 Earthquake Design Criteria for Bridges for the State of California. The California
criteria were developed for California conditions and subsequently modified to allow for
their use in other areas of the United States where possibly damaging seismic motions can
occur. The impact of the application of the specification, particularly in regions outside
California, on design complexity, design and construction costs, and construction complexity
was unknown at the time of adoption.

Consequently, a study entitled "Bridge Seismic Design Criteria" was initiated in 1977
to: (1) evaluate current criteria used for seismic design; (2) review recent seismic
research findings for design potential and use in a new specification; (3) develop new and
improved seismic design criteria; and (4) evaluate the impact of the proposed criteria on
construction and costing.

DRAFT SEISMIC DESIGN GUIDELINES

A working draft of preliminary seismic design guidelines for highway bridges has been
developed by a Project Engineering Panel of Applied Technology Council (ATC) of Palo Alto,
California. Leading experts in seismic design from universities, private consulting firms
and AASHTO were assembled by ATC to serve on the Project Engineering Panel. The Panel has
provided continued guidance into the development of the guidelines. The purpose of the
guidelines is to establish design and construction criteria for bridges subject to earthquake

motions in order to minimize the hazard to life and provide the capability for brid
to survive during and after an earthquake, with essential bridges to remain functic

The design earthquake motions specified in the guidelines are selected so ther
low probability of their being exceeded during the normal lifetime expectancy of th
The probability of the elastic design force levels being exceeded in 50 years is ir
of 80 to 90 percent. However, the design earthquake ground motion by itself does n
mine risk; the risk is also affected by the design rules and analysis procedures us
nection with the design ground motion. Bridges and their components which are desi
resist these motions and which are constructed in accordance with the design detail
tained in the guidelines may suffer damage, but should have low probability of coll
to seismically induced ground shaking.

The principle used for the development of the guidelines are:

1) Small to moderate earthquakes should be resisted within the elastic range
structural components without damage.

2) Realistic seismic ground motion intensities are used in the design procedu

3) Exposure to shaking from large earthquakes should not cause collapse of al
of the bridge. Where possible, damage that does occur should be readily detectable
accessible for inspection and repair.

A basic premise in developing the bridge seismic design guidelines is that the
applicable to all parts of the United States. The seismic hazard exposure varies f
small to rather high across the country. Therefore, for purposes of design, four S
Performance Categories (SPC) are provided to which bridges are assigned. Additiona
bridges are classified as to their relative importance--either as an essential brid
all others. Differing degrees of complexity and sophistication of seismic analysis
design are specified for each SPC. Category D bridges include those designed for t
level of seismic performance with particular attention to methods of analysis, desi
quality assurance. Category C bridges include those where a slightly lower level o
performance is required and therefore the potential for damage is slightly greater
SPC D. Seismic Performance Category B bridges include those where a lesser level o
performance is required and a minimum level of analysis and specific attention to s

design details are provided. Category A bridges include those where no seismic analysis is required, but attention to certain design details for superstructure support is provided.

DRAFT GUIDELINES EVALUATION

The adequacy and impact of the draft seismic design guidelines for highway bridges is to be evaluated during the next year by redesigning selected bridges and then comparing both general design and construction feasibility and costs with that of the originally designed bridges. Based on the evaluations by selected State highway departments and private consultants, the guidelines will be revised and then presented to AASHTO for their consideration and adoption. The guidelines will be finalized in early 1981.

ANALYSIS AND DESIGN OF CRACKED REINFORCED CONCRETE

NUCLEAR CONTAINMENT SHELLS FOR EARTHQUAKES

Peter Gergely

Cornell University, Ithaca, N.Y.

and

Richard N. White

Cornell University, Ithaca, N.Y.

ABSTRACT

Cracking reduces the shear stiffness of reinforced concrete cylindrical containment shells. An extensive experimental program on specimens carrying biaxial tension and cyclic shear produced highly nonlinear shear-slip curves, relatively high sliding shear strength, and low shear stiffness. The effects of this behavior on seismic response and design are discussed.

KEYWORDS: Concrete; containment vessels; cracking; dynamic analysis; hysteresis; nuclear structures; reinforced concrete; seismic effects; shear; stiffness; testing.

INTRODUCTION

One of the critical loading conditions in the design of cylindrical reinforced concrete containment vessels is the combination of internal pressure and membrane shear from seismic forces. Vertical and horizontal cracks form as a result of the pressure and this leads to a significant reduction in shear stiffness of the structure. The seismic shear forces must be transmitted across horizontal cracks by the combination of interface shear transfer, dowel action of the reinforcement, and axial forces in inclined steel (if inclined steel is used). The seismic loading also produces inclined cracks which further reduce shear stiffness.

The cyclic shear transfer mechanism across cracks (also called sliding shear) has been studied at Cornell University with an extensive series of tests on uniaxially or biaxially tensioned reinforced concrete specimens subjected to simultaneous cyclic shear. The accompanying analytical studies have investigated the dynamic behavior of typical containment structures with nonlinear shear-slip characteristics at cracks.

Typical PWR (pressurized water reactor) containment vessels in the U.S. have a cylindrical wall thickness of about 50 in. (about 1.25 m) and usually a hemispherical dome with a thickness of about 30 in. (about 0.75 m). The thickness of the steel liner is usually 1/4 or 3/8 in. (6 or 10 mm) and it is anchored to the concrete on a grid of about 12 in. (0.3 m). The wall is commonly reinforced with two orthogonal layers of #18 (57 mm diameter) bars near each face (Fig. 1). Inclined (diagonal) bars are often added, especially near the base of the wall, to help carry seismic membrane shear forces. One of the principal purposes of the investigation has been to evaluate the seismic response of containment shells without diagonal bars.

FAILURE MODES

The primary objective in the design of containment shells is to preserve the integrity or leak-tightness of the liner. Normally the stresses in the liner are low; for example, for a peak seismic shear stress of 200 psi (1.4 MPa) in a 50 in. (1.27 m) thick concrete shell the stress in a 3/8 in. (9.5 mm) liner is only 1.4 ksi (9.8 MPa), assuming linear elastic behavior. However, when internal pressurization causes horizontal and vertical cracks, the shear stiffness of the concrete decreases and the stress in the liner increases. Two main types of structural failure modes are possible that conceivably could lead to excessive strains in the liner: diagonal tension failure (controlled by the reinforcing), and sliding shear failure (controlled by both concrete and reinforcing).

Sliding Shear Failure

Seismic shear must be transmitted across horizontal cracks that may be open to about 0.030 in. (0.75 mm) as a result of internal pressurization. The transfer mechanism involves interface shear transfer of concrete (also called aggregate interlock or shear friction) and the dowel action of vertical bars. High-level cyclic shear tends to degrade the concrete surfaces and produce relatively large slips. Dowel action of the bars must then carry an increasing share of the total shear force and splitting along the bars may develop. As both of these mechanisms develop, the shear stiffness decreases, which in turn affects the dynamic response of the structure and increases the stresses and strains in the liner. The sliding shear behavior of cracked concrete has been the subject of an extensive research project at Cornell University which is summarized below.

Diagonal Tension Failure

Prior to diagonal cracking the stresses in the orthogonal reinforcing bars are produced by pressurization and by tensile forces generated at the cracks by the interface shear transfer mechanism. Seismic shear may cause diagonal cracks and an equal increase of force in the orthogonal steel, as shown in Fig. 2a, when the cracks are at 45°.

Thus the force in the orthogonal steel is composed of the membrane force due to pressurization and the increase in force caused by seismic shear after the formation of diagonal cracks. It is customary to design the steel on a factored load basis with the steel operating at 0.9 yield stress.

When the factored seismic shear is larger than the capacity of sliding shear transfer, inclined steel must be used for the excess shear which is not carried by the orthogonal steel acting in conjunction with the concrete.

Diagonal cracking changes the load transfer mechanism as can be seen in Fig. 2b. Diagonal compression acts between the diagonal tension cracks and this involves compressive stresses across the pre-existing orthogonal cracks (not shown); this stress intensity is equal to the shear stress for 45° cracks. Thus the width of the orthogonal cracks tends to decrease as the seismic shear stress increases, which enhances the interface shear capacity and reduces the tension in the orthogonal bars.

The equilibrium conditions described above can be used to design the reinforcement. Strains beyond yield; especially in the diagonal bars, do not necessarily represent any impending failure of the shell. A compatibility analysis could result in a prediction of deformations and forces in the concrete, the reinforcement, and in the liner; however, such an analysis hinges on the thorough knowledge of the behavior of all constituent elements and their interaction.

The element shown in Fig. 2a is taken at the maximum tangential shear stress position of the cylindrical shell. The elastic shear stress distribution around the cylinder is sinusoidal but yielding or softening at the peak stress region would clearly increase the stresses elsewhere around the circumference. This force redistribution represents an excess capacity which is currently not considered in design.

Diagonal tension failure obviously begins in a region of high shear. However, it is not clear how diagonal cracks can propagate around the cylindrical shell because the shear forces are theoretically zero at points 90° from the points of maximum shear force. A complete strain and force analysis of the entire vessel may be necessary to evaluate the failure mechanism, which would involve a combination of membrane and radial shear.

TEST PROGRAM

Extensive research at Cornell University on uniaxially loaded specimens with large bars [1] and recent biaxially tensioned specimens [2] with #6 (19 mm) bars have shown that relatively high cyclic shear can be transmitted across cracks but the shear distortions are appreciable.

---

(1) Jimenez, R., Gergely, P., and White, R.N., "Shear Transfer Across Cracks in Reinforced Concrete," Report 78-4, Structural Engineering, Cornell University, August 1978, 357 pp.
(2) White, R.N., Perdikaris, P.C., and Gergely, P., "Strength and Stiffness of Reinforced Concrete Containments Subjected to Seismic Loading: Research Results and Needs," Structural Mechanics in Reactor Technology, Berlin, August 1979.

Fig. 3 summarized the results of one series of tests on 6 in. (15 mm) thick biaxially stressed specimens. The steel ratio ρ refers to the smaller amount of steel; in the other direction twice as much steel was used. The steel stress f_s is the applied initial stress of 0, $0.3f_y$, $0.6f_y$, or $0.9f_y$, where f_y is the yield stress, nominally 60 ksi (414 MPa). Specimens were loaded with either monotonically increasing shear to failure, or with fully reversing shear. Shear was applied after the specimens were cracked by applying biaxial tension to the bars. It is seen that the reduction due to cyclic shear is about 15-20%; usually 10 cycles were applied at 50 psi or 0.34 MPa load increments beginning at 125 psi (0.86 MPa). All specimens failed by diagonal tension indicating that sliding shear failure is probably not a primary failure mode.

However, the shear distortions in cracked concrete are quite high and appreciable slip occurs along cracks and from the effects of diagonal cracking, when the shear is reversed. A set of typical hysteresis curves is shown in Fig. 4 for a specimen stressed biaxially to $0.6f_y$. The effective shear rigidity is only a small fraction (less than 10%) of that for uncracked concrete after a few shearing cycles. This low stiffness affects the dynamic behavior of the containment shell. It should be noted that recent tests on larger specimens 2 ft. *0.61 m) thick and 5 ft. (1.52 m) square at the Portland Cement Association have given similar results.

The splitting of concrete along bars has not yet been fully assessed. In the uniaxial tests [1] splitting occured in some specimens with #11 or #14 bars (36 and 43 mm) but this behavior depends strongly on the orientation of the bars and a slight unintentional component of the tensile bar force paralled to the crack can initiate splitting. In the tests at the Portland Cement Association cracks were observed along nearly every bar but only a few of these penetrated the entire thickness. It is not known whether a splitting plane developed through all bars parallel to the concrete surface.

DYNAMIC ANALYSIS

The hysteretic shear-slip relationships found in the experimental research were used in dynamic analyses of containment structures to evaluate their effects on dynamic response,

in particular on the maximum shear stress and maximum slip at a horizontal crack,
tional inertia was included and idealized hysteretic shear-slip curves were assigne
horizontal crack [3].

A number of idealizations were used in which the main parameters were the numb
horizontal cracks and the shear modulus between cracks. Typical results for an art
ground acceleration scaled to 0.4g maximum ground acceleration show that the peak s
stresses are reduced by about 30% for nonlinear stiffness characteristics as compar
uncracked elastic idealization. The maximum shear stress ranged between about 200
psi (1.38 and 2.07 MPa) for the several cases analyzed so far. The maximum slips a
cracks were only about 0.02 in. (0.51 mm). The displacement time history near the
vessel had more high-level peaks in the nonlinear analysis than in the uncracked lin
ysis. In all analyses the liner was neglected.

The experimental and analytical work is continuing and additional parameters, i
the effect of diagonal bars, are being studied.

ACKNOWLEDGMENTS

This study was supported by the Nuclear Regulatory Commission and the National
Foundation; however, any opinions, findings, conclusions, or recommendations express
are those of the authors and do not necessarily reflect the views of the NRC and NSF
contributions of several current and past graduate research assistants, including R.
J.K. Smith, P. Perdikaris, and C.H. Conley, are acknowledged.

---

(3) Smith, J., Gergely, P., and White, R.N., "The Effects of Cracks on the Seismic
of Reinforced Concrete Nuclear Containment Vessels," Report No. 368, Structural Engi
Cornell University, April, 1977.

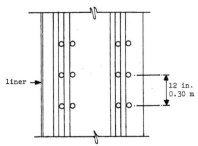

liner →

12 in.
0.30 m

Fig. 1 Typical cross section of wall with #18 (57 mm) bars.

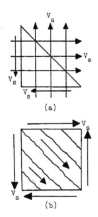

V_s

V_s

V_s

V_s

(a)

V_s

V_s

V_s

V_s

(b)

Fig. 2 Free-body diagrams for diagonally cracked elements: (a) Forces
in orthogonal steel caused by seismic shear, (b) Diagonal
compression forces.

VI-7

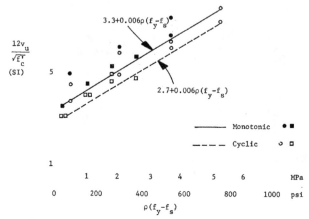

Fig. 3 Diagonal tension strength of biaxially tensioned small scale specimens.

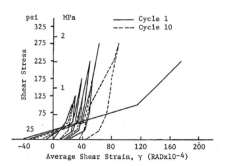

Fig. 4 Typical shear distortion curves for a specimen tensioned biaxially to $0.6f_y$.

VI-8

COMPARISON OF THE MEASURED AND COMPUTED RESPONSES OF THE YUDA

DAM DURING THE JULY 6, 1976 AND JUNE 12, 1978 EARTHQUAKES

Ryuichi Iida

Norihisa Matsumoto

Satoru Kondo

Public Works Research Institute

ABSTRACT

Two accelerograms were recorded at the foundation and dam crest of the Yuda Dam, an arched gravity dam, during the July 8, 1976 and June 12, 1978 earthquakes. The frequency response functions for dam crest to foundation were calculated from these accelerograms. The ground acceleration was used as an input, and the dynamic response was calculated using two-dimensional FEM analysis to compare with the measured response.

The essential conclusions may be enumerated as follows:

1) The displacements at the crest of Yuda Dam were calculated from recorded accelerograms using a digital filtering procedure in which the cutoff frequency of the weighting function was slightly lower than the fundamental frequency of the dam.

2) According to the frequency response functions obtained from two accelerograms, the performance of the dam body may be considered to be linear elastic. The water level of the reservoir influences the frequency response function.

3) The computed response of the dam crest using the foundation accelerogram as an input from the two-dimensional FEM analysis coincides with the measured one. Comparing Westergaard's method and Chopra's method in the evaluation of hydrodynamic pressure, Chopra's method gives the response which contains more low frequency components than Westergaard's method. However, there is not much difference between them. Westergaard's added mass when computed by neglecting the compressibility of water provides a satisfactory approximation of the hydrodynamic pressure.

KEYWORDS: Arched gravity dam; dynamic analysis; earthquake accelerogram; frequency response functions; hydrodynamic pressure.

1. INTRODUCTION

In the design of concrete dams subject to earthquake loads, linear elasticity of concrete can be assumed when the strain of concrete does not exceed certain established values. Because significant parameters such as elastic constants and damping ratio can be determined by laboratory and field tests, the most critical point in analysis procedures of concrete dams lies in the evaluation of reservoir effects. One method of accounting for the reservoir effects is to represent the complete dam/reservoir system as an assemblage of finite elements. This approach requires much computational effort and is not very practical for a dam/reservoir interaction problem. Westergaard's added mass concept [1] is another method used to evaluate the dam/reservoir effects. In this theory, the dam is assumed to be rigid, and the hydrodynamic pressure is represented as an effective mass added to the structure. A third approach is Chopra's substructure concept [2] in which results of the continuum-reservoir solution, including hydrodynamic interaction, are used as input. At present, these latter two methods are considered to provide the most practical design procedures for the dynamic analysis of concrete dams. These two methods can be best verified by comparing analyzed results with actual behavior of a full-scale dam.

The accelerograms of the July 8, 1976 earthquake (simply described as the 1976 earthquake hereafter) and the June 12, 1978 earthquake (simply described as the 1978 earthquake hereafter) were recorded at the foundation and crest of Yuda Dam. The maximum values of acceleration at foundation for the 1976 earthquake and the 1978 earthquake are about 5 cm/s and 25 cm/s^2 respectively, being five times different from each other. The water level of reservoir is El. 219.8 m (76 percent of dam structural height) during the 1976 earthquake and El. 231.9 m (89 percent of dam structural height) during 1978 earthquake. As the effects on dam body due to the difference of input acceleration amplitude and water level are of interest, the recorded accelerograms were analyzed and the results of two-dimensional FEM analysis were compared with the measured response.

2. GENERAL DESCRIPTION OF YUDA DAM

Yuda Dam is an 89.5-m high arched gravity dam which was completed in November 1964 by the Ministry of Construction of the Japanese Government (Figure 1). The location map of Yuda Dam is shown in Figure 2, in which two epicenters of the 1976 and 1978 earthquakes are shown, too.

The general information and layout data for the dam and reservoir are shown in Table 1. The plan and sections through the dam at different locations along the axis are shown on Figures 3 and 4. The multipurpose dam provides for flood control, municipal water, irrigation, and power.

Two strong-motion accelerometers were installed at Yuda Dam. One is located at the center of the dam, at the crest, and the other in the tunnel on the left abutment (see Figure 3). Both accelerometers are at the same elevation and of the electromagnetic type. The sensitivity of the accelerographs is above 90 percent in the frequency range of 0.3 ∼ 30 hertz. They are capable of measuring high-frequency components of vibration with good accuracy.

3. ANALYSIS OF THE 1976 EARTHQUAKE

3.1 The 1976 Earthquake and Its Accelerograms

The earthquake motions were recorded at 8:48 p.m., July 8, 1976. The epicenter of this earthquake was located offshore from Iwate Prefecture at 40°14'N. latitude and 142°26'E. longitude, and the hypocenter at 30 km. The distance from the epicenter to the damsite was 168 km. The magnitude of the earthquake was 5.9 on the Richter scale. The intensity scales in the vicinity of the damsite were Intensity Scale IV (the Japan Meteorological Agency). The reservoir water level was El 219.8 m when the earthquake occurred.

The recorded accelerograms at the foundation and at the crest of the dam are shown on Figure 5.

The maximum values of acceleration in the horizontal upstream/downstream, cross canyon and vertical direction are 4.6, 5.7, and 4.5 cm/s^2 respectively at the foundation, and 58.4, 10.7, and 7.5 cm/s^2 respectively at the crest. At the crest of the dam, the vibration is amplified, especially in the upstream/downstream direction. Assuming the time acceleration histories of the foundation and crest to be f(t) and g(t), respectively, Fourier spectra F(ω) for the foundation and G(ω) for the crest will be given by:

$$G(\omega) = \int_{-\infty}^{\infty} g(t)e^{-\omega t} \, dt \tag{2}$$

and the acceleration frequency response function $Z(\omega)$ for the dam crest to foundation is expressed as follows:

$$Z(\omega) = G(\omega) \, / \, F(\omega) \tag{3}$$

Since Yuda Dam is a three-dimensional structure, the earthquake ground motion in the upstream/downstream direction, for example, affects the crest motion not only for the upstream/downstream direction but also for the cross canyon and vertical ones. For simplicity, however, the frequency response function was calculated for each component, neglecting these couplings of motion. Figure 6 shows the Fourier amplitude spectra of the 1976 earthquake. Figure 7 is the frequency response function for dam crest obtained from Eq. 3. The results shown in Figures 6 and 7 indicate that the first fundamental frequency of the dam is about 5 Hz, two horizontal input accelerograms in the upstream/downstream and cross canyon direction have similar frequency components, most of which are lower than the fundamental frequency of the dam, and the frequency components of the vertical input accelerogram scatters in wider band than those of horizontal accelerograms.

The displacement response for the dam crest relative to the foundation will be calculated later in this report. In order to compare the computed response with the measured one, it is necessary to obtain the displacement from the digitized measured accelerograms. The original accelerograms have baseline error, so that it should be eliminated from original digitized accelerograms.

The baseline correction procedure, which is briefly summarized in the flow chart shown on Figure 8, was applied to the original accelerograms. This procedure is similar to one proposed by M. D. Trifunac [3]. The main feature of this procedure is the digital filtering procedure and its transfer function $H(f)$ is expressed as follows [4]:

$$H(f) = \begin{cases} 1 & f < f_c \\ \dfrac{f_T - f}{f_T - f_c} & f_c \leqq f \leqq f_T \\ 0 & f > f_T \end{cases} \qquad (4)$$

where: f_T = filter roll-off termination frequency

        f_C = cutoff frequency

The weighting function h(t) associated with H(f) is given by:

$$h(t) = \int_{-\infty}^{\infty} e^{i\omega t} H(\omega) \frac{d\omega}{2\pi} \qquad (5)$$

where: ω = $2\pi f$

        ω = circular frequency

        f = natural frequency

For discrete or time sampled data, the weighting function h_n is given by:

$$h_n = \frac{\cos 2\pi_n \lambda_C - \cos 2\pi_n \lambda_T}{2 \lambda_R (\pi_n)^2} \qquad (6)$$

$$n = 0, \pm 1, \pm 2 \ldots \pm N$$

$$\lambda_T = \lambda_C + \lambda_R$$

where: λ_C = f_C / f_S

        λ_R = $(f_T - f_C) / f_S$

        f_S $\cdot$ / Δ_t

        Δ_t = equally spaced time interval

In Eqs. 4 through 6, f_C and f_T should be determined according to the frequency range of the long-period noise introduced into the data by the warping of the record paper and the digitization process. Because the recording speed of the electromagnetic-type accelerographs is 4.0 cm/s, which is faster than that of standard strong-motion accelerographs, the accelerogram of Yuda Dam contains higher frequency errors than that of standard accelerographs, and therefore, a higher cutoff frequency will be required. Since the ground motion mus

long-period displacement, a high-frequency cutoff was not desirable. Also, low-fre

error components are not all filtered out when $f_C = 0.05$ hertz.

The fundamental frequencies of the dam in the three different directions are h:

4.5 Hz, so frequency components of relative accelerograms less than 2.5 Hz will be :

Figure 9 shows the relative displacements for dam crest relative to foundation obta:

the procedure shown in Figure 8 using $f_C = 2.5$ and $f_T = 3.0$. The maximum displacem

crest is approximately 0.05 cm.

3.2 Dynamic Analysis

A two-dimensional finite element analysis was conducted to determine the dynami

response of the Yuda Dam to the July 8, 1976 earthquake using the computer program :

by P. Chakrabarti and A. K. Chopra [5]. Prior to the analysis, two procedures were

this computer program -- one to obtain the acceleration response and the other to ut

Westergaard's added mass approach.

Yuda Dam is a curved concrete gravity dam, as shown on Figure 3. It was design

trial-load method of analysis and should be analyzed as a strictly three-dimensional

ture. However, a two-dimensional analysis would be permissible, because (1) Yuda Da

thick compared to a normal arch dam, and (2) the three-dimensional analysis requires

more computer effort to compute the several different cases required and thus is une

cal.

Material properties used in the analysis are shown in Table 2. The unit weight

elastic modulus of concrete were determined from laboratory tests. The damping rati

cent of the critical) was determined from the field forced-vibration tests of other a

[6, 7].

The hydrodynamic effects of the reservoir on the dam were evaluated using two m

1) Westergaard's added mass concept, and

2) Chopra's substructure concept.

In Westergaard's approach, the following assumptions were made:

* Water is assumed to be linearly compressible and its internal viscosity is

neglected.

* The motion of the dam/reservoir system is considered as two-dimensional.

VI-14

* The upstream face of the dam is vertical.

* The reservoir bottom is horizontal, the water extends to infinity in the upstream direction, and the hydrodynamic pressure decreases to zero at infinity.

* The dam is rigid.

With these assumptions, the small amplitude motion of the reservoir is governed by the two-dimensional wave equation.

$$\frac{\partial^2 \phi}{\partial m^2} + \frac{\partial^2 \phi}{\partial y^2} = \frac{w}{g\,k}\frac{\partial^2 \phi}{\partial t^2} \tag{7}$$

where ϕ = velocity potential

  w = unit weight of water

  g = acceleration of gravity

  k = bulk modulus of elasticity of water

If the fundamental period of the reservoir is less than that of the ground motion, the hydrodynamic pressure for Eq. 7 for above boundary conditions is

$$\sigma = -\frac{8\alpha\,wh}{\pi^2}\cos\frac{2\pi\,t}{T}\sum_{1,3,5}^{n}\frac{1}{n^2 C_n}\sin\frac{n\,\pi\,y}{2\,h} \tag{8}$$

where T = period of earthquake ground motion

  α = acceleration amplitude of ground motion

  h = depth of the reservoir

  y = depth from the water surface

$$C_n = \sqrt{1 - \frac{16\,wh^2}{n^2\,g\,k\,T^2}} \tag{9}$$

In Chopra's approach, the above first four assumptions are the same. However, the dam is considered to be deformable. The complete system is considered as two substructures — the dam, represented as a finite element system, and the reservoir, as a continuum of infinite length in the upstream direction governed by the wave equation.

Eigenvalues and mode shapes were calculated at two sections — one at the maximum section, and the other block 5 section (Figure 10). In computing Westergaard's added mass, the fundamental period of the ground motion was assumed to be 2.0 seconds, because the predominant period of the displacement is about 2.0 seconds.

VI-15

According to the results of the analysis of the recorded accelerograms, the natural frequency of the crest of Yuda Dam in the upstream/downstream direction is 4.9 hertz, whi coincides with the fundamental frequency of block 5 sections. The maximum section has a lower fundamental frequency than the results of recorded accelerograms. Although Yuda Da should be treated as a three-dimensional structure, the block 5 section could be consider to represent the whole structure, thus allowing the analysis of Yuda Dam as a two-dimensi problem. For this study, the response of the dam will be obtained using block 5 section.

Next, the acceleration and displacement response of block 5 section subjected to the recorded ground motion of Yuda Dam is described. High-pass filtered accelerograms were u as the input accelerogram.

The dynamic response of Yuda Dam was calculated for the following seven cases:

Case No.	Foundation Motion	Hydrodynamic Pressure
-	Horizontal	Reservoir empty
2	Horizontal	Westergaard's theory
3	Horizontal	Chopra's theory
4	Horizontal plus vertical	Reservoir empty
5	Horizontal plus vertical	Westergaard's theory
6	Horizontal plus vertical	Chopra's theory ($\alpha* = 0.0$)
/	Horizontal plus vertical	Chopra's theory ($\alpha* = 0.85$)

$*$ α = a refraction coefficient at the bottom of the reservoir, or

$$\alpha = \frac{k-1}{k+1} \quad \text{where} \quad k = \frac{C_r \cdot W_r}{C \cdot W}$$

where W_r = unit weight of the foundation rock

C_r = P-wave velocity of the foundation rock

W = unit weight of water

C = wave velocity of water

The calculated accelerograms in the horizontal and vertical direction at the crest of the dam are shown on Figures 11 and 12, respectively. The calculated displacements are shown on Figures 13 and 14. Included on Figure 11 through 14 are the measured results for the convenience of comparison.

Some of the characteristics drawn from Figures 11 through 14 are:

1) The accelerations and displacements calculated by assuming the reservoir empty were smaller than the measured ones.

2) The effects of vertical ground motions were small. There is not much difference between case No. 6 and No. 7 in which the refraction coefficient at the bottom of the reservoir is different from each other. The fundamental modes of the gravity dam are horizontal bending motions. Although the vertical components higher than 15 Hz do contribute to the response of the dam, the input vertical accelerogram high frequency components.

3) The displacement and acceleration results, using Westergaard's and Chopra's approach, were in good agreement with the measured results.

The two assumptions of Westergaard's theory -- (1) that the period of ground motion is longer than that of the reservoir, and (2) that the dam is rigid, may be permissible for the 1976 earthquake. As to (1), the predominant periods in the stream direction of acceleration and displacement are 0.5 and 2.0 sec., respectively.

As for assumption (2), the relative displacement of the dam below the water level was quite small when compared to the absolute displacement of the foundation.

Comparison of the computed and measured responses in the time domain were described above. Next, the comparison in frequency domain corresponding to Figure 10 will be shown. Figure 15 shows the frequency response functions of FEM analysis.

Figure 15 indicates that whereas the horizontal input acceleration contributes the horizontal response acceleration for wide band frequency in Westergaard's theory, it does not contribute except around 7.8 Hz of resonance frequency of the reservoir in Chopra's theory. The vertical input acceleration contributes the vertical response for dam crest only in the frequency band higher than 15 Hz. The maximum values of response magnification ratio in the horizontal upstream/downstream direction in Figure 15 are from 40 to 50, although the maximum values of measured response are about 30 to 40 in Figure 7. This is because, in Figure 7, the peaks of frequency response function are truncated by the application of Parzen's spectral window. The acceleration response function of the measured accelerogram in the

upstream/downstream have peaks at 5.0, 9.0, 10.5, 11.8 and 15.0 Hz, which is similar to that
of Westergaard's theory than that of Chopra's theory. The frequency response function of
Chopra's theory contains fewer high frequency modes compared to the measured frequency
response function.

4. ANALYSIS OF THE 1978 EARTHQUAKE

4.1 The 1978 Earthquake and Its Accelerograms

This earthquake (= 1978 Miyagi Prefecture Earthquake, Magnitude 7.4) occurred on June
12, 1978, with its epicenter at 38°09'N latitude and 142°10'E longitude, and its hypocenter
40 km deep. The damsite and location of the epicenter are shown in Figure 2. The distance
between damsite and epicenter is 169 km, which is approximately equal to that of the 1976
earthquake. The plot of digitized accelerograms recorded at Yuda Dam during this earthquake
is shown in Figure 16. The ratios of maximum values of acceleration for dam crest to foundation are 7.4 in the upstream/downstream, 1.9 in the cross canyon, and 1.4 in the vertical
direction.

Fourier spectra for recorded accelerograms are shown in Figure 17. Frequency response
functions from Eq. 3 are shown on Figure 18. The frequency characteristics of the foundation accelerograms are similar to those of the 1976 earthquake, with low frequency components from 2 to 3 Hz in the horizontal direction and rather high frequency components in the
vertical direction. The input acceleration levels in the 1976 and 1978 earthquakes are five
times different from each other. Furthermore, the water level of the reservoir is El. 219.8
m (76 percent of the dam structural height) for the 1976 earthquake and El. 231.9 m (89 percent of the dam structural height) for the 1978 earthquake. It is of interest how these two
conditions influence the frequency response function.

There is a good coincidence between the frequency response functions of the upstream/
downstream component on Figures 7 and 18 as for the shape and amplitudes in the band about
5 Hz, although the peak frequency of the 1978 earthquake shifts to the lower frequency. The
peak frequency and amplitude differ in the band from 8 to 25 Hz. The amplitude of the frequency response function of the 1976 earthquake is much greater than that of the 1978 earthquake in the band higher than 15 Hz. Table 3 shows the natural frequencies of modes obtained
from FEM analysis of Section 3.2 (the reservoir effect is evaluated by Westergaard's theory).

The natural frequencies up to the third mode in Table 3 correspond with the peak frequencies of Figures 7 and 18. A peak in Figures 7 and 18 has some peaks, because the dam is an arched gravity dam having the adjacent natural frequencies of symmetrical and asymmetrical modes.

Assuming the reservoir and dam to be two-dimensional, the resonant period T (sec) of the reservoir will be expressed as follows from Eq. 9.

$$T = \sqrt{\frac{16 \ w \ h^2}{g \ k}} \tag{10}$$

Now consider El. 174.0 m of the foundation of No. 5 block used in the analysis as the average elevation of the bottom of the reservoir, then 7.8 Hz and 6.2 Hz will be obtained from Eq. 10 for the 1976 and 1978 earthquake, respectively, so that the frequency which gives the peak amplitude in Figure 18 shifts to the lower part as much as 0.5 to 1.0 Hz compared to Figure 7. The shape and amplitude of the frequency response functions of the two earthquakes coincide in the band below 15 Hz, and therefore it may be concluded that the performance of the Yuda Dam is linear elastic in these two earthquakes.

There is also a good coincidence of the shape and amplitude between the frequency response functions of the cross canyon component. The frequencies which give peak amplitude in the 1978 earthquake are lower than those of the 1976 earthquake, because of the difference of the water level of the reservoir.

The frequency response function of the vertical component of the 1976 earthquake does not have definite peaks, although that of the 1978 earthquake has them at 10.2, 13.7, and 15.2 Hz. This may be related to the accuracy of digitalization of the accelerograms. The recording speed of the accelerographs was 4.0 cm/s, and the sensitivity of acceleration amplitude is the same in all components of the 1976 earthquake.

Accordingly, the amplitude of the original recorded accelerograms except the upstream/ downstream component of dam crest is small, and their accuracy is inadequate in the frequencies higher than 15 Hz. On the contrary, the recording speed of accelerographs of the 1978 earthquake was 10.0 cm/s and the sensitivity of the acceleration amplitude of the foundation components is eight times that of the dam crest components. Therefore the accuracy is the same for all components of the 1978 earthquake.

4.2 Dynamic Analysis

The acceleration and displacement response was computed using the foundation accelero-grams of the 1978 earthquake as an input [8]. The method of analysis, material properties and the analyzed section are the same with those described in Section 3.2. The dynamic response was calculated for the following five cases:

Case No.	Foundation Motion	Hydrodynamic Pressure
1	Horizontal	Westergaard's theory
2	Horizontal	Chopra's theory
3	Horizontal plus vertical	Westergaard's theory
4	Horizontal plus vertical	Chopra's theory ($\alpha^* = 0.0$)
5	Horizontal plus vertical	Chopra's theory ($\alpha = 0.85$)

α^* = refraction coefficient at the bottom of the reservoir.

Since the case of the empty reservoir resulted in a smaller response compared to the measured one in the analysis of the 1976 earthquake, this condition was omitted in the following analysis.

The acceleration and displacement responses for dam crest are shown in Figures 19 through 22. Displacement responses obtained from the measured accelerograms using the same procedure as before are included in Figures 21 to 22 for the convenience of comparison.

The foundation accelerograms presents the S-wave reached at t = 4.5 s. The acceleration amplitude is large from t = 6.0 to 9.0 s. The acceleration amplitude is small after t = 9.0 to t = 16.0 s. The second main shock was reached at t = 16.0 s. Examination of this input acceleration shows the amplitude of the measured response to the upstream/downstream direc-tion to be large in the time from t = 6.0 to 10.0 s and from 17.0 to 19.0 s. Analysis shows that every case presents a large amplitude in the time from 6.0 to 10.0 sec., while the amplitude is large in the time from 12.0 to 14.0 s when the measured response shows the decay of vibration. Comparing case 4 and 5 in which the vertical input was considered as well as the horizontal one, case 5 of the large refraction coefficient gave the larger response.

As for the vertical foundation accelerogram, the S-wave component is not definite, and the acceleration amplitude is flat for both the foundation and the dam crest. Since cases No. 1 and 2 do not consider the vertical input acceleration, the computed vertical acceleration in Figure 20 is the response to the horizontal input, and consists of low frequency components, which is not coincident with the measured response. The vertical acceleration response of case 3 to 5 shows close coincidence with the measured one from the standpoint of frequency characteristics, because vertical foundation acceleration is included as the analysis. However, there is not much difference in vertical displacement responses, because the high frequency components of acceleration do not influence the displacement response much.

As described before, the water level of the reservoir during the 1978 earthquake was El. 231.9 m. Figure 23 shows the frequency response functions which corresponds to this water level. The natural frequencies for the water level of El. 231.9 m shift to the lower frequency as much as 0.5 Hz compared to those for the water level of El. 219.8 m.

5. CONCLUSIONS

The main results of the work presented in this paper can be summarized as follows:

1) The displacements at the crest of Yuda Dam were calculated from recorded accelerograms using a digital filtering procedure in which the cutoff frequency of the weighting function was slightly lower than the fundamental frequency of the dam.

2) According to the frequency response functions obtained from two accelerograms, the performance of the dam body may be considered to be linear elastic. The water level of the reservoir influences the frequency response function.

3) The computed response of the dam crest using the foundation accelerogram as an input from the two-dimensional FEM analysis coincides with the measured one. Comparing Westergaard's method and Chopra's method in the evaluation of hydrodynamic pressure, Chopra's method gives the response which contains more low frequency components than Westergaard's method. However, there is not much difference between them. Westergaard's added mass when computed by neglecting the compressibility of water provides a satisfactory approximation of the hydrodynamic pressure.

REFERENCES

[1] Westergaard, H. M., "Water Pressures on Dams During Earthquakes," Transactions, ASCE, 98, pp. 418-433 (1933).

[2] Chakrabarti, P., Chopra, A. K., "Earthquake Analysis of Gravity Dams Including Hydro-dynamic Interaction," Earthquake Engineering and Structural Dynamics, Vol. 2, pp. 143-160 (1973).

[3] Trifunac, M. D., "Baseline Correction of Strong-Motion Accelerograms," Bulletin of Seismological Society of America, Vol. 61, No. 5, pp. 1201-1211, (1971).

[4] Ormsby, J. F. A., "Design of Numerical Filters with Applications to the Missile Data Processing," Journal of Association for Computing Machinery, Vol. 8, pp. 440-446, (1961).

[5] Chakrabarti, P., Chopra, A. K., "A Computer Program for Earthquake Analysis of Gravity Dams Including Hydrodynamic Interaction," Report No. EE-RC 73-7, EERC, University of California, (1973).

[6] Okamoto, S., "Introduction to Earthquake Engineering," University of Tokyo Press, (1971).

[7] Rouse, G. C. and Boukamp, J. G., "Vibration Studies of Monticello Dam," Research Report No. 9, Bureau of Reclamation (1967).

[8] Matsumoto, N., "A Comparison Between Measured and Computed Response of Yuda Dam During the July 8, 1978 Earthquake," Northern Japan, REC-ERC-78-4, Bureau of Reclamation, (1978).

Table 1 General information and layout data for Yuda Dam and Reservoir

RESERVOIR		DAM	
Reservoir area	6.3 km²	Maximum structural height	89.5 m
Total storage capacity	114 x 10⁶m³	Upstream slope (at crown)	0%
Maximum water elevation	239.0 m	Downstream slope (at crown)	40%
Flood season water surface elevation	222.0 m	Crest width	4.5 m
Normal water surface elevation	236.5 m	Base width	33.0 m
SPILLWAY AND OUTLET WORKS		Crest length	332.0 m
		Axis radius	138.0 m
		Central angle	110°
Orifice	Two gates		
Maximum discharge	600 m³/s	**GEOLOGY OF FOUNDATIONS**	
Overflow spillway	Six gates	Granite, partly Andesite	
Maximum discharge	3240 m³/s		

Fig. 1　Yuda Dam

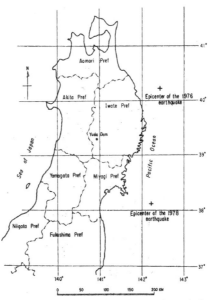

Fig. 2　Map of Northern Japan showing Yuda Dam site and the locations
of the two epicenters of the 1976 and 1978 earthquake

VI-23

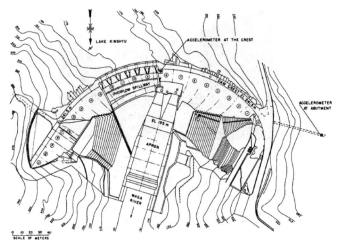

Fig. 3 Plan of the Yuda Dam

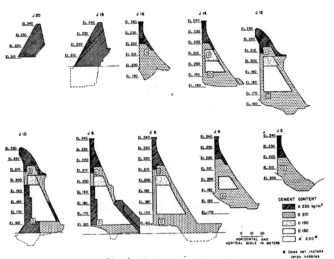

Fig. 4 Sections through Yuda Dam

VI-24

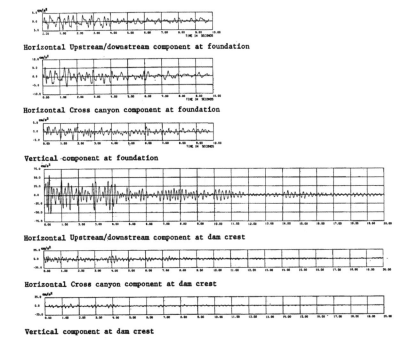

Horizontal Upstream/downstream component at foundation

Horizontal Cross canyon component at foundation

Vertical component at foundation

Horizontal Upstream/downstream component at dam crest

Horizontal Cross canyon component at dam crest

Vertical component at dam crest

Fig. 5 Plot of digitized accelerograms recorded at Yuda Dam during the 1976 earthquake

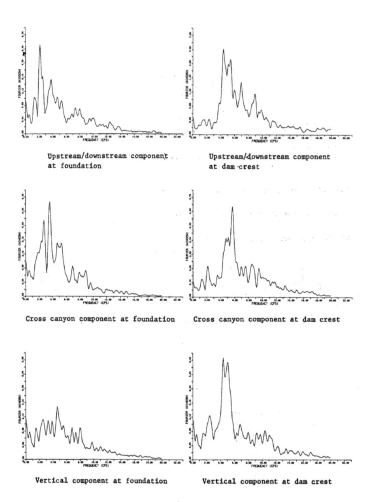

Upstream/downstream component
at foundation

Upstream/downstream component
at dam crest

Cross canyon component at foundation

Cross canyon component at dam crest

Vertical component at foundation

Vertical component at dam crest

Fig. 6 Fourier spectra of recorded accelerograms of the 1976 earthquake

Fig. 7   Frequency response functions for dam crest obtained from recorded accelerograms of
the 1976 earthquake

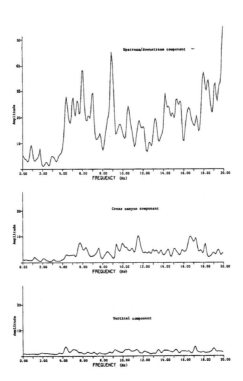

Fig. 8 Flow chart
         displaceme

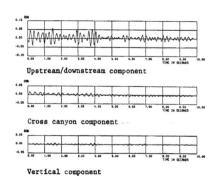

Upstream/downstream component

Cross canyon component

Vertical component

Fig. 9 Relative displacement on the dam crest during
         the 1976 earthquake

VI-28

Table 2 Material properties used in the dynamic analysis

Property	MKSA
Unit weight of water	1.0 t/m³
Unit weight of concrete	2.3 t/m³
Elastic modulus of concrete	2.5 x 10⁶ t/m²
Poisson's ratio of concrete	0.2
Damping ratio	3%

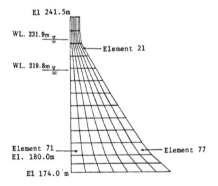

Fig. 10 Finite element idealization for the section through block 5

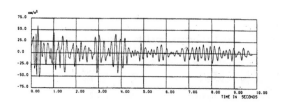

Measured response

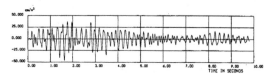

Case No. 1; reservoir empty, horizontal input

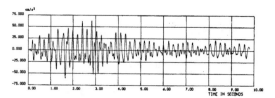

Case No. 2; reservoir by Westergaard's theory, horizontal input

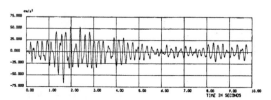

Case No. 3; reservoir by Chopra's theory, horizontal input

Fig. 11 Comparison of measured and analytical acceleration responses in the upstream/downstream direction at the crest of the dam
(the 1976 earthquake)

VI-30

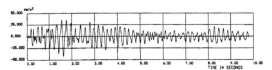

Case No. 4; reservoir empty, horizontal and vertical input

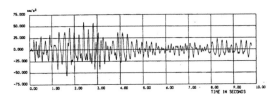

Case No. 5; reservoir by Westergaard's theory, horizontal and vertical input

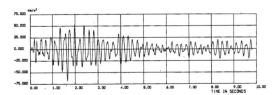

Case No. 6; reservoir by Chopra's theory (α=0.0), horizontal and vertical input

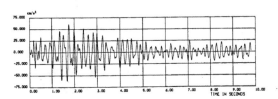

Case No. 7; reservoir by Chopra's theory (α=0.85), horizontal and vertical input

Fig. 11 Comparison of measured and analytical acceleration responses in the
upstream/downstream direction at the crest of the dam
(the 1976 earthquake) (Continued)

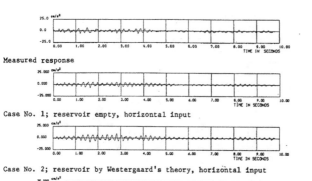

Measured response

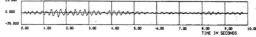

Case No. 1; reservoir empty, horizontal input

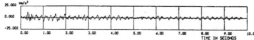

Case No. 2; reservoir by Westergaard's theory, horizontal input

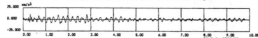

Case No. 3; reservoir by Chopra's theory, horizontal input

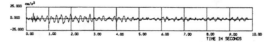

Case No. 4; reservoir empty, horizontal and vertical input

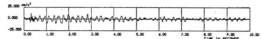

Case No. 5; reservoir by Westergaard's theory, horizontal and vertical input

Case No. 6; reservoir by Chopra's theory (α=0.0), horizontal and vertical input

Case No. 7; reservoir by Chopra's theory (α=0.85), horizontal the and vertical input

Fig. 12 Comparison of measured and analytical acceleration responses in
vertical direction at the crest of the dam (the 1976 earthquake)

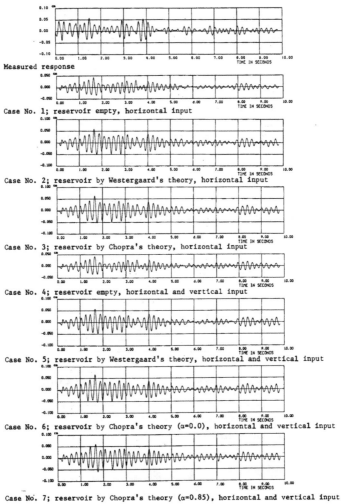

Measured response

Case No. 1; reservoir empty, horizontal input

Case No. 2; reservoir by Westergaard's theory, horizontal input

Case No. 3; reservoir by Chopra's theory, horizontal input

Case No. 4; reservoir empty, horizontal and vertical input

Case No. 5; reservoir by Westergaard's theory, horizontal and vertical input

Case No. 6; reservoir by Chopra's theory (α=0.0), horizontal and vertical input

Case No. 7; reservoir by Chopra's theory (α=0.85), horizontal and vertical input

Fig. 13 Comparison of measured and analytical displacement responses in the upstream/downstream direction at the crest of the dam (the 1976 earthquake)

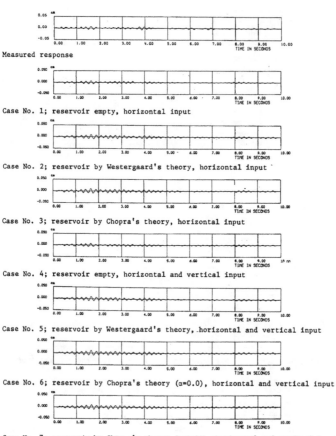

Measured response

Case No. 1; reservoir empty, horizontal input

Case No. 2; reservoir by Westergaard's theory, horizontal input

Case No. 3; reservoir by Chopra's theory, horizontal input

Case No. 4; reservoir empty, horizontal and vertical input

Case No. 5; reservoir by Westergaard's theory, horizontal and vertical input

Case No. 6; reservoir by Chopra's theory (α=0.0), horizontal and vertical input

Case No. 7; reservoir by Chopra's theory (α=0.85), horizontal and vertical input

Fig. 14   Comparison of measured and analytical displacement responses in the vertical direction at the crest of the dam (the 1976 earthquake)

VI-34

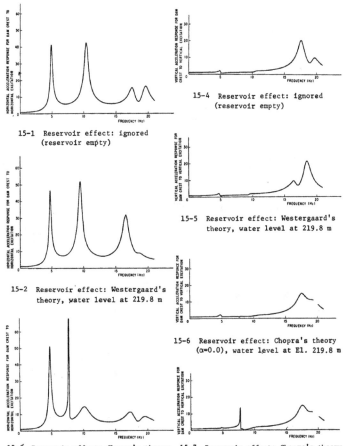

15-1 Reservoir effect: ignored
      (reservoir empty)

15-2 Reservoir effect: Westergaard's
      theory, water level at 219.8 m

15-3 Reservoir effect: Chopra's theory,
      water level at El. 219.8 m

15-4 Reservoir effect: ignored
      (reservoir empty)

15-5 Reservoir effect: Westergaard's
      theory, water level at 219.8 m

15-6 Reservoir effect: Chopra's theory
      (α=0.0), water level at El. 219.8 m

15-7 Reservoir effect: Chopra's theory
      (α=0.85), water level at El. 219.8 m

Fig. 15 Frequency response functions for dam crest acceleration obtained from
         FEM analysis when the water level is El. 219.8 m

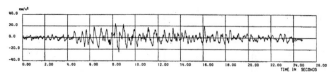

Horizontal Upstream/downstream component at foundation

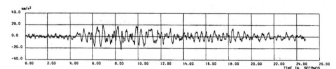

Horizontal Cross canyon component at foundation

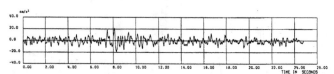

Vertical component at foundation

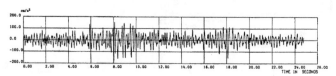

Horizontal Upstream/downstream component at dam crest

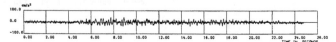

Horizontal Cross canyon component at dam crest

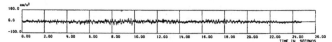

Vertical component at dam crest

Fig. 16 Plot of digitized accelerograms recorded at Yuda Dam during the
1978 earthquake

Table 3 Natural Frequencies
of Block 5 Section

Mode	Block 5 section
	Reservoir, Westergaard's theory
1	4.65
2	9.28
3	15.01
4	17.51
5	24.76

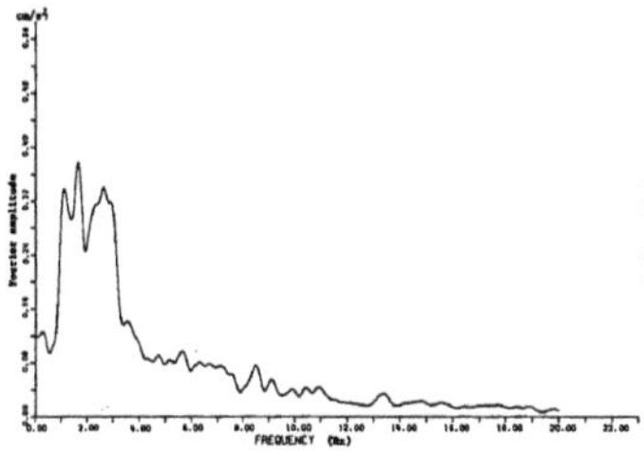

Upstream/downstream component
at foundation

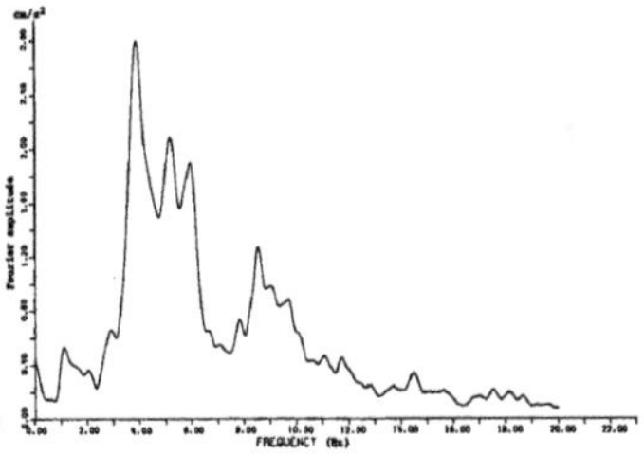

Upstream/downstream component
at foundation

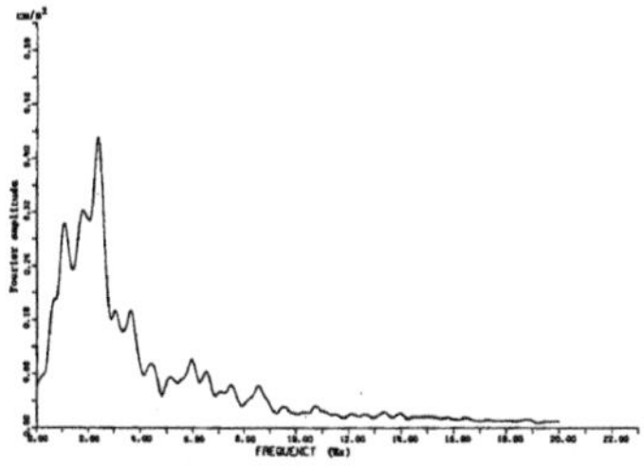

Cross canyon component at foundation

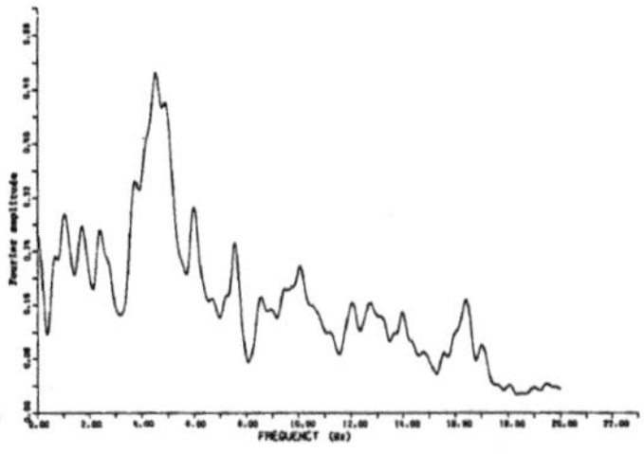

Cross canyon component at foundation

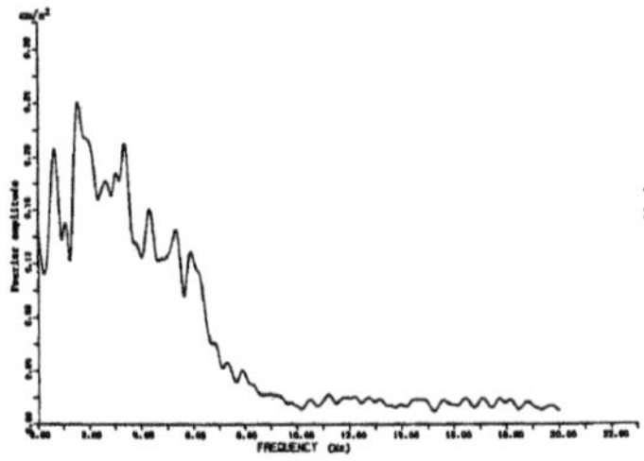

Vertical component at foundation

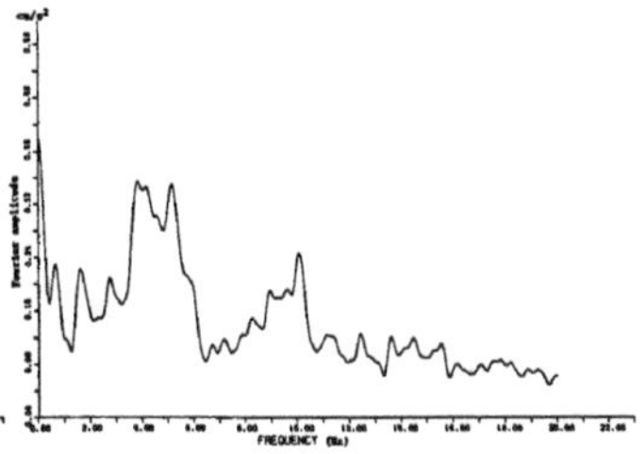

Vertical component at foundation

Fig. 17 Fourier spectra of recorded accelerograms of the 1978 earthquake
VI-37

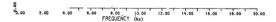

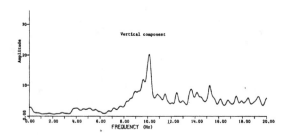

Fig. 18 Frequency response functions for dam crest obtained from recorded
accelerograms of the 1978 earthquake

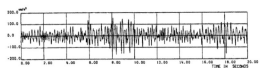

Measured response

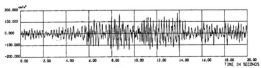

Case No. 1; reservoir by Westergaard's theory, horizontal input

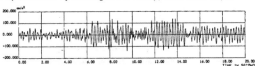

Case No. 2; reservoir by Chopra's theory, horizontal input

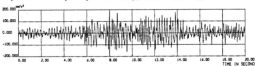

Case No. 3; reservoir by Westergaard's theory, horizontal and vertical input

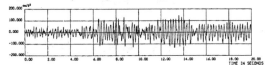

Case No. 4; reservoir by Chopra's theory (α=0.0), horizontal and vertical input

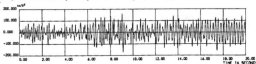

Case No. 5; reservoir by Chopra's theory (α=0.85), horizontal and vertical input

Fig. 19 Comparison of measured and analytical acceleration responses in the
upstream/downstream direction at the crest of the dam
(the 1978 earthquake)

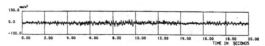

Measured response

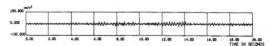

Case No. 1; reservoir by Westergaard's theory, horizontal input

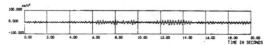

Case No. 2; reservoir by Chopra's theory, horizontal input

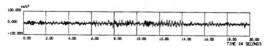

Case No. 3; reservoir by Westergaard's theory, horizontal and vertical input

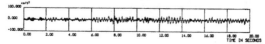

Case No. 4; reservoir by Chopra's theory (α=0.0), horizontal and vertical input

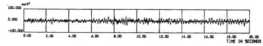

Case No. 5; reservoir by Chopra's theory (α=0.85), horizontal and vertical input

Fig. 20 Comparison of measured and analytical acceleration responses in the vertical direction at the crest of the dam (the 1978 earthquake)

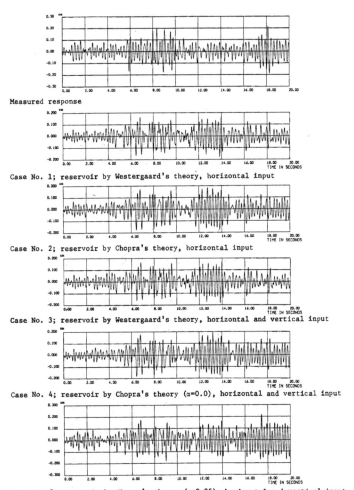

Measured response

Case No. 1; reservoir by Westergaard's theory, horizontal input

Case No. 2; reservoir by Chopra's theory, horizontal input

Case No. 3; reservoir by Westergaard's theory, horizontal and vertical input

Case No. 4; reservoir by Chopra's theory (α=0.0), horizontal and vertical input

Case No. 5; reservoir by Chopra's theory (α=0.85), horizontal and vertical input

Fig. 21 Comparison of measured and analytical displacement responses in
the upstream/downstream direction at the crest of the dam
(the 1978 earthquake)

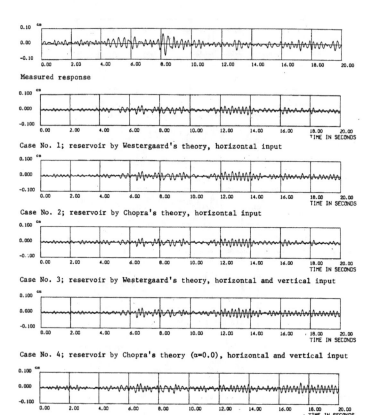

Measured response

Case No. 1; reservoir by Westergaard's theory, horizontal input

Case No. 2; reservoir by Chopra's theory, horizontal input

Case No. 3; reservoir by Westergaard's theory, horizontal and vertical input

Case No. 4; reservoir by Chopra's theory (α=0.0), horizontal and vertical input

Case No. 5; reservoir by Chopra's theory (α=0.85), horizontal and vertical input

Fig. 22 Comparison of measured and analytical displacement responses in the
vertical direction at the crest of the dam (the 1978 earthquake)

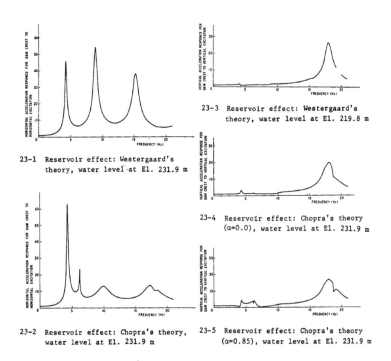

23-1 Reservoir effect: Westergaard's
theory, water level at El. 231.9 m

23-3 Reservoir effect: Westergaard's
theory, water level at El. 219.8 m

23-4 Reservoir effect: Chopra's theory
(α=0.0), water level at El. 231.9 m

23-2 Reservoir effect: Chopra's theory,
water level at El. 231.9 m

23-5 Reservoir effect: Chopra's theory
(α=0.85), water level at El. 231.9 m

Fig. 23 Frequency response functions for dam crest acceleration obtained from FEM
analysis when the water level is El. 231.9 m

REPORT ON

THE 1978 MIYAGI-KEN-OKI EARTHQUAKE

Makoto Watabe

Yutaka Matsushima

Yuji Ishiyama

Tetsuo Kubo

Yuji Ohashi

Building Research Institute

Ministry of Construction

ABSTRACT

The damage features and the causes of damage due to the Miyagi-Ken-Oki Earthquake June 1978 are summarized as follows:

1) Earthquake Ground Motions

a) Damage spread over an area wider than those that experienced a magnitude average of 7.4. b) This fact may be explained by the fact that the source mechanism of this earthquake has been reported as the rebound of subducting plate actions, which is illustrated in Figure 1. c) According to the accelerograms as shown in Figure 3 obtained during this earthquake in Sendai City, approximately 100 km from the reported epicenter, peak acceleration was 0.2 - 0.3 g which caused progressive failure to low rise R.C. structures. Predominant periods for acceleration response spectra are 0.3 to 0.4 second and 1.0 second. Figure 4 shows the world's largest peak structural response (1040 cm/sec^2) in the N-S component of the accelerogram from a recently developed SMAC-M located on the ninth floor of the R.C. Tohoku University Building, which was only slightly damaged. d) Subsoil conditions and damage features are very closely correlated as may be seen in Figure 12.

2) Damage of Structural Members

The main causes of structural damage are as follows: a) Insufficient concrete strength. b) Insufficient reinforcing of hoops against shear force produced in R.C. columns. c) Insufficient ductilities or strengths of connections in R.C. or steel structures. d) Poor quality of construction. e) Lack of horizontal or vertical uniformities, i.e., torsional eccentricities and sudden changes of stiffness weight ratios along the height, as are shown in Figure 13. f) Deteriorations of structural members (especially in the case of wooden houses).

3) Damage of Non-Structural Members

Damage of non-structural members caused various problems in this earthquake: a) Ove turning of concrete block and masonry fences due to this earthquake took the lives of 14 people, or 50% of the total death casualties. b) Damage of non-structural walls due to t earthquake remind us of the importance of earthquake resistant considerations in the tota design of a building (Photo. 1). Insufficient rigidity of main structure and imperfect connection details of these walls to the main structures were two major causes of damage. c) Investigations of the overturning of furniture suggests stochastic evaluation procedur for overturning phenomena in earthquakes.

4) Overall Views

While much earthquake damage was observed and caused by this earthquake, less than o half percent of the total number of houses and buildings in Sendai City suffered severe d age. In view of this fact as well that of the peak structural response in acceleration w was recorded at more than 1G value on the top floor of a building only slightly damaged, standard level of earthquake resistant capacities of the building structures in Japan pro to be satisfactory. The two major tasks for further improvement of earthquake resistant struction: i) overall earthquake resistant consideration of the total design of building and ii) avoidance of poor quality of construction will be the two major tasks.

1. EARTHQUAKE GROUND MOTIONS

The source mechanism of the earthquake developed so far [4] reveals that it is a thrust-
type earthquake due to the subduction of the Pacific Plate (see Figure 1). An unusually
small attenuation of the seismic wave for this earthquake might be associated with this
fact. It is demonstrated in Figure 2 where the peak acceleration values obtained by the
strong motion accelerographs are plotted in terms of epicentral distances. In Figure 2 the
averaged attenuation relation obtained from the past 75 earthquake accelerograms is represented by a dotted line. The solid thick line represents the mean relation between peak
acceleration values (in horizontal component) and the epicentral distances due to this earthquake. Comparison of these two lines reveals that the intensity of the ground motions due to
this earthquake was extraordinary stronger than those estimated by past earthquake data, in
terms of magnitude and epicentral distance of earthquake. In Figure 3 and Figure 4 are shown
accelerograms recorded at the ground floor and the top (ninth floor) of the Structural Engineering Building, Tohoku University. The instruments which produced these accelerograms are
SMAC-Ms with magnetic tape recording systems, the development of which was brought about
through the leadership of our staff in the Building Research Institute. Among these accelerograms, it should be noted that peak acceleration in the structural response of the N-S component at the top floor of this building is the largest value (1,040 gals) ever recorded in
world history. According to these accelerograms, peak acceleration values in ground motions
are 259 gals, 203 gals and 153 gals in N-S, E-W, and U-D component, respectively. Through
integration of these accelerograms in terms of time, the time histories for velocity and
displacement can be obtained, the peak values of which are tabulated in Table 1. In Figure 5
and Figure 6 the resulting time histories of the N-S components at ground level, and top
floor level, are illustrated as examples. Acceleration response spectra of the N-S component
accelerogram at ground level are shown in Figure 7. In Figure 8, tripartite response spectra
of the same accelerogram are shown as well. These response spectra suggest that the predominant periods are 0.3 to 0.4 second and 1.0 second. With 5 percent as the fraction of critical damping, the largest response value (approximately 1,000 gals) is found at the period of
1.0 second, where the response amplification factor becomes 3.86. The fundamental period of
the building where the SMAC-M is installed on the top floor is about 0.7 second in the N-S
direction evaluated by ambient vibration. However, the predominant period of accelerogram

in the N-S component is 1.0 second. Therefore during this earthquake, the fundamental period became larger by 1.43 times than that evaluated by the ambient vibration. Assuming that this building responded linearly or elastically, the peak acceleration of the structural response at the top floor is calculated as 1,350 gals. The difference between this calculated value of 1,350 gals and the 1,040 gals recorded at the top floor may be due to the non-linear behavior of this structure which was subjected to slight damage on the upper stories. It is interesting to note that the elastic displacement response value of 29.7 cm is very close to the actual displacement response value of 25.9 cm, which is evaluated through a double intergration of the relevant accelerogram in terms of time.

Utilizing the N-S component accelerogram at the ground level of the Tohoku University, synthetic accelerograms are generated for three types of subsoil conditions, i.e., bed rock, medium hard soil and soft soil; the latter two of which are representative of local soil conditions in Sendai City. In generating the three synthetic accelerograms, the following procedures and assumptions are used:

i) Deconvolution and convolution of the accelerograms is based on Haskel's transfer matrix method. [1]

ii) Initial values for shear wave velocity v_s (m/sec) are estimated by the following equation [2] in terms of N value and depth H(m) of subsoil;

$$
V_s(m/sec) = 68.79 \cdot N^{0.171} \cdot H^{0.199} \cdot
\begin{bmatrix} 1.0 \\ \text{(alluvial)} \\ \\ 1.303 \\ \text{(delluvial)} \end{bmatrix}
\begin{bmatrix} 1.0 \\ \text{(clay)} \\ 1.086 \\ \text{(fine sand)} \\ 1.135 \\ \text{(coarse sand)} \\ 1.153 \\ \text{(sandy gravel)} \\ 1.448 \\ \text{(gravel)} \end{bmatrix}
\quad (1)
$$

iii) Damping factor Q is assumed to be independent from frequency, and the initial value of Q_o is assumed as 25.0.

iv) Shear rigidity G and damping factor Q are dependent on strain level of subsoil; starting with initial values G_o and Q_o, these values are modified in each step of time increment where the mean shear strain is assumed to be 0.45 times of the maximum shear strain in each subsoil layer.

The results of these synthetic accelerograms are shown in Figure 9. Shown in Figure 10 are the tripartite response spectra of the three accelerograms in Figure 9, with a 5 percent critical damping fraction. Housner's Spectral Intensity values for four accelerograms, the original one at Tohoku University, one on bed rock, one on soft soil, and one on hard soil are 117.0 cm, 96.3 cm, 99.0 cm, and 136.3 cm, respectively, with a 20 percent critical damping.

Incidentally, directions of "principal axes" [3] of the accelerograms at Tohoku University are N5.5°E, E5.5°S and vertical for major, intermediate, and minor axes respectively.

2. DAMAGE STATISTICS AND THEIR DISTRIBUTION

Damage due to this earthquake was concentrated to Miyagi-Prefecture, the capital city of which (Sendai City) suffered 500 million dollar damage, about one-third of the total loss of 1,800 million dollars in the whole Miyagi-Prefecture.

1) Damage Statistics

Twenty-eight were killed and 10,247 injured. Among the deaths, 3 lost their lives by heart attack from the shock of the ground shaking, 5 inside houses, and 16 due to overturned concrete block or masonry fences and gates. About 46.4 percent of the deaths and 90.8 percent of the injuries were concentrated inside Sendai City.

Also in Sendai City, the 42.2 percent total loss of 500 million dollars was to domestic residences. Further breaking down this category, 64.4 percent is for loss of the houses themselves, 19.5 percent for the loss of furniture and 8.7 percent for the loss of fences and gates.

2) Damage Distribution

Damage was clearly concentrated in the area of soft soil conditions and in hillsides. In Figure 11, the distribution of overturned concrete block fences is illustrated. In addition to this distribution, damaged wooden houses and collapsed R.C. buildings are shown in Figure 12. From these figures and Figure 9 of the accelerograms for local soil conditions,

VII-5

the importance of local soil and geographical features for earthquake engineering
seen. The damage distribution map of steel structures shows similar characteristic
local subsoil relationships.

3. DAMAGE FEATURES OF R.C. BUILDINGS

Building Research Institute investigations on R.C. buildings have been made of
institutions: school buildings by the Tokyo University group; a building in Oroshi
where several suffered severe damage; by Tohoku University; throughout Sendai City;
buildings around Oroshimachi; wall R.C. apartment buildings; steel composite tall a
buildings; and precast, pre-stressed concrete buildings.

1) Damaged R.C. Structures

Damage to R.C. structures in the hard soil area, located in the central part o
City, was only slight, while R.C. structures which were subjected to severe damage
centrated in the soft soil area in the eastern and southern parts of the newly deve
in Sendai City. In school buildings, mean damage ratio (collapse: 100%, half colla;
non damage : 0%) is 7 percent (after University of Tokyo group). In the Oroshimach;
where the highest damage ratio was reported, the mean damage ratio for R.C. structu:
6 percent. The mean damage ratio for walled R.C. apartment buildings was zero. Pr;
prestressed concrete structures were subjected to very little damage. Only in the ;
area of the eastern part of the City, tall (9 - 14 stories) steel composite R.C. ap;
buildings suffered shear crack damage in non-structural walls, but no serious damage
observed in main structural members (Photo. 1).

Representative location and fracture patterns can be observed in the columns o:
of ground level floors with shear failure and direction of destructive shear forces
sively in the N-S direction.

2) Structural Features of Damaged R.C. Buildings

The damaged R.C. buildings had one or more items listed as follows:

   a) Open frame structural system in the N-S directions.

   b) Longer span length or wider floor area per column.

   c) Extreme eccentricity (eccentric ratio is more than 0.3) such as the one
sided exterior walls (Figure 13).

d) Discontinuity of stiffness weight ratio such as a soft first-story structural system (Figure 13).

e) Insufficient spacing of column hoops which were constructed before the revision of reinforcing details of column hoops in 1971, revealed by the damage from the Tokachi-Oki Earthquake in 1968.

f) Low strength and rigidity of concrete materials.

3) Examples of Damaged R.C. Buildings

a) "O" Building (Figure 14 and Photo. 2). Two columns, among four, fractured at ground level. This building is a typical "soft first-story" type on soft subsoil layers. Input earthquake energy seemed to concentrate at columns on the first-story. If the redundancy of the structure was larger (number of columns was more than four), this kind of collapse would not have occurred. (Item of structural feature: a, b, d, and e)

b) "M" Building (Figure 15 and Photo. 3). One by three spans 3 story R.C. structure was situated on soft soil with eccentric core at one end. The other end of the first-story is open framed in the N-S direction. Shear fracture occurred in this open frame at the columns, the shear span ratio of which is less than 2.5. (Item of structural feature: c, d, and e)

c) "P" Building (Figure 16 and Photo. 4). Two by three spans 3-story R.C. structure on soft subsoil layers collapsed at the first floor. The thickness of the second floor concrete slab was more than 30 cm (one foot) and spacing of hoops was insufficient. The strength of concrete has been estimated as one-third of the strength required. (Item of structural feature: a, c, e and f)

d) "T" Building (Figure 17 and Photo 5). One by five spans 3-story R.C. structure on soft subsoil layers had a long span and some amount of eccentricity. Eight among 12 columns at the first floor were completely collapsed. (Item of structural feature: a, b, c, e, and f)

4. DAMAGE FEATURES OF STEEL AND OTHER TYPES OF BUILDINGS

1) Steel Buildings

The amount of damage to steel buildings from this earthquake has been the highest i number and in degree of damage to which steel buildings have ever been subjected in Japa history. Damage features of steel buildings which are examined herein are summarized as follows:

  a) Amount of damage, i.e., very severe damage (collapse, or too large a deform to repair).

  b) Cause of damage, i.e., effect of poor construction, especially unskilled welding.

  c) Effect of poor structural design.

  d) Fracture of diagonal bracing systems.

  e) Serious damage to claddings.

As for item a), one steel structure collapsed during the Izu-Oshima Earthquake of January 1978 due to very poor quality of construction. In addition, poor quality of str tural design was noted as a factor leading to the collapse. Damages due to items b) and were also clearly evident. Fractures of diagonal bracings have been observed in various past earthquakes. The problem of item e) became a serious one in this earthquake. Seve claddings, not only of steel sheet mortar finish but also of ALC (Autoclaved Lightweight Concrete), and PC (Precast Concrete) panel walls, fell down. The collapse of a steel-constructed bowling alley with a long span of 34 meters, from vertical ground motion sho be noted here.

2) Damages of Wooden Houses and Other Damage

The destruction inflicted on those wooden houses that were severely damaged was cau by a landslide of sorts. Some other wooden houses suffered damage because of eccentrici or discontinuity of the stiffness weight ratio. It should be noted however that the woo houses with the wall strength required by the existing Building Standard of Japan surviv even in this earthquake.

Toppled concrete block fences which took human lives, without exception, did not sa the reinforcing details required by the Japanese Building Standard.

5. OVERTURNING OF FURNITURE

1) Theoretical Consideration on Overturning of Bodies by Earthquake Motions

The acceleration, a, which will initiate the rocking motion to a prismatic body is given by the following formula:

$$a = \frac{B}{H} g \tag{2}$$

where, B and H are the breadth and height of the body, respectively, and g is the acceleration of gravity.

The acceleration which exceeds the value given by the Eq. 2 does not always overturn a body, because this equation only indicates the necessary condition for overturning. Furthermore, it is a well-known phenomenon that the larger the body is, the more difficult it is to overturn, where the breadth-height ratio of the body remains constant, but this equation depends only on the breadth-height-ratio and not on the dimension. In order to explain this phenomenon, let us suppose that the body has the velocity, v, during earthquake motions and suddenly the rectilinear motion changes to rocking motion. From the conservation of the momentum around 0, we can have: (see Figure 18)

$$M v = I_o \omega \tag{3}$$

where, M is the mass of the body, I_o is the moment of inertia around 0, ω is the angular velocity of the body. In case of the prismatic body:

$$I_o = \frac{4}{3} M r^2 \tag{4}$$

where, r is the length from the center of gravity to 0. From Eqs. 3 and 4:

$$\omega = \frac{4 v}{3 r} \tag{5}$$

Then the kinematic energy, KE, at the beginning of rock motion.

$$KE = \frac{1}{2} I_o \omega^2$$
$$= \frac{3}{8} M v^2 \tag{6}$$

VII-9

The potential energy, PE, when the center of gravity reached to the perpendicular line of 0:

$$PE = \frac{M g r \alpha^2}{2} \tag{7}$$

From Eqs. 6, 7 and the relation $r \alpha = b$:

$$v = 2 \sqrt{\frac{g b \alpha}{3}} \tag{8}$$

If we use the units of centimeters and seconds for length and time, respectively, Eq. 8 will be:

$$v = 25.6 \frac{B}{\sqrt{H}} \tag{9}$$

From Eq. 9, it is possible to estimate the response velocity of the body overturned during earthquake motions. Eq. 9 also shows that the velocity required to overturn the body is proportional to the square root of the dimension, where the breadth-height-ratio remains constant.

2) Estimation of the Velocity from the Field Survey on the Overturning of Furniture

After the earthquake, we surveyed overturned furniture, and interviewed the inhabitants concerning the following items.

a) Type of furniture overturned

b) Dimension of the furniture

c) Direction of the overturning

d) Floor number where the furniture was placed

e) Outline of the building in which the furniture was placed

Most of the buildings surveyed were common wooden residential houses built with conventional Japanese post and beam construction. The total number of buildings surveyed was 138; 130 were two-story wooden houses, 2 were one-story wooden houses, 5 were medium- and low-rise reinforced concrete buildings and one was a high-rise steel frame building.

The result of the survey is shown in Figure 19, where (a)⊕, (b)◎, (c) ●, (d) o and (e) + show the buildings where the response velocity of furniture on the first floor was (a) more than 80 kine, (b) from 80 to 60 kine, (c) from 60 to 40 kine, (d) from 40 to 20 kine and (e) less than 20 kine, respectively. The velocity was determined by Eq. 9 and the magnification factor of the second floor velocity to the first floor was assumed to be 2.0. From Figure 19, the following can be seen:

i) In the western part of Sendai City, where several reinforced concrete buildings collapsed, much of the furniture overturned and the first floor velocity was more than 80 kine, whereas in the eastern part of Sendai City, almost no furniture overturned and the velocity was less than 20 kine.

ii) Dotted part of Figure 19 shows the areas where ground surfaces are covered by alluvium, and velocities in this area were very high compared to the other areas.

iii) In the area 120 km away from the epicenter, almost no furniture overturned. But along the Abukuma River, the overturning was observed up to Shiroishi City and Fukushima City which is more than 120 km from the epicenter. This is due to the alluvium deposits along the river.

iv) Differences of the overturning phenomena among the types of buildings are not totally explainable, because of the scarcity of data concerning types of buildings other than the wooden ones.

v) In the western part of Sendai City, where no furniture overturned in low-rise buildings, much of the furniture nevertheless overturned in high-rise buildings, especially on upper floors. We should note carefully the fact that while high-rise buildings will have less base shear than low-rise buildings, the response velocity will be considerably magnified on the upper floors of high-rise buildings, and overturning furniture may cause special harm to human life.

6. CONCLUDING REMARKS

As outlined in the abstract, the standard level of earthquake resistant capacities in the districts investigated were shown to be fairly satisfactory in this earthquake. However, during this earthquake the importance of total systems design for earthquake resistant qualities has been emphasized here. The importance of construction qualities is also emphasized, therefore some methods to avoid poor quality of construction should be established. These will be the two major tasks for future earthquake engineering in Japan.

REFERENCES

[1] Ohsaki, Y., "Dynamic Characteristics and One-Dimensional Amplification Theory of Soi
     Deposits," Research Report 75-01, University of Tokyo, August 1975.

[2] Ohta, Y., "Establishing Correlations Between the Indices of Soil Properties and Shea
     Wave Velocity," Shizen Saigai Shiryo Kaiseki 4, 1977.

[3] Watabe, M., Penzien, J., "Simulation of Three Dimensional Earthquake Ground Motions,
     International Journal of Earthquake Engineering," 1975.

[4] Seno, T., Shimazaki, K., Somerville, P., Sudo, K., Eguchi, T., "Rupture Process of
     Miyagiken Oki Earthquake of June 12, 1979," Physics of the Earth and Planetary Inter
     (in press).

Tab. 1 Maximum Values of Vibrations At TOHOKU UNIV.

		acceleration		velocity		displ.	
		max.(cm/sec²)	time	max.(cm/sec)	time	max.(cm)	time
G.L.	NS	259.23	7.56	36.17	10.8	-14.53	10.56
	EW	202.57	3.10	27.57	2.96	9.11	3.82
	UD	153.04	4.18	11.92	10.44	3.18	10.02
9FL	NS	1,039.98	15.20	150.41	14.96	-25.87	15.20
	EW	-523.92	14.08	-72.76	14.34	16.61	3.40
	UD	-355.56	7.32	22.92	15.48	7.09	8.18

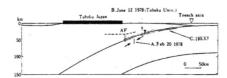

Fig. 1 Source Mechanism of This Earthquake
Hypocenter is located on the boundary of subducting plate

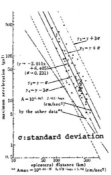

Fig. 2 Relation Between
Max. Accel. and Epicen
Distance

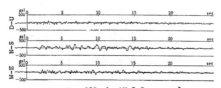

Max.259gals (N-S Component)

Fig. 3 Accelerograms at Ground Level
Installed at Tohoku University

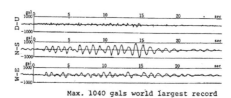

Max. 1040 gals world largest record

Fig. 4 Accelerograms at 9th floor
Installed at Tohoku University

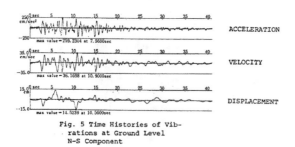

ACCELERATION

VELOCITY

DISPLACEMENT

Fig. 5 Time Histories of Vibrations at Ground Level
N-S Component

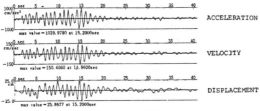

ACCELERATION

VELOCITY

DISPLACEMENT

Fig. 6 Time Histories of Vibrations at 9th Floor

N-S COMPONENT

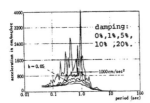

Fig. 7 Response Spectra of
Accelerogram N-S Component
at Ground Level Univ.Tohoku

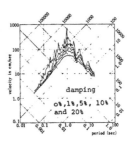

Fig. 8 Tripartite Response
Spectra for Fig.-7

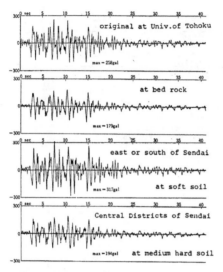

Fig. 9 Synthetic Accelerograms for Different
Local Soil Conditions (N-S Component)

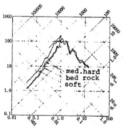

Fig. 10 Response Spectra
for Fiq.-9 in Tripartite.
5% damping

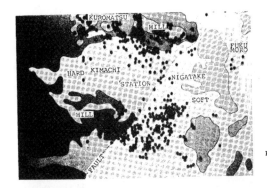

Fig. 11 Damage Distribution
of overturned concrete
block fences, in Sendai

Fig. 12 Damage Distribution
of wooden house and collapsed
R.C. buildings plus Fig.-11.

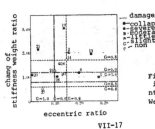

Fig. 13 Damage Feature
in Relation with Ecce
ntricity and Stiffness
Weight Ratio

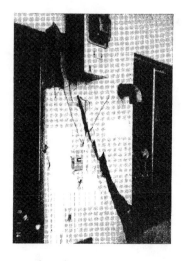

W-SIDE

PLAN

E-SIDE

Fig. 14 Plan &Elev.of
"O" Building.

Photo. 1
Damage of Non-structural
Walls in R.C.&Steel Compo
site Apartment Build.

Phto. 2 Collapse of "O"Building.
due to soft first story

VII-18

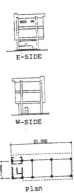

E-SIDE

W-SIDE

Plan

21.000

7.000

Photo. 3 "M" Building;collapsed due to torsion

Fig. 15.
Plan&Elev.
of "M"Build.

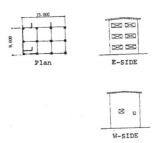

15.000

9.000

Plan

E-SIDE

W-SIDE

Fig. 16 Plan&Elev.of
"P" Building.

Photo. 4 "P"Building:

collapsed due to torsion &poor construction

VII-19

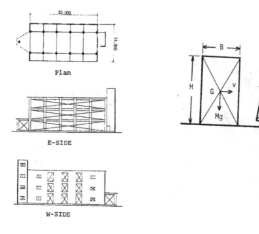

Plan

E-SIDE

W-SIDE

Fig. 17   Plan and elevation
           of the "T" building

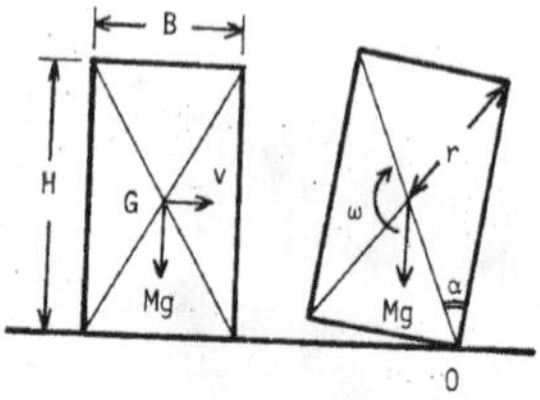

Fig. 18   Rocking of a body

Photo. 5   The "T" building

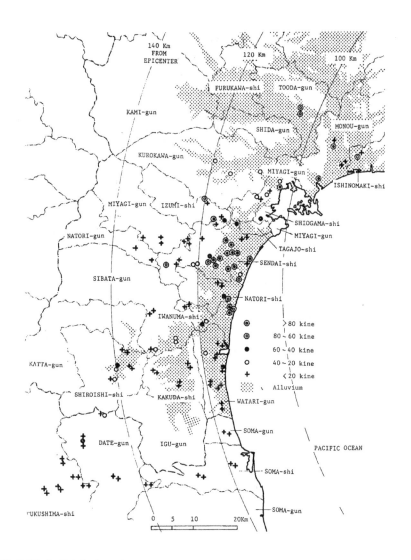

Fig. 19 Velocity Distribution Estimated by Overturning of Furniture
in THE 1978 MIYAGI-KEN-OKI EARTHQUAKE

DISASTROUS GROUND FAILURES IN A RESIDENTIAL AREA OVER A

LARGE-SCALE CUT-AND-FILL IN THE SENDAI REGION

CAUSED BY THE EARTHQUAKE OF 1978

Hitoshi Haruyama

Motoo Kobayashi

Geographical Survey Institute

Ministry of Construction

ABSTRACT

This paper examines the relationship between the disastrous ground failures in a hilly residential area and the land properties, due to the Miyagi-Ken-Oki earthquake of 1978.

We measured the change of landforms by comparisons of two maps based on aerial photographs taken before and after major area earthworks. As the first step, we prepared a landform map at a scale of 1/2,500 with a 2 meter contour interval, from a time just before the earthworks for residential development (1957) and after the earthworks (1978).

Then we constructed a thickness isopleth map of the artificial fill with a 2 meter interval and many profiles across the earthworks.

Then we examined the interrelationships between the distribution of ground cracks, the damage to destroyed houses, the retaining walls, the maps described above, and the original landforms.

KEYWORDS: Artificial fill; disaster prevention; earthquake disaster; ground failure; slope steepness.

1. INTRODUCTION

The damage caused by the earthquake of 1978 in the Sendai region presented new problems of disaster prevention in an urbanized region. One of these problems is disastrous ground failure in hilly residential areas.

This paper is based on a study which is linked to co-operative research projects on the Miyagi-Ken-Oki earthquake which are encouraged by the Science and Technology Agency.

We selected two severely damaged areas for study, namely, the Midorigaoka, and the Kuromatsu districts in the Sendai region.

2. MEASUREMENT OF ARTIFICIAL FILL BY AERIAL PHOTOGRAMMETRY

2.1 Landform Maps for the Time Before and After the Earthworks

The earthworks for residential sites in Midorigaoka and Kuromatsu districts were carried out from 1957 to 1966.

Landform maps at a scale of 1/3,000 had been made before the time of the earthworks, but with an insufficient accuracy to the contours. Consequently, for this study we constructed a new landform map at a scale of 1/2,500 using aerial photographs that were taken in 1957 by the Sendai Municipal Office. For more recent conditions, we used a national large-scale map of 1/2,500 scale which was prepared in 1977-78 by the Geographical Survey Institute.

In this mapping, in order to enhance the accuracy of the 2 meter contours, the number of points plotted was as much as twice the number used in ordinary mapping. The mapping area covered 2 km^2 of the Midorigaoka district and 3 km^2 of the Kuromatsu district.

2.2 Accuracy of the Isopleth Map of Artificial Fill Thickness

By overlaying the two landform maps, the points where contours of the same altitude intersected each other could be located and lines drawn to connect the intersect points in order to determine the boundaries of cut and fill.

Next, to depict the thickness isopleth of artificial fill, we connected the points of intersection of a given contour of (A) meters on the landform map in 1957 and the (A + 2) meters contour of the landform map in 1977-78.

Similarly, we connected the points of intersection of (A) and (A + 4) meters, (A + 6) meters .... (At 2^n) meter contours (Figure 2).

Concerning the accuracy of the thickness isopleths, as map contours generally ha maximum permissible error of 1/2 of the contour interval, and height points have a pe sible error of 1/3 of the contour interval, we may suppose that the thickness isoplet have a contour error of ± 2 meters or, at the lowest estimate, ± 1.34 meters of heigh

Boring survey data in the Midorigaoka and Kuromatsu districts were obtained from earthquake damage reconstruction plan by the Sendai Municipal Office and the Miyagi P ture Office, 34 boring pits having been located on artificial fill.

The mean value of the difference between the thickness of the fill, as shown by boring data and the isopleth map at the same point, was ± 1.3 meters. Two values wit the 34 boring pit survey data had a difference of 4 to 5 meters compared with the val the thickness isopleth. Except for these two values the mean value of the difference ± 1.1 meters. The thickness isopleths for the excavated areas were not estimated.

2.3 Distribution Map of the Disastrous Ground Failure

The acceleration of strong-motion in Sendai City and its environs during the ear quake is shown in Figure 4. The extent of casualties, and the ground failure in Senda Izumi City are shown in Table 1.

From the data on the distribution of cracks, the collapse or swell of retaining and destroyed, and semi-destroyed houses which prepared by the Sendai and Izumi Munici Offices and complemented by our field survey, we depicted the overlay situation on the thickness isopleth map of artificial fill, and the landform map in the time of 1957 p ously mentioned.

In order to examine quantitatively the relationships between the disastrous grou ure and the land conditions, we constructed a grid of 5 mm units on the maps at a scal 1/2,500, and measured the situation on unit areas of 1.56 hectares (12.5 meter x 12.5 dimensions.

3. RELATION OF THE DISASTROUS GROUND FAILURE TO
THE LAND CONDITIONS IN HILL RESIDENTIAL AREAS

3.1 Characteristics of Landform and Geology of the Surveyed Areas

The Central part of Sendai City is located on terraces consisting of Upper Pleistocene
sand and gravel layers having a thickness of several meters underlain by Pleiocene tuff
(See Figure 3).

This material proved to be rather stable, so that the damage of the earthquake was
relatively small.

In the last twenty years, the city area has expanded, and urban construction has spread
over the softer ground of an alluvial plain in the eastern part of Sendai, and over the hilly
areas of northern and south-western Sendai where the landform has been extensively changed
by earthworks. In these newly developed areas, the earthquake damage was conspicuous.

The hills in Sendai and its environs have a rather uniform summit level of about 100 to
200 meters, declining to the east.

Aobayama Hill in southwestern Sendai consists of horizontally-bedded Pleiocene sandstone
tuff, silstone, etc., covered by Middle Pleistocene gravel and aeolian volcanic ash. On
Aobayama Hill, there has been extensive development of residential and educational districts
namely Midorigaoka, Aoyama, Matsugaoka, and Yagiyama.

Midorigaoka district is on the southeastern part of Aobayama Hill, whose eastern margin
is a flexure scarp that is part of the Rifu-Nagamachi structural line bordering the alluvial
plain. Several hundred meters west of the scarp the Dainenji reverse fault runs sub-parallel
to it, the intermediate zone being uplifted like a horst.

Nanakita Hill occupies the northern part of Sendai and the adjacent part of Izumi City.
The hill consists of Upper Miocene sandstone, taffaceous siltstone and Lower Pleiocene sandstone, tuff, without overlying gravel. The relief of the hill is relatively small but it is
rather dissected. The area has been developed widely for residential districts, namely
Kuromatsu, Asahigaoka, Nankodai, Tsurugaya.

3.2 Relation of the Types of Artificial Fill and Disastrous Ground Failure

In former years, before large-scale cuts-and-fill of hilly regions for residential
districts were practiced in Sendai City, some houses were built on the slope of hills,
usually with rather small modifications of the ground surface. Such houses, for instance,

those in the Kano-Honcho neighborhood northeast of Midorigaoka, that are located on the surface of the flexture scarp with a slope of 10 degrees, were only slightly damaged by the earthquake.

Representative landform profiles of the destroyed houses and earth cracks are shown Figure 5. The distribution maps of thickness isopleths and disastrous ground failures a shown as Figures 6 and 7. Areas of artificial fill are divided into two main types: "h side filled" and "valley filled." In general, the thickness of the fill at the former is less than 10 meters. On the later type along the valley line, where the fill is thic the maximum thickness reaches 21 meters inside the surveyed areas. However, the damage to ground failure is not proportional to the thickness of the fill, but rather it is rel to a complex of fill thickness conditions and the original slope steepness of the buried landform.

In many cases, cracks break out in the part of the fill area where there are steep slopes on the buried landform, often running sub-parallel to the direction of the hillsi valley-head, or valley-sides of the buried land surface.

On the hillside filled type, at the place where there was a small valley in the buri landform slopes, ground failure occurred as echelon type cracks and the mass of fill mate sloped down slightly. Also on the hillside-filled type, ground failure was more severe o buried lower slopes than over buried upper slopes.

At the boundary zone between the filled area and the cut area, considerable ground failure also occurred. This zone was only roughly located, but from the appearance of damage it is conjectured to be the area of cut and fill. It seems that a filled area responds more as soft ground to earthquake vibrations than does the cut area.

3.3 Several Attributes of Land Condition Related to the Disastrous Ground Failure

Disastrous ground failures in the artificial fills and their relation to the land condition were examined as follows.

Frequency distribution for thickness of artificial fill in each or all districts as measured at the intersection of the grid is shown in Figure 8. There are slight differen between the districts but most of the fill is less than 4 meters thick as a whole.

The ratio relationship of destroyed houses to the thickness of fill by types is shown in Figure 9. The ratio is calculated as follows: fully-destroyed houses + 1/2 semi-destroyed houses/all houses. A unit consists of a 5 mm diameter circle.

In general, the greater the thickness of the fill, the greater the ratio of destroyed houses. It is possible to see a tendency for an increase in the ratio of destroyed houses as the slope of the original landform increases.

Inter-relations seem to exist between the thickness of the artificial fill, the slope angle of the original landform, and the ratio of destroyed houses for a part of Midorigaoka district (Figure 10). Two groups can be distinguished, a group on the upper right and a group on the lower left. The latter group where the thickness of fill is less than 8 meters and the slope of the original landform is less than 25 degrees is one with a high percentage of destroyed houses.

There is a relationship to a specific catchment area, which is an indicator of a comparatively large water gathering area on the original land surface. In the area of low slope angle on the original landform, and high value for the specific catchment areas, ground failures commonly occurred. (Figure 11).

About the disasters on the artifical fill areas in the hilly region, there are other problems to be examined such as the situation for weathering status of the basement rock and the existence of debris on the slope of the original landform.

The characteristics of the earthquake disaster in the artificial fill areas in the Sendai region are instructive regarding possible future disasters in other urbanized regions, and suggest that more intensive land condition surveys may contribute to avoiding such disasters.

CONCLUSIONS

From our analysis we conclude:

1) Severe damage due to ground failure is concentrated on the areas of artificial fill, rather than on the excavated areas. However, it is not definitely proportional to the thickness of the fill.

2) The damage is related to the original slope steepness underneath the fill, and to the type of fill situation.

3) Damage in the artificial fill area is concentrated at places where the original land surface had a high ratio of specific catchment areas.

ACKNOWLEDGMENT

We must acknowledge the help and assistance in the preparation of much data by the Earthquake Affairs Office of Sendai City and the Management Division of Izumi City. We appreciate the cooperative aid and advice by Mr. S. Hatano and Mr. H. Sunage of the Geographical Survey Institute.

REFERENCES

[1] Asano, T., "Transformed Hill Surfaces Around Sendai," Ann. Tohoku Geogr. Assoc., Vol. 31, No. 2, 1970.

[2] Okutsu, H., "Damage Caused by Miyagi-Ken-Oki Earthquake and the Corresponding Ground Geological Specification," Tsuchi To Kiso (Soil and Foundation), Vol. 26, No. 12, 1978.

[3] Kawakami, F., Asada, A., and Yanagisawa, E., "Damage to Embankments and Earth Structures," Tsuchi To Kiso (Soil and Foundation), Vol. 26, No. 12, 1978.

[4] Speight, J. G., "A Parametric Approach to Landform Regions," Inst. Br. Geogr. Spec. Publ. No. 7, 1974.

[5] Tamura, T., Abe, T., Miyagi, T., "Earthwork for Residential District in Hills and Damage Caused by Earthquake," Sogo Toshi Kenkyu No. 5, 1978.

[6] Tohoku University, "The Science Reports of the Tohoku University," Second Series (Geology), Vol. XXV, 1953.

[7] Institute of Geology and Palentology, Tohoku University, "Phenomena and Disasters Associated with the Miyagi-Ken-Oki Earthquake of 1978 in the East-central Part of Northern Honshu, Japan," Contribution No. 80, 1979.

[8] Nakata, T., "Damage of Houses and Landform Condition in Sendai and Its Environs, Caused by Miyagi-Ken-Oki Earthquake," Chiri (Geography), Vol. 23, No. 9, 1978.

Fig. 1   Surveyed areas

   1. Kuromatsu District ⟶

   2. Midorigaoka District

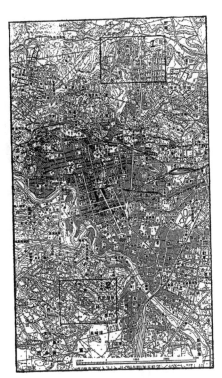

Fig. 2   Contour maps changed by
          large scale cuts-and-fills,
          a) Landform in 1957, b)
          Landform in 1978, c) Thick-
          ness isopleth of artificial
          fills

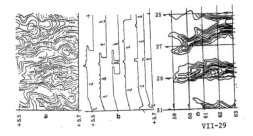

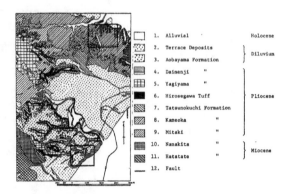

	1.	Alluvial		Holocene
	2.	Terrace Deposits	}	
	3.	Aobayama Formation		Diluvium
	4.	Dainenji	"	
	5.	Yagiyama	"	
	6.	Hirosegawa Tuff		Pliocene
	7.	Tatsunokuchi Formation		
	8.	Kameoka	"	
	9.	Mitaki	"	
	10.	Nanakita	"	}
	11.	Hatatate	"	Miocene
	12.	Fault		

Fig.3 Geological map of Sendai
(by The Science Reports of the Tohoku University Second
Series (Geology) Vol.XXV 1953

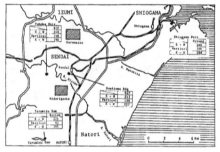

Fig.4 Acceleration of Strong Ground Motion around
Sendai

		Sendai City		Izumi City	
	All	Midorigaoka	Kuromatsu	All	Kuromatsu
Dead person	13	0	0	2	0
Serious wound p.	170	–	–	18	–
Slight wound p.	9,130	–	–	801	–
Destroyed house	769	36	29	94	17
Semi-destroyed h.	3,481	128	47	305	38
Partly-destroyed h.	74,000	–	–	11,856	769
Retaining wall crack swell	482	159	22	} 1,398	–
" collapse	67	13	7		–
Landcrack,Subsidence	199	119	8	2,328	–
Mudflow	25	1	–	104	–

Table 1. Damages due to Miyagi-ken-oki earthquake of 1978
VII-30

Fig. 5 Representative profiles showing the relation between ground conditions
and disasters

1) Destroyed house   4) Crack

2) Semi-destroyed house   5) Collapse of retaining wall

3) Non-destroyed house   6) Present ground surface

                            7) Original ground surface

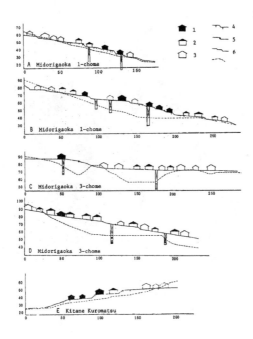

Fig.6 Isopleth map of thickness of artificial fill and the
disastrous ground failure (Main Part of Midorigaoka
district)

1) Destroyed House   4) Collapse, swell of retaining wall
2) Semi-destroyed House   5) Isopleth of thickness of artificial
3) Crack   fill in 2 meter interval

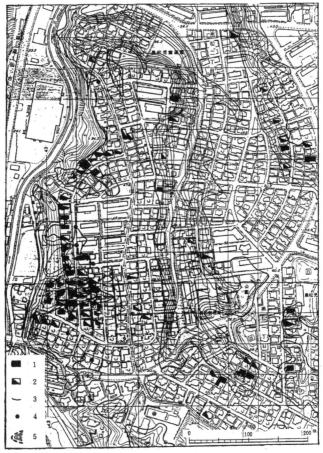

Fig.7 Isopleth map of thickness of artificial fill and the
disastrous ground failure (A part of Kuromatsu district)

1. Destroyed House   4. Collapse, swell of retaining wall
2. Semi-destroyed House   5. Isopleth of thickness of artificial
3. Crack   fill in 2 meter interval

VII-33

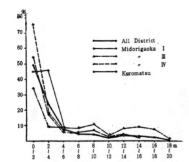

Fig.8 Frequency distribution of thickness of
artificial fill in each districts

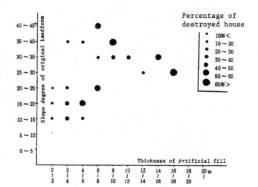

Fig.10 Interrelation between destroyed houses ratio,
thickness of artificial fill and slope degree
of original landform (Midorigaoka 1-chome)

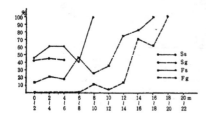

Fig.9 Relation of destroyed houses ratio and
thickness of artificial fill for some types

VII-34

Fig.11 Interrelation between disastrous ground failure, slope
degree of original landform, and specific catchment
area (Midorigaoka 1-chome)

■ Destroyed house
▲ Semi-destroyed house
● Crack

+ Collapse, swell of retaining wall
N Not damaged
5 Thickness of artificial fill
Ⓑ Valley filled type
5 Hill-side filled type

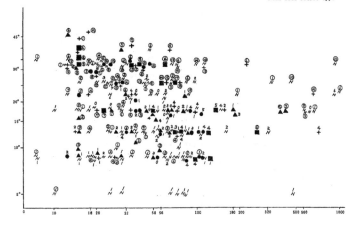

DAMAGE FEATURES OF CIVIL ENGINEERING STRUCTURES

DUE TO THE MIYAGI-KEN-OKI EARTHQUAKE OF 1978

Tadayoshi Okubo

Masamitsu Ohashi

Toshio Iwasaki

Kazuhiko Kawashima

Ken-ichi Tokida

Public Works Research Institute

Ministry of Construction

INTRODUCTION

At 17h 14m (JST), June 12, 1978, a destructive earthquake took place offshore Miya
and caused extensive damage to buildings, highway facilities, river dykes, water supply
sewage systems, electrical and gas supply systems, and others.

Major observed damage characteristics shows: (i) the influence of subsoil condition
great, and (ii) losses in commercial and industrial sectors and residential housings com
a high percentage of the total damage amount.

This paper describes the outline of the earthquake, the earthquake history in the
affected area, the geological and subsoil conditions, the recorded ground motions and da
features to civil engineering structures which were caused by the earthquake, espepially
bridge damage.

OUTLINE OF EARTHQUAKE

Origin Data

Origin data announced by the Japan Meteorological Agency (JMA) [2] is as follows:

Origin time	17h 14m, June 12, 1978 in Japanese Standar
Epicenter	38°09' N, 142°12' E
Focal Depth	30 km
Magnitude	7.4 on the Richter Scale

This earthquake is named the Miyagi-ken-oki Earthquake of 1978 by JMA and was accom
panied by a foreshock which occurred eight minutes prior to the mainshock, and numerous

aftershocks. The number of felt aftershocks announced by JMA is 11, between zero and 24 hours after the mainshock, 8 between 24 and 48 hours, 3 between 48 and 72 hours, 2 between 72 and 96 hours, and 1 between 96 and 120 hours.

The same area was also struck by an earthquake of magnitude of 6.7 on the Richter Scale on February 20, 1978, and suffered minor damage. The epicenter of this earthquake was located at 38°45' N and 142°12' E, about 60 km north of the epicenter of the June 12 Earthquake.

Seismic Intensity

The intensities of ground motions at various sites announced by JMA are shown in Figure 1 together with the location of epicenter. Numerals in the figure are seismic intensities on the JMA intensity scale. A comparison of JMA and Modified Mercalli scales is shown in Table 1.

Source Mechanism

Observation of aftershocks was carried out by JMA, Tohoku University, and other institutions. Figure 2 shows a distribution of aftershocks reported by T. Masuda [3] in the period from June 13 to 30. In the figure, epicenters of the February 20 Earthquake and the mainshock of the June 12 Earthquake are also shown.

Seno, et al. [10] proposed a fault model as shown in Figure 3. In this model, the area of the fault plane is 80 km long and 30 km wide and a dip angle of 20 degrees. From an analysis of surface wave amplitudes, they concluded that the relative movement of the fault plane was about 1.7 m.

GEOLOGY AND SOIL CONDITIONS

Topographic Features of Tohoku Region

As shown in the topography of Tohoku Region in Figure 4, three approximately parallel mountains running north to south characterize the topography of that region. Plains and basins are located between these mountains.

The Kitakami Mountains in the north and the Abukuma Mountains in the south which form the east line of the mountains mainly consist of Paleozoic and Mesozoic formations. The Ohu Mountain Range is the central range and is called the "backbone range in Tohoku." This Range mainly consists of Neocene sediments and Tertiary and Quaternary volcanic rocks.

Dewa Mountains and Ashai Mountains form the west line of the mountains. The formations o
Dewa Mountains are similar to those of Ohu Range. The Asahi Mountains mainly consist óf
Paleozoic and Mesozoic systems. Plains and basins between the three mountains mainly con
sist of Pleistocene and Holocene formations.

Neocene Systems in the Vicinity of Sendai

Table 2 shows Neocene stratigraphy in the vicinity of Sendai. The table is cited fro
the "Data Book of the Nature in Tohoku Region" published by the Tohoku Regional Constructi
Bureau, Ministry of Construction [5] and is modified by the authors [9].

Quaternary System and Topography in the Vicinity of Sendai

Concerning the Quaternary System in the vicinity of Sendai, there are many existing
studies but some discrepancies exist among them. In this section, the Quaternary System a
topography are sketched, drawing primarily from "Ground in the Coastal Area of Sendai Bay
(1965)," published by the Planning Bureau of the Ministry of Construction, and the Miyagi
Prefecture [6].

In the early Quaternary, the Sendai area was exposed above sea level and was subjected
to erosion which resulted in forming an eroded plain. Also in Quaternary, geological forma
tions and topography are characterized by the alternative repetition of glacial and inter-
glacial epochs which resulted in the forming of alluvial fans and terraces.

Figure 5, Figure 6, and Table 3 show a geological map, a typical geological section,
and the Quaternary stratigraphy in the vicinity of Sendai, respectively. Sendai area can
be topographically classified into five zones as follows (Figure 6).

1) Mountain Zone: This zone is located west of Sendai Plain. Its elevation is more
than 150 m. It mainly consists of andesite and aglomerate.

2) Hilly Zone: This zone is located north, west and south of Sendai and north of
Sendai Plain. Its elevation is about 60 to 210 m around Sendai, 20 to 40 m west of Natori
and 10 to 40 m west of Shichigahama. It is mainly consists of Neocene mud stone, tuff,
shale, and partially Triassic slate.

3) Terrace Zone: This zone is located in the valleys of Nanakita River, Hirose River
Natori River, Abukuma River, Naruse River, Yoshida River and Kitakami River. It consists o
four or five terraces, varying with the location. Sendai City has five terraces, i.e.,

(i) Aobayamà Terrace (elevation 100 to 210 m), (ii) Dainohara Terrace (elevation 40 to 90 m), (iii) Kamimachi Terrace (elevation 25 to 65 m), (iv) Nakamachi Terrace (elevation 25 to 60 m), and (v) Shimomachi Terrace (elevation 25 to 50 m).

4) Plain Zone: This zone is located east of Shimomachi Terrace. Its elevation is 1 to 15 m. Stratigraphy is shown in Table 3 and Figure 6.

5) Sand Dune Zone: This zone is located along the coastal line and makes two lines. Between the dune lines, marshes develop.

The most noticeable fact from the viewpoint of the earthquake damage is the existence of Nagamachi-Rifu Fault. It is considered that this Fault moved during late Pleistocene and early Holocene. As shown in Figure 5, the depth of baserock from the ground surface is about 5 to 7 meters in the west of this Fault but sharply increases up to 30 to 35 meters slightly east of the Fault and reaches about 60 meters in the coastal zone.

The Quaternary system mainly consists of Pleistocene deposits (diluvium) in the west of the Fault, but in the east consists of diluvium and Holocene deposit (alluvium). As shown in Table 3, the alluvium in this area contains soft layers. It is believed that these differences caused differences in the earthquake ground motions and in the severity of structural damage.

RECORDED EARTHQUAKE ACCELERATIONS

During the Miyagi-ken-oki Earthquake of June 12, 1978, 191 strong motion accelerographs recorded earthquake accelerations. The maximum ground acceleration was recorded at Kaihoku Bridge, about 80 kilometers from the epicenter and was 287 gals. Minimum acceleration was recorded in Nagoya, about 600 kilometers from the epicenter and was 6 gals.

Recorded maximum accelerations in three directions in the Tohoku Region are shown in Figure 7. From the figure, it can be estimated that the maximum ground acceleration near Sendai and Ishinomaki was approximately 240 to 300 gals. Along the Pacific Coast of Iwate Prefecture, it was about 100 to 170 gals.

Maximum ground accelerations in three perpendicular directions are plotted against the epicentral distances in Figure 8. Acceleration records on the ground and on a pier top at Kaihoku Bridge site are shown in Figure 9. Although very high accelerations (more than 500 gals in the longitudinal direction) were triggered on the pier cap, Kaihoku Bridge did not sustain any structural damage.

VII-39

Next, Figure 10 shows a record at a pier cap on the Date Bridge located near Fukushima City (epicentral distance of 160 km). The maximum accelerations on the pier cap were 480 gals in the longitudinal direction, and 320 gals in the transverse direction. This bridge suffered moderate damage to bearing supports and to a truss member above the fixed bearing. Since frequency properties change at 10 sec after the start of the record as shown in Figure 10, it is supposed that the bearings at the pier top might have been damaged at that time.

DAMAGE TO HIGHWAY BRIDGES

Features of Bridge Damage [1], [8]

Figure 11 shows the locations of severely damaged highway bridges (black circles), places where liquefaction was observed (white circles), and the outline of geological conditions in Miyagi Prefecture.

Numerals near black circles coincide with the number of bridges listed in Table 4, which describes briefly the type of damage to highway bridges. It is seen from Figure 11 that most of the major bridge damage and liquefaction took place in alluvial lands along large rivers such as Kitakami, Naruse, Yoshida, Natori, and Abukuma.

Sendai Bridge [4]

Sendai Bridge, completed in 1965, is located in the south part of Sendai City, and crosses the Hirose River as a part of National Highway No. 4. A general side view is shown in Figure 12. Superstructures are 9-span, simply supported, composite steel-plate girders, with span length of 9 x 33.840 m, total length of 310 m, and width of 19 m. Substructures are T-shape columns (6.1 m high) founded on rigid well foundations (9 to 18 m deep) embedded rather stiff sands. Bearing supports are the line bearing type. Since this highway which connects the Kanto and Tohoku regions is an important one, Sendai Bridge carries very heavy traffic (54,000 cars daily). As shown in Figure 12, the lower half of the column height is embedded into the higher river bed at three piers (P6 to P8). Column bases are above the surface of the lower river bed at the other piers (P1 to P5).

Due to the earthquake (the epicentral distance to the bridge is $\Delta = 120$ km), all of the nine pier columns sustained damage. Piers 1 through 4 cracked horizontally at the column bases, and surface concrete pieces separated severely from the columns near the bases. Piers 5 through 8 had similar damage near the haunches which connect columns and beams of

the piers. Pier 6 which has the lowest free height sustained the severest cracking at both
sides (see Figure 13). concrete pieces separated at the haunch and reinforcing bars buckled.
Near the haunch, the volume of reinforcing bars as well as the concrete sectional area
changes rapidly. It is estimated that relative displacements between adjacent girders were
1 to 2.5 cm on the pier caps and that displacements at the pier caps of Piers 1, 2 and 6 were
11 to 18 cm. The girders and the bearing supports did not sustain any damage.

Figure 14 shows temporary frame works supporting the girders near Pier 6. Since the
bridge is very important, damaged piers were repaired without stopping traffic even for a
short time. Figure 15 illustrates an example of permanent repair work at Pier 6. Figure 16
is a picture of Pier 6 after the repair work was finished. The thickness of added concrete
was 50 to 70 cm, and vertical reinforcing bars were fixed using epoxy adhesive the well foundation, lateral bars were fixed to the columns, and chemical resin was placed into small
cracks. It took only one month to completely repair all the damage to this bridge.

Kin-noh Bridge [7]

Ken-noh Bridge, completed in 1956, is on National Highway No. 346 which crosses the
Kitakami River. As shown in Figure 17, the superstructures of the bridge are single-span
steel plate girder, 5-span simply supported steel trusses, and 9-span Gerber-type steel plate
girders, from left to right. The total length and the width are 575.5 m and 6.0 m, respectively. Substructures are RC columns on caisson foundations for the truss spans, and RC
columns on footing foundations with RC piles for the Gerber plate girder spans. Soils are
of soft silts and sands, and a firm sand layer exists approximately 30 m below the ground
surface. During the June earthquake one suspended girder of this bridge fell down (see
Figure 18). The Kin-noh Bridge is the only one that fell down during the June Earthquake.

The bridge was damaged three times by three different earthquakes, namely the Northern
Miyagi-ken Earthquake of 1962 (M = 6.5, $\Delta \fallingdotseq$ 15 km), two Miyagi-ken-oki Earthquakes of
February 20, 1978 (M = 6.7, $\Delta \fallingdotseq$ 80 km) and the one of June 12, 1978 (M = 7.4, $\Delta \fallingdotseq$ 110 km).

Due to the 1962 Earthquake the side blocks of the bearing supports (oval line bearings)
of Gerber girders failed, and concrete near the fixed bearing supports on the right-bank
abutment cracked. After the 1962 Earthquake, repair work to add stiffening plates was undertaken at three piers (P8, P9, and P10) as shown in Figure 19.

Anchor bolts of the bearing stiffening plates were cut off during the Earthquake of February 20, 1978. Side blocks of bearing supports, which did not sustain damage during the 1962 Earthquake and therefore were not repaired, also failed during the February Earthquake (see Figure 20).

During the Earthquake of February 20, 1978, most bearing supports at the truss girders also failed as did the bearing supports at the Gerber girders. As for the truss spans, fixed bearing supports of pin-type were most severely damaged on Pier 6 (see Figures 21 and 22). Figure 21 shows the upstream support where four anchor bolts pulled out and bent counter-clockwise. It is supposed from Figure 21 that the bearing would have rocked severely, rotated, and translated. Figure 22 is the downstream support in which set bolts were sheared off. Most anchor bolts of the fixed bearing supports on the other piers were also pulled out. Movable bearing supports of the pin-roller-type also failed. Figure 23 shows the protrusion of all the rollers at the upstream movable bearing on Pier 5.

Since there were only four months after February 20 Earthquake, repair work on these bearings was still being undertaken at the time of the June 12 Earthquake. Accordingly, all the girders were able to move freely without restraint by bearing supports during the June Earthquake.

Due to the June Earthquake a suspended girder between Piers 7 and 8 fell down onto the river bed, as shown in Figures 17 and 18. The superstructure moved toward the right-bank side by 55 cm on the top of Pier 8 (see Figure 24). All the Gerber span between Pier 8 and the Right Abutment moved toward the right-bank. The girder moved 10 cm toward the right on the right-bank abutment, and the end of girder collided into the parapet of the abutment (see Figure 25). The asphalt pavement of the backfill heaved due to the collision (see Figure 26).

Truss girders were also heavily damaged during the June Earthquake. Figures 27 and 28 are pictures taken after the June Earthquake at the same places as Figures 21 and 22, respectively. Anchor bolts of the upstream fixed bearing at Pier 6 were severely pulled out by about 20 cm at most (compare Figures 29 and 19), presumably due to the rocking and translation motions of the bearing, and some concrete underneath the lower bearing plate was taken out and the bearing sunk by 2.5 cm. As for the downstream fixed bearing on Pier 6 (Figure 28), a deformed bar which had been used as a temporary set bolt after the February Earthquake was sheared off again. The key of the upper shoe dislodged from the sole plate, and the sole

VII-42

plate deformed. As for pin-roller-type movable bearings, some rollers had rolled out of the shoes during the February Earthquake. Figure 29 shows the condition after the June Earthquake at the upstream movable bearing on Pier 5 where the rollers completely rolled out.

As for pier columns, only the right-bank side of Pier 8 sustained heavy cracks (see Figure 30). It is estimated that these cracks would have taken place when the superstructure collided with the right abutment and the reaction toward the left bank applied to the pier.

To grasp the causes of damage to the Kin-noh Bridge, the Miyagi Prefecture is conducting comprehensive studies which include field surveys and dynamic analyses, and the cooperation of the Public Works Research Institute.

Maiya Bridge

Maiya Bridge, completed in 1928, crosses Kitakami River 4 km downstream from the Kin-noh Bridge. Superstructures are 3-span Gerber-type steel truss girder with a total length of 181.4 m and width of 5.3 m. Two abutments have footing foundations, and two piers have well foundations. Due to the June Earthquake an upper chord member was broken off as shown in Figure 31. Due to the breakage, the lower chord drooped considerably (see Figure 32). The chord member was made of steel channels and broke during the sudden sectional change as shown in Figure 31. At this bridge a pin came out of the fixed pin-type hearing.

Toyoma Bridge

Toyoma Bridge, completed in 1945, crosses Kitakami River 6 km downstream from Maiya Bridge. The superstructures are Gerber-type RC T-shape girders, with a total length of 306 m. The pier columns are on well foundations. Due to the June Earthquake many heavy cracks occurred at the mid-point of several girders, and at webs above the bearing supports. A heavy crack also broke out at the base of the pier column closest to the right bank.

Yanaizu Bridge

Yanaizu Bridge, completed recently in 1974, crosses Kitakami River 6 km downstream from Toyoma Bridge. The superstructures are 6-span steel truss girders, with a total length of 450 m and width of 8.5 m. The substructures are of RC columns on steel-pipe pile foundations. A lower chord was damaged at the fixed pin-type bearing, as shown in Figure 33. Due to seismic forces set bolts between the upper shoe and the lower chord were cut off, welding between the upper shoe and sole plate detached, the web plate deformed, and paint chips flaked off. Substructures and bearing supports of this bridge did not sustain any damage.

Kimazuka Bridge

Kimazuka Bridge, completed in 1931, crosses Naruse River in Kashimadai Town. The superstructures are 19-span simple supported steel plate girders, with a total length of 236 m, and width of 4.5 m. Two abutments are on footing foundations, and 18 piers are on well foundations. Due to the June Earthquake, one pier cap was severely damaged near the bearing (see Figure 34), and one girder was almost dislodged from the pier cap (see Figure 35). The girder moved in the transverse direction.

Eia Bridge [7]

Eai Bridge, completed in 1932, crosses Eai River in Furukawa City. The superstructures are 9-span simply supported steel plate girders, with a total length of 155 m and a width of 7.5 m. Two abutments are on pile foundations, and each of 8 piers rest on two separate well foundations with diameters of 2.5 m and a depth of 7 m. At the time of this earthquake, the embedment of well foundations was almost half of the initial depth due to scoring effects of stream. Located close to the epicenter of the Northern Miyagi Earthquake of 1962 (epicentral distance was approximately 15 km), the bearings of this bridge had been damaged in 1962. Due to the June, 1978 Earthquake lower beams of the eight pier columns were severely cracked, and concrete pieces separated from the beams (see Figure 36). The largest opening of the cracks was 20 mm, and reinforcing bars were exposed.

Yuriage Bridge [7]

Yuriage Bridge, completed rather recently (1972), crosses over the Natori River near its mouth. The superstructures are of 3-span continuous PC box girders with a center hinge (cantilever erection) and 7-span simply supported post-tension PC beams (T-shape) with a total length of 541.7 m and width of 8 m. Two abutments are on steel pipe pile foundations, two piers in the lower river bed are on pneumatic caisson foundations, and 7 piers are on well foundations. Due to the June Earthquake the nine pier columns sustained many cracks (almost all around) mostly at the level of the ground surface. Especially Pier 1 (first pier from the left-bank) which sustained numerous heavy cracks (see Figure 37), and concrete pieces separated from the column. Stoppers on a single-roller-type movable bearing on Pier 1 were damaged, guide pieces of the bearing failed, and the roller almost rolled down from the lower shoe (see Figure 38).

A simply supported PC beam on Pier 6 moved 6 cm downstream. The ends of one handrail had been inserted into the ends of another handrail just above a pier. The length of insertion was 8 cm. During the Earthquake the handrail ends completely came out of the adjacent handrail ends. It is understood from this that the two adjoining beams vibrated relatively with a separation of at least 8 cm in the longitudinal direction.

A number of ground cracks and sand boils were observed on the higher river bed near the right bank. The subsoils made of mostly sands are loose at the surface (about 5 m deep) and medium to dense underneath. A hard layer exists approximately 70 m below the surface.

Date Bridge

Date Bridge, completed in 1963, crosses over the Abukuma River near Fukushima City, with an epicentral distance of approximately 160 km. Superstructures are 4-span continuous steel truss girders, with a total length of 288.0 m and width of 7.0 m. The two abutments are on steel-pipe pile foundations, and the three piers are tall RC columns on caisson foundations embedded into gravel and sand layers.

Due to the June Earthquake a lower chord member buckled just at the fixed bearing on Pier 2 (see Figure 39). Several pins at the fixed bearing and one of the movable bearings were sheared off and came out of the shoes. The substructures did not sustain any damage.

A strong-motion accelerograph is installed on the cap of Pier 2, and triggered a complete time history of the acceleration at the pier cap (see Figure 10).

LESSONS FROM BRIDGE DAMAGE DUE TO THE MIYAGI-KEN-OKI EARTHQUAKE

In view of the damage to highway bridges during the Miyagi-ken-oki Earthquake of June 12, 1978, the following lessons can be derived.

1) Damage to superstructures concentrated on bearing supports and the adjoining portions. Most damage to substructures were cracks and separations of concrete at pier columns and abutments.

2) Damage to bearing supports were most frequently observed. It is advisable to more carefully investigate design practices relating to bearing supports and to design and develop

better bearings which may better resist seismic disturbances. It seems, however, that the breakage of bearing supports served to reduce failure of bridge girders, and failure of substructures. Therefore, it is not always advantageous to design bearings that are too strong.

3) Because of the resultant extensive damage to the whole bridge structure, the fall of bridge girders should be avoided.

4) A number of older bridges such as the Kin-noh, Maiya, Toyoma, Kimazuka, and Eai Bridges sustained relatively severe damage. In most of these bridges, either Gerber-type or simply supported design types are used, the width of pier caps is narrower, and no special consideration to prevent falling girders is introduced. It seems very important to retrofit these older bridges by widening pier caps, installing devices to prevent girders from falling, etc.

5) In view of the damage to pier columns at the Sendai and Yuriage Bridges which were recently constructed according to the current specifications, it is recommended to consider the ductility of pier columns when designing short reinforced concrete columns. In this respect further experimental and analytical investigations are necessary, and future seismic specifications should include an appropriate regulation on ductility of pier columns.

DAMAGE TO ROADS

Damaged road sites and highway bridges amount to 2,350 and 255, respectively, and about 92% of this damage took place in Miyagi Prefecture. The damage to roads was the cracking of pavements, subsidence of road embankments, slope failures, falling stones, etc.

Table 5 shows damage to National Roads in Tohoku Region, in which the number of damaged sites in Miyagi Prefecture amounts to about 70% of the total. As for the type of damages, pavement cracking and road subsidence constitute about 66% of the total. It should be noted that embankment settlement behind bridge abutments and other structures frequently took place and was concentrated in the areas of soft foundations.

DAMAGE TO RIVERS

Damage to rivers were mainly slope failure, longitudinal and transverse cracking and subsidence of river dykes. As for rivers controlled by the national government, the total length of four rivers and the length of river dykes under the control of Tohoku Regional

Construction Bureau are 313 and 422.3 kilometers, respectively. Length of damaged portions of dykes amounts to 34.1 kilometers, i.e., and about 8% of total dyke length. These damaged portions were located mainly in the area of soft foundations.

Weirs, conduits and culverts across the dykes and bank revetment works also suffered damage, though not severe. Two weirs located near the estuaries of the Kitakami and Abukuma Rivers suffered damage of local buckling of gate leaf guides made of H-shape steel. As for conduits, culverts and bank rivetments, cracking of concrete took place.

DAMAGE TO UTILITIES

Damage to utilities concentrated in the Miyagi Prefecture.

Electricity was shut off immediately after the Earthquake and 419,000 families lost electrical power. Repair work was energetically carried out and within 24 hours, 323,000 families recovered power and within 50 hours all domestic electricity was back in service. One week after the earthquake, the electric supply for industrial use was totally restored.

Two hydro-thermal power plants ceased functioning because of breakdowns of boiler tube systems, circuit breakers and transformers. Hydroelectric plants also suffered damage to intake facilities, water channels, tanks, and retaining walls; however, they remained functional.

Trunk transmission lines suffered cracking of retaining walls at tower foundations, however, such damage did not hinder transmission. Fifteen substations suffered severe damage to transformers, circuit breakers and so on, and caused shutdown of electricity in the Prefecture.

In distribution systems, toppling, snapping, and tilting electric poles, snapped wires and tilted transformers on poles took place in the area of soft foundations and reclaimed land.

1,704 cases of damage to potable water supply systems took place; 2 for water storage facilities, 50 for purification facilities and 1,652 for pipeline systems resulting in a suspension of water supply to 87,740 families in 55 municipalities. Repair work was rapid; 49% back in service by June 13, 67% by June 14, and 78% by June 15, and about 10 days after the earthquake, the water supply was totally restored.

In Sendai City, office buildings of the Water Supply Bureau, the intake, purifying and pumping facilities of purification stations, the water supply and delivery pipes all suffered some damage. After the Earthquake, leakage of water increased and on June 13, the estimated leakage amounted to about one third of supply. Suspension of water supply was caused by broken pipelines and which was concentrated around reclaimed lands in hilly areas and in coastal zones.

In Miyagi Prefecture, 7 cities have gas supply systems and 6 sites suffered damage of the system. In Sendai City, a cylindrical tank of 17,000 cubic meters storage capacity collapsed and burned but the gas manufacturing plant did not cease to function. Leakage of gas took place at 533 places in the supply pipe systems. Therefore, gas supply was shut off. It took 31 days for total recovery. Shiogama City had 168 leaks, Ishinomaki City 97, and Furukawa City 23. This damage consisted of breaking, cracking, dis-socketing and loosening of supply mains, branches and home deliveries. Damage to gas supply systems was concentrated in the area of soft ground, but occurred in reclaimed lands in hilly areas too.

Sewage systems also suffered damage. This damage was: (i) disfunctioning of the treat-ment and pumping stations due to the electrical shutdown, (ii) damage to power supply facil-ities, pumps, independent power generators, pressure pipes, and discharge pipes, (iii) cracking, buckling, and dissocketing of pipe ducts and manholes.

Concerning telecommunication systems, 669 poles were tilted and damaged, 28.2 kilometers of telephone wires was damaged, wires were severed in 4,538 places, 905 phones were crushed, failure of joints between manhole and embedded duct totalled 162, breaking of embedded pipes and ducts totalled 103 places and damage to pipes attached to bridges took place at 27 sites. This damage was repaired within a week after the Earthquake. Other than damage described above, telephone exchange stations in the damaged area were loaded by numerous phone calls into the area by worried relatives and friends.

CONCLUSIONS

Similar to the lessons of past earthquakes, the Miyagi-ken-oki Earthquake indicates that damage to structures and facilities largely depends upon ground condition. It also shows that some design codes and specifications should be re-examined and revised along with construction standards. To lessen not only the structural damage but functional damage as well, more research would be beneficial.

ACKNOWLEDGMENT

In surveying the damage caused by the Miyagi-ken-oki Earthquake of 1978, much cooperation and valuable information were provided by the personnel of the Ministry of Construction,
the National Land Agency, the Miyagi Prefecture, and other related organizations. The
authors wish to express their sincere thanks to the staff members at these organizations.

REFERENCES

[1] Iwasaki, T., Kawashima, K. and Tokida, K., "Interim Report of the Miyagi-ken-oki
    Earthquake of June 1978," Technical Note of PWRI, No. 1422, Public Works Research
    Institute, Ministry of Construction, 1978.

[2] Japan Meteorological Agency, "Report on the 1978 Miyagi-ken-oki Earthquake," Technical
    Note of the Japan Meteorological Agency, No. 95, 1978.

[3] Masuda, T., "Aftershocks of the 1978 Miyagi-ken-oki Earthquake," Abstract of Annual
    Meeting, Seismological Society of Japan, 1978.

[4] Ministry of Construction, "Damage Due to the Miyagi-ken-oki Earthquake of 1978," Tohoku
    Regional Construction Bureau of the Ministry of Construction, 1978.

[5] Ministry of Construction, "Data Book on the Nature of the Tohoku Region," Planning
    Division of Tohoku Regional Construction Bureau, Ministry of Construction, 1961.

[6] Ministry of Construction and Miyagi Prefecture, "Ground in Coastal Area of Sendai
    Bay," Planning Bureau of the Ministry of Construction and the Miyagi-ken, Vol. 10,
    1965.

[7] Miyagi Prefecture, "Damage Due to the Miyagi-ken-oki Earthquake of 1978," 1979.

[8] Okubo, T. and Iwasaki, T., "Summary of Recent Experimental and Analytical Seismic
    Research on Highway Bridges," Workshop on Research Needs of Seismic Problems Related
    to Bridges, San Diego, USA, 1979.

[9] Okubo, T. and Ohashi, M., "Miyagi-ken-oki," Japan Earthquake on June 12, 1978 --
    General Aspects and Damage, U.S. National Conference of Earthquake Engineering --
    1979, August 22-24, 1979," Stanford, California, USA.

[10] Seno, T., Sudo, K. and Eguchi, T., "Fault Mechanism of the 1978 Miyagi-ken-oki
     Earthquake, Abstract of Annual Meeting, Seismological Society of Japan, 1978.

Table 1. Comparison between Intensity Scale of Japan Meteorological Agency and Modified Mercalli Scale

Maximum Accelerations (gals)	0.5	1	2	5	10	20	50	100	200	500	1000
Japan Meteorological Agency Seismic Intensity Scales	0	0.8 I	2.5 II	III	III	25 IV	80 V	250 VI 400 VII			
Modified Mercalli Scale	I	I 2.1 II	5 N	10 V	21 M	44 VI	94 VII	202 K	432 X,XI,XII		

Table 2. Stratigraphy of Tertiary in the Vicinity of Sendai (After "Data Book of the Nature in Tohoku")

Era	Group	Formation	Thickness	Lithofacies	Remark
Pliocene	Sendai G.	Dainenji F.	30-130	Unconsolidated, sand stone, silt stone, tuff, limonite	Stage 6 Marine Deposit
		Yagiyama F.	20-30	Alternate of tuff and silt stone	Stage 6 Continental Deposit
		Hironosegawa tuff F.	15-100	Pisolite	"
		Kitayama F.	1-10	Tuff, silt stone	"
		Tatsuno-kuchi F.	30-60	Alternation of tufaceous sand stone and silt stone	Stage 6 Marine Deposit
		Kameoka F.	15-50	Silt stone, sand stone, tuff	Stage 6 Continental Deposit
Miocene	Akiu G.	Mitaki Basalt	200	Basalt	
		Shirasawa F.	330	Shale, white sandy tuff	Stage 5 Continental Deposit
		Yumoto F.	150	Massive tuff breccia, trachitic green tuff	"
	Natori G.	Tsunaki F.	130	Agglomerate, tuff, sand stone	Stage 4 Marine Deposit
		Hatatate F.	130	Alternation of conglomerate and shale	Stage 3 Marine Deposit
		Moniwa F. Tsukinoki F.	60-130	Sand stone, conglomerate tuff	Stage 2 Marine Deposit Stage 2 Continental Deposit
		Takadate F.	60-250	Andesite	Stage 1 Continental Deposit

Table 3. Quaternary Stratigraphy in the Vicinity of Sendai
(After "Ground in the Coastal Area of Sendai Bay")

Epoch	Formation	Thickness (m)	Facies	Sedimentation Circumstance	Bearing Capacity (t/m²)
Holocene	Fukanuma F.	0-5	Fine to coarse sand	Sand dune	5-10
	Kasuminome F.	1-5	Sandy loam	Alluvial fan and flood plain	3-7
			Coarse sand with gravel		15-20
	Fukudamachi F.	1-10	Loam to peat	Lacustrine	2-5
			Sandy loam	Swamp	3-8
	Iwakiri F.	10-30	Fine to coarse sand	Shallow sea	20-25
			Silty loam		10-20
			Clay		3-7
			Sandy loam		3-7
Pleistocene	Shimomachi terrace F.	10-35	Sand and gravel	River terrace	20-35
	Wakabayashi clay F.		Clay	Alluvial fan	15-20
	Gamou F.				

VII-50

Table 4 List of Highway Bridges Severely Damaged in Miyagi
and Fukushima Prefectures

| No. | Bridge | Route | Characteristics of Bridges | | | Year Com-pleted | Outline of Damages |
			Total Length (m)	Width (m)	Superstructure		
1	Sendai	National Highway No.4	310.0	19.0	Composit Steel Plate Girder	1965	Horizontal Cracks at All Pier Columns
2	Abukuma	National Highway No.6	571.6	6.0	Steel Warren Truss, Simple Steel Pl. Girder	1932	Horizontal Cracks at Pier Columns
3	Ono	National Highway No.45	247.3	5.5	Simple Steel Pl. Girder	1936	Movement of All Girders, Failure of Bearing Supports, Failure of Slabs
4	Ten-noh	National Highway No.45	367.7	6.0	Steel Gerber Pl. Girder, Steel Langer Truss	1959	Vertical Crack at a Pier Column
5	Toyama	National Highway No.342	306.0	5.3	RC T-beam	1945	Cracks at Beams and Pier Columns
6	Kin-noh	National Highway No.346	575.5	6.0	1-Simple Steel Pl. Girder, 5-Steel Truss, Steel Gerber Pl. Girder	1956	Fall of a Suspended Girder, Failure of Bearing Supports
7	Turiage	Miyagi Prefectural Highway	541.7	8.0	3-Hinged PC Girder, 7-Simple PC Post-Tension T-beam	1972	Cracks at Pier Columns, Liquefaction
8	Kimazuka	Miyagi Prefectural Highway	236.0	4.5	19-Simple Steel Pl. Girder	1931	Failure of Bearing Supports, Failure of Expansion, Liquefaction
9	Eai	Miyagi Prefectural Highway	155.0	7.5	9-Simple Steel Pl. Girder	1932	Cracks at Pier Columns
10	Maiya	Miyagi Prefectural Highway	181.4	5.5	Gerber Truss Girder	1928	Break of Upper Chord, Movement of Truss Girder
11	Yanaizu	Miyagi Prefectural Highway	450.0	8.5	1-Simple Steel Truss 2-Cont. Steel Truss 3-Cont. Steel Truss	1974	Failure of Bearing Supports, Crack and Buckling at Lower Truss Chord
12	Date	Fukushima Prefectural Highway	288.0	7.0	4-Cont. Steel Truss	1963	Failure of Bearing Supports, Buckling at Lower Truss Chords, Crack of Pavement

Table 5   Damage of National Road in Tohoku Region

(a) Classification of Damage in accordance with Prefectural District

| Prefecture | Route No. | | | | | | | Total |
	4	6	13	45	47	48	108	
Fukushima	2	6	3					11
Miyagi	58	6		36	2	6	8	116
Iwate	10			28				38
Yamagata			2					2
Total	70	12	5	64	2	6	8	167

(b) Classification of Damages in accordance with Types

| | Route No. | | | | | | | Total |
	4	6	13	45	47	48	108	
Bridge Failure	8	2		4		1		15
Pavement Failure	47	7	2	49	1	1	3	110
Slope Failure	1	2		5				8
Falling Stones				2		1		3
Others	14	1	3	4	1	3	5	31
Total	70	12	5	64	2	6	8	167

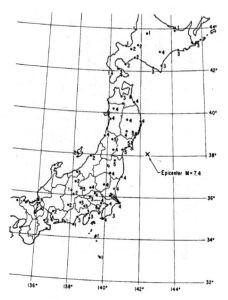

Fig. 1   Distribution of Seismic Intensities

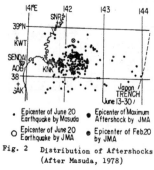

Fig. 2   Distribution of Aftershocks
(After Masuda, 1978)

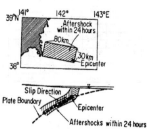

Fig. 3   Fault Model of Miyagi-ken-oki
Earthquake of June 12, 1978
(After Seno et al, 1978)

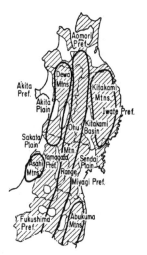

Fig. 4   Topographic Feature
of Tohoku Region

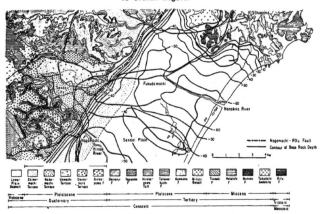

Fig. 5   Geological Map in the Vicinity of Sendai (After "Ground in the
Coastal Area of Sendai Bay")

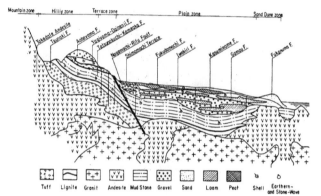

Fig. 6   Typical Geological Section of Sendai Area (After "Ground in the Coastal Area of Sendai Bay")

Fig. 7   Distribution of Maximum Accelerations in Gals

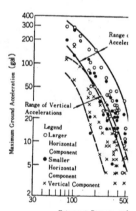

Fig. 8   Relation between Epicentral Distance and Maximum Acceleration

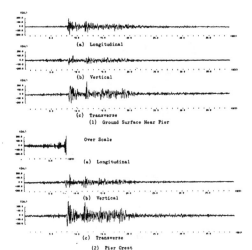

(a) Longitudinal

(b) Vertical

(c) Transverse

(1) Ground Surface Near Pier

Over Scale

(a) Longitudinal

(b) Vertical

(c) Transverse

(2) Pier Crest

Fig. 9   Strong-Motion Acceleration Records at Kaihoku Bridge
and Ground Nearby

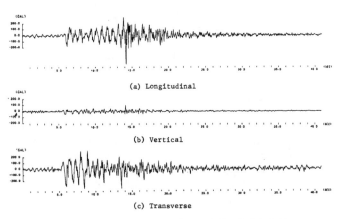

(a) Longitudinal

(b) Vertical

(c) Transverse

Fig. 10   Strong-Motion Record at Cap of Pier 2, Date Bridge

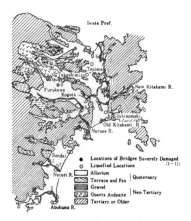

Fig.11   Geological Features and Locations
          of Major Bridge Damages and
          Liquefaction Sites

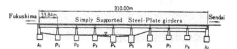

Fig.12   General View of Sendai Bridge

Fig.13   Failure of Pier 6,
          Sendai Bridge

Fig.14   Temporary Frames Supporting
          Girders near Pier 6, Sendai
          Bridge

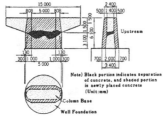

Note) Black portion indicates separation of concrete, and shaded portion is newly placed concrete (Unit:mm)

Column Base

Well Foundation

Fig. 15 Damage and Repair Work of Pier 6, Sendai Bridge

Fig. 16 Pier of Sendai Bridge after Repair Work Completed

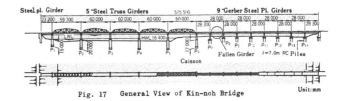

Steel pl. Girder 5 ‾Steel Truss Girders 575 5½0 9 ‾Gerber Steel Pl. Girders

Caisson

Fallen Girder $l = 7.0$m RC Piles

Fig. 17 General View of Kin-noh Bridge

Unit:mm

Fig. 18 Fall of a Suspended Girder, Kin-noh Bridge (During the Earthquake of June 12, 1978)

Fig. 19 Bearing Stiffening Plates to Resist Transverse Movement, Kin-noh Bridge (Added after 1962 Earthquake)

Fig. 20 Failure of Side Block of
Oval Line Bearing on Pier 11,
Kin-noh Bridge (Just after
February 20, 1978)

Fig. 21 Pull-out of Anchor Bolts at
Upstream Fixed Pin-type
Bearing on Pier 6, Kin-noh
Bridge (Just After February
20, 1978)

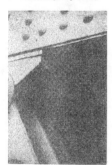

Fig. 22 Failure of Set Bolts at downstream Fixed Pin-type Bearing
on Pier 6, Kin-noh Bridge
(Just After February 20, 1978)

Fig. 23 Protrusion of Rollers at the
Movable Pin-roller-type
Bearing on Pier 5, Kin-noh
Bridge (Just after February
20, 1978)

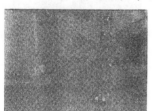

Fig. 24 Movement (55 cm) of Plate
Girder at the Downstream
Support on Pier 8, Kin-noh
Bridge (After June 12, 1978)

Fig. 25 Failure of the Right-Bank
Abutment, Girder Moved Toward
Abutment by 10 cm, Temporary
Support is seen.
(After June 12, 1978)

Fig.26　Heaving of Asphalt Pavement at the Backfill of the Right Abutment, Kin-noh Bridge (After June 12, 1978)

Fig.27　Pull-out of Anchor Bolts and Settlement of the Shoe at the Upstream Fixed Bearing on Pier 6, Kin-noh Bridge. (After June 12, 1978)

Fig.28　Failure at Downstream Fixed Bearing on Pier 6, Kin-noh Bridge (After June 12, 1978)

Fig.29　Failure of Upstream Movable Bearing on Pier 5, Kin-noh Bridge (After June 12, 1978)

Fig.30　Cracks of Sides of Columns of Pier 8, Kin-noh Bridge (After June 12, 1978)

Fig.31　Breakage at Upper Truss Chord, Maiya Bridge

Fig.32   Drooping of the Lower Truss
         Chord, Maiya Bridge

Fig.33   Damage to Lower Flange of Truss
         Chord near a Fixed Pin-type
         Bearing, Yanaizu Bridge

Fig.34   Failure of Concrete near a
         Bearing, Kimazuka Bridge

Fig.35   A Girder almost Dislodging
         from a Pier Cap, Kimazuka
         Bridge

Fig.36   Cracks at Pier Columns, Eai
         Bridge

Fig.37   Cracks at Pier 1, Yuriage
         Bridge

Fig.38   Failure of Guide Piece at
          Movable Bearing on Pier 1,
          Yuriage Bridge -

Fig.39   Buckling of Lo
          Member Above t
          Bearing on Pie

DAMAGE TO RIVER DYKES CAUSED BY THE MIYAGI-KEN-OKI

EARTHQUAKE OF JUNE, 1978

Kazuya Yamamura

Yasushi Sasaki

Yasuyuki Koga

Eiichi Taniguchi

Public Works Research Institute

Ministry of Construction

ABSTRACT

A large number of engineering structures were severely damaged by the Miyagi-Ken-Oki Earthquake of June, 1978. This paper describes the damage to the river dykes in the Kitakami, Naruse, Eai, and Yoshida Rivers by this earthquake.

Field surveys and laboratory investigations were performed on subsoils underlying the dykes and following conclusions drawn

    i) River dykes in the Miyagi prefecture were extensively damaged by this earthquake.

    ii) The seismic resistance of river dykes is affected by the stability of supporting subsoils during an earthquake: either by a tendency toward liquefaction, or by the bearing capacity of the subsoils.

    iii) The damage of river dykes has a close relationship to the micro-topographical features. Dykes on a "former river bed" have a high potential to be damaged by earthquakes. It is suggested by the analyses on Eai River that the Land Form Classification Map for Flood Control Planning is useful in predicting river dyke damage.

    iv) Re-liquefaction may or may not occur at the same place. It was observed not to occur at a layer which had been compacted by previous earthquakes.

KEYWORDS: Earthquake-resistant dykes; river dyke damage; soil liquefaction; soil relationships.

1. INTRODUCTION

An earthquake of Richter magnitude 7.4 occurred off the Pacific coast of the Miyagi prefecture in northern Japan at 17:14 on June 12, 1978. The seismic intensities according to the Japan Meteorological Agency are V at Sendai, Fukushima, Ofunato, and Shinjo, and IV at Tokyo, Mito, Akita, etc. This earthquake caused extensive damage to diverse engineered structures such as buildings, bridges, river dykes, road embankments, etc.

This paper describes the outline of damage to river dykes caused by the earthquake, and the result of investigations into the failure and settlement of dykes. In general the damage to dykes is closely related to subsoil conditions, and the field survey and investigation were performed from this point of view.

2. AN OUTLINE OF THE DAMAGE TO RIVER DYKES

Table 1 shows the number of sites where restoration works were carried out after the Miyagi-ken-oki earthquake in each prefecture of the Tohoku region through the assistance of the Japanese government. This table shows that the total number of damage points to the river facilities amounts to 176, which is next to the number of the damaged sites for highway facilities. It is also shown that most of the damage is concentrated in the Miyagi prefecture.

Some distinctive features can be seen on the damage caused by this earthquake as compared with that of Izu-Oshima-Kinkai Earthquake (1978, M = 7). A large number of slope failures occurred, and highways at Izu Reninsula were cut off by the Izu-Oshima-Kinkai Earthquake, because the mountainous area was struck by this earthquake. On the other hand the Miyagi-ken-oki earthquake hit the plain area and the bridges and earth structures such as river dykes on flood plain deposits were extensively damaged.

Table 3 shows the number of damaged sites for river dykes, revetment and others. Under "others," the damage to transverse pipes in dykes and water gates is included.

Figure 1 illustrates the damaged sites of river dykes in Miyagi prefecture. The rivers controlled by the regional bureau of the Ministry of Construction in this area are the Kitakami River, Naruse River, Natori River, and Abukuma River, and severe failures, cracks and

settlements of dykes were observed along these rivers. The Nanakita River was also extensively damaged. Most of these damage points are located on alluvial plain deposits where
the epicentral distance is less than 120 km.

Table 4 shows the length of damaged dykes along the main four rivers in Miyagi prefecture. It is seen in Table 4 that the total length of damaged dykes shown in Table 3 reaches
20.2 km (4.8% of total dyke length 422 km), and if the slight damage which was repaired by
the maintenance works budget are included, the damaged length increases to 7% of the total
dyke length.

The most severe failure and settlement were observed at the left side bank of the
downstream of Kitakami River (5.2 ~ 10 km), the right side bank of Yoshida River (13.2 ~
18.2 km), a branch river of Naruse River, and the right side bank at a distance of 4 km,
and the left side bank at a distance of 5 km from the mouth of the Natori River. Table 5
shows the length of embankments where the cracks or settlements took place along Yoshida
River and Kitakami River. At the right side bank of Yoshida River the settlement over 1 m
was seen as long as 242 m. As shown in Photo 1 the dykes at this site were so extremely
destroyed that the repair work was performed by recompaction of the dykes. In this area
the cracks were observed at the top and slope of the embankments, the maximum settlement
of the top of the dykes was 1.5 m, and the width of maximum crack was 60 cm.

3. MICROTOPOGRAPHICAL FEATURES AND DAMAGES TO RIVER DYKES

In this section the relationships between the microtopographical features and the
damage to the river dykes are discussed. For the area along large rivers throughout Japan a
Land Form Classification Map for Flood Control Planning has been compiled. This map provides
the microtopographical features near the rivers.

Table 6 demonstrates the relationship between the number of sites, the length of damaged
dykes, and the topographical features determined by using the map described above. This
table includes only rivers directly controlled by the Japanese government. If a damaged site
is located in two microtopographical regions the site is weighted as 0.5. Table 6 denotes
that 54% of the total number of the damaged dykes are constructed on the "flood plain" and
that 14% belongs to the "former river bed," and 60% of the total length of damaged dykes
occurred on the "flood plain." Dykes on "reclaimed land" are lower in number. This fact
suggests that the dykes on soft ground were more severely damaged.

The Land Form Classification Map for Flood Control Planning was compiled using the topographical maps surveyed on and after 1911, and so the history of the river is partially included in the map. However, in general, the geotechnical characteristics of the subsoils of the river dykes which have a lengthy history are not well known, so the map can be useful in estimating the geotechnical characteristics of the soil.

Figure 2 illustrates the ratio of the total damaged length of the Eai River dykes to the whole length of dykes in each topographical classification, whereas Table 6 showed the percentage of the length (or number) of damaged dykes in each topographical classification to the total length (or number) of damaged dykes. For a river flowing in an alluvial plain the length of dykes located in the "flood plain" is so much longer than that of any other classification that the percentage of the damaged length of dykes on "flood plain" to the total damaged length is the highest as shown in Table 6. But Figure 2 shows that the highest damage ratio to be the microtopographical classification "former river bed." This indicates that dykes on a "former river bed" easily become unstable under seismic loading. In Figure 2 the damage caused by the North Miyagi Earthquake in 1962 (M = 6.5) is shown and the ratios of damage on former river bed sites in both earthquakes are shown to differ. The details of this will be described later. The sites examined to construct Figure 2 include not only the places rebuilt during the disaster repair work but the places repaired by the maintenance work as well.

4. LIQUEFACTION OF SUBSOILS UNDER RIVER DYKES

Sand and water spouting was observed on the ground surface just after the Miyagi-Ken-Oki Earthquake. This indicates that liquefaction took place in the ground. It is reported [3] that liquefaction occurred at 30 sites. Since sand spouting was observed at several points where the river dykes were damaged, it is believed that liquefaction is one of main causes for the damage of the river dykes.

There were extensive cracks and failures to the dykes on the Eai River by the North Miyagi Earthquake of April 30, 1962 (M = 6.5). It is estimated that the maximum acceleration of the surface ground near Eai River (0 ~ 27 km) was almost the same for both earthquakes of 1962 and 1978.

The geological profiles of this area are shown in Figure 4. The alluvial deposit is piled up on tuff bed rock for a thickness of 20 ~ 60 m. The thickness of the alluvial deposit is about 60, 16 ~ 20, and 40 m for upper, middle and lower stream areas, re tively. Figure 5 illustrates the damage distribution of river dykes by two earthqu described above by denoting the damaged spots with letter A for 1962 earthquake and letter B for 1978 earthquake. It indicates that the intense damage is concentrated upper and lower areas of Eai River where thicker layers of alluvial soils are depos while no damage took place in the middle stream area in both earthqukes, where the deposit is less than 20 m in thickness. The microtopographical condition underlayt dykes was examined by using the Land Form Classification Map for Flood Control Plan

The dykes of Eai River were classified into five groups according to the class of microtopographical features on this map: most of the dykes lie on the "flood pl the "natural levee"; the rest are on the "former river bed," the relatively higher low land surface," or the "hill." From this classification, the damaged sections of dykes belong to the "former river bed," "flood plain" or "natural levee." Table 7 s relationship between the microtopographical features and the extent of damage. It c seen in this table that the damage length ratio on the "former river bed" site in th earthquake was much smaller than it was in 1962, while the damage length ratios of tl classification remained nearly the same.

The total length of dykes damaged by the 1962 earthquake was about 1.5 km and tl the 1978 earthquake was about 1 km. It is estimated that some of these damaged dykel destroyed by liquefaction from the sand spouting observed near the dykes.

In Table 8 are shown six sites where liquefaction occurred in each earthquake oɪ Three places out of these six seem to be located on the "former river bed," and site very near the former river bed, so it appears that a sandy layer may exist in these ɪ sites.

It is interesting that the liquefaction occurred during the 1962 earthquake and liquefaction took place during the 1978 earthquake at Ninofukuro, Gokenyashiki, Hayaɪ and Eai, although the acceleration at the ground surface was almost same for both eaɪ as shown in Figure 3.

The boring logs at Ninofukuro and Hayamado are shown in Figure 6. The alternaɪ sandy and clayey layers are seen. It can be estimated from the geotechnical data oɪ in 1979 that the sand layers from the depth of 8.5 to 10 m at Hayamado and from the of 3.3 to 7.5 m at Nonofukuro were liquefied.

Dynamic triaxial tests were performed on undisturbed soil samples taken from the layers described above. The relative density Dr_1 at the time of 1962 can be calculated from the following equation by substituting the liquefaction stress ratio R $(= \frac{\tau}{2\sigma_c'}$, τ: shear stress, σ_c': effective confining pressure) obtained from the dynamic triaxial tests and the relative density Dr_2 in 1978.

$$Dr_1 \leqq (L/R_2) \times Dr_2 \tag{1}$$

where L is the stress ratio which was developed in the ground during the 1962 earthquake, which was calculated by using the equation proposed by Iwasaki, et al. [4]. The results are shown in Table 9.

The settlement of the estimated liquefaction zone in Figure 6 was calculated from the following equation by substituting the void ratio in 1962 and the recent one.

$$\Delta H = \frac{e_1 - e_2}{1 + e_1} \times H \tag{2}$$

The results are shown in Table 10. This table indicates that the settlement due to dissipation of pore water pressure developed by the liquefaction during the 1962 earthquake is 0.7 cm at Hayamado and 42.5 cm at Ninofukuro. According to the survey just after the 1962 earthquake, the settlement of the dyke at Ninofukuro was about 50 cm while no settlement of the dyke was observed at Hayamado, although the revetment was extensively damaged.

Also while the result presented above is based on some assumptions, it nevertheless suggests that where a sandy ground is compacted by an earthquake as a result of liquefaction, little or no liquefaction will occur at the same place during a subsequent earthquake with a similar order of ground motion.

In Figure 2 of the previous section, the damage ratio of dykes on the former river bed was 41% for the 1962 earthquake, and it decreased to about 6% for the 1978 earthquake. It is reasonable to note that such a phenomenon can be attributed to the change of liquefaction characteristics of subsoils underlying the dykes due to seismic loading. However re-liquefaction was observed at the left side bank of Natori River (2.8 km) after the earthquake of June, 1979, while sand spouting was observed just after the earthquake in February, 1978, so investigation on reliquefaction should be continued.

5. FAILURE OF RIVER DYKES ON THE SOFT CLAY SUBSOILS

The Yamazaki area of the Yoshida River dykes (right side bank 15 ~ 18 km) was severely damaged. The cross section of the dykes after the earthquake is shown in Figure 7 and the settlement of the embankments and the large mouthed cracks were observed. The maximum settlement of the embankment in Yamazaki area was 2 m and the maximum width of cracks at the top of the levee was 60 cm.

The dykes in this area were constructed in the 1920's. The right side bank was moved outside the river and both side banks were enlarged by the construction work of 1949.

Figure 8 illustrates the distribution of the earthquake disaster in dykes along Yoshida River and the position of former dykes. In Figure 8 the extent of damage of dykes is denote by A, B, C corresponding to the amount settlement of the tops of the dykes. This figure indicates that only dykes along the right bank were damaged by the earthquake and that no extensive failure was observed to dykes along the left bank. Figures 9 and 10 show soil profiles beneath the dyke in the cross-section and longitudinal directions, which were obtained from a boring and sounding survey after the earthquake. These figures show that the ground of this area consists of an alternation of sandy and clayey strata. As far as the sounding data obtained by the Datch cone penetration test and Swedish sounding are concerned, it was difficult to clearly point out geotechnical difference in subsoil condition between the right bank and the left bank. It is recorded that the construction work of the dyke was hard because of the settlement and foundation failure due to the low strength of the layer. It leads to the estimation that the dykes had low stability during the earthquake.

In Figure 11 former dykes are shown by dotted lines. Figure 11 also shows the new dyke of larger scale which was constructed outside the river from 1949 to 1954, and the distance between new and old dykes which reaches 80 ~ 100 m on the right bank. On the other hand, the new dyke for the left bank was built up at almost the same place as the old one. This historical aspect of the dyke may be associated with the variations of disaster severity by the earthquake on the right side and left side dykes. However, further investigation is necessary to pursue this possible relationship.

6. CONCLUSIONS

The following conclusions were obtained from the investigation:

1) The river dykes in Miyagi prefecture were extensively damaged by this earthquake.

2) The seismic resistance of river dykes is affected by the stability of supporting subsoils during an earthquake: either by a tendency toward liquefaction, or by the bearing capacity of the subsoils.

3) The river dyke damage has a close relationship to the microtopographical features. Dykes on a "former river bed" have a high potential to be damaged by earthquakes. It is suggested by the analyses of the Eai River that the Land Form Classification Map for Flood Control Planning is useful in predicting river dyke damage.

4) Re-liquefaction may or may not occur at the same place. It was observed to occur at a layer which had been compacted by previous earthquakes.

ACKNOWLEDGMENTS

The authors would like to express their heartiest appreciation to the staff members of the river control section, the disaster prevention section of the river bureau, the Geographical Survey Institute, the Tohoku regional construction bureau, the construction office of the downstream area of Kitakami River, and to the Ministry of Construction for their help in carrying out the damage survey and field investigations.

REFERENCES

[1] The Land Classification Map for Flood Control Planning, Geographical Survey Institute.

[2] Sasaki, Y., Taniguchi, E., and Funami, K., "The Damage of River Dykes by Earthquakes and the Hysterisis of Subsoils," The annual meeting of JSCE, 1979 (in Japanese).

[3] Yamamura, K., Ihasaki, T., Sasaki, Y., Koga, Y., Taniguchi, E., and Tokida, K., "Ground Failures and Damage to Soil Structures from the Miyagi-Ken-Oki, Japan Earthquake of June 12, 1978," U.S. National Conference on Earthquake Engineering, 1979.

[4] Iwasaki, T., Tatsuoka, F., Tokida, K., and Yasuda, S., "A Practical Method for Assessing Soil Liquefaction Potential Based on Case Studies at Various Sites in Japan," Proceedings of the Fifth Japan Earthquake Engineering Symposium, 1978 (in Japanese).

Table 1 Number of sites where restoration works were carried out
with assistance of the Japanese government after the
Miyagi-Ken-Oki earthquake

Prefecture	River	Coast	Sabo	Road	Bridge	Total
Miyagi	167	14	1	549	70	801
Iwate	8	4	-	108	16	136
Akita	-	-	-	3	-	3
Yamagata	1	-	-	6	1	8
Fukushima	-	-	-	1	2	3
Total	176	18	1	667	89	951

Table 2 Number of Sites where restoration works were carried
out with assistance of the Japanese government after
the Izu-Oshima-Kinkai earthquake

	River	Coast	Sabo	Road	Bridge	Total
Number	45	-	-	637	4	687

Table 3 Number of sites where river facilities were repaired
after the Miyagi-Ken-Oki earthquake

Epicentral distance	Directly controlled by the Japanese government			With assistance of the Japanese government			Total
	Dyke	Revetment	Others	Dyke	Revetment	Others	
Δ= ∿ 80Km	8	-	-	-	7	-	15
80 ∿ 90	3	6	1	-	14	2	26
90 ∿ 100	14	-	1	5	6	2	28
100 ∿ 110	35	7	1	33	26	3	105
110 ∿ 120	5	4	1	10	36	4	60
120 ∿	-	-	1	5	11	3	20
Total	65	17	5	53	100	14	254

Table 4 Damaged dykes in Miyagi prefecture directly controlled by the Japanese government

number in ( ) shows the number of sites

	Controlled length	Length of dykes			Damaged dykes (disaster restoration works)		Damaged dykes (maintenance works)		Damaged dykes (total)	
		Complete dyke	Provisional dyke	Total	Length	Ratio	Length	Ratio	Length	Ratio
Kitakami	137.5km	78.5km	104.6km	183.1km	(19) 8.9km	4.9%	(35) 3.1km	1.7%	(54) 12.0km	6.6%
Naruse	82.4	22.2	111.0	183.2	(26) 7.7	5.8	(45) 2.6	2.0	(71) 10.3	7.7
Natori	29.4	28.6	3.4	32.0	(17) 3.0	9.4	(52) 2.8	8.8	(69) 5.8	18.1
Abukuma	63.7	20.5	53.5	74.0	(3) 0.5	0.7	(14) 1.2	1.6	(17) 1.7	2.3
Total	313.0	149.8	272.5	422.3	(65) 20.2	4.8	(146) 9.6	2.3	(211) 29.8	7.1

Table 5 Length of damaged dykes

(m)

		Yoshida River	Kitakami River
Crack		6,219	4,469
Settlement	1m ～	242	0
	0.5 ～ 1m	321	1,866
	0.2 ～ 0.5m	343	2,368
	Total	906	4,234

Photo 1 Damage of Yoshida River dyke

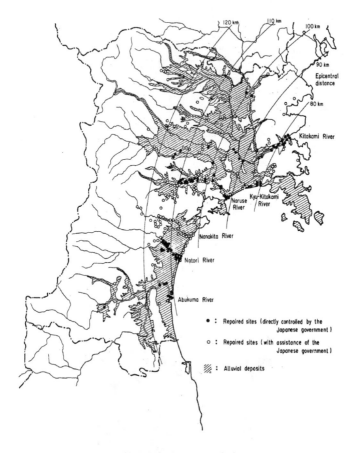

Fig. 1 Damaged sites of river dykes by the Miyagi-Ken-Oki earthquake

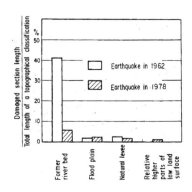

Fig. 2 Microtopographical features and rate of damage
to dykes

Table 6 Microtopographical features and damaged site
and length

	Natural levee	Relative higher parts of low land surface	Dune	Former river bed	Flood plain	Reclaimed land	Total
Sites of damaged revetment	(13.1) 8.5	(7.7) 5	(6.2) 4	(13.8) 9	(53.8) 35	(5.4) 3.5	(100%) 65
Length of damaged dyke	(8.6) 1.73	(8.0) 1.61	(1.3) 0.27	(4.4) 0.89	(60.0) 12.12	(17.8) 3.59	(100%) 20.21
Sites of damaged revetment	(47.1) 8	(17.6) 3	(11.8) 2	-	(17.6) 3	(5.9) 1	(100%) 17
Length of damaged revetment	(26.9) 0.35	(9.2) 0.12	(39.2) 0.51	-	(21.5) 0.28	(2.3) 0.03	(100%) 1.30
Total damaged sites	(20.1) 16.5	(9.8) 8	(7.3) 6	(11.0) 9	(46.3) 38	(5.5) 4.5	(100%) 82
Total damaged length	(9.7) 2.08	(8.0) 1.73	(3.6) 0.78	(4.1) 0.89	(57.6) 12.40	(16.8) 3.62	(100%) 21.51

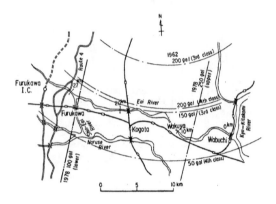

Fig. 3 Estimated acceleration at the ground surface near
Eai River by two earthquakes in 1962 and 1978

Table 7 Microtopographical feature and damage to dykes
in Eai River

Micro-topographical feature	Total length of dykes		Damaged dykes by the 1962 earthquake			Damaged dykes by the 1978 earthquake		
	Site	Length	Site	Length	Rate of damage	Site	Length	Rate of damage
Former river bed	9	km 1.23	2	km 0.5	% 40.7	1	km 0.07	% 5.7
Flood plain	47	22.17	4	0.43	1.9	4	0.52	2.3
Relative higher parts of low land surface	16	8.07	–	–		1	0.01	0.1
Natural levee	40	25.45	7	0.61	2.4	6	0.47	1.8
Mountain, hill, platform	6	0.84	–	–		–	–	
Total	118	57.76	13	1.54		12	1.04	

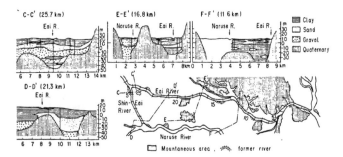

Fig. 4 Geological Profiles Near Eai River

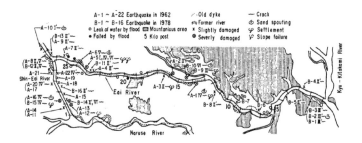

Fig. 5 Damaged Points by the Two Earthquakes in 1962 and 1978

Table 8 Liquefaction of subsoils of Eai River dykes

	Site	Microtopographical feature	Kilopost	Earthquake causing liquefaction
a	Wabuchi (B-1∿3)	Former river bed	Right side bank $0.6^{Km}+125\sim1.0^{Km}+46$	Only the 1978 earthquake
b	Ninofukuro (A-1)	Flood plain	Right side bank 14.0∿14.0+100	Only the 1962 earthquake
c	Kamiyaji (B-9, 10)	Natural levee	Left side bank 14.4+70∿14.6+10	Both
d	Gokenyashiki (A-2)	Former river bed	Left side bank 15.8∿15.8+100	Only the 1962 earthquake
e	Hayamado (A-6)	Former river bed	Left side bank 23.8∿24.2	Only the 1962 earthquake
f	Eai (A-10)	Flood plain	Right side bank 27.0+100∿27.2	Only the 1962 earthquake

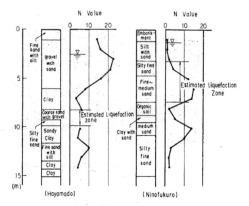

Fig. 6 Columnar section at Hayamado and Ninofukuro

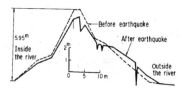

Fig. 7 Cross section of damaged dyke (Yoshida River,
right side bank 18.0 km)

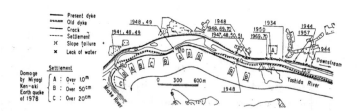

Fig. 8 Distribution of Earthquake Disaster along Yoshida
River and Historical View of Slope Failure and
Leak of Water by Flood

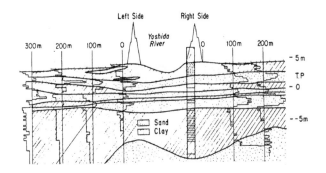

Fig. 9 Cross section of the ground
(Yoshida River)

Table 9 Estimation of relative density

Sample No.		Dynamic triaxial test		α_{max} = 150 gal	
		R	Dr (%)	L	Dr (%)
Hayamado	S1-6 (-9m)	0.215	50.2	0.206	48.0
	S1-8 (-12.7m)	0.310	56.9	0.210	39.0
Ninofukuro	S4-1 (-1.7m)	0.38	97.2	0.175	45.0
	S4-3 (-4.7m)	0.42	68.0	0.239	39.0

Table 10 Estimation of settlement

Site	e_{max}	e_{min}	before liquefaction		After liquefaction		Liquefaction layer		Settlement (cm)
			Dr (%)	e	Dr(%)	e	Depth (m)	Thickness(m)	
Hayamado	0.985	0.612	48.0	0.806	50.2	0.798	8.35 ~ 9.90	1.55	0.7
Ninofukuro	1.754	0.898	39.0	1.420	68.0	1.172	3.25 ~ 7.40	4.15	42.5

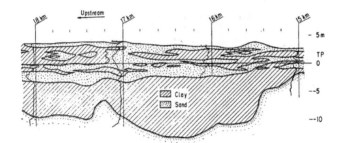

Fig. 10   Longitudinal Section of the Ground
(Yoshida River)

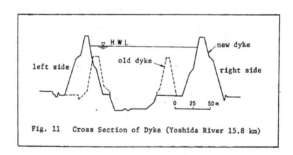

Fig. 11   Cross Section of Dyke (Yoshida River 15.8 km)

DAMAGE TO PORT STRUCTURES BY THE

1978 MIYAGI-KEN-OKI EARTHQUAKE

Hajime Tsuchida

Setsuo Noda

Structures Division

Port and Harbour Research Institute

ABSTRACT

The 1978 Miyagi-ken-oki earthquake (M = 7.4) caused damage to port facilities. From field investigations and analyses, the following lessons concerning earthquake engineering were learned:

1) Strong-motion accelerograms were recorded by an observation network of the Port and Harbour Research Institute. The largest peak ground acceleration of 280 gals was recorded at Shiogama Port.

2) Port facilities were damaged seriously when a backfill liquefied. Otherwise, they suffered only lightly.

3) Current procedures for estimating liquefaction potential can lead to an appropriate judgement for actual sites.

4) A relationship between the seismic coefficient and the maximum ground acceleration for gravity quaywalls: $e = \frac{1}{3} \left(\frac{\alpha}{g}\right)^{1/3}$ agrees with the investigations of this earthquake.

5) The above-mentioned relationship may be equally applicable to a stability calculation for sheetpile quaywalls with anchor plates.

KEYWORDS: Earthquake damage; seaport damage characteristics; site liquefaction; stability analysis; strong motion accelerograms.

1. INTRODUCTION

Ports and fishing ports in Miyagi prefecture and Fukushima prefecture were the 1978 Miyagi-ken-oki earthquake. In order to collect engineering information assist in the reconstruction plan of the damaged facilities, staffs of the Burea and Harbours, the Second District Port Construction Bureau, and the Port and Har Institute were immediately sent to the damaged ports [1,2]. In this paper the a sent characteristics of damage on port facilities, strong-motion accelerograms i analyses of liquefaction at the actual sites, and comparison of the maximum groun tions and the seismic coefficients by the stability analyses of the gravity $q_{ua y}$ sheetpile quaywalls with anchors.

2. DAMAGE OF FACILITIES IN PORTS AND FISHING PORTS

2.1 Outline of Damage

The public facilities in ports and fishing ports were damaged in both the Mi Fukushima prefectures. The number of the damaged facilities and the cost of dama$ ports are shown in Table 1. 98% of the total cost corresponds to damage in Miyag; ture. Table 2 shows the detail of damage in Miyagi prefecture. This table shows corresponding to 90% of total cost happened in Ishinomaki port, and 95% of the co: port corresponds to damage of the quaywalls, except the apron. This cost correspo of sum amount of damage to the port facilities. Damage on port facilities in Fuku fecture occurred in Shoma port. Damage in fishing ports occurred mainly at Ishino Yuriage in Miyagi prefecture. Consequently damage in ports and fishing ports was trated Ishinomaki port and both Ishinomaki and Yuriage fishing ports. Damage to t ports was not so severe.

Damage cost comparisons to the port facilities from this earthquake to those recent earthquakes follows: 22 billion Yen in the 1964 Niigata earthquake (M = 7. billion Yen in the 1968 Tokachi-oki earthquake (M = 7.8), 0.5 billion Yen in the 1 hanto-oki earthquake (M = 7.4), and 3.1 billion yen in the 1978 Miyagi-ken-oki ear (M = 7.4). It is judged that this earthquake was not small from view point of the damage.

2.2 Damage to Gravity Quaywalls

Many gravity type quaywalls and revetments were damaged by the earthquake. Typical aspects of the damage were: sliding by gravity wall toward sea, tilting and settlement of the wall, and settlement, cracking and opening of the joint of the apron as shown in Figure 2. This damage is similar to the damage on gravity structures during past earthquakes. Swelling of the wall toward the sea was less than 10 cm in most cases and settlement of the apron occurred to the same extent. Most of the gravity facilities did not lose their function and were used after the earthquake without any trouble. However most of the gravity quaywalls in Yuriage fishing port were damaged seriously as shown in Figure 3. They were useless immediately after the earthquake because the walls swelled 122 cm in maximum and the concrete pavements of the aprons crashed seriously.

2.3 Damage to Steel Sheetpile Quaywalls

The extent of damage to steel sheetpile quaywalls may be broadly divided into the following three groups:

a) Quaywall not damaged or damaged very slightly. Repair is not necessary.

b) Slight repair necessary because settlement and cracking in the pavement of apron occurred. The main parts of quaywalls such as steel sheetpile, tie rod, and anchor not damaged.

c) The front line of the quaywall moved measurably toward sea, the main part of the quaywall needs repair.

Five steel sheetpile quaywalls were classified as category (c). Figure 4 shows the plane of the major part of Ishinomaki port which is located at the west side of the mouth of the Kitakami river. Both the Nakajima port wharf walls of -10 m water depth, and the Hiyori wharf (-9 m) slide toward the sea (57 cm maximum) and the wall in Shiomi wharf slid by 119 cm at maximum. Damage to these quaywalls were also classified category (c). As shown in Figure 5, the type of anchor used with these quaywalls was steel sheetpile. The Ishinomaki fishing port is located on the east side of the mouth of the Kitakami river. At the Junbi wharf (-7 m) a wall with a single vertical pile anchor slide (34 cm maximum). At the back of these quaywalls settlement of the ground, and sand at cracks and joints in the pavement was observed. Therefore it was recognized that the backfilling sand had liquefied.

Maximum swelling of the quaywall (-3.5 m) at Yuriage fishing port was 106 cm and the extent of the damage belong to category (c). Although many traces of sand boiling were found around the port, they were not found any place inside the port.

The extent of damage to quaywalls, except the above-mentioned, were classified into either category (a) or (b). The general state of damage was the cracking of the pavement just above the anchor, and the swelling and the settlement of the ground in front of and in the rear of the anchor, respectively. In these cases the marks of the sand liquefaction were not found on the ground. When the tie rods were dug up for inspection in several damaged sites, the soil condition around the anchors were observed but the marks of liquefaction were not found. The anchor type used with these slightly damaged quaywalls was limited to the plate shown in Figure 7, or the vertical steel pipepile.

At Nakano wharf in Sendai port the steel sheetpile quaywalls with two kinds of anchor adjoined each other. This was the quaywall with the steel sheetpile anchor shown in Figure 7, and with the coupled-piles anchor as shown in Figure 8. Cracking, settlement, and swelling of the pavement just above the anchor was found in the former, while nothing happened to the latter during this earthquake. This fact shows the different earthquake resistance of the quaywalls caused by the anchors.

When the backfill sand liqufied, the steel sheetpile quaywalls suffered severely. This fact was similarly observed at Niigata port in the 1964 Niigata earthquake and Hanasaki port in the 1973 Nemuro-hanto-oki earthquake. When the backfill sand did not liquefy, damage to the steel sheetpile quaywalls was minimal. This judgement was based on external appearance. Judging from the change of the ground surface in front and in the rear of the anchor, it was also estimated that the lateral resistance of anchor almost reached the maximum. If the anchor breaks down, the sheetpile quaywalls immediately collapse. Even if the sheetpile deforms slightly, considerable stress remains in sheetpile and tie rod, and it possibly decreases the stability of the structure. While, many gravity type quaywalls slid and settled slightly, it was held that the stability of the structure did not decrease significantly. Therefore we cannot absolutely judge the extent of damage from the change of appearance such as swelling of the wall front toward the sea and the settlement of the wall capping.

2.4 Steel Pipe Pilled Wharf

Steel pipe piled wharves have never been damaged by past earthquakes, nor were they during this one. The gravity quaywalls and the steel sheetpile quaywalls were severely damaged in the Ishinomaki port and the Yuriage fishing port, however, the steel pipe piled wharves neighboring them were undamaged except for the retaining wall behind the wharves. Figure 9 shows the -3.5 m wharf in Yuriage fishing port which was reconstructed in front of the gravity type quaywall. It is thought that the steel pipe pile resisted the large horizotal force through the access bridge during the earthquakes, and at the Ohte wharf in Ishinomaki port. Judging from the foregoing, it was confirmed again that the steel pipe pile wharf are very earthquake-resistant.

3. STRONG-MOTION EARTHQUAKE RECORDS IN PORT AREAS

3.1 Maximum Ground Acceleration in Ports

The earthquake triggered 34 strong-motion accelerographs in the recording network appeared by the Port and Harbour Research Institute [3]. The ports where the observation stations exist are shown in Figure 10 with Arabic numerals. The maximum accelerations in each port are also listed up in Table 3. Abbreviated names of stations in Table 3 show the name of the site and the type of accelerographs, that is, "S" indicates SMAC-B2 and "M", ERS-B (or ERS-C), respectively. The largest peak ground acceleration in the port areas during this earthquake was recorded at Shiogama port as 280 gals which is larger than the 233 gals recorded at Hachinohe port in the 1978 Tokachi-oki earthquake. The ground condition of the observation site at Shiogama port is fill from surface to -3.4 m (N < 3), tuff clay to -10.6 m, coarse sand with silt (N > 3) to -13.75 m, sandy tuff to -14.90 m, and tuff under -14.9 m (N < 50). The maximum acceleration of 170 gals was recorded at Ofunato port which is almost the same epicentral distance as the Shiogama port.

3.2 Major Wave Forms and Their Response Spectra

Figures 11 and 12 show the main aspects of the acceleration records at Shiogama port and Ofunato port, respectively. Figure 13 shows the displacements (the EW components) which were calculated by integrating the acceleration records. The method of integration was proposed by one of the authors. [4] The filter cut-off frequency employed in the integration is shown as Fc in Figure 13.

The response spectra of the accelerograms (EW component) at Shiogama port and Ofunato port are shown as solid lines in Figures 14 and 15. Spectra indicated by dotted lines were obtained from the records of past earthquakes. These earthquakes were rather small (M < 6.4), while the earthquake of February 20, 1978 is not included. According to Figure 15, spectra calculated by accelerograms in rocky ground are similar in shape. Spectra in this earthquake (Figure 14), however, are different in shape because the surface layer of about 15 m in thickness responded in a different manner for very strong acceleration.

3.3 Predominant Direction of Ground Motion

A predominant direction did not clearly appear from the orbit of the two horizontal components of the acceleration records. However, orbits of the displacement records did show the predominant direction of the ground movement at the stations which were located close to the source region of rather simple ground condition. Orbits at Shiogama, Ofunato, Miyako and Ishinomaki are shown in Figures 16 through 19. An acceleration record at Ishinomaki was observed at Kaihoku-bashi which is located close to Ishinomaki port and it is offered by the Public Work Research Institute. [5] Displacements were obtained from the calculation by the source model in which some parameters proposed by Seno [6] and the temporary values are as shown in Table 4. In this calculation of the displacement, the amplification of the surface layers was not considered. The orbits by this method of calculation are also indicated in Figures 16 through 19. Comparing the two kinds of orbits, they agree well on the predominant directions of ground motion, but the absolute quantity of displacement are different. The predominant directions by the observed records in each port, and the fault plane of the earthquake, are shown in Figure 20.

4. CASE STUDY OF LIQUEFACTION

4.1 Damage of Hiyori Wharf in Ishinomaki Port

There were steel sheetpile quaywalls with -9 m and -10 m water depths at Hiyori wharf as shown in Figure 4. At the part of the wharf with a -9 m water depth, it is clear that subsoil behind the wall liquefied. This judgement is based on the evidence that the pavement settled remarkably, and sand arising from cracked pavement and from the openings between the wall capping and the pavement. However, at that part which was of -10 m water depth the wall swelling toward the sea and the settlement of the ground behind the wall did not occur, it was judged that the backfill did not liquefy. Both the quaywalls at the Hiyori wharf are adjacent to each other, and it is supposed that the intensity of earthquake motion were almost the same at the base of these sites. Therefore, the reason why liquefaction happened at one site and not at the other was investigated.

4.2 Subsoil Condition and Damage of Quaywall

N-values of the subsoil at four sites along the wall front of the Hiyori wharf are shown in Figure 21. The data at CB18, 14 and 10 were obtained by soil exploration after the earthquake, and it is judged that they suffered the disturbance due to ground vibration and liquefaction. However comparison of N-values before and after the earthquake at Shiomi wharf did not show any difference between them. Therefore, it was assumed that this fact was also applicable to the subsoil at Hiyori wharf. The values at No. 5 is the result of standard penetration test prior to the earthquake. It was judged that the small N-values at CB-18, 14 and 10 were caused by the ejector for making sand piles in foundation improvement work to lower layer. Figure 21 shows the swelling of the wall front line toward the sea and the settlement of the asphalt pavement 3 m behind the capping. From this figure it is apparent that the quaywall was damaged at -9 m water depth and to the adjacent part of it. Figure 22 shows the grain size distribution curves of soil at the depth of -1.0 to 1.4 and -3.0 to 3.4 m at CB18, 10 and No. 5, respectively. In Figure 22 the values at CB18 and 10 were also obtained from the result before the earthquake. Judging from Figure 22, the grain size distribution of the sand at any places in this wharf were almost same.

4.3 Investigation of Subsoil Condition Against Liquefaction

It was clear in the previous section that N-values of the subsoil were different in the liquefied and non-liquefied sites. In order to make this fact clear, N-values at each site are plotted in Figure 23. This figure shows the clear different of N-values between them. This difference may be the reason why the subsoil did or did not liquefy.

Dynamic response during the earthquake was estimated from the CB14 site which was modeled into the six lumped mass system. Consequently, it was estimated that the maximum acceleration was about 250 gals at ground surface and the average value of the maximum shear stress in the -4.7 m to -13.9 m sandy layer was almost 0.7×10^{-3}. The number of cyclic waves was six when the shear stress reached half the maximum value. As the grain size distribution at the site are similar to the Bandai-jima site in Niigata port, the liquefaction potential at this site was estimated by the results of dynamic tri-axial test for Bandai-jima sand. According to this investigation the site was judged to have a tendency to liquefy during the earthquake. Although some input data were possibly inexact in this investigation, the estimation agreed reasonably with the fact of liquefaction at the site.

In the design of port structures the possibility of liquefaction is commonly based on the grain size distribution and N-value [7,8]. If the grain size distribution for uniform soil is in the range of (B) in Figure 22, it is considered the subsoil might possibly liquefy under certain conditions. It is in the range of (A), the soil is quite certain to be in the bounds of possibility of liquefaction. The relationship between the critical N-value for liquefaction and the maximum ground acceleration are based on the field experience in past earthquakes and the laboratory tests. If the N-value of the subsoil below the ground water table is smaller than the critical one, it is judged that the subsoil may liquefy during an earthquake. Where the maximum acceleration at ground surface is 250 gals as in the previous calculation, the critical N-value becomes 14. When the critical N-value was compared to those of sites at the Hiyori wharf, it was larger than those at liquefied sites and smaller than those of non-liquefied sites except in very shallow layers. Therefore this procedure for liquefaction estimation was considered to be one of some use. However, as the result in this method of estimation depends upon the maximum acceleration of ground surface, input data must be carefully selected for calculating the ground response.

5. STABILITY ANALYSES OF GRAVITY QUAYWALLS AND SHEETPILE QUAYWALLS

5.1 Facilities

The stability of facilities in several ports were analysed using the current design standard [7] and the results were compared to the extent of damage and the severity of ground motion during the earthquake. The port facilities investigated number 60 quaywalls in 11 ports, that is, 36 gravity quaywalls in 10 ports, and 24 steel sheetpile quaywalls in 7 ports. The classification of these facilities are listed in Table 5 in terms of type of structure and extent of damage. The classification of extent of damage in the table is the one proposed by Kitakima and others for considering the damage records of port facilities since the 1964 Niigata earthquake [9,10]. The steel sheetpile quaywalls which were damaged by liquefaction of the backfill was not discussed in this chapter.

5.2 Estimation of Maximum Ground Acceleration in Ports

The large values of ground acceleration were recorded around Sendai, Shiogama and Ishinomaki cities. The earthquake triggered 34 strong-motion accelerographs of the network of the Port and Harbour Research Institute. Two of them were installed on the ground of Shiogama port and Ofunato port. When earthquake resistant design is investigated through analyses of damaged structures, the severity of the earthquake motion acting on the structures must be estimated first of all. Therefore, the maximum ground accelerations at the ports, except the two ports above, was estimated by calculating the ground response during the earthquake. The epicenter location was given by the Tohoku University as shown in Figure 1 and the ground motions were calculated by the SHAKE program. As the incident wave form the acceleration records which were used and were observed on the rocky site at Ofunato port and Kaihoku Bridge. The source region and the maximum acceleration of base rock motion in each port were determined using the same procedure which was described in the previous report [11]. Maximum ground accelerations in each port were estimated as shown in Figure 24. The expression $\big|$ indicates that the estimated acceleration exists in the range. As the facilities in Kinkazan, Hagihama, Onagawa and Okatsu were directly built on the rocky ground, the ground accelerations were assumed to be same as those at the base rock.

5.3 Stability Analysis of Gravity Quaywalls, and Relationships Between Seismic Coefficient
     and Maximum Ground Acceleration

The stability of the gravity quaywalls and bulkheads were analysed using current design procedure, and the seismic coefficients which correspond to the severity of the seismic effect were estimated by the way reported in the previous paper [11]. The outline of the method is shown in Figures 25 and 26. For example, the stability of the sliding, the overturning, and the bearing capacity at the base of structures for each port facility were analysed, and the seismic coefficients which give the unit value of the safety factor were obtained. Judging from the characteristics of the damage to facilities, the causes of damage were estimated for the above-mentioned factors. If the facility was damaged due to one of these factors, the actual seismic coefficient corresponding to the seismic effect may be larger than the calculated seismic coefficient and it is plotted by the expression $\uparrow$ as shown in Figure 26.1. If the facility was not damaged, the actual seismic coefficient may be smaller than the calculated one and it is plotted by the expression $\downarrow$. Considering these

VII-87

results, the seismic coefficient acting on the facilities during the earthquake in Shiogama port is judged to fall in the range of the dotted lines. In case of Kinkazan port in Figure 26.2, the seismic coefficient is estimated to exist in a more limited range. In case of Onagawa port in Figure 26.3, the upper boundary is only estimated because no facilities were damaged. The seismic coefficients or their ranges were determined in 10 ports in this manner.

The relationships between the maximum ground acceleration and the seismic coefficient at each port were estimated and the results are shown in Figure 27. In the figure, the curved line indicates the proposed relationship which was obtained by the same procedure to the gravity quaywalls in the past earthquakes, and the straight line indicates the relationship for the rigid body on the solid foundation. Even though the range of the estimated values in Shiogama, Matsushima and Hagihama ports range somewhat widely, the results of the estimation show similar characteristics to the proposed curve. However, the maximum ground accelerations in Ofunato and Okatsu port are smaller than 200 gals, and it seems more suitable that the straight line indicates a better relationship than the curved line. As arrows point downward in Sendai and Ishinomaki ports, the expected seismic coefficients in these ports do not always contradict the curved line. Consequently, it is concluded that in this earthquake the relationship between the acceleration and the seismic coefficient for gravity type quaywalls is similarly expressed by the relationship determined by damage records in past earthquakes.

5.4 Stability Analysis of Steel Sheetpile Quaywalls, and Relationships Between Seismic Coefficient and Maximum Ground Acceleration

The current earthquake-resistant design suggestions for steel sheetpile quaywalls are analyzed for the following items, that is, the embedded length and stress of the sheetpile, the stress of the tie rod, and the stability of the anchor. In this chapter the critical seismic coefficients which provide the unit value of the safety factor were also estimated. In the stability analysis the active and passive earth pressure on the wall and the lateral resistance of the anchor piles are the most important items, but the technical level of the design for the latter is not as reliable when compared to the former. Therefore, investigation was limited to the two items but was not considered for all items in the case of the gravity quaywalls.

VII-88

All of the critical seismic coefficients of the bending stress of the sheetpile and the tensile stress of the tie rod were larger than 0.25. These results are not in contradiction with the fact that the sheetpile and the tie rod did not break down in the actual event.

The critical seismic coefficients for the embedment of sheetpile were compared to the maximum ground acceleration in Figure 28. As no damage due to the lack of the embedment length occurred during this earthquake, only the upper limit of the seismic coefficient was estimated. The results of the stability analyses in past earthquakes also show a similar tendency in the figure. It might be concluded that the seismic coefficients on the necessary length of embedment of sheetpile did not increase from increasing the ground acceleration. But, the results were not sufficient to make clear the relationship between them.

As the typical damage to the steel sheetpile quaywalls was the swelling of the wall toward sea due to the movement of the anchor, the most important item in the analyses was the stability of the anchor. The relationship between the maximum ground acceleration and the seismic coefficient for the anchor plate, which is the most common type of anchor, was estimated in Figure 29. The results from 4 ports and those from previous investigations are also shown in the figure. According to the figure the results from 3 ports, except for the Yuriage fishing port, are close to a straight line. The horizontal resisting capacity of the anchor is expressed as the difference between the passive earth pressure in front of the anchor and the active earth pressure in the rear, and the stability of the gravity quaywalls are mainly influenced by the difference in pressure. Judging from these facts, the relationship between the maximum ground acceleration and the seismic coefficient for the anchor is possibly similar to that for the gravity quaywalls. The stability of the anchor by the single vertical pile and the sheetpile was investigated next. The stability of these anchors appears to result from the lateral resistance of flexible members in the soil, however, this is not altogether clear in the present stage. If the swelling of the wall is considered to be equivalent to the lateral displacement of the anchor head, actual damage is ten times larger than the estimated one. This fact has been pointed out [10] and remains a technical problem to be solved. Stability of the coupled-piles anchor are determined mainly by the pulling resistance of the coupled piles. Since the seismic coefficient was obtained only from the upper limit at 2 ports, the relationship between the maximum ground acceleration and the coefficients can not be discussed.

6. CONCLUSIONS

The 1978 Miyagi-ken-oki earthquake (M = 7.4) caused damage to port facilities. As a result of the field investigations and the analyses, the following lessons concerning earthquake engineering were learned.

1) Strong-motion accelerograms were recorded by the observation network of the Port and Harbour Research Institute. The largest peak ground acceleration was recorded at Shiogama Port as 280 gals.

2) Port facilities were damaged seriously when the backfill liquefied. Otherwise, they suffered only lightly.

3) Current procedures for estimating liquefaction potential can lead to appropriate judgements for actual sites.

4) The relationship between the seismic coefficient and the maximum ground acceleration for gravity quaywalls: $e = \frac{1}{3}(\frac{\alpha}{g})^{1/3}$ agrees with the investigations of this earthquake.

5) The above-mentioned relationship may be equally applicable to a stability calculation for sheetpile quaywalls with anchor plates.

REFERENCES

[1] Tsuchida, Noda, et al., "The Damage to Port Structures by the 1978 Miyagi-ken-oki Earthquake," Technical Note of the Port and Harbour Research Institute, in press.

[2] Tsuchida, H., "Damage to Port Facilities Due to the 1978 Miyagi-ken-oki Earthquake," Minato-no-bosai, September 1978.

[3] Kurata, E., Iai, S., Yokoyama, Y., and Tsuchida, H., "Strong-Motion Earthquake Records on the 1978 Miyagi-ken-oki Earthquake in Port Areas," Technical Note of the Port and Harbour Reserch Institute, No. 319, March 1979.

[4] Iai, S., Kurata, E., and Tsuchida, H., "Digitization and Correction of Strong-Motion Accelerograms," Technical Note of the Port and Harbour Research Institute, No. 286, March 1978.

[5] Iwasaki, T., Wakabayashi, S., Kawashima, K., and Takagi, Y., "Strong-Motion Earthquake Records from Public Works (No. 2)," Bulletin of Public Works Research Institute, Vol. 33, October 1978.

[6] Hirasawa, T., "Seismological Characteristics of Miyagi-ken-oki Earthquake," Report to the Technical Meeting of Earthquake Engineering Committee, the Japan Society of Civil Engineering, November 1978.

[7] The Japan Port and Harbour Association, "Engineering Requirements on Port and Harbour Facilities and Its Application," March 1978.

[8] Tsuchida, H., and Hayashi, S., "Estimation of Liquefaction Potential of Sandy Soils," Third Joint Meeting, U.S.-Japan Panel on Wind and Seismic Effects, UJNR, May 1970.

[9] Bureau of Ports and Harbours, First District Port Construction Bureau, and Port and Harbour Research Institute, "Damage to Port Structures by the Niigata Earthquake (First Report)," September 1964.

[10] Kitajima, S., and Uwabe, T., "Analysis of Seismic Damage in Anchored Sheet-piling Bulkheads," Report of the Port and Harbour Research Institute, Vol. 18, No. 1, March 1979.

[11] Hayashi, S., Noda, S., Uwabe, T., "Relation Between the Seismic Coefficient and Ground Acceleration for Gravity Quaywalls," Eighth Joint Meeting, U.S.-Japan Panel on Wind and Seismic Effects, UJNR, May 1976.

Table 1 Damage of Public Port Facilities

Port Managements Body	Number of Damaged Facility	Amount of Money (×1000 Yen)
Miyagi Prefecture	45	3,079,500
Fukushima Prefecture	4	58,500
Total	49	3,138,000

Table 2 Damage of Public Port Facilities in Miyagi Prefecture

( ) : number of facility, unit : 1000 Yen

Port	Quaywall	Quaywall (Apron)	Jetty	Jetty (Apron)	Bulkhead	Bulkhead (Apron)	Road	Dock Railway	Break-water	Sea Wall	Revet-ment	Parapet Wall	Total f Port
Sendai		(2) 79,300					(3) 25,500				(2) 2,800		(104,6
Shiogama				(1) 1,200	(6) 117,100	(2) 13,200	(1) 3,200		(1) 1,000	(1) 8,100		(1)	(1 143,8
Matsushima			(1) 700		(2) 11,800						(1) 1,900		(20,7
Ishinomaki	(3) 2,650,000	(2) 14,100					(4) 30,900	(1) 1,100		(3) 48,500	(1) 31,100		(1 2,775,7
Okatsu					(1) 9,000				(2) 1,600		(1) 1,000		(11,6
Hagihama					(1) 9,500								(9,5
Kinkazan					(4) 10,600								(10,6
Total for Facility	(3) 2,650,000	(4) 93,400	(1) 7,000	(1) 1,200	(14) 158,000	(2) 13,200	(8) 59,600	(1) 1,100	(3) 2,600	(4) 56,600	(5) 36,800	(1)	(Sum To (4 3,079,5

Table 3 Maximum Accelerations in Ports

No. in Fig. 10	Station			Record Number	Max. Acceleration (Gal)		
	Abbreviated Name	Installation Condition	Epicentral Distance (km)		NS	EW	UD
16	Yamashita-dai6-S	on structure	374	S-1187	20	19	9
16	Keihin-ji-S	on ground	376	S-1188	31	25	10
16	Yamashita-hen-S	on ground	374	S-1189	25	15	6
11	Onahama-S	on ground	174	S-1191	44	51	19
6	Aomori-S	on ground	319	S-1192	21	23	9
24	Sakata-S	on ground	220	S-1193	30	34	19
14	Shinagawa-S	on ground	354	S-1194	9	10	5
13	Chiba-S	on ground	345	S-1195	26	29	6
19	Shimizu-miho-S	on ground	478	S-1196	13	13	10
19	Okitsu-S	on ground	476	S-1197	4	4	3
19	Shimizu-kojyo-S	on ground	480	S-1198	15	15	6
24	Akita-S	on ground	249	S-1200	25	24	13
10	Shiogama-kojyo-S	on ground	100	S-1201	270	280	169
7	Hachinohe-S	on ground	273	S-1202	63	61	30
22	Niigata-ji-S	on ground	274	S-1203	25	19	4
8	Miyako-S	on ground	166	S-1204	151	115	50
20	Nagoya-zokan-S	on ground	584	S-1205	6	4	3
12	Kashima-zokan-S	on ground	281	S-1206	40	30	9
3	Tomakomai-S	on ground	500	S-1207	9	8	1
18	Tagonoura-S	on ground	459	S-1209	23	36	8
9	Ofunato-bochi-S	on ground	103	S-1210	141	170	60
9	Ofunato-bo-S	on structure	103	S-1211	350	275	106
4	Muroran-S	on ground	475	S-1217	4	4	1
17	Koken-M	on ground	392	M-216	8	7	
16	Yamashita-hen-M	on ground	374	M-217	25	17	7
16	Yamashita-dai7-M	on structure	374	M-218	1	1	
15	Kawasaki-dai5-ko-M	on structure	364	M-219	34 over	--	
15	Kawasaki-dai5-chi-M	on ground	364	M-220	24	27	
21	Yokkaichi-sekitan-M	on structure	609	M-221	5	15	
19	Shimizu-sekitan-M	on structure	481	M-222	23	20	
20	Inae-yaita-M	on structure	583	M-223	6	7	
5	Hakodate-M	on ground	422	M-224	12	14	7
2	Tokachi-M	on ground	469	M-225	10	10	4
1	Hanasaki-M	on ground	638	M-226	4	4	2

Table 4

Parameters of

Source Model

1.	Width of Fault	80 km
2.	Length of Fault	30 km
3.	Slip Angle	-20°
4.	Slip Vector	N10°E
5.	Slip Direction	N65°W
6.	Slip Displacement	1.7 m
	(1 ~ 6 : after Seno[6])	
7.	Rise Time	2.0 sec.
8.	Rupture Velocity (in length)	11.6 km/s
9.	Rupture Velocity (in width)	3.1 km/s
10.	Velocity of P Wave	6.0 km/s
11.	Velocity of S Wave	3.5 km/s

Table 5 Classification of Facilities

Extent of Damage	Gravity Structure				Sheetpile Structure					
	Block	Caisson	L-Block	Total	Plate Anchor	Sheetpile Anchor	Single pile Anchor	Coupled pile Anchor	Self Standing Sheetpile	Total
0	11	4	0	15	5	1	2	7	0	15
I	17	0	1	18	1	1	2	0	1	5
II	2	0	0	2	1	3	0	0	0	4
III	1	0	0	1	0	0	0	0	0	0
IV	0	0	0	0	0	0	0	0	0	0
Total	31	4	1	36	7	5	4	7	1	24

Table 6 Classification of Extent of Damage

Extent of Damage	Swelling of Wall		Settlement of Apron (cm)	Settlement of Coping (cm)
	Maximum (cm)	Average (cm)		
0	0	0	0	0
I	0~30	under 10	under 20	under 30
II	30~100	10~60	under 50	under 40
III	100~400	60~120	50~100	under 50
IV	over 200	over 120	over 100	over 50

Fig. 1 Location of Epicenter, Source Region and Ports

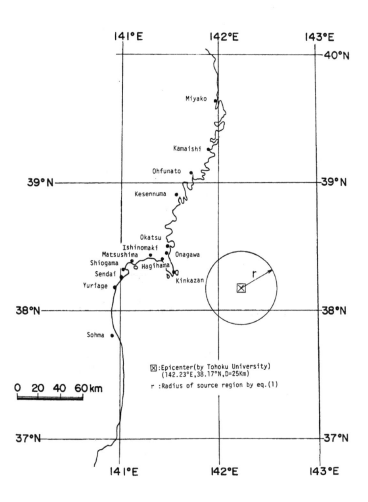

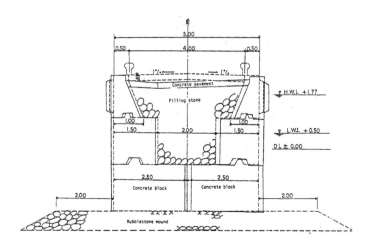

Fig. 2   Damaged Gravity Quaywall (Shiogama Port)

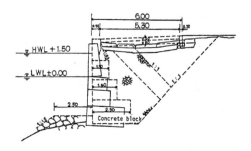

Fig. 3   Damaged Gravity Quaywall (Yuriage Fishing Port)

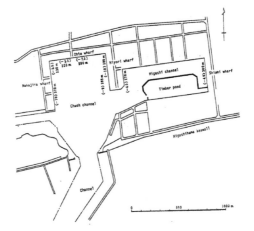

Fig. 4   Plane of Ishinomaki Port

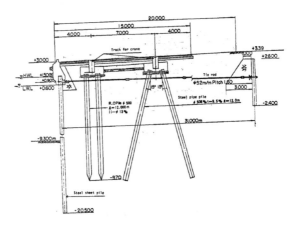

Fig. 5   Hiyori -9 m Wharf (Ishinomaki Port)

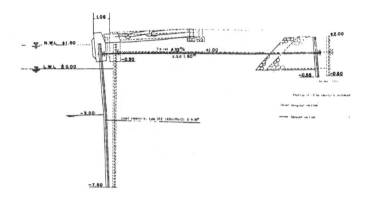

Fig. 6 -3.5 m Wharf (Yuriage Fishing Port)

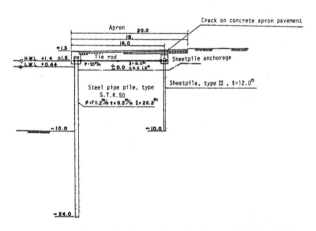

Fig. 7 Nakano -10 m Wharf (Sheetpile Anchor) (Sendai Port)

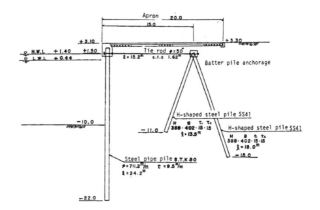

Fig. 8   Nakano -10 m Wharf (Coupled Piles Anchor) (Sendai Port)

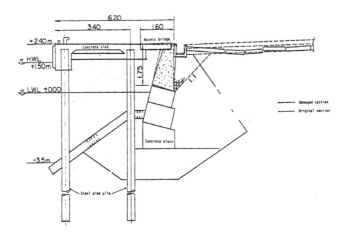

Fig. 9   -3.5 m Wharf (Yuriage Fishing Port)

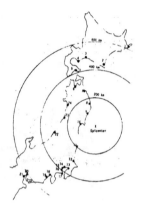

Fig. 10 Location of Observation Stations

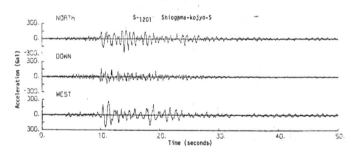

Fig. 11 Accelerograms (Shiogama-kojyo-S)

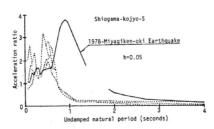

Fig. 12   Accelerograms (Ofunato-bochi-S)

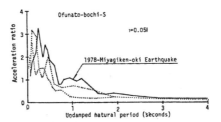

Fig. 13   Displacements of Shiogama-kojyo-S and Ofunato-bochi-S
E-W Component

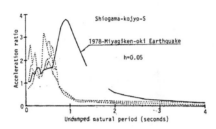

Fig. 14   Response Spectra of Shiogama-kojyo-S

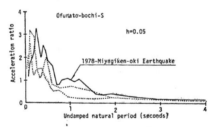

Fig. 15   Response Spectra of Ofunato-bochi-S

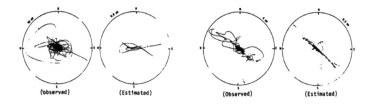

(observed)	(Estimated)

Fig. 16   Locus (Shiogama)

(Observed)	(Estimated)

Fig. 17   Locus (Ofunato)

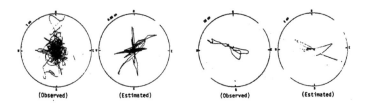

(Observed)	(Estimated)

Fig. 18   Locus (Miyako)

(Observed)	(Estimated)

Fig. 19   Locus (Ishinomaki)

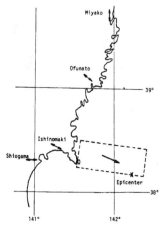

Fig. 20 Predominant Direction of Ground Motion and Location
of Fault

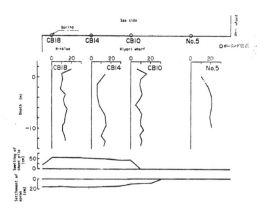

Fig. 21 N-values and Damage (Hiyori Wharf)

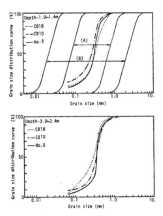

Fig. 22 Grain Size Distribution Curves (Hiyori Wharf)

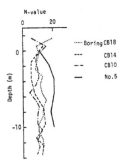

Fig. 23 Comparison of N-value (Hiyori Wharf)

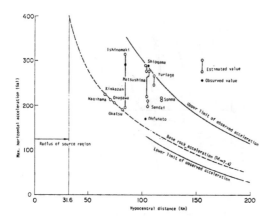

Fig. 24 Maximum Ground Acceleration in Ports

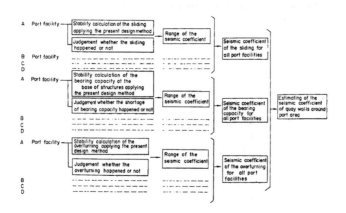

Fig. 25 Procedure for Estimation of Seismic Coefficient

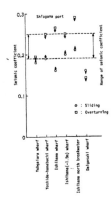

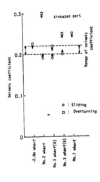

Fig. 26.1 Range of
          Seismic Coefficient
          (Shiogama Port)

Fig. 26.2 Range of
          Seismic Coefficient
          (Kinkazan Port)

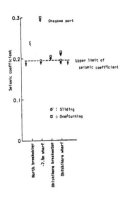

Fig. 26.3 Range of
          Seismic Coefficient
          (Onagawa Port)

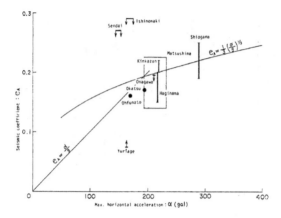

Fig. 27 Seismic Coefficient and Maximum Ground Acceleration
(Gravity Structure)

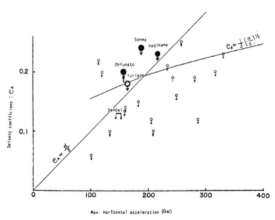

Fig. 28 Seismic Coefficient for Embedment of Sheetpile and Maximum
Ground Acceleration (Sheetpile Structure)

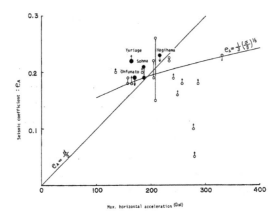

Fig. 29 Seismic Coefficient for Anchor Plate and Maximum Ground
Acceleration (Sheetpile Structure)

VII-109

WOOD DIAPHRAGMS IN MASONRY BUILDINGS

Mihran S. Agbabian

Agbabian Associates

El Segundo, California

ABSTRACT

Reports of damage from past earthquakes indicate that wood diaphragms
masonry walls have suffered little or no damage during excitation. However
masonry walls have suffered degrees of damage varying from minor tension cr
from the diaphragms and complete collapse.

The interaction of a diaphragm with the masonry walls has a critical e
dynamic behavior of the structure. Numerical analyses of plywood, diagonal
sheathed floor and roof diaphragms indicate that highly nonlinear and hyste
tion relationships are needed to describe the response of wood diaphragms.
planned to improve the analytical model by obtaining performance characteri
anchorages to diaphragms as well as deformations of diaphragms under cyclic
Finally, candidate retrofit/strengthening techniques will be incorporated i
diaphragms, and the tests will be repeated in order to develop structural s
survive disastrous earthquakes.

KEYWORDS: Seismic hazard; wood diaphragms; unreinforced masonry; dynamic r
          analytical model.

INTRODUCTION

Wood diaphragms supported on masonry walls are found in buildings located in various seismic zones. Reports of damage from past earthquakes indicate that such diaphragms have suffered little or no damage during earthquake excitations. However, the supporting masonry suffered degrees of damage varying from minor tension cracks to separation from the diaphragms and complete collapse. The diaphragm's response and masonry wall damage may be directly related because of the effect of diaphragm deformation on the out-of-plane motions of the masonry wall. An analytical model is used in this paper that includes a nonlinear hysteretic element which simulates the behavior observed in cyclic tests of wood diaphragms. The results of the analysis are compared with corresponding results obtained with linear elastic and nonlinear elastic assumptions. Forthcoming experiments, where several diaphragms will be statically and dynamically tested, will provide new data for correlation with these analyses.

ANALYTICAL MODEL

A typical one-story unreinforced masonry (URM) building shown in Figure 1 was selected for analysis. The building consists of a wood-diaphragm roof supported on four sides by URM walls. The earthquake ground motion was assumed to be directed normal to the side wall.

The analytical model shown in Figures 2 and 3 assumes that the end walls are rigid. A preliminary study of wall overturning due to seismic loading [1] has indicated that the earthquake input motions will be transmitted from the foundation level (Levels C and D in Figure 2a) to the top of the end walls (Levels A and B) with little modification. The side wall was assumed to crack along the foundation when subjected to the out-of-plane earthquake motions. Since the cracked wall will have little lateral stiffness, only its weight has been included.

The roof diaphragm/wall system is represented as a lumped parameter model and is shown in Figure 2b. The roof diaphragm is divided into several segments (eight segments are shown in Figure 2b), the out-of-plane wall mass is distributed along both edges, and the model is subjected to-earthquake ground motion at its ends. The diaphragm is also assumed to act as a deep shear beam. The lumped parameter model is shown in Figure 3. The earthquake motions at each end are taken to be identical, and due to symmetry only a half model is required.

The nonlinear internal springs shown in Figure 3 represent the shear stiffness of the diaphragm, and the internal dampers represent viscous damping of the roofing material. The earthquake input motion is described by coordinate X_1, and the independent degrees of freedom (DOF) of the diaphragm/wall system are described by coordinates X_2 through X_5. The displacement gages shown connecting the DOF X_2 through X_5 with DOF X_1 are used to monitor the relative out-of-plane motions of the top and bottom of the side wall. This relative motion will serve as an indicator of wall instability. The analyses were performed using the STAR/ III computer program. [2]

DIAPHRAGM STIFFNESS ELEMENT

Cyclic tests [3] show a hysteretic behavior for plywood diaphragms as shown in Figure 4. A review of numerous tests of wood diaphragms [6-20] also shows that the behavior of these diaphragms is highly nonlinear. Most of these tests were, however, noncyclic.

Based on the available test results, a nonlinear hysteretic model for the wood diaphragms has been developed as shown in Figure 5. The overall force-deflection envelope is shown in Figure 5a, and a typical cyclic load path for the spring is shown in Figure 5b. It can be seen that this load path is a multilinear version of that shown in Figure 4. The unloading slope as well as the residual force value may be specified to match experimental data. The test data indicate that the envelope can be adequately represented by a second-order curve of the form

$$F(e) = \frac{F_u e}{\frac{F_u}{K_i} + e} \quad \text{for compression (positive e)}$$

$$F(e) = \frac{F_u e}{\frac{F_u}{K_i} - e} \quad \text{for tension (negative e)}$$

where

$F(e)$ = Spring force

e = Spring deformation

F_u = Ultimate capacity of spring at large e

K_i = Initial spring stiffness

ANALYSIS OF TYPICAL ROOF DIAPHRAGMS

Dynamic analyses using the model described in the previous section were conducted for two types of earthquake ground motions. The inputs correspond to the highest seismic areas of the United States [4] where effective peak acceleration is 0.40 g, as shown in Figure 6 for a 5% damped response. The SOOE component of the 1940 D1 Centro record, representing a distant event, was scaled (1.25 scale factor) to the 0.40 g level, and its response spectrum is shown in Figure 6a. The N69W component of the 1971 Castaic record, representing a nearby event, was scaled (1.80 scale factor) to the 0.40-g level, and its response spectrum is shown in Figure 6b. The scaled displacement, velocity, and acceleration records are shown in Figures 7 and 8 for the scaled El Centro and Castaic records, respectively.

The cases run are shown in Table I. Two basic types of roof diaphragms were analyzed; namely,

° 1/2-in. plywood, unblocked, and roofed

° Diagonal sheeting, roofed

Each type was subjected to the scaled El Centro and Castaic input motions. In addition, each type was analyzed using four material properties:

° Linear elastic

° Nonlinear elastic (envelope curve of Figure 5a)

° Nonlinear hysteretic (Figure 5b)

° Nonlinear hysteretic (Figure 5b) with F_u = one-half the value of F_u of the immediately preceding case

The last case shows the sensitivity of the response to a variation in the ultimate force, F_u. Although it is clear that the roofing material will provide some degree of viscous damping, all of the analyses were performed without viscous damping to clearly show the effect of the nonlinear hysteretic model assumptions. Peak responses for all of the cases are given in Table I.

It can be seen that the nonlinear elastic cases show a general reduction (relative to the linear elastic cases) in peak responses at DOF 5, an attenuation of the peak relative deformation at the midlength of the diaphragm, Spring No. 12, and significant reductions in the maximum shear force at the end wall, Spring No. 1. The nonlinear hysteretic cases show further significant reductions in all response quantities. Typical results for the plywood roof diaphragm are shown in Figures 9, 10, 11, and 12. Figures 9a, b, and c, and 10a, b, an c compare the acceleration and displacement response at DOF 5, respectively, for the first three model types. Figures 11a, b, and c compare the shear force transmitted to the end wal (i.e., Spring No. 1).

The relative deformation between the top and bottom of the masonry wall at the diaphrag midlength (i.e., Spring No. 12) gives an indication of the out-of-plane motions of the side wall. Typical results for this relative deformation are given in Figures 12, 13, and 14. The main effect of the hysteretic model can be seen in these three figures. For example, comparing Figures 12a, b, and c, it can be seen that not only is the peak deformation reduced, but the number of repeated loadings at nearly constant peak values is greatly reduced. It is believed that elastic analyses overestimate the actual response and that con- sideration of the hysteretic characteristics of the wood diaphragms is necessary for more realistic response predictions.

FORTHCOMING DIAPHRAGM AND WALL TESTS

Data on the dynamic response of wood diaphragms and masonry walls are generally lacking. Accordingly, a series of tests is planned to provide needed data and to calibrate/verify the analytical model that has been developed. The test program consists of two related test series: (1) diaphragms and (2) out-of-plane deflection of URM walls. These tests are brief- ly described in an accompanying paper. [5]

FURTHER STUDIES AND CONCLUSIONS

During the forthcoming test program, the analytical model will be refined, calibrated, and verified for use in predicting the dynamic response of wood diaphragms. In addition, stability criteria for the response of masonry walls to out-of-plane motions will be estab- lished. The final goal will be a reliable method of predicting wall/diaphragm failures in seismic areas and the development of retrofit measures to prevent such failures.

ACKNOWLEDGMENTS

The study is being conducted by the joint venture ABK comprising the firms of Agbabian Associates, S.B. Barnes & Associates, and Kariotis, Kesler and Allys. It is authorized by the National Science Foundation (NSF) in support of its ongoing earthquake hazard mitigation program. Dr. John B. Scalzi is Cognizant Program Manager for NSF. John Kariotis and Albin Johnson are designing the test elements, and Robert Ewing and Tim Healey are conducting the analyses.

REFERENCES

[1] Adham, S. A. and Ewing, R. D., "Methodology for Mitigation of Seismic Hazards in Existing Unreinforced Masonry Buildings, Phase 1," R-7815-4610. El Segundo, California: Agbabian Assoc., March 1978.

[2] Agbabian Assoc. (AA)., "User's Guide and Formulation of the Analysis Method for the STARS/III Code," U-4700-4816. El Segundo, California: AA, (in press) 1979.

[3] Young, D. H. and Medearis, K., "An Investigation of the Structural Damping Characteristics of Composite Wood Structures Subjected to Cyclic Loading," Technical Report No. 11. Stanford, California: Stanford University, April 1962.

[4] Applied Tech. Council (ATC)., "Tentative Provisions for the Development of Seismic Regulations for Buildings," ATC-3-06. San Francisco, California: ATC, June. 1978.

[5] Agbabian, M. S., "Mitigation of Seismic Hazards in Existing Unreinforced Masonry Buildings," Proceedings of 11th Joint UJNR Panel Conference, Tokyo, Japan, September 1979.

Additional References for Diaphragm Test Data

[6] Forest Products Lab. (FPL)., "Diaphragm Action of Full-Scale Diagonally Sheathed Wood Roof or Floor Panels." Madison, Wisconson: FPL, April 1955.

[7] Atherton, G. H. and Johnson, J.W., "Diagonally Sheathed Wood Diaphragm. Corvallis, Oregon: Oregon Forest Research Lab., October 1951.

[8] Stillinger, J. R., Johnson, J. W., and Overholser, J. L., "Lateral Tests on Full-Scale Lumber-Sheathed Diaphragms." Corvallis, Oregon: Oregon Forest Products Lab., October 1952.

[9] Stillinger, J. R., "Lateral Tests on Full-Scale Lumber-Sheathed Roof Diaphragms." Corvallis, Oregon: Oregon Forest Research Lab., October 1953.

[10] Johnson, J. W., "Lateral Tests on Full-Scale Lumber-Sheathed Roof Diaphragms of Various Length-Width Ratios. Corvallis, Oregon: Oregon Forest Research Lab., October 1954.

[11] Johnson, J. W., "Lateral Tests on 12-by-60 ft and 20-by-80 ft Lumber-Sheathed Roof Diaphragms." Corvallis, Oregon: Oregon Forest Research Lab., October 1955.

[12] Burrows, C. H. and Johnson, J. W., "Lateral Test on Full-Scale Roof Diaphragm with Diamond-Pattern Sheathing." Corvallis, Oregon: Oregon Forest Research Lab., October 1956.

[13] Stillinger, J. R., "Lateral Tests on Full-Scale Lumber and Plywood-Sheathed Roof Diaphragms," in Symp. on Methods of Testing Building Construction, ASTM-STP-166. New York: American Society of Mechanical Engineering, 1955.

[14] Oregon Forest Labs. (OFL)., "Lateral Tests on Full-Scale Lumber and Plywood-Sheathed Roof Diaphragms." Corvallis, Oregon: OFL, March 1956.

[15] Countryman, D., "Lateral Tests on Plywood Sheathed Diaphragms." Tacoma, Washington: Douglas Fir Plywood Assoc. Lab., March 1952.

[16]   Oregon Forest Research Lab. (OFRL)., "Lateral Tests on Full-Scale Plywood-Sheathed Roof Diaphragms." Corvallis, Oregon: ORFL, October 1953.

[17]   Countryman, D. and Colbenson, P., "1954 Horizontal Plywood Diaphragm Tests." Tacoma, Washington: Douglas Fir Plywood Assoc. Lab., October 1954.

[18]   Douglas Fir Plywood Assoc. (DFPA)., "1954 Horizontal Plywood Diaphragm Tests." Tacoma, Washington: DFPA, January 1955.

[19]   Johnson, J.W., "Lateral Test on a 12-by-60 ft Plywood Sheathed Roof Diaphragm." Corvallis, Oregon: Oregon Forest Research Laboratory, January 1955.

[20]   Tissel, J.R., "Horizontal Plywood Diaphragm Tests." Tacoma Washington: Douglas Fir Plywood Assoc., 1966.

TABLE I. SUMMARY OF DIAPHRAGM DYNAMIC ANALYSIS RESULTS

Diaphragm	Earthquake Input	Material Model Type	K_1 kips/in.	F_m kips	Peak Acceleration at Midlength DOF 5, g	Peak Velocity at Midlength DOF 5, in./s	Peak Displacement at Midlength DOF 5, in.	Peak Relative Deformation at Midlength Spring 12, in.	Peak Shear Force Spring 1, kips
Plywood[1]	Scaled[a] El Centro	Linear Elastic	25	--	2.4	112.0	20.0	15.5	183.0
Plywood[1]	Scaled[a] El Centro	Nonlinear Elastic	25	140	1.48	67.7	17.4	14.2	76.3
Plywood[1]	Scaled[a] El Centro	Nonlinear Hysteretic	25	140	0.62	35.3	7.2	6.0	45.6
Plywood[1]	Scaled[a] El Centro	Nonlinear Hysteretic	25	70	0.45	27.3	8.4	5.9	31.7
Plywood[1]	Scaled[t] Castaic	Linear Elastic	25	--	2.2	98.2	17.4	13.7	166.0
Plywood[1]	Scaled[t] Castaic	Nonlinear Elastic	25	140	2.0	77.5	13.0	10.9	72.3
Plywood[1]	Scaled[t] Castaic	Nonlinear Hysteretic	25	140	0.53	35.4	8.0	6.1	52.3
Plywood[1]	Scaled[t] Castaic	Nonlinear Hysteretic	25	70	0.41	30.5	8.3	5.4	36.5
Diagonal Sheeting[2]	Scaled[a] El Centro	Linear Elastic	8.3	--	1.8	103.0	25.4	23.6	88.6
Diagonal Sheeting[2]	Scaled[a] El Centro	Nonlinear Elastic	8.3	120	0.68	55.9	19.8	17.9	42.0
Diagonal Sheeting[2]	Scaled[a] El Centro	Nonlinear Hysteretic	8.3	120	0.50	43.6	12.0	13.5	34.7
Diagonal Sheeting[2]	Scaled[a] El Centro	Nonlinear Hysteretic	8.3	60	0.38	31.2	12.4	13.9	24.6
Diagonal Sheeting[2]	Scaled[t] Castaic	Linear Elastic	8.3	--	1.66	91.8	19.0	17.9	76.9
Diagonal Sheeting[2]	Scaled[t] Castaic	Nonlinear Elastic	8.3	120	0.99	42.6	13.8	9.5	30.0
Diagonal Sheeting[2]	Scaled[t] Castaic	Nonlinear Hysteretic	8.3	120	0.48	36.4	11.7	9.4	26.5
Diagonal Sheeting[2]	Scaled[t] Castaic	Nonlinear Hysteretic	8.3	60	0.36	28.2	11.8	9.4	22.6

[a] El Centro, 1940, S00E, scaled by 1.25

[t] Castaic, 1971, N69W, scaled by 1.80

[1] 1/2-in. plywood, unblocked, roofed

[2] Diagonal sheeting, roofed

(1 in. = 2.54 cm)

(1 kip = 4450 N)

VII-117

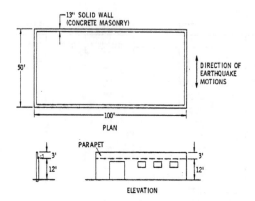

FIGURE 1. TYPICAL ONE-STORY UNREINFORCED MASONRY BUILDING
WITH A WOOD ROOF DIAPHRAGM (1 ft = 0.3+ m)

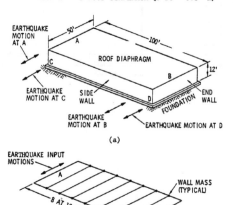

FIGURE 2. DIAPHRAGM/WALL CONFIGURATION AND ANALYSIS MODEL
(1 ft = 0.3+ m)

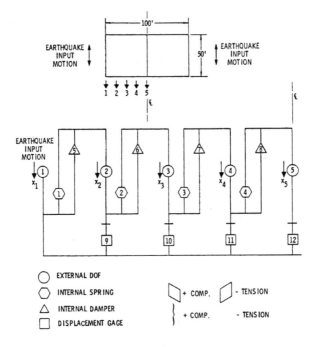

FIGURE 3. LUMPED PARAMETER MODEL (8 SEGMENTS — HALF MODEL)
(1 FT = 0.3⁺ m)

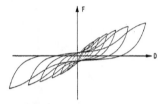

FIGURE 4.   TYPICAL CYCLIC LOAD DEFLECTION DIAGRAM FOR
PLYWOOD DIAPHRAGMS

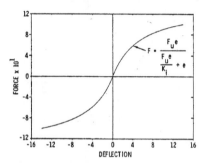

(a)   Force-deflection envelope of model

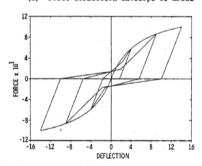

(b) ´Typical cyclic load-deflection diagram for model

FIGURE 5.   LOAD DEFLECTION MODEL FOR WOOD DIAPHRAGMS˙

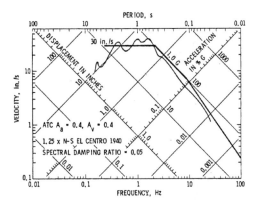

(a) Comparison of ATC spectra for Los Angeles and N-S El Centro 1940
scaled by a factor of 1.25

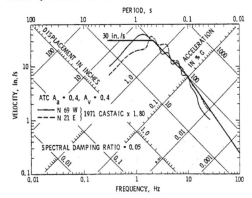

(b) Comparison of ATC spectra for Los Angeles and Castaic 1971
scaled by a factor of 1.80

FIGURE 6.   RESPONSE SPECTRA FOR LOS ANGELES SHOWING SELECTED INPUT MOTIONS
(1 in. = 2.54 cm)

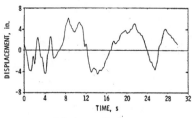

(a) Input displacement

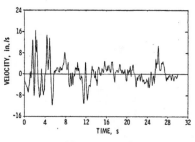

(b) Input velocity

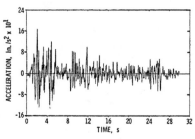

(c) Input acceleration

FIGURE 7. EARTHQUAKE INPUT MOTION, EL CENTRO SCALED BY 1.25
(1 in. = 2.54 cm)

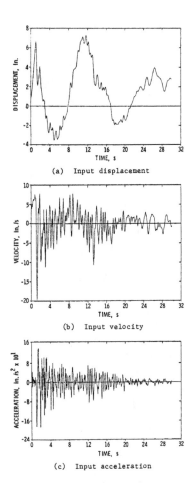

(a) Input displacement

(b) Input velocity

(c) Input acceleration

FIGURE 8.   EARTHQUAKE INPUT MOTION, CASTAIC SCALED BY 1.80
(1 in. = 2.54 cm)

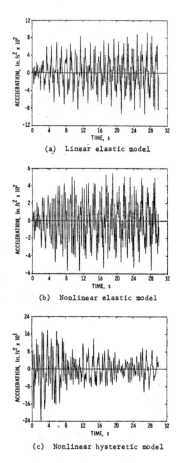

(a) Linear elastic model

(b) Nonlinear elastic model

(c) Nonlinear hysteretic model

FIGURE 9. ACCELERATION RESPONSE AT MIDLENGTH OF PLYWOOD DIAPHRAGM,
EL CENTRO INPUT (1 in. = 2.54 cm)

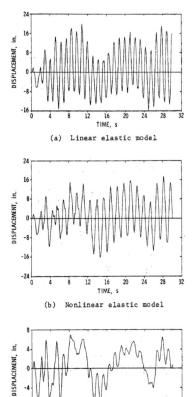

(a) Linear elastic model

(b) Nonlinear elastic model

(c) Nonlinear hysteretic model

FIGURE 10. DISPLACEMENT RESPONSE AT MIDLENGTH OF PLYWOOD DIAPHRAGM,
EL CENTRO INPUT (1 in. = 2.54 cm)

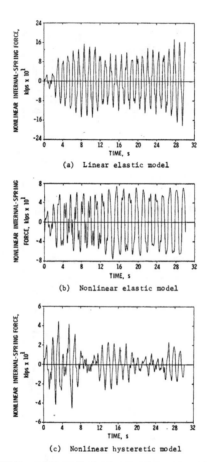

(a) Linear elastic model

(b) Nonlinear elastic model

(c) Nonlinear hysteretic model

FIGURE 11. SHEAR FORCE AT END WALL FOR PLYWOOD DIAPHRAGM,
EL CENTRO INPUT (1 kip = 4450 N)

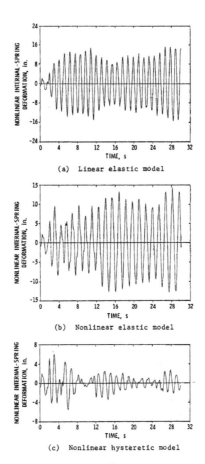

(a) Linear elastic model

(b) Nonlinear elastic model

(c) Nonlinear hysteretic model

FIGURE 12. RELATIVE DEFORMATION BETWEEN THE TOP AND BOTTOM OF THE MASONRY
WALL AT PLYWOOD DIAPHRAGM MIDLENGTH, EL CENTRO INPUT
(1 in. = 2.54 cm)

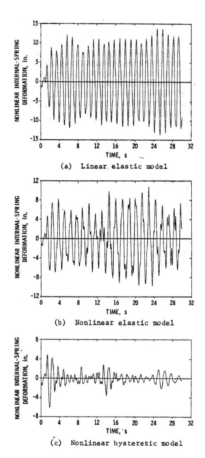

FIGURE 13. RELATIVE DEFORMATION BETWEEN THE TOP AND BOTTOM OF THE MASONRY
WALL AT PLYWOOD DIAPHRAGM MIDLENGTH, CASTAIC INPUT
(1 in. = 2.54 cm)

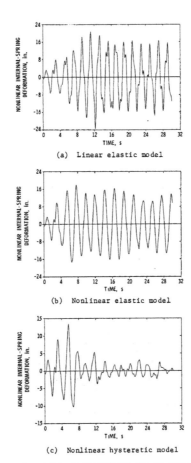

(a) Linear elastic model

(b) Nonlinear elastic model

(c) Nonlinear hysteretic model

FIGURE 14. RELATIVE DEFORMATION BETWEEN THE TOP AND BOTTOM OF THE MASONRY
WALL AT DIAGONAL SHEATHED DIAPHRAGM MIDLENGTH, EL CENTRO INPUT
(1 in. = 2.54 cm)

MITIGATION OF SEISMIC HAZARDS IN

EXISTING UNREINFORCED MASONRY BUILDINGS

Mihran S. Agbabian

Agbabian Associates

El Segundo, California

ABSTRACT

Analytical and experimental investigations to determine resistance of structur
various seismic zones of the United States are combined in an ongoing study. The s
being carried out by a joint venture of three consulting engineering firms in Los A
under the sponsorship of the National Science Foundation.

A survey was initially made of unreinforced masonry (URM) buildings in seven g
areas in the United States. The selection of types of structures for investigation
ered construction materials, size, distribution of walls around the perimeter and wi
building, connection details, and the application or absence of seismic design crite

The structures selected for study consist of:

a) Rectangular, six-story industrial building

b) Rectangular, four-story public school

c) Irregular, four-story plus basement public building

d) U-shaped, four-story apartment building

e) Rectangular, six-story and three-story office buildings

f) Rectangular, one-story and three-story industrial buildings

A methodology for earthquake hazard mitigation is being developed that will be appli
the broad range of buildings surveyed. Concurrent with these analyses, tests are pl
masonry walls in the out-of-plane and in-plane directions, anchorage of walls to dia
and wood and steel diaphragm characteristics. Tests of repaired and/or retrofitted
will also be carried out.

The effect of the following rehabilitation techniques is being investigated:

° Strengthening of masonry walls

° Adding or improving anchorages

° Repair and strengthening of diaphragms

° Amelioration of foundation settlement

° Addition of shear walls

° Removal of upper stories

° Parapet renovation

° Bracing of nonbearing partitions

The final result of this analytical and experimental study will be the development of a methodology for earthquake hazard mitigation that will include (a) methods for determining seismic hazard and seismic input, (b) methods for establishing physical properties of unreinforced masonry, (c) recommended analytical methods, and (d) recommended retrofit methods.

KEYWORDS: Seismic hazard; unreinforced masonry; retrofit methods; hazard mitigation; diaphragms; anchorages; dynamic response.

INTRODUCTION

Unreinforced masonry (URM) buildings exist·in all seismically active areas of t
United States. The potential for injury or death resulting from failure of these bu
during earthquakes is of great concern to government officials and structural engine
yet, the high cost of rehabilitating the buildings to meet current earthquake—resist
standards has slowed down or deferred the execution of the practical steps that are r
for hazard mitigation.

Research for determining realistic hazard—mitigation requirements and cost—effec
methods of retrofit is being carried out by several organizations. The purpose of su
research is to develop procedures which improve the earthquake resistance of URM stru
to acceptable levels but within the economic constraints that may be tolerated by pro
owners.

This paper reports on an ongoing study that combines analytical and experimental
tigations for the mitigation of seismic hazards in existing unreinforced masonry buil

SURVEY OF EXISTING BUILDINGS

The three geographic areas selected for survey of existing URM buildings correspo
seismic zones identified by effective peak accelerations* (EPA) of 0.10 g, 0.20 g and
The geographic areas covered are: St. Louis, the Southern Carolina area, and the New
area in the lowest EPA category; the Pacific Northwest, the Salt Lake City area, and M
in the intermediate EPA category; and California in the highest EPA category.

Construction materials, size, distribution of walls around the perimeter and with
building, and connection details were all considered in the selection of types of stru
for investigation.

Materials

In pre-1940 buildings, floors and roofs consist of concrete or boarded wood-frame
struction. Interior vertical elements are plastered wood-framed walls. In post-1940
ings, wood-framed floors and roofs are sheathed in plywood. Floors and roofs are fram

---

* EPA's are defined by the ATC-3 document entitled "Tentative Provisions for the Devel
of Seismic Regulations for Buildings," Applied Technology Council, June 1978.

steel joists or beams, and steel decking is topped with concrete. Steel deck roofs generally are covered with insulation board and roofing. Interior partitioning often terminates in the ceiling rather than connecting with the framing at the floor levels.

Size

The size of the structures observed ranged from very large industrial buildings more than 600 ft (183 m) long to residences with an end wall of about 15 ft ($\sim$ 5 m). The height of the surveyed buildings ranged from 150 ft (46 m) high to a single story of 15 ft (5 m). Post-1940 URM buildings are generally no more than two or three stories high.

Distribution of URM Walls

URM walls in the perimeter of industrial buildings have relatively uniform penetrations. Fire walls having very few penetrations are used as building dividers. Multiple housing, public buildings, schools, and churches do not have consistent fire-wall subdivisions. Commercial buildings such as offices and retail stores have few or no URM elements at the lowest level facing the public ways. In most buildings, these elements are not anchored into a frame that could develop significant resistance to lateral displacement. In many cases, the framework supports several stories of masonry above the street level.

Anchorage of Horizontal Elements and Vertical URM Walls

Beams and joists are usually anchored to the URM walls where wood framing is combined with URM bearing walls. However, a remarkable number of surveyed structures have no apparent anchorage at all. In pre-1940 buildings, the concrete floor was generally poured after completion of the masonry walls below the floor, and the connection depended on the interlock of frictional surfaces. In recent construction practice, the concrete frame/floor system was first completed and the masonry was infilled into the finished frame.

STRUCTURES SELECTED FOR STUDY

Characteristics of the seven types of structures that best represent that total United States inventory of existing URM buildings are shown in Table 1.

CATEGORIZING FAILURE MODES

Failure or damage of URM buildings past earthquakes will provide data for the initial models that will be used for numerical analysis. Observed damage or failure has been a result of:

1) Wall collapse due to (a) inadequate anchorage between walls and floors or roof diaphragms, (b) out-of-plane bending failure, (c) in-plane shear or flexural failure, and (d) excessive deflection of the diaphragm system

2) Diaphragm failures in (a) shear, (b) shear connection to walls or other resisting elements, (c) horizontal-shear connnection chords (if any), and (d) chords (if any)

3) Excessive deflection of the diaphragm system, causing failures of interior walls

4) Differing dynamic response of component parts of complex buildings

5) Conditions due to previous strains on structure, such as foundation settlement, deterioration, or previous shaking

6) Effect of infilled URM wall on building frames

EVALUATION OF SEISMIC HAZARD

Seven major U.S. cities have been selected for investigation. Figure 1 summarizes the design criteria for these cities as provided in the ATC-3 document. Response spectra for these seismic zones are shown in Figure 2.

Since URM buildings are constructed of undesigned elements that have hysteretic or aperiodic response, they cannot be analyzed by spectral response methods. Accordingly, the spectra shown in Figure 2 have been used as the bases for selecting an ensemble of time-history motions for each area as input for dynamic analysis of the structures.

ANALYTICAL AND EXPERIMENTAL METHODS

Nonlinear dynamic response of the following URM building components will be obtained:

° End walls rocking on their foundations (overturning)

° Stability of walls subjected to out-of-plane and out-of-phase motions

° Diaphragm response (hysteretic behavior)

° Torsional response (plan irregularity)

° Complete structures (all component responses included)

Linear and nonlinear static finite element analyses will be carried out for correlation with tests on panels, core specimens, and anchorages, as well as analysis of perforated walls and piers.

The test program consists of four related test series: (1) diaphragms, (2) out-of-plane deflection of URM walls, (3) in-plane URM strength, and (4) anchorages.

Diaphragms

The first series will test five basic 20 ft by 60 ft (6.1 m by 18.3 m) diaphragms (including repaired and retrofitted specimens) by static displacement and by dynamic, in-plane shaking. These tests are being conducted to (a) study the behavior of diaphragms under earthquake loading, (b) assess the effect that various kinds of retrofitted strengthening will have on the response of the diaphragms, and (c) correct, refine, and verify the mathematical models of typical building diaphragms.

One diaphragm will be made of steel decking, two of plywood, one of board sheathing applied diagonally, and one of board sheathing applied straight across the diaphragm frame. The diaphragms will be retrofitted for further testing.

Each candidate diaphragm will be subjected to a series of tests in which the amplitudes of displacement or motion will be sequentially increased to produce progressively more severe levels of excitation. At each level, the diaphragm will be subjected to a testing sequence of (1) quasi-static loading, (2) sine-sweep testing, (3) sine-beat-and-dwell at the resonant frequency, and finally (4) dynamic testing using a properly scaled replica of the horizontal motion of a selected earthquake.

Quasi-static loading will develop stiffness values, which will be used to analyze the subsequent sinusoidal and earthquake type of motions to the same deflection limits developed in the tests. The sine-sweep sequence will reveal the first-mode resonant frequency, which will be examined in detail by the sine-beat-test dwelling at the resonant frequency to determine the amplitude gain factors for a particular diaphragm.

The prescribed excitation will be applied to the ends of the diaphragm as shown in Figure 3. For the quasi-static testing, the reaction forces will be developed at pillars positioned at the one-third points on the diaphragm. For the dynamic tests, the reaction constraint will be removed and only inertia will resist the input motion. Along the 60-ft (18.3 m) sides of the diaphragm, separately supported lead weights will be attached to equal the inertial loads experienced by the diaphragm owing to the wall systems and adjunct roof sections. A typical test configuration for dynamic testing is depicted in Figure 4.

Out-of-Plane Walls

The second series will dynamically test approximately 30 masonry wall sections 6 ft by 10-to-16 ft high (1.8 m by 3.0 to 4.9 m) by out-of-plane and out-of-phase motions applied the top and bottom of the walls. Repaired walls will also be tested.

Testing unreinforced masonry wall sections in the out-of-plane direction is expected t reveal why actual walls fractured by out-of-plane motions often do not collapse in real earthquakes. It is possible that the fractured walls remain statically stable and thus are still capable of supporting compressive loads without collapse. The dynamics of the phenom enon will be investigated by numercial analysis prior to testing, and the test program will be used to correct, refine, and verify the computer analysis.

A test specimen will consist of a 6-ft wide (1.8-m) section of masonry wall constructe on a fabricated base. The specimen will be as high as is required to satisfy a preselected height-to-thickness ratio for the particular method used in the construction of the wall se tion. The section will be preloaded vertically by weights to simulate additional stories above the test section. Two different materials (and their construction methods) and two different height-to-thickness ratios for each of the materials will be treated. The speci-mens will be tested with up to four levels of vertical preloaded weights. After the char-acteristics of the unreinforced sections have been determined, a series of six retrofitted specimens will be tested and evaluated.

The specimen and base will be fastened in a test fixture having a low-friction base support that can be displaced by servohydraulic actuators in the out-of-plane direction but without rotation. The top of the wall will be restrained either in a pinned condition or fixed against rotation and moved independently of the base by additional actuators. A typical test configuration is depicted in Figure 5.

The input to the base of the specimen will range from the minimum design ground motion to a suitably scaled replica of a selected earthquake displacement record. The input to the top of the wall will be determined from the assumed transfer function between the foundatior and the overhead floor or roof of the modeled building. It is expected to be nearly sinu-solidal and to be substantially different from the base motion.

In-Plane Walls

The purpose of these tests is: (1) to ascertain methods for determining the strength properties of existing URM with an accuracy equivalent to the analysis techniques, and (2) to determine the strengthening gained by certain types of retrofit. The tests proposed will include brick, concrete block, and hollow clay tile masonry of thicknesses nominally 12 in. (0.3 m), 8 in. (0.2 m), and 8 in. (0.2 m), respectively. The wide variation in workmanship and quality will be typed as good, medium, and poor quality masonry.

Anchorages

Values have been established by industry for various types of anchors in new masonry. Since poor anchorage of walls to floor and roof diaphragms has been the cause of many failures in earthquakes, the placement of new anchors in existing URM buildings is an important part of any retrofit program. Therefore, the frequently used types of anchors that may be installed in existing masonry will be tested in order to determine shear and pull-out values.

Two conditions will be considered: (1) an anchor that penetrates a wall and is fastened with a plate or washer on the outside of the wall, (2) an anchor that cannot be installed completely through the wall, such as when the wall is on a property line.

ASSESSING AND SELECTING RETROFIT METHODS

The methodology is expected to focus on investigative and analytical processes that enable a designer to determine whether retrofitting is needed and, if so, the extent of retrofitting required. From that point, the details and costs of retrofitting will influence the decision to proceed with the hazard mitigation or to remove the building.

Depending on the geometry of the existing URM structure, the accessibility of its various components, and the methods of original construction, one or more of several techniques can be considered for rehabilitation:

Strengthening of masonry walls

Adding or improving the anchorage of walls to diaphragms

Repair and strengthening of diaphragms

Amelioration of foundation settlement

Addition of shear walls

Removal of upper stories

Parapet Renovation

Bracing of nonbearing partitions

All of these methods will be assessed for both adequacy and relative costs. If
innovative methods of retrofitting evolve during the study, they too will be considered.

CONCLUSION

The methodology will be developed from the sum total of this research. It will be a
manual of practice applicable to all seismic zones of the United States. When the methodology is complete, the structures selected for study will be evaluated. Applicable retrofit methods will be designed for the structures as required. The cost of retrofit required
for these structures will then be estimated.

ACKNOWLEDGMENTS

The study is being conducted by the joint venture ABK comprising the firms of Agbabian
Associates, S. B. Barnes & Associates, and Kariotis, Kesler and Allys. It is authorized by
the National Science Foundation in support of its ongoing earthquake hazard migitation program. Dr. John B. Scalzi is Cognizant Program Manager for NSF. John Kariotis conducted the
survey of buildings, Albin Johnson is designing the test elements, and Robert Ewing is
conducting the analyses.

TABLE 1. STRUCTURES SELECTED FOR STUDY

370 x 70 (113 x 21)	Wood	Wood	Minimal
140 x 200 (43 x 61)	Wood	Wood	Floors not anchored to walls
112 x 173 (34 x 53)	Concrete on steel beams	Concrete on steel beams	Floors cast on walls
100 x 180 (31 x 55)	Wood	Wood	Minimal
75 x 150 (23 x 46)	Concrete	Concrete	Floors cast on walls
60 x 150 (18 x 46)	Wood	Wood	Minimal
160 x 240 (49 x 73)	N.A.	Steel with light-gage steel decking	Minimal
75 x 150 (23 x 45)	Concrete in steel decking	Light-gage steel, steel joists	Minimal

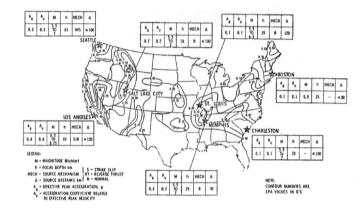

FIGURE 1.   SUMMARY OF SEISMIC INPUT DATA FOR SEVEN MAJOR U.S. CITIES
             (Adapted from ATC, 1978)

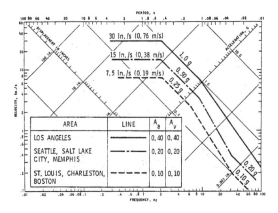

FIGURE 2.   ATC 5% DAMPED RESPONSE SPECTRA FOR 3 SEISMIC ZONES

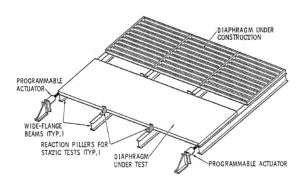

FIGURE 3.   QUASI-STATIC TEST SETUP FOR DIAPHRAGM TESTING

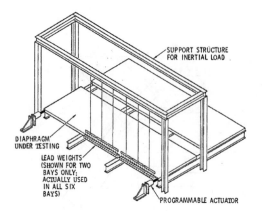

FIGURE 4.   TEST SETUP FOR DYNAMIC TESTING OF DIAPHRAGMS

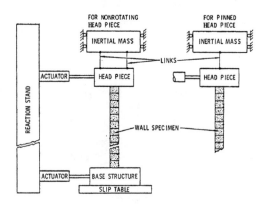

FIGURE 5.   SCHEMATIC TEST SETUP FOR OUT-OF-PLANE WALL TESTING

DEVELOPMENT OF A UNIVERSAL FASTENER FOR

WOODEN BUILDING ROOF FRAMES

Tatsuo Murota

Yuji Ishiyama

Building Research Institute

Ministry of Construction

ABSTRACT

A new type of joint fastener to be used in reinforcing wood building roof frames was developed at the Building Research Institute. This fastener, its behavior and strength developed in tension tests are described and shown in this paper. The structural performance of this fastener is also compared to those of conventional joint fasteners such as cramp irons.

KEYWORDS: Joint fasteners; roofing damage; structural performance; tension tests; universal fasteners; wooden roofs.

1. INTRODUCTION

In observing the types of damage to dwellings caused by the latest severe storms in Japan, i.e., the Hachijojima Typhoon, October 1975, the Okierabu Typhoon, September 1977 al., it is worth noting that the most common damage in the past severe storms, i.e., the tial loss of roofing, has decreased in number but increased in the removal of roof struct members such as rafters, purlins, roof posts, bed boards, etc.

The increase in this new type of damage to wood frame dwellings occurs despite large decreases in the total dead weight of roof in newer dwellings compared to conventional Ja ese roofing, joint connections of roof structural members often remain unchanged.

The Japanese conventional roofing material of wood frame dwellings is clay-tile and total roofing weight including rafters is usually 95 kg/m^2 when tiles are bedded on soil 60 kg/m^2 when tiles are set directly on bed boards. On the other hand in place of clay-t roofings sheet metal roofings represented by galvanized iron sheet roofing have increased rapidly for these ten years. Dead weight of those roofings is 15 kg/m^2 at most.

According to regulations in the Building Standard Law of Japan, roofs high above gro level by h meters shall resist to suction wind force of 0.5 x 60 $\sqrt{h}$ kg/m^2. Supposing ave age Japanese single- or two-story wood frame dwellings, average roof height is about 4 or and then suction forces to which roofs shall resist are 60 or 80 kg/m^2, respectively. Co paring those values with dead weight of conventional and modern roofing described above, weight of clay-tile roofings with or without bed soil is larger than or nearly equal to w pressure, but in case of sheet metal roofings, however, dead weight is far smaller than w pressure.

This leads to the conclusion that in structural design of roofs, timber connections roof frames in clay-tile roofing dwellings must not necessarily be designed to resist ten forces but in case of light weight roofings such as galvanized iron sheet, the action of siderably large tensile forces must be supposed in the design of roof frame connections.

Therefore it is recommended in Japan that major timber connections shall be mortise-tenon joints reinforced by cramp irons or nails. However it is often difficult to nail o crimp effectively without cracking timbers to be connected because diameter of crimps or nails is usually too large comparing timber size, and therefore the above recommendation has not been easily put into practice and consequently timber connections in roof frames are often ineffective to resist tensile forces.

The new type of roof damage described before is considered to be caused under these conditions. Under this present situation an idea of applying joint fasteners used in 2/4 construction is being examined in recent years. Several types of joint fasteners have already been designed 2/4 fasteners but however it is predicted that those fasteners will not easily spread because types of connections and dimensions of timbers are too many in number in Japanese conventional timber construction and it is difficult for manufacturers to prepare many kinds of fasteners.

Under these circumstances Building Research Institute has developed a new joint fastener having special merits as follows;

    i) having simple figures but applicable to almost all types and dimensions of timber connections

    ii) low cost

    iii) able to be manufactured easily in situ

    iv) able to be used for reinforcing timber connections in existing dwellings.

The following sections description of this fasteners and some results of experiments will be shown.

2. DESCRIPTION OF FASTENER

Material of this fastener is a galvanized iron sheet strip of 0.8 mm thickness and 20 mm width having nail holes along the center line (Figure 1). By cutting the material, ordinary straight type strip fasteners are obtained and by bending it along 45° line as shown in Figure 1 and cutting off at appropriate length, joint fasteners to be used at joints where two members cross perpendicularly are obtained. Some examples of application are also shown in the figure. In order to connect timbers by this fastener N38 nails or longer ones are used. Special tools for bending and cutting materials has also been developed.

3. DESCRIPTION OF TENSION TEST

In order to know the tensile strength of timber connections reinforced by this joint fastener, tension tests were conducted. Two types of test specimens shown in Figure 2 were tested. One is a purlin-roof post or roof post-roof joist connection model for testing straight type fastener and the other is a rafter-purlin or purlin-collar connection model for testing bent type fastener.

Major factors related to the tensile strength of this connection are (a) detail of timber connection, (b) kind of timber, (c) roof slope, (d) thickness of fastener and (e) length and number of nails. As for the factor (a), butt joint are adopted for test specimen in order to exclude the other strength than that of fastener. For factor (b), western hemlock was used for test specimens because western hemlock is used most frequently now in Japan. Concerning the other factors, a few conditions were selected respectively as shown in Tables 1 and 2.

In case of purlin-roof post connection, blind mortise-and-tenon joint reinforced by cramp irons (6φ, 120 mm length) and full mortise-and-tenon joint reinforced by nails (N100) were also tested for comparison (Table 3 and 4).

4. RESULT OF TENSION TEST

4.1 Straight Type Fastener

    i) Behavior and Failure Mode

        Figure 3 shows tensile force (P) - displacement (δ: see Figure 2) relations obtained by the test. From this figure it is observed that;

    a) plastic displacement is observed from the beginning

    b) maximum strength P_{max} occurs at nearly the same displacement of 7 to 15 mm (Table 1)

    c) strength decreases gradually at displacement larger than those stated above.

        Final mode of failure was pull-out of nails and no rupture of fasteners was observed. But at large displacement small checks or shakes of timbers were observed around nails.

    ii) Maximum Tensile Strength

        Maximum tensile strength P_{max} obtained by test is shown in Table 1. It was observed that effect of thickness and number of fasteners on tensile strength were rather weak and P_{max} seemed to depend mainly on number and length of nails.

Rough estimates of P_{max} were as follows:

    150 kg for 2-N38 nails

    200 kg   3-N38

    300 kg   4-N38

    400 kg   6-N38

    250 kg   3-N50$_F$

    500 kg   6-N50$_F$

iii) Allowable Tensile Strength

According to the timber construction design regulation prepared by the Architectural Institute of Japan, the allowable strength of timber connections shall be decided as the value does not exceed both 1/2 to 2/3 of the maximum strength and the strength at 1.5 to 2 mm joint displacement. Following the idea, $\text{Min}(1/2 \cdot P_{max}, P_{\delta=2mm})$ is also shown in Table 1. From these values rough estimate of allowable strength of straight type fasteners are led as follows

    70 kg for 2-N38 nails

    110 kg   3-N38

    150 kg   4-N38

    210 kg   6-N38

    120 kg   3-N50$_F$

    360 kg   6-N50$_F$

4.2 Bent Type Fastener

i) Behavior and Failure Mode

Figure 4 show typical P-δ relations of bent type fasteners. Two major differences are observed comparing with Figure 3; (a) initial rigidity is generally far lower, (b) decrease of strength after the strength reached the maximum is more intense.

Failure mode of bent type fastener was also pull out of nails and no rupture of fasteners was observed. Small checks or shakes around nails were also observed.

ii) Maximum Tensile Strength

Maximum tensile strength P_{max} is shown in Table 2. From these values it is inferred that the maximum tensile strength seemed to be effected mainly by the roof slope; maximum strength decreased with the increase of roof slope. Effect of number and thickness of fasteners was not observed clearly.

iii) Allowable Tensile Strength

Following the same ideas in the preceding section, and neglecting the effect of roof slope, allowable tensile strength of bent type fastener nailed by N38 can be obtained as follows:

      50 kg for 2-N38 nails

      100 kg   4-N38

4.3 Crimped Joint

Test results of crimped joints are shown in Table 3 and also in Figure 4 concerning P-δ relations.

Initial rigidity of cramped joints was very high compared with straight fasteners. The maximum strength was as much as that of straight fastener nailed by 6-N50$_F$. It will be noted that when cramp axes are not buried into grooves, the failure becomes very brittle.

4.4 Nail Reinforced Full Mortise-and-Tenon Joint

Table 4 shows that test result of nail reinforced full mortise-and-tenon joint. The maximum tensile strength of this joint is as high as 600 kg for N100 and more than 800 kg for 2-N100 and and behavior is very ductile.

As for full mortise-and-tenon joints not reinforced by nails, tenons were pulled out from mortises at one or two millimeter displacement. Tensile force necessary to pull out tenons scattered very widely such as 170 kg to 400 kg. Accuracy of carpentry is considered to be the reason.

5. CONCLUDING REMARKS

Supposing that rafters, purlins, roof posts, collars and other roof frame members are arranged at ordinary intervals (rafter; @ 450 mm, purlin; @ 900 mm, post; @ 1800 mm), tensile forces to act on those member connections can be estimated from the regulation in the

Building Standard Law of Japan; 20 to 30 kg for rafter-purlin connections, 80 to 120 kg for purlin-roof post or purlin-collar connection and 160 to 230 kg for roof post-roof joist connection.

Tension test results show that the strength of new type fasteners depends mainly on number and diameter of nails, and therefore by selecting appropriate nails the fastener can safely resist to such an extent of tensile forces.

When this fastener is adopted for reinforcing timber roof frame connections, the reinforcement standard shown in Table 5 will be recommended.

Table 1 Straight type fastener test specimens and test results

Name of test specimen	Number	Fasteners Thickness (mm)	Length*	Nail**	Test result P_{max} (kg)	δ_{pmax} (mm)	P_{2mm} (kg)	P_a*** (kg)
6H2F0.6DN38	double	0.6	6	4-N38	293	11.6	185	146
7H3F0.6DN38	"	"	7	6-N38	419	12.6	236	210
7H3F0.6DN50	"	"	7	6-N50$_F$	553	11.8	322	276
6H2F0.8DN38	double	0.8	6	4-N38	316	11.9	225	158
7H3F0.8DN38	"	"	7	6-N38	437	10.0	287	218
7H3F0.8DN50	"	"	7	6-N50$_F$	563	12.9	392	281
6H2F1.0DN38	double	1.0	6	4-N38	332	12.9	191	166
7H3F1.0DN38	"	"	7	6-N38	429	6.9	295	214
7H3F1.0DN50	"	"	7	6-N50$_F$	473	11.8	363	236
6H2F0.8SN38	single	0.8	6	2-N38	145	14.4	68	68
7H3F0.8SN38	"	"	7	3-N38	214	15.3	116	107
7H3F0.8SN50	"	"	7	3-N50$_F$	233	8.3	158	116

* Length of fasteners is shown by the number of nail holes contained in each fastener.

** Number of nails for each timber connected.

*** P_a = Min($0.5 \times P_{max}$, P_{2mm}).

Table 2 Bent type fastener test specimens and test results

Name of test specimen	Fastener***		Nail*	Roof slope	Test result			
	Number	Thickness (mm)			P_{max} (kg)	δ_{pmax} (mm)	P_{2mm} (kg)	P_a** (kg)
2/10D0.6	double	0.6	4-N38	2/10	259	12.9	112	112
3/10D0.6				3/10	294	14.4	104	104
4/10D0.6				4/10	229	9.4	101	101
2/10D0.8	double	0.8	4-N38	2/10	301	15.2	112	112
3/10D0.8				3/10	264	13.4	99	99
4/10D0.8				4/10	297	12.9	129	129
5/10D0.8				5/10	249	11.6	120	120
10/10D0.8				10/10	202	13.3	67	67
2/10D1.0	double	1.0	4-N38	2/10	277	13.1	125	125
3/10D1.0				3/10	294	10.4	139	139
4/10D1.0				4/10	257	12.0	127	127
2/10S0.6	single	0.6	2-N38	2/10	153	17.1	38	38
3/10S0.6				3/10	147	17.0	45	45
4/10S0.6				4/10	104	13.8	33	33
2/10S0.8	single	0.8	2-N38	2/10	164	18.0	42	42
3/10S0.8				3/10	160	15.2	50	50
4/10S0.8				4/10	163	13.9	57	57
5/10S0.8				5/10	148	15.5	50	50
10/10S0.8				10/10	131	16.6	31	31
2/10S1.0	single	1.0	2-N38	2/10	153	15.7	44	44
3/10S1.0				3/10	152	13.7	47	47
4/10S1.0				4/10	138	14.3	53	53

* Number of nails for each timber connected.

** See Table 1.

*** Fastener is 10 cm long and contains 5 nail holes.

Table 3 Test specimens and test results of butt joints reinforced by crimp irons

Name of test specimen	Cramping	Number of cramps	Test result	
			P_{max} (kg)	pmax (mm)
CO-1			496	3.7
CO-2	ordinary	2	439	4.1
CO-3			436	3.4
CG-1			492	5.6
CG-2	buried in groove	2	491	9.7
CG-3			401	4.4

Table 4 Test specimens and test results of full mortise-and-tenon joints reinforced by nails

Name of test specimen	Nail	Test result	
		P_{max} (kg)	pmax (mm)
N1	N100	639	29.6
N2-1	2-N100	812	9.9
N2-2	2-N100	969	17.8
NO-1		400	1.4
NO-2	none	292	2.3
NO-3		172	0.5

Table 5 Reinforcement standard

Joint connection	Average roof height (m)	Tensile force estimated (kg)	Reinforcement	
			Fastner	Nails necessary for each timber connected
rafter-purlin	4	20	bent type, single*	2-N38
	7	30		2-N38
purlin-roof post	4	80	bent type, double	4-N38
purlin-collar	7	120		6-N38
post-joist	4	160	straight type, double	6-N38
			none	2-N100
	7	230	straight type, double	6-N50$_F$
			none	2-N100

* Fastner shall be placed on both sides of rafter alternatively.

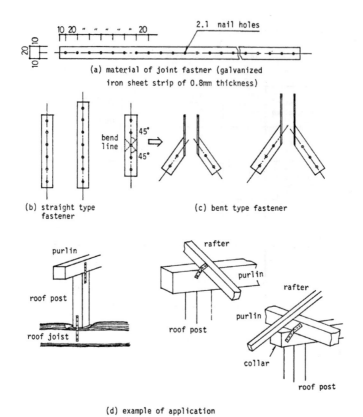

(a) material of joint fastner (galvanized
iron sheet strip of 0.8mm thickness)

2.1 nail holes

bend line

(b) straight type fastener

(c) bent type fastener

(d) example of application

Fig.1 Material and basic types of joint fastner and examples of application

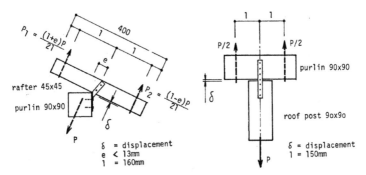

$P_1 = \frac{(1+e)P}{2l}$

400

rafter 45x45

purlin 90x90

$P_2 = \frac{(l-e)P}{2l}$

P

e

δ

δ = displacement
e < 13mm
l = 160mm

purlin 90x90

P/2   P/2

roof post 9ox9o

δ

P

δ = displacement
l = 150mm

(a) bent type fastner test specimen   (b) straight type fastner test specimen

Fig.2 Test specimen

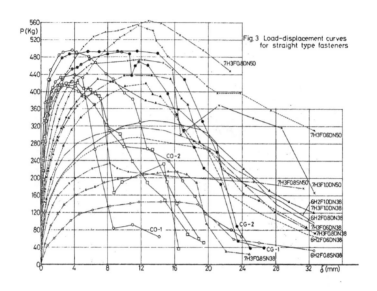

Fig. 3 Load-displacement curves for straight type fasteners

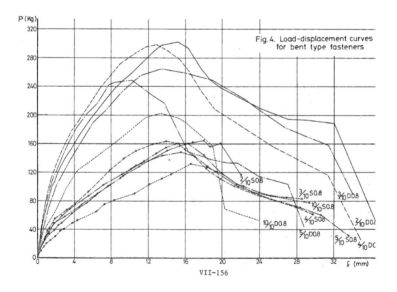

Fig. 4. Load-displacement curves for bent type fasteners

THE NASA/MSFC EXPERIMENTAL FACILITIES

AT HUNTSVILLE, ALABAMA

John B. Scalzi

National Science Foundation

Washington, D.C.

and

George F. McDonough, Jr.

Nicholas C. Costes

George C. Marshall Space Flight Center

National Aeronautics and Space Administration

Huntsville, Alabama

ABSTRACT

An evaluation of National Aeronautics and Space Administration (NASA) facilities and supporting capabilities for earthquake engineering research was made during a site visit and workshop held at George C. Marshall Space Flight Center (MSFC), Alabama, on February 22, 23, and 24, 1979. Workshop participants included twenty-six earthquake engineering specialists from the academic community, industry, and government. The workshop was sponsored by the National Science Foundation and NASA, and it was hosted by MSFC. The chairmanship and over-all direction of the workshop was assumed by the Earthquake Engineering Research Institute.

The results of the workshop indicate that the NASA/MSFC facilities and supporting capabilities offer unique opportunities for conducting earthquake engineering research. Specific features that are particularly attractive for large-scale static and dynamic testing of natural and man-made structures include the following: large physical dimensions of buildings and test bays; high loading capacity, wide range and large number of test equipment and instrumentation devices; multichannel data acquisition and processing systems; technical expertise for conducting large-scale static and dynamic testing; sophisticated techniques for systems dynamics analysis, simulation and control; and capability for managing large-size and technologically complex programs.

This paper will describe the pertinent facilities at MSFC inc
original purpose and utilization and potential uses, both in their
modification, to support seismic testing of structures.

KEYWORDS: Large scale testing; test facilities; structures soils,

1. INTRODUCTION

The George C. Marshall Space Flight Center (MSFC) of the National Aeronautics and Space Administration (NASA) is located within the U.S. Army Redstone Arsenal (RSA), a 60 square mile (96 square kilometers) facility adjacent to Huntsville, Alabama. The Arsenal is strategically located roughly 100 miles (160 km) from Nashville, Tennessee and Birmingham, Alabama and 200 miles (320 km) from Memphis, Tennessee and Atlanta, Georgia. It has its own ship docks, railroads and an airport in addition to close proximity to public transportation, including a modern airport and the interstate highway system. Barges played a prominent role in the early Apollo Program in carrying launch vehicles to Cape Canaveral for launch; the Arsenal airstrip was used in the shipping of stages and payloads as well as the delivery of the first Space Shuttle to MSFC piggyback aboard a 747 aircraft for modal testing. The location of MSFC within the RSA came about because the nucleus of MSFC was made up of scientists and engineers of the U.S. Army Missile Command who were transferred to NASA when it was formed in 1960. Dr. Wernher von Braun, who led the group which launched the first U.S. earth satellite in 1958, was the first Director of MSFC. The U.S. Army continues to conduct development and testing of military missiles and rocket systems ar RSA as well as training of crews for their operation worldwide; whereas the activities of MSFC are limited to peaceful uses of space and the used of space technology in the solution of problems in the civil sector. Seismic testing for earthquake hazard mitigation falls into the latter category of our charter.

Most of the NASA full-scale structural testing of large space and launch vehicles has been performed at MSFC, primarily because this Center was assigned responsibility within NASA for the largest vehicles: the Saturn moon rockets and Skylab. Facilities developed for these large systems recently have been used to test the Space Shuttle. Types of tests include static and dynamic structural load, vibration and acoustics, thermal vacuum and pressure proof tests. These tests were necessary to prove vehicle integrity and contributed greatly to the success of the programs conducting them. Because of the large size facilities required for the Saturn, which was 360 ft (110 m) and weighted 6.5×10^6 lbs (2.9×10^6 kg) when fully loaded, MSFC now has the largest and best equipped facilities of this kind in the United States. Recent completion of the Shuttle tests allows these facilities to be made available for other purposes, an important one of which can be testi of structures for earthquake hazard mitigation. This paper will describe the major facili ties which could be used for this purpose.

2. MAJOR FACILITIES

Figure 1 shows an overview of the MSFC Test Area as it is configured for Space Shuttle testing. Building 4550 is the Saturn V dynamic test stand recently converted for Shuttle vibration testing; a smaller test stand used for Saturn I dynamic testing is adjacent to it Building 4670 was originally the first stage booster engine test stand on which were test fired simultaneously the five engines of Saturn V; other engine test stands are throughout the area. Building 4619, which is the headquarters of the MSFC structural test activities, is the building of primary interest for earthquake testing since·it contains three of the most important elements needed for the tests envisioned in early phases of these tests. These elements are the Structural Test Tower, the Modal Test Equipment, and the Structural Test Data Acquisition System.

2.1 Structural Test Tower

The Structural Test Tower is located in a large test bay area of Building 4619 called the Load Test Annex, which is 170 ft (51.5 m) by 160 ft (49 m) by 155 ft (47.3 m) high and which is served by an access door 60 ft (18.3 m) by 75 ft (22.9 m). The Annex is served by a crane system utilizing two trolleys each with two hooks. Each trolley has a 30 k* (13,600 kg) lifting capacity and maximum height capacity of 106 ft (32.3 m).

The Tower, shown in Figure. 2 with a test specimen and load reaction fixtures in positio was completed in 1963 at cost of approximately $4.5 million. Its original purpose was test- ing of individual stages of the Saturn launch vehicle for critical load conditions; however, it was designed for greater capability, both in size and applied loads, than needed for Saturn so that it would be useful· for a wide range of research and development testing for post–Apollo launch vehicles. The tower is approximately 140 ft (42 m) high and can apply vertical and horizontal loads to structures ranging in height from 40 ft (12.2 m) to 115 ft (35 m) and having maximum lateral dimensions of 80 ft (24.4 m) by 50 ft (15.2 m) for

---

* k = kilopound or kip

structures assembled under the crosshead. These limitations are due to the spacing of the legs and the height of the crosshead. Larger structures can be tested within the Load Test Annex but outside the tower without loss of horizontal loading capability; however, vertical loading by the crosshead is lost.

The foundation of the Test Tower, shown in Figure 3 during its construction, is based on bedrock 26 ft (7.9 m) below the floor. The main slab is about 11.5 ft (3.5 m) thick and has 2,356 2 3/4 in (7 cm) diameter high-strength anchor bolts (ASTM A-354 steel) with sleeves so that there can be a flush floor. The bolts are on 18 in (0.5 m) centers in an 78 ft^2 (7.2 m^2). The crosshead is driven vertically using four roll ramps which move on four supported by 7.5 ft (2.3 m) wide reinforced concrete footings which surround the floor slab and which extend to bedrock. These continuous footings eliminate the need for uplift anchors for the stand.

Vertical loads are applied through the load reaction head shown in Figure 4, which weighs 4 million lbs (1.8 million kg). This crosshead consists of 8 ft (2.4 m) deep intermediate beams framing into six 20 ft (6.1 m) deep girders which in turn frame into two 20 ft (6.1 m) deep box girders. The top and bottom plates are 4 in (10 cm) thick with an area of 78 ft^2 (7.2 m^2). The crosshead is driven vertically using the four roll ramps which move on four stationary 14 in (350 mm) diameter double acme threaded stems. These ramps have a capacity of 6 million lbs (2.7 million kg); the stems run almost the full height of the tower. After the crosshead is put in position, it is bolted to the columns of the tower legs to serve as strongback in resisting vertical loads. Compression capability is 30 million lbs (13.6 million kg); tension capability is 2 million lbs (907,000 kg). Horizontal loads are applied through five horizontal box plate girders spanning two of the tower legs which are spaced at 20 ft (6.1 m) intervals up to 100 ft (30.3 m). Biaxial loading can be applied by utilizing special test equipment (STE) designed to provide lateral reaction. The massiveness of the crosshead is paralleled by that of the tower legs, each of which weights more than 600,000 lbs (272,000 kg) and in the horizontal girders.

A wide variety of hydraulic actuators is available for use with the structural test tower. A summary of these actuators along with other equipment is given in Section 3 of this paper.

2.2 Modal Special Test Equipment

Also located in Building 4619 is the modal special test equipment (modal STE) us
perform vibration tests, with three dimensional excitation, on the liquid oxygen tank
Space Shuttle External Tank. The modal STE consists of an air-bag support system whic
vides essentially a free-free boundary condition for the tank and an enclosure cage 78
(24 m) high and 52 ft (15.9 m) in diameter. As shown in Figure 5, the modal STE is ca
of being canted up to 13°, a requirement for simulating Space Shuttle takeoff.

The air bag suspension can be tailored to many configurations and uses. It consi
33 Firestone Model 321 air bags with a maximum load capabity of 1.9 million lbs (864,0
a vertical suspension frequency range of 0.7 to 5 Hz, a horizontal suspension frequenc
of 0.2 to 1 Hz, a test range displacement of $\pm$ 3.0 in (37 mm) either horizontal or ver

2.3 Vertical Ground Vibration Test Facility, Building 4550

The Vertical Ground Vibration Test Facility, was initially used for Saturn V moda
tests. It is 360 ft (110 m) high, 100 ft (30 m) by 120 ft (37 m) in plan, and has a ma
test bay 74 ft (22.5 m) square. The internal configuration of the building is variable
meet the requirements of the test being conducted; hence, it was possible to test the S
Shuttle configurations successfully in a facility designed for a much different vehicle
Figure 6 shows the mated Orbiter-External Tank (ET) - Solid Rocket Booster (SRB) config
tion in place for six-degree-of-freedom dynamic tests; hydrodynamic supports were used.
second configuration, Orbiter-ET, was tested using a cable support system. The roof of
Building 4550 can be removed over the main test by as can a large section of the buildi
sidewall 74 ft (22.5 m) by 145 ft (4.4 m), which facilitates moving large structures in
the building using the two large derricks 440 k (200,000 kg) and 385 k (175,000 kg). T
foundation is 8 ft (2.4 m) thick and has a load capacity of 12 million lbs (5.5 million

The hydrodynamic support system of Building 4550 is a state of the art device whic
lizes four identical supports with a total load capacity of 8 million lbs. (3.63 millio
It has negligible damping due to lateral resistance and is suitable for use with a wide
of loadings by changing its spring constants through pressure changes in the gas reserv

Building 4550 has full utility service as well as full instrumentation and control
capabilities with 600 channels.

VIII-6

2.4 Structural Test Facility for Hazardous Tests

Building 4572 was built and used for structural testing of the SRB. It is suitable for hazardous testing because it is set up for remote control from an adjacent building. The building has the capability of reacting loads as great as 3.1 million lbs (1.4 million kg) through the end walls and through beams embedded in the concrete floor.

As shown in Figure 7, access to the building for large test articles is through the top using a 100 k (45,000 kg) gantry crane outside the building. The building is designed for hydrostatic tests and has high pressure water supply through 40 in (1.2 m) and 13 in (.4 m) lines.

The building itself is 48 ft (14.8 m) by 152 ft (46.3 m) by 36 ft (11 m) high with a clear test area of 41 ft (12.5 m) by 94 ft (28.7 m) by 32 ft (9.8 m) high. One end wall is 12 ft (3.6 m) by 40 ft (12.2 m) by 29 ft (8.8 m) high reinforced concrete with embedded girders; the other is 7 ft (2.1 m) by 40 ft (12.2 m) by 28 ft (8.7 m) high with an embedded steel load ring. The floor is 5 ft (1.5 m) reinforced concrete with tiedown provisions to WF 10-60 to WF 10-112 embedded beams.

2.5 Static Test Stand

Bldg. 4670, shown in Fig. 8, was originally used as a hold down for the first stage of the Saturn V as its engines were test fired generating 7.5 million lbs (3.4 million kg) of vertical thrust and 370k (170,000 kg) lateral thrust. The stand consists of four concrete piers, 200 ft (61 m) high keyed into bedrock at 46 ft (14 m) depth and tied together at the top by a 20 ft (6 m) deep steel load platform. The interior spacing of the piers is 61 ft (18.6 m) and they taper from 48 ft (14.6 m^2) at the base to 10 ft (3 m^2) at the top.

Because of its original use the facility has extremely large storage and pumping capacity for fluids and high pressure gases as well as an 880 k (400,000 kg) and 770 k (350,000 kg) derricks. It also has complete instrumentation in a nearby blockhouse.

3. TEST SUPPORT EQUIPMENT

A large amount and great variety of large scale test equipment exists at MSFC to support the facilities listed in the previous section and other facilities within the Center. Most of this equipment is either portable, such as shakers, or capable of supporting tests at remote locations, such as the Structural Test and Data Acquisition System (STDAS).

3.1 Vibration Test Equipment

For elastic range vibration testing MSFC has 11 hydraulic and 28 electrodynamic shakers ranging in payload from 90 lbs (40 kg) to 200 lbs (90,000 kg) with maximum stroke of 6 in (22.5 cm) or less, and operating at frequencies from as low as 0 Hz to as high as 2500 Hz. These shakers are supported by three digital computers with Fourier analyzer and multipoint control and one computer with Fourier analyzer capable of simultaneous control of eight shakers. All systems can perform shock tests as well as sine and random testing with automatic cutoff at present tolerances.

3.2 Hydraulic Actuators

Eighty-one portable actuators are available at MSFC; their capabilities are summarized in Table I.

3.3 Structural Test and Data Acquisition System (STDAS)

The STDAS is a state of the art multiprocessor data system, developed for the Shuttle program, but capable of handling any system requiring up to 6,000 data channels including the simultaneous support of two separate tests. Most transducers, passive or active, can be handled by the system including strain and displacement gages, load and pressure cells, and thermocouples. The system is divided into three main parts: the Static Input Unit (SIU), the Data Selector Unit (DSU) and the Central Facility (CF). The STDAS as now configured has 24 SIU's, 3 DSU's, and 1 CF.

The SIU interfaces the transducer with the data system; it scans the signals, digitizes the analog signal, and transmits it up to 1600 ft (490 m) to a DSU. It provides excitation, calibration and signal conditioning as well as noise filters, automatic gain-ranging and other features and includes a Mod. Comp II/05 digital computer which, among other capabilities stores channel addresses and performs data averaging. This computer is hardware and software compatible with the DSU and CF computers. Portability, simplicity of operation and ruggedness are key features of the SIU; it is totally transportable and operates unattended in remote locations in an environment of 40°F (5°C) to 75°F (9°C) and relative humidity of 40 to 60% while maintaining accuracy of 0.15%. Control and display capabilities exist at the SIU to supplement the remote capability.

The DSU is a secondary computer subsystem which performs data reduction (conversion to engineering units), recording and display for as many as 9 SIU's (2000 channels) and

VIII-8

transfers data to the CF through modems. The DSU operator verifies the data, chooses and implements storage modes and updates displays. Mass data storage of 1.28 million 16-bit words is available on disc cartridge as well as magnetic tape. Graphic display includes an operator Cathode Ray Tube (CRT) plus three annunciator CRT's (9 more can be added if required) plus a high speed line printer. A ModComp IV/25 computer with a core memory of 64,000 16-bit words serves as the control center; it transfers data to and from the CF at the rate of one megabit per second. In short, the operator at the DSU console can exercise full control over the data acquisition as well as monitor test limit values vs. measured and display real-time, quick-look data to assess test performance.

The CF, as its name implies, is the hub of the DSU-SIU network, since DSU's can only communicate through the CF. At this station the test director and analysts monitor the conduct of the test and assess the validity of the data being acquired through the DSU's. Testing is conducted from the CF, if multiple DSU's are utilized, or from the DSU if only a single DSU is being used. Simultaneous control from one or more DSU's and the CF is not possible. A key feature of the CF is the ability to control DSU's as far as 3 miles (5 km) distant. Incorporated in the CF are two ModComp IV/25 computers similar to those in the DSU's with a full range of peripherals, some of which are switchable between processors. The CF performs post-test analysis as well as pre-test operations and can provide reduced test data in several formats suitable for reports or other final use. Hence, test data can be made available in final form within no more than a few hours, and usually much less, after completion of the test.

The STDAS has served as an excellent example of how time and manpower can be saved in test data handling through automation. The real-time interaction with the tests by analysts and other technical experts allows tests to be repeated at once if data is faulty or results questionable as well as identifying additional test conditions that would be useful.

4. MINOR FACILITIES

In this category are facilities of large capability which probably would be of limited usefulness in seismic testing and those of lesser capability which might fit well into a comprehensive earthquake program, especially in support of other tests which might be conducted at MSFC. Discussions of these facilities will be brief; however, much more detailed information is available on request from the authors or in the references listed at the end of this paper.

A small centrifuge is located in Building 4487. Under uniform rotation, it can handle a specimen up to 500 lbs (230 kg), the c.g. of which is 6 ft (1.8 m) from the center of rotation; maximum acceleration is 100 g. In a second mode a specimen of 110 lbs (50 kg) can be subjected to a 20 g centrifugal acceleration; to this motion, either sinusoidal oscillation up to 28 g at frequencies of 5 - 2000 Hz, or 28 g random at frequencies of 20 - 2000 Hz can be superimposed. Full data acquisition equipment, including television is available.

A small six degree-of-freedom (6 DOF) shake table, shown in Figure 9, is located in Building 4663. The table is 17 ft (5.2 m) by 13 ft (4 m), has a capacity of 23,000 lbs (10,500 kg) and a horizontal stroke limit of 4 ft (1.2 m). As shown in Figure 9, six hydraulic actuators, operated in pairs, drive the table under the control of a large hybird (analog/digital) computer system in an adjacent room. This system has had extensive use in simulating docking of spacecraft, vehicle dynamics on the lunar surface and several other space applications. Its use for simulating earthquake loadings appears to be limited, however, because its fully combined 6 DOF capabilities are limited and can be achieved only for small payloads and low frequencies.

The Neutral Buoyancy Space Simulator, Building 4706, is a very large facility consisting of a water-filled tank 75 ft (22.9 m) in diameter and 40 ft (12.2 m) deep. By appropriate adjustment of weights or floats attached on test subjects or equipment, a state of zero gravity is achieved which is used for astronaut training on zero-gravity space tasks such as extra-vehicular activity (EVA). Included in the facility are full instrumentation capabilities. Possible use in earthquake testing would be in soil-fluid-structure interaction, such as tests related to offshore structures.

A Geotechnical Research Laboratory is located in Building 4481. It was originally established primarily for scientific and engineering studies related to the lunar surface exploration program including wheel soil interaction for the Lunar Roving Vehicle (LRV). Its primary use today is in defining basic soils experiments for Spacelab missions and extending the lunar surface studies to the planets. Its instrumentation capabilities are available to support soil and geotechnical field and laboratory tests at MSFC.

5. POTENTIAL OF MSFC FACILITIES FOR EARTHQUAKE ENGINEERING

It is beyond the scope of this paper to define or propose a series of tests or other uses of the MSFC facilities in support of earthquake engineering; however, in examining facilities, it was useful to evaluate what types of tests might be conducted on what sizes and types of specimens. We were guided by the comments and suggestions of the participants in the NSF-NASA Workshop in February of this year as well as other statements of the earthquake engineering community.

Although a wide variety of tests could be conducted in the facilities at MSFC, only those which utilized the unique capabilities were considered. The most important of these are physical dimension and loading capability with efficient remote data acquisition, an important second-order capability. Based on this, several strawman tests were evaluated in detail; two of these, a structural test and a soil test, will be discussed in the following paragraphs to indicate what is feasible and what modification to facilities might be required.

5.1 Multi-Story Steel Frame Cyclic Test

Biaxial loading of a 6-story building, with 12 ft (3.6 m) stories, which is 4 bays long by 3 bays wide, with 20 ft (6.1 m) bays can be accomplished in the Structural Test Tower. Column and beam sizes would be typical of existing MSFC rigid frame buildings. With minor modification, the present load reaction structure and the main towers will be sufficient. The structure could be loaded along one side and one end either at top panel points or all panel points to achieve maximum cyclic deflections of top panel points of $\pm$ 24 in (61 cm). The existing Load Control System would be sufficient to synchronize the required 42 actuators and existing hydraulic pumps are sufficient to drive these actuators; however, additional long-stroke 200 k (90,000 kg) actuators would be required. Instrumentation is available: load cells, pressure transducers, electrical deflection instruments and strain gages. One DSU of the STDAS would be adequate to support the test. This is essentially a no-facility-modification test, the main cost of which is the test specimen.

5.2 Dynamic Soil Behavior in Large Test Bins

The Modal STE can readily be modified to support tests of dynamic soil behavior performed in cylindrical test bins 5 ft. (1.5 m) in diameter and 10 ft. (3 m) high and containing carefully prepared soil test specimens weighing approximately 24.5 kips (11,000 kg), which would serve as pilot tests for larger scale dynamic soil investigations using specimens approximately 15 ft (4.5 m) in diameter and 30 ft (9 m) high, or 30 ft (9 m) in diameter and 15 ft (4.5 m), as desired by the geotechnical community. The hydrodynamic support system of Building 4550 could be used in the large-scale tests. With only slight modification of existing hardware, the bins can be tested in torsion; with more elaborate modifications, four degrees of freedom can be provided. All actuators, pumps, controls, instrumentation and data handling are available. The main costs of such tests lie in the fabrication of the bins and equipment for the placement and removal of test soil and, for the multi-degree of freedom tests, modifications to current loading and lateral support systems.

6. SUMMARY

The Marshall Space Flight Center has developed facilities and equipment of unique capability which have been used successfully in the space program and show substantial potential for use in earthquake engineering research as well. Current program schedules indicate that some of these facilities can be made available for extensive use. Furthermore, preliminary studies indicate that only small to moderate modifications of these facilities are needed to tailor them to earthquake testing needs. Thus a large capability, which would be costly and time consuming to reproduce, can be made available to the earthquake engineering community at small cost relative to the value of the facility.

An additional asset of MSFC, equally important but less tangible than the facilities, is the experience of its staff in full-scale testing, the supporting analysis, and working with outside investigators to reach solutions to extremely complex problems. Full-scale testing has been one of the key factors in the success of the space program and it should prove equally valuable to earthquake engineering.

REFERENCES

[1] G. F. McDonough, "Dynamic Testing of Saturn Launch Vehicles," Sixth International Symposium on Space Technology and Science, Tokyo, 1965.

[2] R. E. Scholl, Editor, "An Evaluation of NASA Test Facilities and Supporting Capabilities for Earthquake Engineering Research," EERI (1979).

[3] G. E. Shofner, Jr., "Unique Structural Testing Facility – Load Test Annex," ASCE National Water Resources Meeting (1970).

TABLE 1
VIBRATION EQUIPMENT CHARACTERISTICS

Type of Shaker	Number Available	Force (tons)	Maximum Stroke (mm)	Operating Frequency Range (Hz)
Electrodynamic*	4	13.6	25	5 - 2,000
Electrodynamic*	4	9.1	225	5 - 800
Electrodynamic*	2	6.8	25	5 - 2,000
Electrodynamic*	4	3.2	25	5 - 2,000
Electrodynamic*	14	0.5	150	0 - 2,500
Hydraulic	8	22.7	25	0 - 350
Hydraulic	3	45.5	25	0 - 350

*Shock capacity - 2,000g; spectrum - 4 - 10 kHz.

FIGURE 1
AERIAL VIEW OF NASA/MSFC

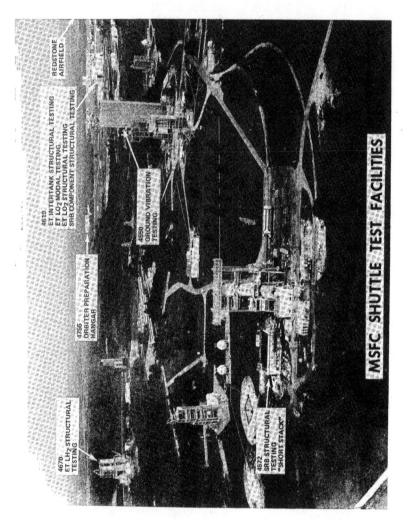

FIGURE 2
STRUCTURAL TEST TOWER WITH TEST SPECIMEN
AND LOAD REACTION TEST FIXTURE

FIGURE 3

STATIC TEST TOWER FOUNDATION CONSTRUCTION

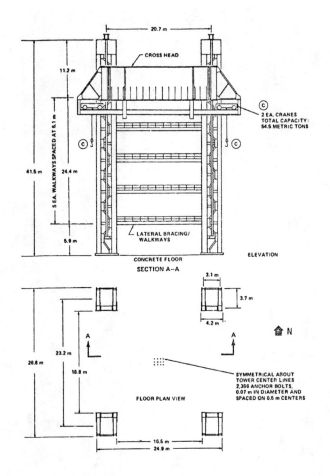

FIGURE 4
BUILDING 4619 -- LOAD TEST ANNEX MAIN TEST TOWERS

FIGURE 5   SPECIAL TEST EQUIPMENT WITH TEST SPECIMEN
VIII-18

FIGURE 6 VERTICAL GROUND VIBRATION TEST FACILITY: INTERIOR VIEW OF
BUILDING 4550 SHOWING SPACE TRANSPORTATION SYSTEM (SPACE SHUTTLE ORBITER,
EXTERNAL TANK, AND TWO SOLID ROCKET BOOSTERS)

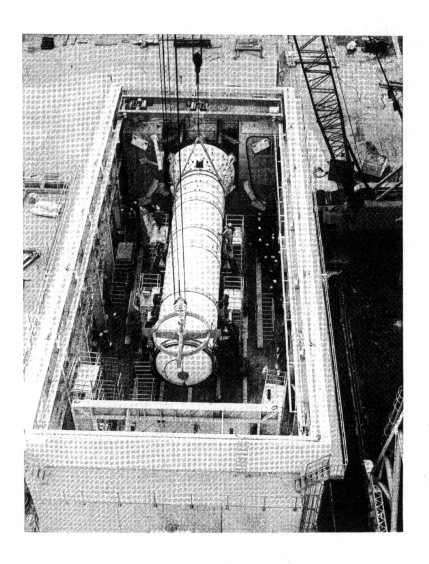

FIGURE 7 STRUCTURAL TEST FACILITY FOR HAZARDOUS TESTS:
AERIAL VIEW OF BUILDING 4572 WITH ROOF REMOVED
VIII-20

FIGURE 8 OVERALL VIEW OF THE S-1C STATIC TEST STAND

VIII-21

FIGURE 9 SIX-DEGREEE-OF-FREEDOM SHAKING TABLE

EARTHQUAKE HAZARDS REDUCTION

RESEARCH SUPPORTED IN 1978

Charles C. Thiel

William A. Anderson

Michael P. Gaus

William Hakala

Frederick Krimgold

Shih Chi Liu

John B. Scalzi

Division of Problem-Focused Research Applications

National Science Foundation

ABSTRACT

Research activities supported by the National Science Foundation in 1978 are reviewed. Abstracts of 153 awards are presented to describe the diversity of research activities underway as part of the National Earthquake Hazards Reduction Program.

KEYWORDS: Earthquake design; earthquake hazards research; geotechnical engineering; seismic design standards.

The objectives of the Earthquake Hazards Mitigation Program are to develop methods and techniques that can provide effective protection for man, his work and institutions from life loss, personal injury, property damage, social dislocations, and economic and ecological disruption associated with potential or realized earthquake hazards. The Program is structured in three principal components—siting, design and policy. These are described below.

Siting

The Siting category of the Earthquake Hazards Mitigation program provides research support for the elucidation of the physical basis of earthquake energy generation and the transmission and propagation of the generated shock waves through various geologic and soil conditions; with the impact of earthquake ground motion on structures; and with development of criteria and guidelines for the mitigation of potential impacts on the built environment. The specific objectives of the Siting program are to:

a) Develop a comprehensive data base on the nature of earthquake motion at typical construction sites and for representative structures;

b) Establish the physical basis for characterizing the nature of earthquake motions and the dynamic forces generated by such motions and other natural hazards;

c) Develop capabilities for prediction of the magnitude and frequencies of ground motion;

d) Develop a methodology for qualitative and quantitative estimates of local or regional risk associated with earthquake and other types of hazards and combined hazards.

e) Develop a comprehensive and unified program to improve geotechnical engineering practices applicable to soil dynamics, foundation design, failure and instability, and other aspects of earthquake ground motion; and

f) Identify procedures for integrating information on natural hazards into land use planning, urban and coastal zone planning, offshore engineering and siting procedures.

Design

The specific objectives of the Design program are to:

(a) Improve the characterization of earthquake and natural hazard loadings necessary for the economical design of structures subject to dynamic loading;

b.) Develop new methods of analysis and design of buildings and structures of all types which will take into account nonlinear and inelastic behavior of materials;

c) Develop methods to assess the hazard potential and risk assessments applicable to existing structures and facilities, and devise innovative methods for improving performance within economically acceptable bounds;

d) Obtain information for engineering analysis and design by observing the damage of facilities following actual earthquakes, and incorporate this information into standard design practice;

e) Develop improved computational capability for dynamic analysis of structures and facilities and improve user access to any computer software which is developed;

f) Develop model standards and design criteria for design of structures and facilities subjected to earthquake and natural hazard loadings; and

g) Conduct detailed studies of the behavior of smaller nonengineered structures and secondary components of buildings to improve recommended minimum analysis and design guidelines.

Policy

The specific objectives of the Policy program are to:

a) Expand the base of knowledge on alternative social adjustments to earthquakes;

b) Identify the social, economic, political, legal, and related factors which facilitate or hinder the adoption of both social and technological solutions to earthquake hazards;

c) Facilitate the beneficial use of earthquake hazards mitigation measures by devising effective techniques for disseminating information to the public and to decision-makers at the local, State and national levels; and

d) Investigate measures which will reduce possible negative social, economic, and political consequences of earthquake predictions and warnings.

One of the most important objectives of the Program is the timely and widespread dissemination of the results of supported research to potential users. The name and mailing address of the Principal Investigator and Grantee Institution is contained in each project description.--Persons wishing to obtain information on project findings including project reports, monographs, journal articles, technical reports, and other such relevant materials should write to the Principal Investigator at the Grantee Institution to determine what

information is available and at what, if any, cost it may be obtained. The Grantee Insti
tion may charge a nominal amount for the duplication and mailing of such materials to cov
costs. The Principal Investigator may furnish information on how interested persons may
acquire reports and other materials as appropriate from the National Technical Informatio
Service (NTIS) of the Department of Commerce in lieu of furnishing the report or other
material directly. NTIS is the central point in the United States for the pulic sale of
Government-funded research and development reports and other analyses prepared by Federal
agencies, their contractors, and grantees. The Principal Investigator may also cite journ
or other publications where project information may be looked up instread of furnishing a
copy of the article.

PROGRAM ACTIVITIES

The following sections present abstracted information for each of the awards made by
the Problem-Focused Research Applications Division during calendar year 1978. During fisc
year 1978 (October 1, 1977 through September 30, 1978) total program expenditures were
approximately $17.3 M; expenditures for fiscal year 1979 (October 1, 1978 through September
30, 1978) are estimated at $18.3 M.

Information on each award is presented in the following format:

Structural Stability of Columns;[1] Walter G. Lewis;[2] Lewis & Day, Inc., 6822
Westcott Drive, Richmond,Virginia 22325;[3] Award #77-0017.[4]

1) Title of the Specific Grant.

2) Principal Investigator: the chief scientist or administrator who
is responsible for the research plan and fiscal expenditures as an NSF-
sponsored awardee.

3) Institution Conducting the Research: any college, university, laboratory, industry, or other organization, whether operating on a profit or
non-profit basis, as well as State governments and Federal organizations.

4) Award Number.

Data is presented in the following sections for awards made in each quarter of the yea
Prior to the last quarter of the year, abstracted information for each award was very brief
Starting with this quarter, a more detailed summary of each project is given. This informa
tion has been extracted from a quarterly announcement "Recent Awards" which is available to

all those interested in timely notification of activities undertaken by the Program. If you desire to receive this publication please notify one of the authors or write: Professional Assistant, Division of Problem-Focused Research, Room 1134, National Science Foundation, Washington, D.C. 20550.

FUNCTIONAL DAMAGE AND REHABILITATION OF LIFELINES

IN THE MIYAGI-KEN-OKI EARTHQUAKE OF 1978

Kazuto Nakazawa

Eiichi Kuribayashi

Tadayuki Tazaki

Takayuki Hadate

Ryoji Hagiwara

Public Works Research Institute

Ministry of Construction

INTRODUCTION

The Miyagi-ken-oki Earthquake of June 12th, 1978 with a magnitude of 7.4 brough
disasters to Sendai City, population six hundred thousand, and the adjacent area. T
aster is deemed one of the biggest earthquake disasters since the Kanto Earthquake o
Since then earthquake disasters in modernized prefectural cities have been experienc
Fukui Earthquake of 1948 and the Niigata Earthquake of 1964. However, Sendai City w
greatly urbanized than Fukui City and Niigata City was in those days.

In this survey, the facts connected with earthquake disaster prevention measure
investigated cooperatively with the organizations listed later. This paper aims pri
a successful analysis of the functional losses and the rehabilitation of lifeline sy

It is organized as follows:

1. Functional Damage to Lifeline Facilities

    1.1 General Description

    1.2 Transportation Facilities

    1.3 Water Supply Facilities

    1.4 Sewer Facilities

    1.5 Medical Facilities

    1.6 Energy Supply Facilities

    1.7 Telecommunication Facilities

1. FUNCTIONAL DAMAGE TO LIFELINE FACILITIES

1.1 General Description

This earthquake is characterized by intense damage to lifeline facilities and their functional interruptions in Sendai, a large city of 629,000 population, and its adjacent area.

After the earthquake, Sendai City and the area experienced suspension of electric power, water and gas supply, railway operation, traffic confusion by the interruption of traffic signals, fall of telephone lines, etc. The various damage brought great inconvenience to urban activities.

Lifeline facilities fulfill their function when they are organized as a network. The experience of the earthquake shows the possibility of functional paralysis to whole lifeline systems by partial damage to lifeline facilities, and the seriousness of the influence of functional damage on urban activities.

VIII-29

1.2 Transportation Facilities

1) Roads

After the earthquake occurrence at 5:14 p.m. of June 12, the automatic control system at the traffic control center of Sendai Central Police Station sustained electrical failure, and 260 traffic signals in Sendai City could not be operated. That was right during the rush hours in the evening, and in such situations, the traffic in Sendai City became confused. After the recovery of electricity in the traffic control center at 8 p.m., the disposition of policemen for traffic control, and emergency vehicles for electric-power supply to traffic signals, traffic congestion was gradually dissolved.

Elsewhere, traffic through major roads had to be suspended or restricted by structural damage to national and prefectural highways and the Tohoku Expressway. (see Table 1-1 through 1-4).

2) Railways

After the earthquake, all the trains under the control of the Sendai Railway Control Bureau were suspended. Therefore many commuters and travelers were deprived of means of transit. After inspection and temporary repairs, the Tohoku Main Line south of Sendai and Joban Line re-opened on June 14, and the Tohoku Main Line north of Sendai on June 15. As for local lines in and around Sendai, their re-opening was completed on July 8.

3) Harbors

Ishinomaki Industrial Harbor, Sendai Harbor, etc. suffered such damage as cracks in aprons, breakdown of loading machines, etc. But by emergency measures, harbor functioning was restored.

4) Airports

The Sendai Airport was slightly damaged by small cracks in the runways. Take-offs and landings were suspended for a while, but normal operation was soon resumed. For the security of long-distance transport, Nippon Airways added additional flights between Sendai and Tokyo [June 13 (Tuesday) to June 14 (Thursday), June 17 (Saturday) to June 18 (Sunday)].

1.3 Water Supply Facilities

Water supply facilities were damaged in 54 cities, towns and villages of the Miyagi Prefecture.

In Sendai City, water supply was suspended for about 7,000 houses after the earthquake. The number of houses with suspended water service decreased to 2,000 on June 14, and that decreased to 700 on June 15. After that, water supply was increased day by day. As for other small cities, due to breaks in the water pipes, water supply was suspended at all 16,000 houses in Shiogama City and for 10,500 houses in Izumi City. The functional recovery of water supply facilities is shown in Table 1 to 5.

Damage to industrial water supply facilities caused more than 20 factories to suspend water supply from 24 to 79 hours.

1.4 Sewer Facilities

In Miyagi Prefecture, construction of public sewerage facilities has advanced in 16 cities and towns.

After the earthquake, sewer plants and pump yards in Sendai City and other cities around Sendai were out of service due to electricity failures and structural damage. In this situation, sewage had to be discharged into a river without full treatment.

As a result of temporary repairs, operation of these damaged facilities was restored within a few days except for the Kooriyama Pump Yard (recovered at 4:25 p.m. of June 23) in Sendai City.

1.5 Medical Facilities

Medical activities after the earthquake were more or less hampered because of electricity failures and damage to equipment and fittings at hospital and clinics in Sendai City and its neighboring districts.

However, harm to hospital patients was avoided by emergency measures, and serious confusion to medical activities was prevented through operation among undamaged and just slightly damaged hospitals.

1.6 Energy Supply Facilities

1) Electric Power Facilities

Total electric power demand in the Tohoku District (consisting of Anomori Prefecture, Iwate Prefecture, Akita Prefecture, Miyagi Prefecture, Yamagata Prefecture and

Fukushima Prefecture) was 4,900 MW immediately before the earthquake. After the earthquake, the electric power loading reduced 1,500 MW, and the electric power supply reduced 1,000 MW in Miyagi Prefecture and other districts of Tohoku. Reductions in electric power are shown in Table 1-6 and Figure 1-2.

As a result of functional damage, the number of houses without electricity reached 681,600 (419,100 houses in Miyagi Prefecture, 103,000 in Fukushima Prefecture, 95,800 in Yamagata Prefecture, and 58,600 houses in Iwate Prefecture, etc.). As restoration proceeded, the number of houses without electricity decreased to 280,000 at the end of June 12, and that decreased to 12,000 on June 13. Electric power facilities were completely recovered and functioning by the early morning of June 14.

2) Gas Supply Facilities

Earthquake damage struck gas supply facilities in four cities of Sendai (including Izumi City and Tagajo City), Shiogama, Furukawa and Ishinomaki. Damage in these areas was slight except for Sendai City. After inspection and restoration, gas supply facilities were completely functioning on June 13 in Ishinomaki City, on June 16 in Shiogama City, and on June 18 in Furukawa City.

However, in Sendai City, gas production facilities at Minato Plant and Haramachi Plant were more severely damaged, and worse, gas pipes were severed at many places. Gas supply in Sendai City was wholly suspended by such serious damage. As repairs were made, gas supply was resumed at some places in Sendai City on June 16. After inspection and repair to all gas pipes, from main pipe lines to twig pipes inside every house, gas was supplied block-by-block. It was on July 9 that gas supply resumed in every house in Sendai City. The restoration of gas supply facilities is shown in Table 1-7 and Figure 1-3.

Propane-gas supply by pipe was also suspended to 7,861 of the 16,266 houses in Miyagi Prefecture after the earthquake. Propane-gas supply to 4,444 houses was resumed on June 14, and the supply to the rest of the 3,417 houses was completed gradually thereafter.

1.7 Telecommunication Facilities

3,845 of the 1,738,000 subscription telephones in four prefectures of Miyagi, Fukushima, Iwate, Aomori were damaged by snapped wires, damage to telephone sets, etc. They were entirely restored on June 20 as shown in Table 1-8. As for long-distance transmission lines, 24,000 of the 70,000 circuits into, out of, and through Sendai City were down as shown in Figure 1-4. This was caused by damage to coaxial cables, tipping of micro antennas, etc. All the circuits were restored on June 14 as shown in Table 1-9. Short-distance transmission lines were damaged as shown in Table 1-10, but all the circuits were quickly restored on June 13. Further, as telephone inquiries poured into Sendai City after the earthquake, an extraordinary congestion of calls existed until June 14 as shown in Figure 1-5.

Public telephones were totally unusable for several hours due to electrical power failures, and telegraph service was knocked out at the Sendai Central Telegraph Office due to transmission line damage. This function was soon recovered. By June 15, three days after the earthquake, 64,000 telegrams reached Sendai City. This is 20 times more than normal.

Wireless facilities for emergencies connecting the prefecture and cities, towns and villages were just slightly damaged in Miyagi Prefecture. There was only one town that could not communicate by wireless.

Broadcasting, TV and radio stations could not broadcast for a brief period after the earthquake due to electrical failures and damage to micro-circuits. Their functions were soon restored as shown in Table 1-11 and Table 1-12.

2. CHARACTERISTICS OF DAMAGE TO LIFELINE FACILITIES, THEIR FUNCTIONS, AND REHABILITATION

2.1 Relationship of Damage, Topography and Geology

The sites where structures suffered damage correspond roughly to the areas of topographically- or geologically-severe conditions. For example, Figure 2-1 shows the distribution of razed houses and failed slopes. The razed houses are concentrated near alluvium, the eastern suburb of Sendai City, and to the hilltop, northern and south-western suburbs of the city. The latter area also suffered slope failures. The distribution of the damage roughly coincides with that of the water supply facility and of the gas supply facility shown in Figure 2-4.

Through the experience of past earthquakes, we have learned that structures on alluvium are likely to suffer greater damage than those on diluvial layers. The earthquake proved that by and large this is still accurate. Some geologists assert that the back marsh behind the ocean ridge on the Pacific Ocean is vulnerable to earthquake effects. However, little evidence of this was observed.

The hills in the western part of Sendai City (i.e. Aoba-yama, Yogi-yama and Dainenji), and the western part of Natori City (i.e. Minowa and Noda-yama) are covered with a 5 m thick loam surface layer a 10 to 20 m thick gravel layer underneath, both of which are called the Aoba-yama-Layers. The Tertiary layer of sandstone spreads beneath them. The average N-value of the loam layer at 2.5 m beneath the surface is 5 and the gravel layer ranges from 20 to 35. In the Midorigaoka District, Sendai City, where residential lots were developed in the late 1950's and early 1960's, much slope failure took place under the same geological conditions as above. The residential lots had been developed by cutting the hilltops and filling the valleys. As shown in Figures 2-2 and 2-3, there were the fills more than 10 m thick with N-values of less than 10. Slope failure and settlement occurred mostly at these fills due to the earthquake, which damaged residential houses, water supply, and gas supply pipes.

As for the residential lots of the northern part of Sendai City, and Izumi City, there are no loam and gravel layers. Hard rock and soft rock layers alternate. The residential lots had been developed through the use of bulldozers with ripper attachments. Figure 2-4 shows the damage distribution of water supply and gas supply pipes at Nanko-dai Housing Complex, Izumi City. The damage sites are concentrated in the areas where there used to be valleys that had been artificially filled.

2.2 Damage and Function of Lifeline Facilities After the Earthquake

Table 2-1 shows the individual loss ratios of lifeline facilities in Miyagi Prefecture. The individual loss ratio is defined as the ratio of the loss caused by the earthquake to the existing assets in the affected area. The existing assets can be estimated from the population density.

$$\text{Individual Loss Ratio} = \frac{L \; P}{W \; P_D}$$

where

        L = loss valued in money,

        W = national wealth,

        P = national population and

        P_D = population of the quake-affected area.

The national wealth of 1978 is not available, so the 1970 data are used here.

The individual loss ratio to harbor facilities shown in the Table is the highest. The highway and gas supply facilities figures are also high. Water supply and electric power facilities have relatively low ratios while the telecommunication facilities are the lowest. The list individual loss ratios provide a suggestion of the relative amounts of damage. It can be used as an index to quantify the damage to the various facilities.

The loss ratios also roughly coincide with the amount of functional damage. In Ishinomaki Port, the most heavily damaged harbor, the harbor function was maintained after the earthquake by the emergency repairs by utilizing wharves damaged only slightly. As for the harbor facility the functional damage was not minor, but the reserve facilities were well prepared.

Many highways were blocked by the earthquake damage. As of October 5, 1978, about 4 months after the earthquake, two highway sections and four highway bridges were still out of service. Many bridges suffered considerable damage as shown in Appendix II, which caused discontinuity of highway networks in Miyagi Prefecture. The gas supply facility suffered great functional damage and the individual loss ratio was relatively high. It took about one month to resume gas supply. The area where gas supply stopped after the earthquake encompassed Sendai and adjacent cities. The water supply facility sustained less functional damage than the gas supply, as the individual loss ratio to the water supply was also less. The suspension of water supply in Sendai was over on June 20, eight days after the earthquake. Islands of Shiogama City experienced the longest suspension which was eleven days. During the suspended period, water wagons supplied water to the residents. Functional damage to electric power was not severe. The domestic power supply was resumed on June 14. However, the suspension did influence other lifeline facilities. For example, traffic signals did not function because they did not have electric power. This caused traffic congestion in central Sendai City. The Traffic Control Center of Sendai Central Police Station, which

controls area traffic signals in Sendai, was disordered by the electric power suspension.
It could not collect data on the traffic just after the earthquake which would be useful to
analyze. Tohoku Electric Power Company, Ltd. asked industrial firms to reduce the consump-
tion of electric power until the power plants and transformer stations recovered. This
influences industrial productivity to some extent. The telecommunication facility suffered
little functional damage. However, the demand for telephone calls increased after the earthquake, which caused difficulties and delays in placing them.

As shown above, functional damage and individual loss ratios correlate to each other
quite well.

Lifeline facilities can be classified into terminal facilities (e.g., power plants,
filtration plants, centrals) and distribution facilities (e.g., electric wires, conduits,
telephone lines). Highways are distribution facilities and harbors are solely terminal
facilities respectively, so that we have taken electric power, water supply, gas supply, and
telecommunication facilities as the examples for the classification.

Electric power facilities suffered damage mainly to the power plants and transformer
stations, the terminal facilities. This caused a decrease of electric power supply for
several days, although the individual loss ratio was low. This teaches us that the terminal
facility is important to preserve functions after the earthquake.

On the other hand, water supply, gas supply and telecommunication facilities sustained
more damage to their distribution facilities than to this terminals. The telecommunication
facility was well prepared with reserved networks, which held functional damage to the minimum. As for the conduits, valves were in place to isolate the damaged sections from the safe
networks. Moreover some water leakage could be tolerated in order to keep the pipes flowing.
This helped to lessen functional damage.

Valves were not in place in the gas pipe network. The explosiveness of gas also delayed
the resumption of supply.

Figure 2-5 illustrates the difficulty in restoration of the function. The valves in the
pipe networks are also considered as a reserve facility. The upper and right direction indicates the difficulty of resumption. In this earthquake, electric power was considered to be
classified as IV, water supply and telecommunication was III and gas supply was I in the
figure.

2.3 Characteristics of the Disaster in Urban Areas

This earthquake may be classified as an urban-type disaster. Dwellings and lifeline facilities damage was concentrated in urban areas and functional damage due to the disorder of lifeline facilities followed. The Niigata Earthquake disaster of 1964 of Japan was similar in nature.

While all earthquakes shake the ground, not all of them cause damage to inhabited areas. Urban-type earthquake disasters represent the opposite case. Concentration of population and facilities accentuates the disaster. Table 2-2 shows the comparison of this earthquake and the Kinkasan-oki Earthquake of 1936, whose epicenters were about 5 km away from each other. The Kinkasan-oki Earthquake caused less damage than this earthquake did. As shown in the Table the population had increased by 1.6 times between the two earthquakes. The population of Sendai City had increased from 260,000 in 1944 to 620,000 in 1975, although the population of Miyagi Prefecture in 1944 which was 1,260,000 had changed very little from 1936. By definition, urban-type disasters are more severe when an earthquake hits an urbanized area of high population density.

Typical examples of urban-type earthquake disaster were observed at the newly developed residential lots in Sendai and adjacent cities. The residential lots have been developed by cutting hilltops, filling valleys, and by embanking on farm lands, which were usually soft and saturated. On the other hand, the old city of Sendai, except downtown Naga-machi, which is spread over the terrace of the Hirose River, mostly consists of diluvial gravel. The old city on the terrace suffered minor damage. The increase of population in turn increased the pressure to spread the building of residences to the area of steep slopes and soft ground which are relatively vulnerable to earthquakes. The damage to lifeline facilities in the vulnerable area greatly increased disruption in citizens' lives.

The impact on, and the response of, residents to earthquakes was likely to have been less than what would be expected in a metropolitan area such as Tokyo or Osaka.

Table 2-3 shows the tenure of houses as of 1975 National Census. In the central part of Tokyo occupant-owned houses do not form a high percentage of the total. However in Sendai it is comparatively high. Moreover, it is said that the citizens of Sendai have a greater tendency toward domiciliation. The disaster showed that they were eager to maintain order and calm. Panic, such as rushing to food stores, did not take place. Relatively few complaints about the suspension of electric power, water, and gas supply were made to the relevant organizations.

2.4 Public Facilities and Private Goods

Public facilities are, by definition, designed to possess a uniform minimum strength for public safety. Structures should continue to incorporate design improvements to resist the severe circumstances of nature.

Private facilities, primarily residences, also relate to the public so they should have similar design resistivity. The prescription for average buildings in the Standard Building Law contains the example of avoiding nuisances to neighbors. However, private facilities are apt to be designed as inexpensively as possible, because they tend to emphasize economy compared to public facilities. Therefore the earthquake damage to private facilities is likely to be more influenced by natural circumstances. For instance, the damage to dwellings was concentrated in areas such as developed lots on soft ground and hilltops.

The residents, whose houses were damaged, suffered the economic burden of having to reconstruct them. Prefectural and municipal governments aided them indirectly by reconstructing roads and sabo dams, and strengthening steep slopes in the damaged areas. They also recommended loans for reconstruction. However, the responsibility of reconstruction was on the residents in the end.

2.5 Rescue and Rehabilitation

National, prefectural and municipal governments, and public corporations set up temporary organizations for rescue and rehabilitation. They functioned well, since they had earlier experience with many disasters, such as floods.

The Department of Public Works and the Police play an important role in communication of data, because they shoulder the main responsibility for it in flood reconstruction.

3. LESSONS FOR EARTHQUAKE DISASTER PREVENTION MEASURES

3.1 Data Collection on the Disaster

It is important after the earthquake to collect disaster data as quickly and accurately as possible. It is invaluable in determining methods of reconstruction and in analyzing the disaster later.

The earthquake caused much functional damage to several lifeline facilities, but the functional damage has not been quantified. Adding up the functional damage to whole facilities and then comparing them has not been possible so far. The quantification of such functional damage is important in order to relate it to the total damage to the whole system.

3.2 Resumption of Functions

Road functions were somewhat disturbed for a few days after the earthquake due to the damage to them. However it did not greatly hamper the transportation of goods. There were no shortage of subsistence commodities.

The disaster was not big enough to apply the ministerial notification of The Designation of Charge-free Vehicles on Toll Roads.

Gas supply was stopped for about one month after the earthquake. One reason for this was that the gas pipe network was continuous within the supply area. The supply could not be resumed until entire networks of damaged pipes were repaired.

In Sendai the pipe network was divided into eight blocks temporarily, in order to be efficiently repaired, and to resume the supply as fast as possible. They were connected again after the repair was over.

Tokyo Gas Co., Ltd. has started a plan to divide the pipe network into several blocks, after the experience in Sendai.

Functional damage to the electric power facilities was not extensive. However it did cause disorder to the traffic signals and water supply pumps. Functional damage to the electric power facilities had similar disruptive influences on other lifeline facilities.

3.3 Utilization of the Earthquake Disaster Experience

Fortunately, few fires occurred during the earthquake. The earthquake of February 20, 1978, described in the preceding chapter might have resulted in the training of people to extinguish fires in earthquakes.

The Nippon Telegraph and Telephone Public Corporation suffered considerable damage by the Tokachi-oki Earthquake of 1968 (M = 7.9). They changed their design criteria from the experience. It avoided routing lines on high embankments and softground as much as possible. This would have contributed to a reduction of the damage from this earthquake.

The damaged residential lots in and around Sendai had been developed on valleys and soft grounds, vulnerable to earthquake effects. Research to measure the failure mechanism has

been launched by universities and local governments. However the problem was that th
vulnerability had not been communicated to the residents before the disaster. Landsli
prone areas and dangerous steep slopes have been legally designated, and the designati
flood areas has been introduced into the by-laws.

The Miyagi Prefectural Government began its investigation into seismic micro-zona
after the experience of the recent earthquake.

APPENDIX

I. An Outline of the Miyagi-ken-oki Earthquake of 1978

I-1 The Earthquake Feature

The general feature of the earthquake is described as follows:

1) Data and time of the occurrence of the main shock: 5:14 p.m.,
June 12 (Monday), 1978,

2) Epicenter and Focal depth: 142°10'E, 38°09'N and h = 40 km,

3) Location of the epicenter: The sea bottom about 60 km off Mt.
Kinka-san in the Miyagi Prefecture,

4) Magnitude on the Richter scale: M = 7.4,

5) Seismic Intensity: Isoseismal Map shown in Figure I-1,

6) Focal mechanism: The focal mechanism shown in Figure I-2 proposed by
T. Seno, K. Sudo and T. Eguchi,

7) Severest after-shock:

Data and time: 8:34 p.m., June 14th (Wedenesday), 1978,

Epicenter and Focal depth: 142°29'E, 38°12'N and h = 40 km,

Magnitude: M = 6.4, and

8) After-shock area: 80 km long and 30 km wide as shown in Figure I-1.

I-2 Fore-sign of the Main Shock

The outline of fore-shocks is described as follows

1) On February 20, 1978, an earthquake with magnitude 6.7 which brought
slight damage in Miyagi and Iwate prefectures occurred at the sea bottom
about 60 km north of the epicenter of the main shock,

2) After the earthquake on February 20, there occurred a few sensible
earthquakes for a month in and around the area, and

3) Eight minutes before the main shock, a fore-shock (M = 5.8) with third
degree of JMA's Intensity Scale was recorded at the meteorological station in
Ohfunato and Miyako, Iwate Prefecture.

I-3 Weather in the Damaged Area Before and After the Earthquake

The average temperature of June, in 1978 in Sendai City was a little high in
comparison to the average while the precipitation was about normal. (Refer to
Table I-1.)

The average temperature and the precipitation of the most severely damaged
Sendai City ranged from 17.6° to 20.3°C and from 0 to 2.5 mm/day, respectively.

1-4 An Outline of the Disaster

The earthquake brought damage and losses mainly to Northern Japan, especially
Miyagi, Iwate, Fukushima and Yamagata Prefectures. Miyagi Prefecture suffered most
severely. Urban functions were temporarily paralyzed by damage to lifelines such
as transporation, water and gas supply, electric power and communication facilities. A considerable number of casualties were caused by the toppling of prefabricated concrete segment fences.

1-5 Historically Damaging Earthquakes off Kinkasan

Since the seventeenth century, twelve earthquakes with a magnitude greater
than 6.2 have occurred off the coast of Kinkasan. (Refer to Figure I-3.)

The earthquakes, and the return periods, are shown in Figure I-3. According
to this record, the return periods range from 23 to 42 years with one exception.

As shown in Figure I-3, the Kinkasan-oki Earthquake of 1936 brought slight
damage to the Miyagi and Fukushima prefectures. The epicenter and magnitude of
the earthquake was similar to that of the Miyagi-ken-oki Earthquake of June 1978.

II. Damage and Losses

II-1 General Description

The injury and loss of life and property caused by the earthquake occurred in
Miyagi, Iwate, Fukushima and Yamagata Prefectures. Injuries were estimated at
eleven-thousand persons and property damage at three hundreds billion yen, respectively. Figures II-1 and II-2 show the losses of property and lives. Based on
Figures II-1 and II-2, the property losses of ¥15,000,000,000, ¥300,000,000,

¥100,000,000 and ¥30,000,000 took place in Miyagi, Iwate, Fukushima and Yamagata Prefectures, respectively. One hundred thousand residents were injured.

Tables II-1 and II-2 describes the classified property losses and casualties. The loss of lives and the amount of damage are detailed in the following sections.

II-2 Losses of Life

The number of dead and injured are shown in Table II-2. The number of lives lost is largest in Miyagi Prefecture.

The dead and injured were mainly due to the collapse of buildings, houses, fences and the drop of fragments of glass. Fence collapse was the cause of about 60% of the deaths in Miyagi Prefecture.

II-3 Buildings and Houses

The damage to buildings and houses mainly occurred in the Sendai plain and Fukushima basin. Damage also occurred on the alluvium along the Kitakami and the Hirose River. The greater part of damage was to wooden dwelling houses.

The damage and the loss is shown in Table II-3.

II-4 Medical and Sanitary Facilities

The loss of medical and sanitary facilities, such as hospitals, medical stations, and solid waste stations, is shown in Table II-4.

II-5 Commerce and Industry

The loss to commerce and industry, such as the loss of goods and factories, is shown in Table II-5.

II-6 Agriculture, Forestry and Fishery

Farmland, agricultural production, roads, sawmills for forestry, and the quay walls of fishing ports were damaged.

These losses are shown in Table II-6.

II-7 Educational Facilities

Education facilities losses were mainly to school buildings. It is shown in Table II-7. In this table the monetary loss to cultural assets is included.

II-8 Facilities for Flood Control and Preservation of Mountainous Areas

The losses to flood control facilities and mountainous areas are mainly those to dykes, seashore, and sabo facilities. The losses are shown in Table II-8.

Countermeasures to prevent secondary disasters such as floods and debris flow were rapidly implemented after the earthquake, which occurred in the rainy season.

II-9 Transporation Facilities

The loss of transportation facilities is shown in Table II-9.

II-10 Water Supply Facilities

As the water supply system in Miyagi Prefecture was damaged, drinking water and water for industry were temporarily suspended.

The loss of the water supply facilities is shown in Tables II-10 and II-11. The damage distribution to pipelines is shown in Figures II-3 and II-4.

These figures show the damage was concentrated at newly developed housing lots from cutting hills and banks on soft ground in Sendai City. Detailed investigations have been conducted by the organizations concerned.

II-11 Gas Supply Facilities

The damage occurred to gas pipes and gas terminals.

Figure II-5 shows damage distribution to buried pipes. The damage distribution is similar to that of the water supply pipeline.

The loss in Miyagi Prefecture is shown in Table II-12.

II-12 Electric Power Supply Facilities

The losses to hydroelectric and thermal power plants and transformer stations are shown in Table II-13.

II-13 Telecommunication Facilities

Telephone calls were suspended because buried coaxial cables in Sendai and Kabuta City were cut completely.

The loss is shown in Table II-1.

II-14 Sewerage Facilities

The damage to sewerage facilities was to pipes, treatment plants, and pumping facilities.

The loss is shown in Table II-14. The damage distribution is shown in Figure II-6.

II-15 Park

The facilities at Kokeshi Island of Matsushima Park in Rikuchu-kaigan Nation Park were damaged. Observatories and lodgings at Miyako and Rikuzentakada cities were slightly damaged.

The loss is shown in Table II-1.

II-16 Other Facilities

Damage to other facilities occurred in a water tank reservoir for fire fighting, office furniture, etc.

The loss is shown in Table II-1.

II-17 Indirect Losses

Indirect losses were caused by the suspension of various industrial activities. These have not yet been evaluated. The losses are primarily in commerce, agriculture, forestry, fishery, buildings, and manufacturing.

III. Rescue and Rehabilitation After the Earthquake

III-1 General Description

The earthquake caused the death of 28 persons, the collapse of many houses, and other serious damage. The damage to railways and roads caused a paralysis of transportation in the affected area. In Sendai City and the adjacent area, traffi confusion, suspension of water, electric power and gas supply, interruption of telephone and sewerage, all caused inconveience both in urban living and to industrial activity.

Each administration proceeded with:

* Rescue of casualties,

* Evacuation of inhabitants from dangerous areas,

* Guaranty of the necessities of life such as food, drinking water, etc.,

* Restoration of basic facilities for living, such as facilities for water supply, electric power, gas supply and communication, and

* Restoration of transportation.

Fortunately, there were only 3 fires that could not be extinguished by residents themselves among the 12 fires which occurred after the earthquake. These were soon extinguished by fire fighters. At a petroleum.industrial area, some oil spilled.from cracks in 3 oil tanks, but there were no fires.

In spite of this extensive damage, greater confusion was prevented by the rapid countermeasure responses of the organizations concerned and the cooperative behaviour of inhabitants.

III-2 Organizational Systems and Their Operations

In Miyagi Prefecture, rescue and rehabilitation centers for emergencies were set up and operated by each organization mentioned below:

1) Rescue and Rehabilitation Center of Miyagi Prefectural Government (opened at 5:30 p.m., June 12 as shown in Figure III-1).

2) Rescue and rehabilitation centers by local public bodies (56 cities, towns and villages).

3) Rescue and rehabilitation centers of:

* Designated government offices in Tohoku District (Tohoku District of Jurisdiction .Police Bureau, Sendai International Trade and Industry Bureau, Şendai Overland Transportation Bureau, Sendai District of Jurisdiction Meteorological Bureau, Tohoku Regional Construction Bureau, etc.), The Self-Defense Forces (Tohoku District General Administration Division, The 22nd General Course Regiment),

* Designated public organizations (Sendai Railways Superintendent Bureau of Japanese National Railways, Miyagi Telecommunication Division of the Nippon Telegraph and Telephone Public Corporation, Miyagi Prefectural Branch of the Japanese Red Cross Society, Tohoku Headquarters of Japan Broadcasing Corporation, Sendai Branch of Japan Transportation Co., Ltd., Miyagi Branch of Tohoku Electric Power Co., Ltd., etc.), and,

* Headquarters of Miyagi Prefectural Police Department.

The Miyagi Prefectural Government requested aid from the Ground Self-Defense Force which put 2,025 personnel and 400 vehicles in 5 cities and 7 towns. Since

June 13, the Disaster Relief Law was applied to Sendai City, Izumi City, and the towns of Hasama Yoneyama, Kogota and Naruse. Each organization carried out rescue and rehabilitation activities with mutual cooperation.

The Miyagi Prefectural Government substituted the Miyagi-ken-oki Earthquake Disaster Reconstruction Section for the Rescue and Rehabilitation Center on July 10, which had been charged with rehabilitation operations.

III-3 Emergency Communications

Emergency communications are vital to carry out refuge, rescue and restoration operations.

The damage to telecommunication facilities by the earthquake was relatively slight, and their functions were restored in an early stage. The emergency telephone networks which were set up greatly contributed to rapid responses in many activities.

Information was relayed to inhabitants by TV, radio, patrol-cars and other information media. This helped promote calm and reduce anxiety.

III-4 Official Information on the Earthquake

The Meteorological Agency announced earthquake information eleven times after the earthquake. Tsunami information was broadcast after the earthquake as shown below:

° Tsunami Warnings: To the Pacific Coast of Tohoku District at 5:21 p.m., and the Pacific Coast of Chiba and Ibaragi Prefectures at 5:24 p.m.

° Tsunami Cautions: To the Pacific Coast of Hokkaido Prefecture at 5:28 p.m.

Order of evacuation were sent to the coastal cities, towns and villages where danger of damage by tsunamis was suspected. Height of tsunamis was observed only between 14 to 22 cm at the Pacific Coast of Tohoku District. The tsunamis brought no damage. Cancellation of the warnings and the caution was announced at the Pacific Coast of Hokkaido at 8:15 p.m. and the Pacific Coast of Tohoku District, Chiba Prefecture and Ibaragi Prefecture at 8:30 p.m.

III-5 Evacuation and Rescue

About 13,650 inhabitants had to be evacuated due to collapse of houses and tsunami warnings. The refugees returned home after the cancellation of the warnings. But the inhabitants in areas where houses were severly damaged stayed away for fear of secondary disasters, as shown in Table III-1.

Even on July 20, more than one month later, 2,558 habitants of 751 families in 7 cities and 13 towns continued living at places of refuge, relative's houses, etc. The rescue of casualties was carried out almost without interruption in cooperation with medical teams who continued to function after the earthquake.

III-6 Fire Fighting Activities

Twelve fires at nine places occurred during the earthquake, but nine fires were extinguished by residents. The three that could not be extinguished in an early stage occurred at the Science Faculty of Tohoku University, Tohoku Pharmaceutical College and at a low-pressure gas holder of Sendai Gas Bureau. These fires did not spread to neighboring buildings.

The following reasons for the slight damage from fires seem accurate:

● The earthquake occurred before the cooking hour of evening meals.

● Heaters were not used, as it was early summer.

● A fore-shock of the 3rd degree of JMA's Intensity Scale in Sendai City occurred about 8 minutes before the main shock, and precautions were taken to prevent fires.

● The experience gained from the earthquake of M = 6.7 on the Richter Scale which occurred offshore Miyagi Prefecture on February 20, 1978 was not forgotten.

III-7 Traffic and Transporation After the Earthquake

1) Roads

In Sendai, Ishinomaki, Shiogama, and Furukawa Cities, 744 traffic signals were damaged. No traffic signals worked due to the electrical failures. Traffic became confused in these urban areas. Owing to the dispatch of policemen for traffic control, and emergency vehicles for electric-power supply to principal intersections, the confusion settled down in about five hours.

Tohoku Regional Construction Bureau, Miyagi Prefectural Government, etc., patrolled main roads and collected information on the damage. Traffic control is shown in Tables 1-1, 1-2 and 1-3.

The damage to the designated national highways and the methods of restoration are shown in Table III-2.

As for Tohoku Expressway of JHPC, traffic was suspended between the Shirakawa and Tsukidate Interchanges at 5:17 p.m. June 12, and between the Ichinoseki and Hirasawa-Maesawa Interchanges at 5:32 p.m., June 12. It was wholly reopened at 7:00 a.m., June 15. However, speed restrictions continued at a few sections because of repair work. Traffic on the expressway resumed normal speed on July 11.

2) Railways

The Japanese National Railways (JNR) sustained damage to railway bridges, tracks, platforms, etc. Railway operation was suspended in and around Miyagi Prefecture. JNR quickly restored the Tohoku line, the most important one in the Tohoku District. Railways on the southern side of Sendai were reopened early on the morning of June 14, and the rest on June 15. All lines were totally restored about one month after the earthquake.

3) Harbors

In spite of damage to facilities at Sendai Harbor, Ishinomaki Industrial Harbor, etc., temporary measures enabled them to keep functioning without interruption. At some fishing ports, unloading was limited by damage to wharves and other facilities.

4) Airports

The damage to Sendai Airport was not severe, and its operation was returned to normal soon after the earthquake.

III-8 Rehabilitation of Lifeline Facilities

1) Water Supply Facilities

Water supply was suspended due to the damage to water supply facilities. Restoration work was started immediately after the earthquake.

The Water Supply Bureau of the Sendai Municipal Government ordered "3rd Deposition" based on "Disposition Plan for Disaster Prevention." They called up all staff to establish the Earthquake Disaster Countermeasures Headquarters at 5:30 p.m., June 12. Four water wagons and 40 water tanks were prepared at 7:00 p.m. Fourteen water wagons from 13 private companies were also prepared by 9:00 p.m. Forty water wagons were dispatched. At Midoriga-oka and Kuromatsuichinebo of Sendai, water supply was suspended because of the possibility of ground failures. Temporary common hydrants were provided. The temporary water supply is shown in Table III-3. The temporary water supply works required the mobilization of 149 water wagons and 702 m^3 of water was supplied.

In Shiogama and Izumi Cities, where the water supply facilities were damaged severely, the Self-Defense Forces were requested to help supply water temporarily. Though restoration work was hastened by cooperation with the troops dispatched from other prefectures (see Table III-4), it took more than 10 days to complete the repairs in these cities.

2) Electric Power Facilities

The degree of functional damage and the repairs to the electric power facilites of Tohoku Electric Power Co., Ltd. are described as follows:

1) Hydroelectric Power Generation Facilities

Waterways and water tanks were damaged at the hydroelectric power plants, but the interference with generating electricity was not serious, and power was restored by 10:48 p.m., June 14.

2) Thermal Power Generation Facilities

Generators mentioned below were turned off after the earthquake.

• Sendai Thermal Power Plant; No. 2, No. 3 generators (generating power of 175,000 KW each)

• Shin-Sendai Thermal Power Plant; No. 1 generator (generating power is 350,000 KW), No. 2 generator (generating power is 600,000 KW)

• Hachinohe Thermal Power Plant; No. 3 generator (generating power is 250,000 KW)

As the result of inspection and repair, their functions were recovered at:

* Sendai Thermal Power Plant; No. 1 generator (12:54, June 16), No. 2 generator (13:45, June 26), No. 3 generator (15:39, July 20)

* Shin-Sendai Thermal Power Plant; No. 1 generator (18:16, June 18), No. 2 generator (13:02, June 19)

* Hachinohe Thermal Power Plant; No. 3 generator (18:54, June 12)

3) Power Transmission Facilities

The power transmission lines were inspected using 3 helicopters. Damage was not serious enough to interrupt the tranmission of electricity.

4) Transformer Facilities

Sendai Transformer Substation suffered serious damage. But as the result of restoration work, main transformers No. 1, No. 3 and No. 4 were restored on June 15, June 14, and June 15, respectively. Zao Main Line, Miyagi Main Line and Shin-Sendai Thermal Power A Line were restored on June 15, June 16, and June 18, respectively. As for Miyagi Transformer Substation, main transformers of No. 1 and No. 3 were restored on June 12, and June 16, respectively.

5) Power Distribution Facilities

Collapse of distribution poles, snapped wires, and the tipping of transformers poles occurred mainly in the southeastern part of Sendai City. The restoration of these power distribution facilities was completed at 6:50 a.m., June 14.

After the earthquake, a great number of people were engaged in restoration work. They tried to restore the supply by sharing among organizations. Some electric power was transmitted temporarily from outside of the damaged district. With such efforts, the function of electric power facilities was restored on the early morning of June 14. The number of people performing repair work is shown in Table III-5.

3) Gas Supply Facilities

The Gas Bureau of Sendai Municipal Government supplies 135,863 houses in Sendai, Tagajo, and Izumi Cities. After the earthquake, gas production was decreased by the burning-down of a low-pressure gas holder at Haramachi Plant, and by the electricity failure at the Minato Plant. When 200 reports of gas leakage were made by inhabitants, the Gas Bureau suspended the gas supply at 6:15 p.m., June 12. After the decision, the Gas Bureau warned users to turn off gas taps through TV and radio.

With the cooperation of the Disaster Countermeasures Headquarters of the Gas Bureau, Sendai Municipal Government, and the Japan Gas Association, restoration work was carried out giving priority of to resupply residential quarters. As the damage to gas pipes was scattered widely (see Table III-6), the supply area (Sendai, Tagajo, and Izumi Cities) was divided into 8 blocks as shown in Figure III-3. Some of them, where many gas pipes were damaged, were subdivided into two to eight sections for efficient restoration work. The procedure for the restoration work is shown in Figure III-4. It took 25,635 man-days to restore gas supply facilities by the middle of July, with the help of 10,583 man-days (June 13 to July 9) from 22 other companies. The detailed inspection of underground structures has been continued.

4)   Telecommunication Facilities

The Nippon Telegraph and Telephone Public Corporation established the Disaster Countermeasure Headquarters immediately after the earthquake to insure the function of telecommunication facilities and to restore damaged facilities. As the electricity failed at 165 telecommunication stations, 9 emergency vehicles for electric-power supply went to the stations to supply electric power temporarily for 40 hours until the electricity was recovered at all stations. Coaxial cables were damaged at 9 sites, and temporary cables were provided. By these temporary measures, the function of telecommunication facilities was recovered rapidly. However, telephone calls poured in from the evening of June 12 til June 14, and an extraordinary congresion of calls arose. A 25 to 75% restriction of calls was placed by restricted call identification equipment. Telephone communication routes for emergencies were established to maintain vital communications.

Public telephones did not work because of the electricity failures. Twenty-nine emergency public telephones in Sendai were temporarily supplied with electric power. At the refuge places, 22 free telephones were installed.

As for telegrams, the function of telegraph facilities were restored soon
after the earthquake, and 150,000 telegrams delivered in the Tohoku District.
Urgent telegrams were given delivery priority.

Temporary restoration work to telecommunication facilities required 5,000 man-
days, including 400 man-days by construction firms.

III-9   Schools and Their Operations

Ninety-nine elementary schools, nine junior high schools, and nine senior high
schools were closed. After the inspection of school buildings, almost all of them
reopened within a few days.

One hundred forty-three schools suspended meal-service for about a week
because of damage to meal service facilities, the suspension of electric power and
gas supply.

III-10 Other Measures for the Emergency

1)   Petroleum Industrial Facilities

Tohoku Petroleum Co., Ltd. has 87 oil tanks at Sendai Oil Factory. Three of
them settled and were cracked by the earthquake causing the leakage of oil. It
overflowed defensive banks to the precinct, and some oil flowed into Sendai Bay.
Oil fences were stretched to prevent the spread of the oil. The inspection of each
tank and oil spill cleanup was completed by June 18. The details of the outflow
and the collection of oil are shown in Figure III-5.

2)   River Facilities

Damaged dykes suffered cracking, settlement, and sliding on the Shin-Kitakami
River, the Naruse River, and the Natori River. The Tohoku Regional Construction
Bureau made field inspections after the earthquake. Facing the rainy season,
temporary restoration works were started on June 15 and were almost completed on
June 30. Some severely damaged dykes were repaired with revetments. The details
of the damage and the restoration of dykes are shown in Table III-7.

IV. Reconstruction

IV-1 General Scope

Reconstruction entails those measures necessary to restore socio-economic order.
It involves the measures needed to repair damaged structures, to construct substitutes,
to review design criteria for earthquake resistivity, and to improve related organizational structures.

IV-2 Organizational System for Reconstruction

Relevant administrations set up the disaster-oriented organizations temporarily for reconstruction.

Miyagi Prefectural Government organized the Earthquake Disaster Reconstruction
Section on July 10, 1978. It controlled the reconstruction jobs in the Government.

IV-3 Reconstruction Works of Public Utilities

Reconstruction Work of roads, rivers, harbors, agricultural facilities
schools and water supply facilities are scheduled under the supervision of the
Government. Table IV-1 and IV-2 show the budgets for reconstruction of water
supply and other public facilities.

IV-4 Long-term Reconstruction of Public Utilities

1) Railroad

The Japanese National Railroad has established the Railroad Structures
Aseismicity Committee, headed by Prof. S. Okamoto of Saitama University, to
review of the aseismic design of railroad structures.

2) Water Supply

Water supply administrators in the quake-affected area have compiled the
data on the damage to water supply facilities.

3) Gas Supply

Japan Gas Association is compiling the data on the damage and reconstruction of gas supply facilities.

Tokyo Gas Co., Ltd. has started research on how to efficiently restore
gas supply functions following earthquakes.

4) Electric Power Facilities

The earthquake resistivity of insulators on transformers had been studied
by analyses and experiments. Insulators had been considered to have enough

resistivity. Nevertheless some insulators suffered damage from the earthquake. The Electricity Enterprise Federation formed a Transformer Aseismicity
Committee to research the cause of the damage and future countermeasures.

5) Telecommunication Facilities

The Nippon Telegraph and Telephone Public Corporation is researching the
causes of the damage and future countermeasures. It is concentrating on the
study of damage to buried coaxial cables.

IV-5 Application of Law Systems and Loans

The Miyagi Prefectural Government administered the legally established
assistance and loan systems shown below.

1) Loans:

Public Welfare Loan

Minor Enterprise Loan

Agriculture and Fishery Loan

Privately-Owned House Loan

Municipal Finance Aid

2) Exemption of Tax and Tuition

3) Petitions for Government Aid in Violent Disasters

V. References

[1] Rika-Nenpyo (Science Yearbook), Tokyo Astronomical Observatory, Maruzen Co., Ltd.,
     1978.

[2] Abstracts of Autumnal Presentation Meeting: The Seismological Society of Japan, A36,
     1978, No. 2.

[3] Ohutsu, H., Topography, Geology and Ground Water In and Around Sendai City, Hohbundo,
     Co., Ltd.

[4] Preliminary Damage Investigation Report on Miyagi-ken-oki Earthquake of June in 1978,
     Public Works Research Institute Investigation Party, Civil Engineering Journal, Vol.
     20, No. 8, 1978.

[5] Kuribayashi, E., Tazaki, T., Hadate, T., The Investigation on the Rescue and Rehabili-
     tation of the Izu-ohshima Kinkai Earthquake of 1978, Tenth Joint Meeting, U.S.-Japan
     Panel on Wind and Seismic Effects, UJNR, 1978.

[6] Kuribayashi, E., and Tazaki, T., An Evaluation Study on the Distribution Characteris-
     tics of Property Losses Caused by Historical Earthquakes, Tenth Joint Meeting, U.S.-
     Japan Panel on Wind and Seismic Effects, UJNR, 1978.

[7] Bureau of Statistics, Office of Prime Minister, Abstract of the National Census of
     1944, 1977.

[8] Ishizaki, K., et al., Flood Hazard Map, Civil Engineering Journal, Vol. 18, No. 5, 1976.

Photo. 1. Damaged Reinforced Concrete Building at Oroshimachi, Sendai

Photo. 2. Damaged Reinforced Concrete Building at Oroshimachi, Sendai

Photo. 3. Damaged Wooden House at Naga-machi, Sendai

Photo. 4. Overturned Fence at Naga-machi, Sendai

Photo. 5. Landslide in the
Development at
Midorigaoka,
Sendai

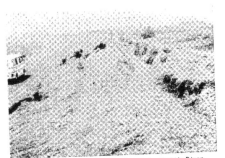

Photo. 6. Subsidence of Dyke Settlement at the Yoshida River

Photo. 7. Fallen Span at Kinno Bridge
on National Highway No. 346

Photo. 8. Temporary Repair
on Kinno Bridge

Photo. 9. Road Offset near the
Residential Development
at Midorigaoka, Sendai

Photo. 10. Sidewalk Subsidence on
Principal Prefectural
Road, Sendai-Izumi

Photo. 11. Damaged Conduit of 400 mm at Mikoyama 4, Sendai

Photo. 12. Inundated National Highway No.
Sendai due to the Damaged Cond

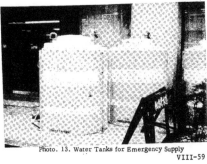

Photo. 13. Water Tanks for Emergency Supply

Photo. 14. Burned Gas Holder at Harano-machi, Sendai

Photo. 15. Broken Elbow Joint of Gas Pipe at Nanko-dai, Izumi

Photo. 16. Broken Insulators at the Sendai Transformer Station

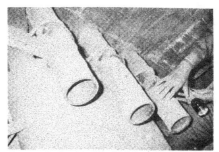

Photo. 17. Cut Site of the Coaxial Telephone Cables at Sendai-
ohashi Bridge on the National Highway No. 4

Photo. 18. Overturned Computer Equipment at the Fukushima
Prefectural Government Office

Table 1-1 Traffic Control on Designated National Highways

Number of the Route	Damaged Sites (Kilometer Post)	States of the Damage	Restriction	Period of the Restriction	
				Beginning of the Restriction	End of the Restriction
4	Tamachi, Shiroishi City (304.5km)	Sinking of the road surface by the bursting of a water pipe	One-side restriction	23:00, June 12	1:30, June 13
"	Tomiya Town, Kurokawa County (371.7km)	Upheaval of the road surface h=20.40cm	"	17:20, June 12	8:00, June 13
6	Kashima Town, Soma County (283.4km)	Sinking of the road surface near an abutment h=10cm	"	18:00, June 12	19:00, June 12
45	Naruse Town, Monou County (Ono Bridge) (38.0km)	Movement of the girder	Traffic suspension	19:00, June 12	18:00, June 17 Traffic restriction of vehicles loaded more than 1 ton continued.
"	Naruse Town, Monou County (38.2km)	Landslide V=400m³	One-side restriction	19:00, June 12	13:00, June 16
"	Kanan Town, Monou County (Tenno Bridge) (58.3km)	Cracks in the pier	Traffic restriction of vehicles loaded more than 1 ton	18:00, June 20	
"	Oofunato, Oofunato City (165.5km)	Cracks in the road surface 70m	"	13:00, June 13	19:30, June 16
"	Sanriku Town, Kesen County (180.6km)	Cracks in the road surface 10m.	"	19:05, June 12	17:00, June 15
45	Sanriku Town, Kesen County (183.6km)	Cracks in the road surface 50m	Traffic restriction of vehicles loaded more than 1 ton	19:05, June 12	17:00, June 15
"	Sanriku Town, Kesen County (184.3km)	Cracks in the road surface 32m	"	17:51, June 12	17:00, June 15
"	Noda Village, Kunohe County (336.7km)	Landslide L=10m V=100m³	"	18:20, June 12	17:00, June 14
48	Miyagi Town, Miyagi County (30.5km)	Falling down of a tree	"	17:30, June 12	19:50, June 12
108	Kanan Town, Monou County (5.8km)	Depression of the road L=6m h=70cm	Traffic suspension	18:00, June 12	16:25, June 13
"	Furukawa City (35.5km)	Leakage of gas	"	17:30, June 12	19:50, June 12

Table 1-2 Traffic Control on the Roads Administrated by Miyagi Prefectural Government from June 12 till July 10.

	Date and Time / Restriction	18:00, June 12	17:00, June 13	17:00, June 14	17:00, June 15	17:00, June 16	17:00, June 17	17:00, June 18	8:00, June 19	～	10:00, July 10	Note
Road	Traffic suspension	16	10	8	8	8	8	8	10		5	
	Traffic restriction of large-sized vehicles	1	2	2	1	0	2	2	2		1	Time restriction from July 7 except for on Sunday
	One-side restriction	12	9	7	7	8	7	7	7		0	
	Speed restriction	2	0	0	0	0	0	0	0		0	
	Total	31	21	17	16	16	17	17	19		6	
Bridge	Traffic suspension	5	3	4	4	4	5	5	5		0	
	Traffic restriction of large-sized vehicles	2	1	1	1	2	2	2	2		3	Traffic suspension of vehicles at Kinno Bridge, Eai Bridge and Komazuka Bridge
	One-side restriction	3	3	3	3	3	3	3	3		3	Traffic suspension of vehicles loaded more than 1 ton
	Speed restriction	2	4	4	3	4	3	3	3		0	
	Total	12	11	12	11	13	13	13	13		6	
Total	Traffic Suspension	21	13	12	12	12	13	13	15		5	
	Traffic restriction of large-sized vehicles	3	3	3	2	2	4	4	4		4	
	One-side restriction	15	12	10	10	11	10	10	10		3	
	Speed restriction	4	4	4	3	4	3	3	3		0	
	Total	43	32	29	27	29	30	30	32		12	

Table 1-3 Suspension of Traffic on the Roads Administrated by Miyagi Prefectural Government
till July 5 after the Earthquake

Name of the Route	Damaged Sites	States of Damage	Restriction	Reopening Time
National Highway Route 113	Zaimokuiwa, Shiro-ishi Town	Falling of stones	Suspension of traffic	22: 00, June 12
" Route 284	Kesennuma City	"	"	19: 00, June 12
" Route 286	Hagurodai, Sendai City	Landslide	"	
" Route 286	Akaishi, Sendai City	Danger of stone falling	"	11: 00, June 20 One-side restriction continued
" Route 346	Kinno Bridge, Towa Town	Falling of the bridge	"	12: 00, June 24 Reopened by a temporary bridge
Principal Prefectural Road Okumatsushima ·Matsu-shima Park Line	Tetaru, Matsu-shima Town	Depression of the road	"	11: 00, June 13 One-side restriction continued
" Taiwa ·Matsushima Line	Dobori Bridge, Taiwa Town	Depression of the abutment	"	22: 00, June 19 Traffic restriction of large-sized vehicles continued
" Yuzawa ·Tsukidate · Shizugawa Line	Onyu, Hanayama Village	Falling of stones	"	16: 00, June 13
" Ishinomaki ·Kashimadai · Oohira Line	Komazuka Bridge, Kashimadai Town	Cracks in the floor plate	"	
" Onagawa ·Shizugawa Line	Kamaya, Okatsu Town	Collapse of the shoulder of the road and landslide	"	12: 00, June 13
Principal Prefectural Road Shiogama · Watari Line	Yuriage Bridge, Natori City	Sinking of the abut-ment	Suspension of traffic	12: 00, June 21 Traffic restriction of large-sized vehicles continued
" Ishinomaki ·Ayukawa Line	Bankoku Bridge, Wataha, Ishinomaki City	Cracks in the pier	"	15: 00, June 29 One-side restriction continued
Prefectural Road Sendai · Yamadera Line	Takinohara, Akiu Town	Falling of stones	"	10: 00, June 13
" Sadayoshi, Sendai Line	Sadayoshi, Miyagi Town	"	"	June 28, Time restriction continued
" Kensennuma·Motoyoshi Line	Kesennuma City	Cracks in the road surface and falling down of a tree	"	16: 00, June 14
" Masusawa · Yoshioka Line	Taiwa Town	Falling of roof tiles	"	11: 00, June 13
" Kurikoma Park Line	Ueda, Kurikoma Town~Komanoyu	Falling of stones	"	16: 00, June 13
" Arikabe Station Line	Kannari Town	"	"	20: 00, June 12 Traffic restriction of large-sized vehicles continued
" Gunsawa ·Hasama Line	Hanayama Village	"	"	16: 00, June 13
" Arikabe · Wakayanagi Line	Gobo Bridge, Waka-yanagi Town	Sinking of the abutment	"	20: 00, June 12
" Wakuya · Sanbongi Line	Keshozaka Over-bridge, Kogota Town	Sinking of the road surface	"	13: 30, June 13

(Table 1-3 continued)

Name of the Route	Damaged Sites	States of Damage	Restriction	Reopening Time
Prefectural Road Kitakami · Kahoku Line	Shunosu, Kitakami Town	Cracks in the road surface	Suspension of traffic	
" Kamaya · Oosu · Okatsu Line	Tachihama, Okatsu Town	Landslide	"	16: 30, June 13
" Towa · Usuginu Line	Yoshine, Towa Town	Falling of stone	"	19: 00, June 13
" Semine · Toyosato Line	Toinuki Bridge, Toyosato Town	Cracks in the abut-ment	"	10: 30, June 22 Traffic restriction of large-sized vehicles continued
" Oosato · Rifu Line	Morisato, Rifu Town	Landslide	"	
" Kasukawa · Hataya Line	Kozunai, Oosato Town	"	"	
" Nenoshiroishi · Shiogama Line	Yaotome Bridge, Yaotome, Izumi City	Sinking of the road surface near the abutment	"	12: 00, June 13
" Kitakami · Kahoku Line	Ookawa, Kahoku Town	Cracks in the read surface	"	12: 00, June 17 Traffic restriction of large-sized vehicles continued
" Shizugawa · Toyoma Line	Uzawa Mountain Pass	Cracks in the road surface and landslide	"	16: 30, June 13
" Naruse · Nango Line	Sunayama, Naruse Town	Cracks in the road surface	"	9: 00, June 22
" Taiwa · Matsuyama Line	Sokawa Bridge, Ai-kawa, Taiwa Town	Cracks in the abutment	"	
Prefectural Road Kashimadai · Naruse Line	Takeya, Matsushima Town		Suspension of traffic	8: 00, July 4
" Sanbongi · Taiwa Line	Oohira Village~ Taiwa Town	Depression of the road	"	12: 30, June 26
Toll Highway Cobalt Line	Whole line	Landslide, collapse of the shoulder of the road and crack in the road surface	"	7: 00, July 1

Table 1-4   Traffic Restriction at Tohoku Expressway

Date	Closing	Reopening
June 12	17:17 (Shirakawa ~ Tsukidate) 17:32 (Ichinoseki ~ Hiraizumi · Maesawa)	18:30 (Shirakawa ~ Kooriyama) 19:15 (Ichinoseki ~ Hiraizumi · Maesawa : up line) 19:50 (Koriyama ~ Fukushima · Iizaka) 20:25 (Ichinoseki ~ Hiraizumi · Maesawa : down line)
June 13	Closing continued at the rest sections because of temporary restoration works.	23:55 (Shiroishi ~ Izumi)
June 14		12:00 (Taiwa ~ Tsukidate) 16:00 (Fukushima · Iizaka ~ Shiroishi)
June 15	Reopening was completed at all sections, but speed restriction continued at some sections	7:00 (Izumi ~ Taiwa)
July 11	Traffic returned to normal conditions at all sections at 12:00	

Note: Maximum Speed

```
             100km/h          80km/h        100km/h            under construction
Shirakawa  ⌒     Zao P.A.  ⌒  Izumi   ⌒   Tsukidate  ⌒
             80km/h          100km/h
Ichinoseki  ⌒  Maesawa  ⌒   Morioka
```

Table 1-5 Suspension and Decrease of Water Supply in Miyagi Prefecture

Date	Number of Local Public Bodies Resumed Water Supply after Suspension	Number of Facilities Resumed Water Supply after Suspension	Number of Local Public Bodies Resumed Water Supply after Reduction
June 12	3	4	
June 13 (AM)	3	7	1
June 13	20	24	
June 14	10	13	1
June 15	6	8	
June 16	1	1	
June 17	2	4	
June 18	2	4	
June 19	1	1	
June 20	2	2	.
June 21	2	2	
June 22	1	1	
June 23	1	1	
Total	54	72	2

Table 1-6 Functional Damage to Electric Power Facilities
(Tohoku Electric Power Co., Ltd.)

Decrease in Electric-power Generation	815,210 KW	⎛ Hydroelectric power 115,210 KW ⎞ ⎝ Thermal power 700,000 KW ⎠
Decrease in Electric-power Supply	1,130,700 KW	⎛ General demand 762,650 KW ⎞ ⎝ Big demand 368,050 KW ⎠
Decrease in Electric-power Reception	377,800 KW	⎛ General demand 105,300 KW ⎞ ⎝ Big demand 272,500 KW ⎠

Table 1-7   Restoration of Gas Supply Facilities in Sendai City
Including Izumi City and Tagajo City (Gas Bureau
of Sendai Municipal Government)

Date	Percentage of Restoration Achievement	Number of Gas-Supplied Houses
June 16	Beginning of gas supply	
19	19 %	26,000
22	32 %	44,000
25	60 %	82,000
29	92 %	125,000
July 6	98.2 %	133,000
9	Restoration was completed except for at specified districts and deserted houses.	

Note : Total number of gas-supplied houses : 135,863 houses (The percentage of pervasion : 72.8%), total length of main pipe lines :
196km, total length of trunk and bough pipes : 1,225km, Total:
1,421 km. Specified districts consists of three dangerous
districts (Midorigaoka, Kitane-Kuromatsu, Kuromatsu in Izumi
City) which have 530 houses.

*Reference

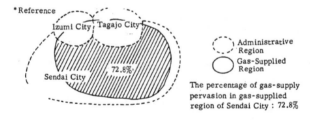

The percentage of gas-supply
pervasion in gas-supplied
region of Sendai City : 72.8%

Table 1-8 Damage to Telephone Systems (The Nippon Telegraph and
Telephone Public Corporation)

Name of the Branch	Number of Telephone Subscribers	Number of the Telephones Suspended	Date of the System Recovered
Miyagi	About 533,000	About 2,660	
(Sendai Station)	(273,000)	(1,160)	All telephone systems were recovered on June 20 except for 15 subscriptions impossible to be restored because of the collapse of houses.
Fukushima	488,000	530	
Iwate	331,000	650	
Aomori	386,000	5	
Total	1,738,000	3,845	

Note: Suspension of telephones was brought about by snapping of wires, damage
to telephone sets, etc.

Table 1-9 Damage to Long-Distance Transmission Lines
(The Nippon Telegraph and Telephone Public Corporation)

a. Micro System

Damaged Section, System	Number of Accommodated Circuits	Number of Damaged Circuits	Time of the System Damaged	Time of the System Recovered	Cause of the Damage
Tokyo ~ Sendai 6GHz (Jujo) (Tsutsujigaoka)	circuits 6,100	circuits 6,100	17:15, June 12	23:18, June 12	Inclination of the wireless-relay phone antenna at Tsutsujigaoka Station.
Sendai ~ Hakodate 4GHz	250	250	"	19:30, June 12	Cords connecting the facilities pulled out at Sendai Wireless Station.
Sendai ~ Ishinomaki 11GHz	1,500	400	"	19:00, June 12	Fuses of the source of electric power for 3 systems gone at Oohira Wireless Station
Ishinomaki ~ Kesennuma 2GHz	320	130	"	19:03, June 12	Recovered naturally Under investigation of the cause.
Tokyo ~ Sendai 4GHz (Jujo)	TV up 6 down 6	TV up 6 down 6	"	17:24, June 12	Board of the source of electric power loosened at Sendai Wireless Station
Total	8,170 TV up 6 down 6	6,880 TV up 6 down 6			

VIII-69

b. Coaxial Cable System

Damaged Section	Number of Accommodated Circuits	Number of Damaged Circuits	Time of the System Damaged	Time of the System Recovered	Damaged Sites
Tohoku II (Fukushima ~ Sendai)	circuits 4,900	circuits 4,900	17:15, June 12	0:03, June 14	Breaking down of the cable at 1 point about 31 km from Sendai, 3 points about 37~38 km from Sendai at 1 point about 42 km from Sendai. The cable was damaged at 5 points.
Tohoku II (Sendai ~ Morioka)	4,600	4,600	"	21:55, June 13	Breaking down of the cable at the culvert about 18 km from Sendai. The cable damaged at 2 points.
Joban (Haranomachi ~ Sendai)	6,900	6,900	"	18:57, June 13	Breaking down of the cable at the culvert about 11 km from Sendai. The cable damaged at 2 points.
Total	16,400	16,400			

Note : The Damage to cables was brought about by sinking of the ground at weak-ground areas and near bridges of directly buried sections.

c. Long Distance Circuits

Date and Time	Total Circuits into, out of and through Sendai			Curcuits into and out of Sendai (The number of circuits is included in the number of the left column)			Note
	Total Number of Circuits	Number of Damaged Circuits	Ratio of Damaged to Total	Total Number of Circuits	Number of Damaged Circuits	Ratio of Damaged to Total	
17:15, June 12	about circuits 70,000	about circuits 24,000	% 34	about circuits 31,000	about circuits 13,000	% 42	
19:30, June 12	"	16,000	23	"	7,700	25	
23:30, June 12	"	8,000	11	"	4,300	14	
12:00, June 13	"	5,500	8	"	3,500	11	
17:00, June 13	"	3,400	5	"	1,600	5	
0:03, June 14	"	0	0	"	0	0	All circuits recovered

Table 1-10 Damage to Short-Distance Transmission Lines
(The Nippon Telegraph and Telephone Public Corporation)

Damaged Section, System	Number of Accommodated Circuits	Number of Damaged Circuits	Time of the System Damaged	Time of the System Recovered	Cause of the Damage
Ishinomaki - Oshika PCM-24	circuits 180	circuits 161	17:15, June 12	0:01, June 13	Relay boards loosened at Oshika Station and Oharahama Station.
Tsutsujigaoka - Nanakita PCM-24	637	48	"	18:50, June 12	Changing-over cords pulled out at Nanakita Station.
Ishinomaki - Kahoku PCM-24	177	112	"	0:45, June 13	An electric-power supply board loosened at Kahoku Station.
Tsutsujigaoka - Shiogama PCM-24	487	24	"	20:15, June 12	Changing-over cords pulled out at Shiogama Station.
Sendai - Natori PCM-24	690	212	"	18:52, June 12	Fuses of the electric-power distribution board gone at Sendai Station.
Oogawara - Kawasaki PCM-24	194	113	"	17:30, June 12	A relay board loosened at Kawasaki Station.
Miyako - Iwaizumi T-12SR	121	12	"	20:30, June 12	A relay board loosened at Taro Relay Station.
Total	2,486	682			

Table 1-11 Functional Damage to TV and Radio Broadcasting
due to the Damage to Micro Circuits

TV	NHK : 94 stations (General : 47 stations, Educational : 47 stations) 24 stations in Miyagi Prefecture 36 stations in Iwate Prefecture 30 stations in Yamagata Prefecture 2 stations in Akita Prefecture 2 stations in Fukushima Prefecture 6 broadcasting companies : 46 stations
Radio	NHK : 3 stations (First : 1 station, Second : 2 stations) 2 broadcasting companies : 3 stations
Short-wave international broadcasting	NHK : 1 station

Table 1-12 Suspension of TV Broadcasting

Name of the Broadcasting Station	Time of the Beginning of Suspension	Time of the Recovery	Suspended Period	Cause of Suspension
Fukushima NHK General	17:15, June 12	17:47, June 12	32 minutes	Damage to relay lines between Tokyo and Sendai (4GHz).
Morioka NHK Educational	"	17:26, June 12	10 minutes 30 seconds	
Yamagata NHK Educational	"	17:24, June 12	9 minutes	
Sakata NHK Educational	"	17:18, June 12	3 minutes 40 seconds	o Sakata NHK and Aomori NHK were recovered by round about routes.
Akita NHK Educational	"	17:24, June 12	7 minutes 55 seconds	o Broadcasting of Fukushima NHK didn't suspend by receiving electric waves directly from a satellite station in Sendai Area.
Aomori NHK Educational	"	17:18, June 12	3 minutes 40 seconds	
Sendai NHK Educational	"	17:24, June 12	9 minutes	
Sapporo NHK Educational	"	17:25, June 12	10 minutes 35 seconds	

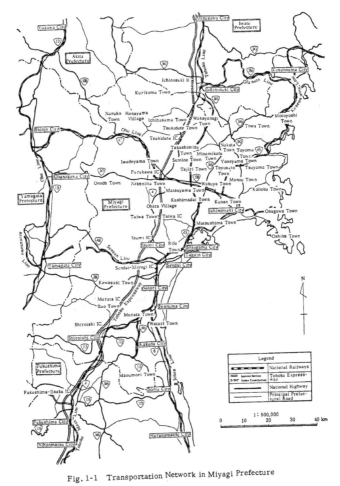

Fig. 1-1   Transportation Network in Miyagi Prefecture

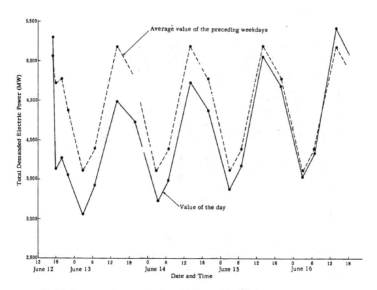

Fig. 1-2 Decrease and Recovery of Total Demanded Electric Power (Tohoku Electric Power Co., Ltd.)

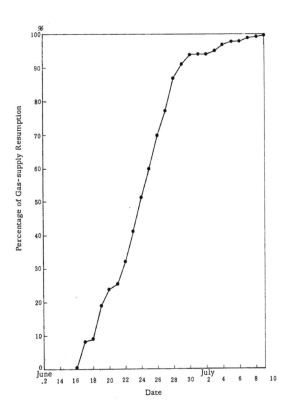

Fig. 1-3 Transition of Gas-supply Resumption
(Gas Bureau of Sendai Municipal Government)

VIII-75

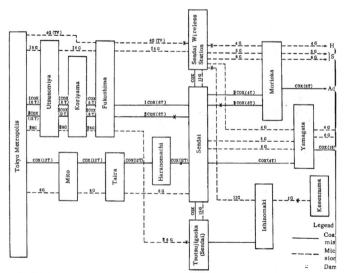

Fig. 1-4 Damage to Long-Distance Transmission Lines (The Nippon Telegraph and Telephone Public Corpor

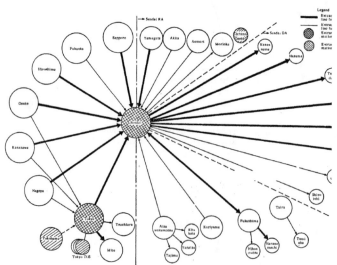

Fig. 1-5 Extraordinary Congestion of Trunk Calls (The Nippon Telegraph and Telephone Public Corporation)

VIII-76

Table 2-1. Comparison of Individual Loss Ratios of Lifeline Facilities

	Loss (L) in million yen	National Wealth[1] in billion yen	$\dfrac{L_K}{W_K}\dfrac{P}{P_D}$
Highway	[2] 10,721	8,121.5	0.076
Harbor	3,746	2,076.5	0.103
Water Supply	1,733	4,502.2	0.022
Gas Supply	947	867.4	0.063
Electric Power	2,960	8,133.7	0.021
Telecommunication	850	4,857.2	0.010

[1] Referred from the National Wealth Survey of 1970.

[2] Excluding National Expressways.

P : 111,933 thousand

P_D : 1,955 thousand

Table 2-2   Comparison of the Earthquake of 1978 to Kinkasan-oki Earthquake of 1936

Earthquake	Date	Magnitude	Epicenter	Damage	Population in the Time of the Earthquake
Kinkasan-oki	Nov. 3, 1936	7.7	$\lambda = 142.2°E$ $\varphi = 38.2°N$	Injured 4, Razed Non-dwelled Building 3, Half-razed Dwelling 2, Half-razed Non-dwelled Building 2, Damage to Highways 35	[1] 1,234,801
Miyagiken-oki	June 12, 1978	7.4	$\lambda = 142°10'E$ $\varphi = 38°9'N$	Refer to Appendix II	[2] 1,955,274

[1] as of 1935

[2] as of 1975 referred from the National Census of 1975.

Table 2-3 Ordinary Households by Tenure of House and One Person
Quasi-households Living in Rented Rooms as of 1975

Area	Ordinary Households Living in Houses						One Person Quasi-households Living in Rented Room
	Total	Owned House	Rented House Publicly Owned	Rented House Privately Owned	Issued House	Rented Room	
Japan	31,321,700	18,151,500	2,343,300	8,547,900	2,056,300	222,600	547,200
Miyagi Pref.	506,100	309,800	28,400	134,300	31,100	2,500	16,800
Tokyo Pref.	3,820,700	1,446,200	382,200	1,695,100	248,100	49,000	102,600
Aichi Pref.	1,580,400	855,300	148,500	448,100	120,900	7,500	19,200
Ohsaka Pref.	2,467,500	999,700	320,900	989,300	143,400	14,200	21,700
Sendai City	186,100	76,400	10,300	80,500	18,200	700	12,000
Central Tokyo	2,920,900	1,064,100	229,900	1,391,100	194,000	41,700	87,500
Nagoya City	611,700	244,100	68,800	246,800	48,700	3,200	12,600
Ohsaka City	882,300	298,100	106,800	427,100	42,600	7,800	9,400

Referred from ; Bureau of Statistics, Office of the Prime Minister : Population of Japan, 1975

Population Census of Japan Abridged Report Series No. 1, 1977.

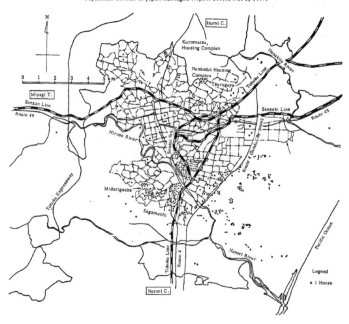

Fig. 2-1 (a) Distribution of Razed Houses in Sendai

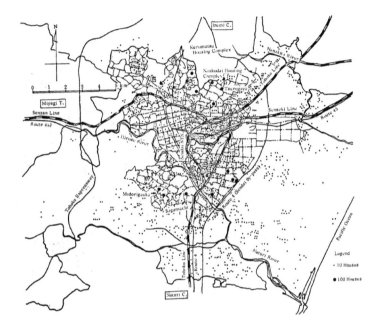

Fig. 2-1 (b)   Distribution of Damaged Fences in Sendai

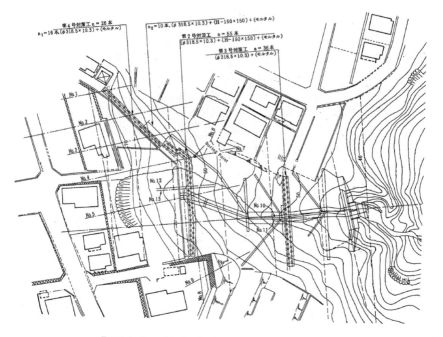

Fig. 2-2　Plan of the Residential Lots in Midorigaoka, Sendai

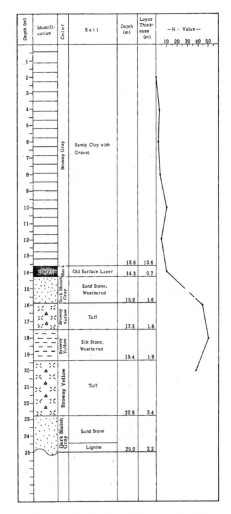

Fig. 2-3　Soil Profile at Midorigaoka, Sendai

VIII-81

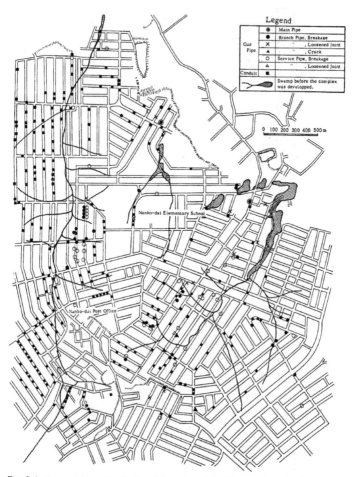

0 100 200 300 400 500 m

Nanko-dai Elementary School

Nanko-dai Post Office

Fig. 2-4   Damaged Sites Distribution of Conduits and Gas Pipes at Nanko-dai, Izumi'City

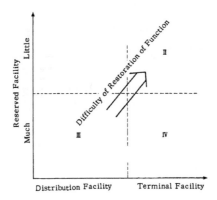

Fig. 2-5   Difficulty of Restoration of Function of
           Lifeline Facilities

Table I-1   Weather Conditions before and after the Main Shock

(a) Sendai Meteorological Bureau

Items / Date	Weather	Temperature (°C)			Precipitation (mm/day)	Relative Humidity (%)	
		Average	Max.	Min.		Average	Min.
June 1	Fine	18.2	23.3	13.1	—	42	29
2	Cloudy, later fine	17.5	23.3	10.5	—	59	36
3	Cloudy	18.9	23.7	14.1	—	60	41
4	Cloudy, later rain	15.4	17.0	14.5	21.5	91	79
5	Fine, occasionally cloudy	19.1	23.4	15.1	0.0	61	44
6	Fine, temporarily rain	17.2	23.9	12.9	1.0	67	40
7	Fine	17.1	22.4	11.3	—	71	48
8	Fine	19.0	22.7	13.8	—	75	65
9	Cloudy	21.8	29.3	16.8	—	69	40
10	Cloudy	20.6	25.9	17.5	—	77	58
11	Cloudy, later rain	19.5	21.7	16.7	0.0	86	73
12	Cloudy	20.3	25.9	17.5	0.0	82	61
13	Cloudy, occasionally rain	18.1	21.0	16.0	3.5	96	82
14	Cloudy, later fine	19.5	24.3	16.0	1.5	90	67
15	Fine, later cloudy	19.8	24.9	16.5	—	86	64
16	Cloudy, temporarily rain	22.0	27.2	18.8	0.0	83	65
17	Cloudy	24.2	29.3	20.6	0.0	79	66
18	Fine	25.0	29.6	20.9	—	74	63
19	Fine, later rain	24.2	29.0	20.8	5.5	81	62
20	Fine, temporarily rain	24.2	28.7	20.9	0.0	81	63
21	Cloudy, later fine	24.1	29.8	20.6	2.0	75	52
22	Rain	19.4	21.8	17.7	25.5	86	77
23	Cloudy, temporarily rain	18.6	19.9	17.8	0.5	90	83
24	Cloudy	19.5	22.3	17.6	—	85	78
25	Cloudy	21.1	26.4	17.6	2.0	85	67
26	Rain, later cloudy	21.3	23.5	19.6	43.0	92	83
27	Rain	18.4	20.9	17.5	47.5	97	95
28	Cloudy, occasionally rain	19.8	23.5	17.4	6.0	86	75
29	Cloudy, temporarily fine	20.6	26.2	17.0	—	83	64
30	Fine	20.4	25.0	17.9	0.0	82	64
Total	—	604.8	735.8	505.0	159.5	2371	—
Average	—	20.2	24.5	16.8	—	79	—

Note: The average temperature and the precipitation of June are 18.2°C
and 160 mm respectively in the year 1941-1970 in Sendai City.

Table I-1 Weather Conditions before and after the Main Shock

(b) Sendai Airport Meteorological Station

Items / Date	Weather	Temperature (°C)			Precipitation (mm/day)	Relative Humidity (%)	
		Average	Max.	Min.		Average	Min.
June 1	Fine	18.0	23.9	10.6	—	46	—
2	Cloudy, later fine	15.5	20.1	8.9	—	75	—
3	Cloudy	17.5	21.9	13.3	0.0	74	—
4	Rain	15.3	16.3	13.6	18.0	96	—
5	Fine	19.9	24.1	13.5	0.0	60	—
6	Fine, temporarily cloudy	17.3	23.9	12.2	1.0	75	—
7	Fine	16.1	20.3	9.7	—	81	—
8	Fine	17.9	21.9	12.1	—	85	—
9	Cloudy, temporarily fine	20.6	27.6	15.7	—	79	—
10	Cloudy	19.2	23.8	16.4	—	88	—
11	Cloudy, later rain	18.2	21.1	16.5	0.5	94	—
12	Cloudy	19.1	23.6	16.4	0.0	89	—
13	Rain, later cloudy	17.6	20.9	15.7	2.5	99	—
14	Cloudy	18.2	21.0	15.7	0.0	93	—
15	Fine, later cloudy	18.8	23.2	16.7	—	93	—
16	Cloudy	21.1	25.9	17.9	—	92	—
17	Cloudy	22.3	27.1	19.5	—	89	—
18	Cloudy	23.2	27.2	19.6	—	86	—
19	Fine, later cloudy	22.9	26.2	19.0	2.0	92	—
20	Fine, temporarily cloudy	23.2	27.9	20.5	—	92	—
21	Cloudy, temporarily rain	23.4	30.8	19.9	2.0	84	—
22	Rain	18.9	21.0	17.9	25.5	93	—
23	Cloudy, temporarily rain	18.3	20.3	17.2	0.5	96	—
24	Cloudy	18.8	21.2	17.3	—	94	—
25	Cloudy	20.0	23.5	17.5	—	95	—
26	Rain, later cloudy	21.1	24.0	19.2	39.5	97	—
27	Rain	18.4	20.9	17.3	43.5	99	—
28	Cloudy	19.4	22.2	17.3	5.5	91	—
29	Cloudy	19.9	24.7	16.3	—	92	—
30	Fine, later cloudy	19.6	23.0	17.4	—	88	—
Total	—	579.7	699.5	480.8	140.5	2607	—
Average	—	19.3	23.3	16.0	—	87	—

Note : The average temperature and the precipitation of June are 18.2°C
and 160 mm respectively in the year 1941-1970 in Sendai City.

Table I-1 Weather Conditions before and after the Main Shock
(c) Ishinomaki Meteorological Station

Date	Weather	Temperature (°C)			Precipi-taion (mm/day)	Relative Humidity (%)	
		Average	Max.	Min.		Average	Min.
June 1	Fine, temporarily cloudy	16.3	21.1	12.2	–	48	36
2	Fine, temporarily cloudy	15.9	20.6	10.8	–	55	39
3	Cloudy, temporarily fine	17.5	22.6	12.7	–	64	41
4	Cloudy, later rain	15.9	17.7	14.2	31.0	86	74
5	Fine	17.6	22.0	14.1	0.0	63	47
6	Fine, temporarily cloudy	16.2	22.8	11.8	0.5	69	40
7	Fine	16.0	20.7	11.2	–	72	53
8	Fine	17.3	21.6	12.5	–	77	60
9	Fine, occasionally cloudy	19.6	23.1	15.8	–	75	58
10	Fine, temporarily cloudy	19.2	23.9	16.1	–	76	60
11	Cloudy, later rain	17.6	19.6	16.0	0.0	89	75
12	Cloudy, later fine	19.8	24.1	16.9	0.0	80	64
13	Rain, later cloudy	17.3	20.5	15.6	3.0	91	85
14	Fine, occasionally cloudy	17.4	21.7	15.4	0.0	86	73
15	Fine, later cloudy	18.8	23.5	15.9	0.0	85	60
16	Cloudy	20.6	25.2	17.7	0.0	83	63
17	Cloudy, temporarily fine	22.2	27.6	19.7	–	80	67
18	Fine	23.0	29.0	19.8	–	79	60
19	Fine, later cloudy	21.4	24.8	19.1	2.5	83	72
20	Cloudy, later fine	21.5	25.1	18.9	1.0	86	71
21	Cloudy, later fine	21.1	27.7	18.4	2.0	81	59
22	Cloudy, later rain	19.3	22.0	17.6	30.5	83	69
23	Cloudy, temporarily rain	17.6	19.3	16.9	0.0	91	82
24	Cloudy	18.2	20.6	16.4	–	88	82
25	Fine	20.4	24.9	16.9	–	84	73
26	Rain, temporarily cloudy	20.3	21.3	18.6	27.0	94	88
27	Cloudy, later rain	18.8	21.1	17.6	21.5	89	78
28	Cloudy	19.2	22.1	17.1	9.5	87	72
29	Cloudy, temporarily fine	19.0	23.7	17.2	–	85	69
30	Fine, temporarily cloudy	19.9	24.4	17.6	–	79	59
Total	–	564.9	684.3	480.7	128.5	2388	–
Average	–	18.8	22.8	16.0	–	80	–

Note : The average temperature and the precipitation of June are 18.2°C
and 160mm respectively in the year 1941-1970 in Sendai Cidy.

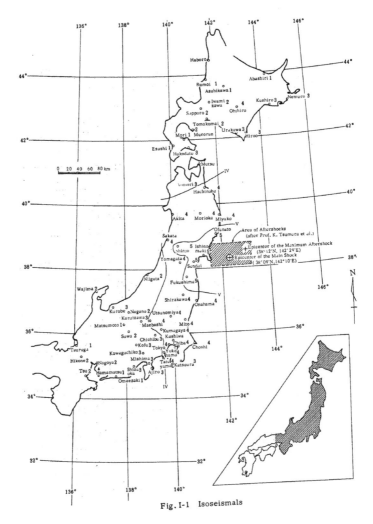

Fig. I-1 Isoseismals

Fault parameters
Length   30km
Down dip width   80km
Fault plane dip direction   N280°E
               dip angle   20°
Seismic moment $2.9 \cdot 10^{27}$ dyn-cm
Dislocation   1.7m
Stress drop   64bars

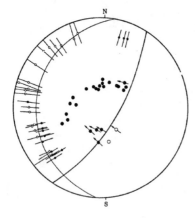

Legend
● : compression
○ : dilatation
-○-: error bar
-○-: direction of S-wave

Fig. I-2   An Example of Focal Mechanism
          Solution (by T. Seno, K. Sudo and
          T. Eguchi)

№	Date	ϕ, λ	M
①	Ⅶ 28, 1616	$\phi = 38.1°$ $\lambda = 142.0°$	M = 7.0
	30 years		
②	Ⅳ 26, 1646	$\phi = 37.7°$ $\lambda = 141.7°$	M = 7.6
	90 years		
③	Ⅲ 20, 1736	$\phi = 38.3°$ $\lambda = 140.8°$	M = 6.2
	34 years		
④	Ⅴ 3, 1770	$\phi = 38.6°$ $\lambda = 142.0°$	M = 7.4
	23 years		
⑤	Ⅱ 17, 1793	$\phi = 38.3°$ $\lambda = 142.4°$	M = 7.1
	42 years		
⑥	Ⅵ 25, 1835	$\phi = 37.9°$ $\lambda = 141.9°$	M = 7.6
	26 years		
⑦	Ⅸ 18, 1861	$\phi = 37.7°$ $\lambda = 141.6°$	M = 6.4
	36 years		
⑧	Ⅱ 20, 1897	$\phi = 38.1°$ $\lambda = 141.5°$	M = 7.8 (7.3)
⑨	Ⅷ 5, 1897	$\phi = 38.0°$ $\lambda = 143.7°$	M = 7.7 (7.2)
	39 years		
⑩	Ⅺ 3, 1936	$\phi = 38.2°$ $\lambda = 142.2°$	M = 7.7
	42 years		
⑪	Ⅱ 20, 1978	$\phi = 38.75°$ $\lambda = 142.2°$	M = 6.7
⑫	Ⅵ 12, 1978	$\phi = 38.15°$ $\lambda = 142.17°$	M = 7.4

Fig. I-3   Historical Damaging Earthquakes
          off Kinkasan

Table II-1   Totaled Losses of Miyagi, Iwate, Fukushima and Yamagata Prefectures.

Classification	Loss in Thousand Yen	Note
[Direct Losses]	275,725,326	
Buildings and Houses	79,239,128	
Medical and Sanitary Facilities	3,618,040	
Commerce and Industry	96,853,864	
Agriculture, Forestry and Fishery	19,307,776	
Educational Facilities	9,456,640	
Facilities for Flood Control and Preservation of Mountainous Area	14,237,936	
Transportation Facilities	22,572,113	
Water Supply Facilities	1,744,948	
Gas Supply Facilities	947,000	
Electric Power Supply Facilities	3,116,862	
Telecommunication Facilities	871,582	
Sewerage Facilities	838,453	
Park	97,667	
Other Facilities	22,823,317	Water tank, etc.
[Indirect Losses]	Obscure	
Commerce	"	Due to impossibility of business
Agriculture, Forestry and Fishery	"	Due to impossibility of cultivation
Construction and Manufacture	"	Due to discontinuity of operation
Others	"	
Total	275,725,326	

Note: By Miyagi, Iwate, Yamagata and Fukushima Prefectural Government Offices.

Table II-2 Contents of Losses of Lives in Four Prefectures.

Classification		Miyagi Pre.	Iwate Pre.	Fukushima Pre.	Yamagata Pre.
Dead		27	0	1	0
Injured	Serious	262	0	3	0
	Slight	10,700	11	27	1
Missing		0	0	0	0
Total		10,989	11	31	1

Table II-3 Damage to Buildings and Houses

Classification		Number of Damaged	Losses in Thousand Yen	Note
Dwelling	Totally Razed	1,377	60,180,765	
	Half Razed	6,178		
	Breakage	126,051		
	Flooding over Floor	3		
	Flooding under the Floor	2		
Non-dwelling		43,532	19,058,363	
Total		177,143	79,239,128	

Table II-4 Damage to Facilities for Medical and Sanitary.

Classification	Number of Damaged	Losses in Thousand Yen	Note
Medical	—	663,495	
Sanitary	—	1,300,977	
Others	—	1,653,568	
Total		3,618,040	

Table II-5 Damage to Commerce and Industry.

Prefecture	Number of Damaged	Losses in Thousand Yen	Note
Miyagi Pre.	53,524	95,753,230	
Iwate Pre.	2,252	562,782	
Fukushima Pre.	—	518,106	
Yamagata Pre.	—	19,746	
Total		96,853,864	

Table II-6 Damage to Agriculture, Forestry and Fishery.

Classification	Number of Damaged	Losses in Thousand Yen	Note
Agriculture		14,108,580	
Forestry		1,102,045	
Fishery	—	4,097,151	
Total		19,307,776	

Table II-7 Damage to Education.

Classification	Number of Damaged	Losses in Thousand Yen	Note
School	1,529	8,528,197	
Cultural Assets		245,940	
Facilities for Education and Welfare	—	682,503	
Total		9,456,640	

Table II-8   Damage to Facilities for Flood Control and Preservation of Mountainous Areas.

Prefecture	Number of Damaged	Losses in Thousand Yen	Note
Miyagi Pre.	River Facilities : 482 Seaside Facilities : 14 Sabo Facilities : 15	13,693,698	
Iwate Pre.	River Facilities : 31 Seaside Facilities : 2	443,870	
Fukushima Pre.	River Facilities : 3 Sabo Facilities : 21	94,980	
Yamagata Pre.	River Facilities : 1	5,388	
Total		14,237,936	

Table II-9   Damage to Transportation Facilities

Classification	Number of Damaged	Losses in Thousand Yen	Note
Roads	2,317	5,614,184	
Bridges	252	5,817,009	
Railways		6,377,883	
Harbors	89	4,000,872	
Others	—	762,165	
Total		22,572,113	

Table II-10   Damage to Water Supply Facilities.

Prefecture	Number of Damaged	Losses in Thousand Yen	Note
Miyagi Pre.	—	1,733,399	
Iwate Pre.	14	10,105	
Fukushima Pre.	—	0	
Yamagata Pre.	2	1,444	
Total		1,744,948	

Table II-11 Losses of Water Supply Facilities in Sendai City

Classification	Losses in Thousand Yen
Receiving and Storing Reservoirs	1,800
Filtration Plants	12,400
Distribution Stations	33,400
Conduits	111,400
Supplying Facilities	13,600
Buildings	5,300
Others	77,100
Total	255,000

Table II-12 Damage to Gas Supply Facilities.

Prefecture	Number of Damaged	Losses in Thousand Yen	Note
Miyagi Pre.	190	947,000	
Total		947,000	

Table II-13 Damage to Hydroelectric Power Plants, Thermal Power Plants and Transformer Stations in Miyagi Prefecture.

Classification	Number of Damaged
Hydroelectric Power Plants	11
Thermal Power Plants	2
Transformer Stations	18
Total	31

Table II-14 Damage to Sewerage Facilities.

Classification	Number of Damaged	Losses in Thousand Yen	Note
Drainpipes	60	578,332	
Pump Yards	19	142,231	
Treatment Plants	23	52,245	
Others	4	65,645	
Total	106	838,453	

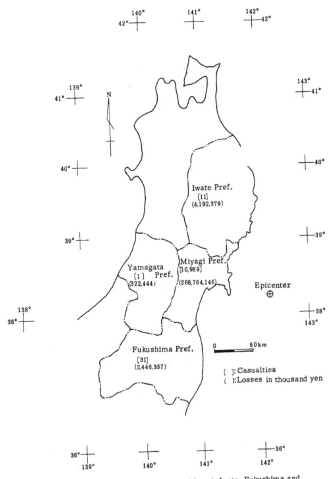

Fig. II-1 Casualties and Losses in Miyagi, Iwate, Fukushima and
Yamagata Prefectures.

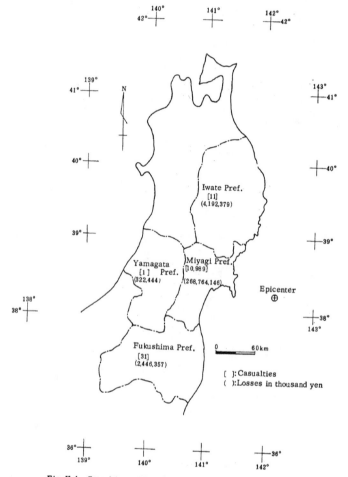

Fig. II-1 Casualties and Losses in Miyagi, Iwate, Fukushima and Yamagata Prefectures.

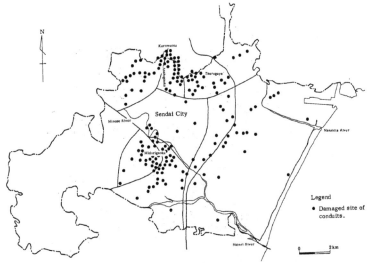

Fig. II-3   Damaged Sites Distribution of Conduits in Sendai City.

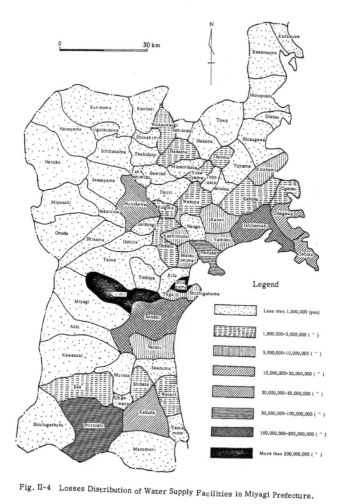

Fig. II-4 Losses Distribution of Water Supply Facilities in Miyagi Prefecture.

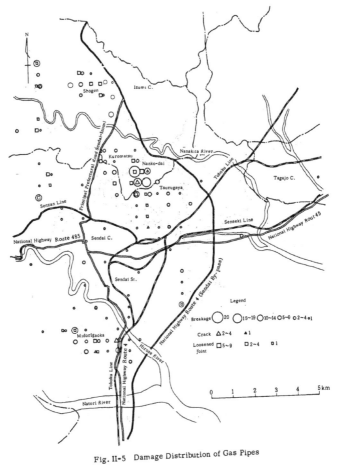

Fig. II-5   Damage Distribution of Gas Pipes

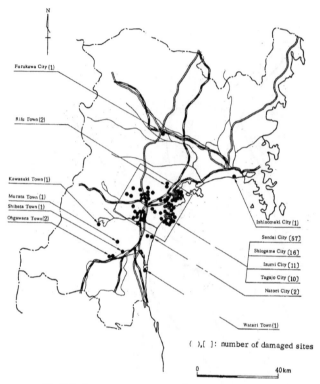

Furukawa City (1)

Rifu Town (2)

Kawasaki Town (1)
Murata Town (1)
Shibata Town (1)
Ohgawara Town (2)

Ishinomaki City (1)

Sendai City (57)

Shiogama City (16)

Izumi City (11)

Tagajo City (10)

Natori City (2)

Watari Town (1)

( ),[ ]: number of damaged sites

0   40 km

Fig. II-6   Damage Distribution of Sewerage Facilities.

Table III-1   Residential Damage and Refuge in Sendai City till June 15 after the Earthquake

Date	Damaged Area	States of Damage	Number of Refugees	Shelter	Note
June 13 0:30	Midorigaoka	Danger of landslides Cracks in roads	About 70 inhabitants	Kano Elementary School	About 80 refugees on the evening of June 13. About 100 refugees on the evening of June 14.
2:00	Futatsusawa	Landslides	20 inhabitants of 4 households	"	About 20 refugees on the evening of June 13. About 20 refugees on the evening of June 14.
"	Minamoto-Shinden	Danger of landslides	11 inhabitants of 3 households	Kitane Assembly Hall	About 8 refugees on the evening of June 13. About 8 refugees on the evening of June 14.
"	Tsurugaya Housing Corporation Terrace	Electricity failures at multistory apartment houses	75 inhabitants	Tsurugaya First Assembly Hall (15 refugees) Tsurugaya Third Assembly Hall (30 refugees) Tsurugaya-higashi Elementary School (30 refugees)	Refuge was ended on the morning of June 13.
"	Takasago Sunny Heights	Bursting of a water tower	About 120 inhabitants	Takasago Elementary School	Refuge was ended on the morning of June 13.
June 13 16:00	Sunaoshi-cho	Landslides	5 inhabitants of a household	A house of an acquaintance	
"	Sunaoshi-cho	Landslides	6 inhabitants of a household	A house of an acquaintance	
"	Komatsushima	Landslides	1 household	A house of an acquaintance	
June 14 12:30	Anyoji	Landslides	1 household	A house of an acquaintance	
"	Kitane-ichinenbo	Collapse of an apartment house by landslides	11 inhabitants of 7 households	Houses of acquaintances	
"	"	Danger of the collapse of an apartment house by landslides	9 inhabitants of 9 households	Houses of acquaintances	
June 15 13:20	Kanisawa	Danger of the collapse of a house	1 household	A house of an acquaintance	

Table III-2 Damage and Restoration Methods in Designated National Highways
(More than 1 Million Yen of the Cost of the Restoration)

Number of the Route	Damaged Sites	Kilometer Post (km)	Damage		Restoration Method
			Main Damage	Other Damage	
4	Kanairo, Nihonmatsu City	249.2	Swelling of the protection blocks of the slope		Setting up of blocks and construction of protection wall
	Shindate, Shiroishi City	303.9	Cracks in the pavement		Pavement of the road surface
	Zao Town, Katsuta County	313.6	Sinking of the block piles		Piling of blocks
	Kooriyama, Sendai City (Chiyo Bridge)	347.3	Sinking of the road surface near the abutment		Pavement of the road surface
	Okino, Sendai City (Chiyo Bridge)	347.5	Damage to the pier		Reinforcement of the pier by wrapping with concrete
	Okino, Sendai City (Chiyo Bridge)	347.7	Sinking of the road surface near the abutment		Asphalt pavement of the road surface
	Furujiro, Sendai City	348.4	Sinking of the road surface near the artificial structure		"
	Oroshimachi, Sendai City	351.7	" "		"
	" "	352.0	" "		
	" "	352.3	" "		
	" "	352.4	" "		"
	Hinode-cho, Sendai City	353.1	Sinking of the road surface near the artificial structure		Asphalt pavement of the road surface
	" "	353.8	" "		
	" "	354.0	" "		
	Kozuru, Sendai City	354.2	" "		
	" "	354.4	" "		
	Tsubamesawa, Sendai City	355.3	" "		
	"	355.5	Landslide at the artificial bank	Sinking of the sidewalk and damage to the guard pipe	Restoration of the artificial bank, pavement of the sidewalk and restoration of the guard pipe
	Nanakita, Izumi City	363.5	Cracks in the pavement		Pavement of the road surface
	Sawada, Furukawa City	397.0	Sinking of the road surface near the abutment		Asphalt pavement of the road surface
	Takashimizu Town, Kurihara County	404.4	Inclination of protection shelf of the artificial bank	Sinking of the sidewalk and inclination of the side blocks	Construction of the protection concrete wall, pavement of the sidewalk and resetting of the blocks
6	Kashima Town, Soma County	283.4	Cracks in the abutment		Reinforcement of the abutment
	Yamamoto Town, Watari County	321.3	Sinking of the road surface		Asphalt pavement of the road surface
	Watari Town, Watari County	321.9	Sinking of the shoulder of the road		Concrete pavement of the road surface
	" "	333.6	Damage to the pier	Damage to the shoe	Reinforcement of the pier by wrapping with concrete and restoration of the shoe

(Table III-2 continued)

Number of the Route	Damaged Sites	Kilometer Post (km)	Damage		Restoration Method
			Main Damage	Other Damage	
45	Fukudamachi, Sendai City	7.3	Landslide at the artificial bank	Sinking of the sidewalk and damage to the protection shelf	Protection of the artificial bank with sheet piles, pavement of the sidewalk and resetting of the protection shelf
	Matsushima Town, Miyagi County	28.3	Inclination of protection shelf of the artificial bank	Sinking of the sidewalk with cracks	Protection of the artificial bank with block piles and pavement of the sidewalk
	" "	28.4	Sinking of the road surface near the abutment		Asphalt pavement of the road surface
	" "	28.5	" "		"
	" "	28.6	Landslide at the artificial bank	Sinking of the sidewalk with cracks	Restoration of the artificial bank and pavement of the sidewalk
	" "	29.1	Inclination of the block piles	Inclination of the protection wall of the sidewalk	Restoration of the protection wall
	Naruse Town, Monou County	37.2	Sinking of the road surface with cracks		Pavement of the road surface
	Naruse Town, Monou County (Ono Bridge)	38.0	Movement of the girder	Damage to the shoe, the joint, the floor plate and the pier	Restoration of the girder, the joint, the floor plate and the pier
	Kanan Town, Monou County (Tenno Bridge)	58.3	Cracks in the pier	Damage to the abutment and the shoe	Reinforcement of the pier and the abutment and renewal of the shoe
	Kanan Town, Monou County	66.0	Cracks in the pavement		Pavement of the road surface
	Shizugawa Town, Motoyoshi County	98.4	Cracks in the artificial bank	Cracks in the protection wall and the pavement	Restoration of the artificial bank with protection piles
	Utatsu Town, Motoyoshi County (Utatsu Bridge)	102.5	Damage to the shoe		Reinforcement of the shoe
	Shimofunato, Oofunato City	165.5	Cracks in the pavement	Sinking of the road surface	Pavement of the road surface
	Noda Village, Kunohe County	336.2	Inclination of the protection concrete wall		Renewal of the protection concrete wall
108	Kanan Town, Monou County	5.0	Inclination of the protection block wall	Sinking of the sidewalk with cracks	Piling of blocks and pavement of the sidewalk
	Kanan Town, Monou County	5.8	Sinking of the road surface	Damage to the block piles	Asphalt pavement of the road surface and piling of blocks
	" "	14.7	Cracks in the stone piles	Cracks in the pavement	Construction of the protection concrete wall and the pavement of the road surface

Table III-3 Temporary Water Supply after the Earthquake till June 21 in Sendai City
(Water Supply Bureau of Sendai Municipal Government)

Date	Water Wagon			Personnel			Number of Cases	Amount of Temporary Water Supply (m³)	Main Water-Supplied Districts
	Total (Wagons)	Water Supply Bureau (Wagons)	Assistance (Wagons)	Total (Personnel)	Water Supply Bureau (Personnel)	Assistance (Personnel)			
June 12	5	5	0	20	20	0	7	7	Tsurugaya Housing Corporation Terrace Asahigaoka, Midorigaoka Jiyugaoka, Futabagaoka, Izumigaoka Kofukugaoka Kitaneichinenbo Tomisawa, Fukurobara Idehana, Arai Izai, etc. Kuromatsu, Saiwai-cho Housing Corporation Apartment House Sakuragaoka, Tsurugaya Apartment House Sunny Heights Dainohara Lions Mansion Nissui Apartment House Toroku Elementary School, etc.
13	29	7	22	72	50	22	165	180	
14	37	7	30	90	60	30	213	230	
15	27	7	20	70	50	20	120	130	
16	17	7	10	50	40	10	60	65	
17	10	6	4	30	26	4	26	30	
18	10	6	4	30	26	4	18	20	
19	8	5	3	18	15	3	18	20	
20	4	4	0	9	9	0	7	14	
21	2	2	0	5	5	0	3	6	
Total	149	56	93	394	301	93	637	702	

Table III-4 Assisting Activities in Temporary Water Supply and Restoration of Conduit Pipes
(Based on Mutual Assisting Plan of Tohoku Local Branch of Japan Water Supply Association)

Prefectural Branch	Assisting City	Assisting Period	Accepting City	Assisting Activities	Dispatch Items
Aomori	Hachinohe City	June 13 - June 16	Shiogama City	Temporary water supply	15 water tanks, 10 vehicles, 20 drivers of related companies, 4 personnel of water supply bureau.
Akita	Akita City	June 14 - June 18	Izumi City	Restoration of conduit pipes	7 personnel of 2 related companies, 2 vehicles.
Yamagata	Yamagata City	June 14 - June 16	Shiogama City	Temporary water supply	A water wagons, 3 personnel of water supply bureau.
	"	June 15 - June 16	Izumi City	Restoration of conduit pipes	4 personnel of a related company
	Yonezawa City	June 14 - June 16	Shiogama City	Temporary water supply	A water wagon, 3 personnel of water supply bureau.
	Kaminoyama City	"	"	" "	A water wagon, 2 personnel of water supply bureau.
	Sagae City	"	"	" "	A water wagon, a personnel of water supply bureau.
Fukushima	Fukushima City	June 14 - June 17	Izumi City	Restoration of conduit pipes	4 personnel of related companies.
	Koriyama City	"	"	" "	4 personnel of related companies.
	Haranomachi City	"	"	" "	4 personnel of related companies.

(Table III-4 continued)

Prefectural Branch	Assisting City	Assisting Period	Accepting City	Assisting Activities	Dispatch Items
Iwate	Morioka City	June 14 - June 18	Izumi City	Restoration of conduit pipes, water-flow test and investigation of water-leakage	June 14-June 17, 16 personnel of water supply bureau, 7 vehicles. June 17-June 18, 4 personnel of water supply bureau, 16 personnel of related companies. 4 vehicles of related companies.
	Mizusawa City	June 14 - June 17	"	Restoration of conduit pipes	A personnel of water supply bureau, 4 personnel of related companies, 2 vehicles.
	Kitakami City	June 14 - June 17	"	" "	A personnel of water supply bureau, 4 personnel of related companies, 2 vehicles.
	Hanamaki City	June 17 - June 18	"	" "	A personnel of water supply bureau, 4 personnel of related companies, 2 vehicles.
Miyagi	Kesennuma City	June 15 - June 18	Izumi City	Temporary water supply	2 water wagons.
	"	June 17 - June 19	Shiogama City	Restoration of conduit pipes	3 parties of related companies.
	Sendai City	June 15 - June 17	Izumi City	" "	2 parties of related companies.
	"	June 17 - June 19	Shiogama City	" " "	3 personnel of water supply bureau, 5 parties of related companies.

Table III-5　The Number of Personnel Required for the Restoration of Electric Power Facilities till June 30 after the Earthquake (Tohoku Electric Power Co., Ltd.)

Items \ Facilities	Personnels of Tohoku Electric Power Co., Ltd. (man-days)	Other Personnels (man-days)	Total (man-days)	Note
Thermal Power Generation	3,433	12,902	16,335	
Hydroelectric Power Generation	219	130	349	
Transformer	1,099	1,878	2,977	Assistance by Tokyo Electric Power Co., Ltd.
Power Transmission	1,107	283	1,390	"
Power Distribution	4,003	3,317	7,320	
Communication	339	194	533	
Total	10,200	18,704	28,904	

Note : Other personnels are from Tokyo Electric Power Co., Ltd., related companies and electrical machinery companies.

Table III-6   Damage to Gas Pipes (Gas Supply Bureau of Sendai Municipal Government)

Damage / Block	Trunk Pipe						Bough Pipe						Branch Pipe						Twig Pipe						Total	
	Severance	Crack	Pulled Out	Slackness of Joint	The Others	Total	Severance	Crack	Pulled Out	Slackness of Joint	The Others	Total	Severance	Crack	Pulled Out	Slackness of Joint	The Others	Total	Severance	Crack	Pulled Out	Slackness of Joint	The Others	Total	Number	Percentage
I	2	1	1	1		5	10		3	1		14	30	3			2	35	31	2		1	3	37	91	16.5%
II			1			1	3	1	1			5	4					4	7	9				16	26	4.7
III			5	1	1	7	81	9	27	3	5	125	71	5		1	2	79	69	6	1		3	79	290	52.6
IV			3			3	8		5			13	4					4	3	1			1	5	25	4.5
V	1					1	2	1			1	4	3			1		4	9	2		1	1	13	22	4.0
VI				1		1							6				1	7	8					8	16	2.9
VII				1		1							3					3	10				1	11	15	2.7
VIII			1			1	22	3	10	1	1	37	6		2		1	9	12				4	16	63	11.4
Tagajo, Fukuda, Fukuzumi											1	1	2					2	1					1	4	0.7
Total Number	3	1	11	4	1	20	127	14	46	5	7	199	129	8	2	2	6	147	150	20	1	2	13	186	552	

Note : According to the data of restoration works from June 12 till July 8.

Table III-7   Damage to Dykes on Rivers and Their Restoration

Name of the River	Length of Administrated Section (km)	Year of Improvement Undertaken	Total Length of Dykes (km)			Length of Damaged Dykes (km)				Note
			Finished	Provisional	Total	Emergency Disaster	Emergency Maintenance	Total	Restoration	
Lower Reaches of Kitakami River	137.5	1911	78.5	104.6	183.1	5.3	4.2	(0.052) 9.5	8.9	Including Eai River
Naruse River	82.4	1921	22.2	111.0	133.2	4.3	6.4	(0.080) 10.7	8.1	
Natori River	29.4	1941	28.6	3.4	32.0	2.3	9.0	(0.353) 11.3	3.0	
Lower Reaches of Abukuma River	63.7	1936	20.5	53.5	74.0	0.1	2.5	(0.035) 2.6	0.6	
Total	313.0		149.8	272.5	422.3	12.0	22.1	(0.081) 34.1	20.6	

Note : ( ) in the column of "Length of Damaged Dykes" gives the ratio of length of damaged dykes to total length of dykes.

Length of "Restoration" is at the point of dykes reported to have been suffered disaster, and its restoration costs more than 1 million yen, and the length is included in "Total".

Numerical values in the column of "Emergency Disaster" gives the length of dykes coped with by emergency temporary restoration, and numerical values in the column of "Emergency Maintenance" gives the length of dykes coped with by the budget for maintenance expenses.

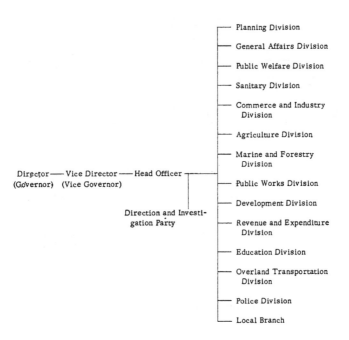

Fig. III-1   Structure of Rescue and Rehabilitation Center of
Miyagi Prefectural Government.

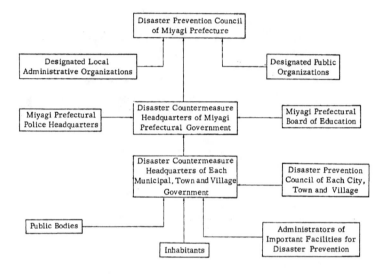

Fig. III-2   Disaster Countermeasure Structure in Miyagi Prefecture.

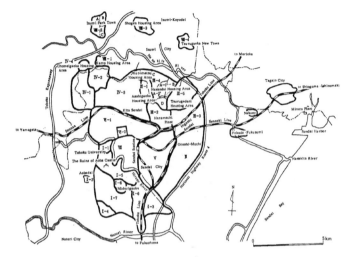

Fig. III-3　Regional Division for Restoration Works by Gas Supply Bureau
of Sendai Municipal Government.

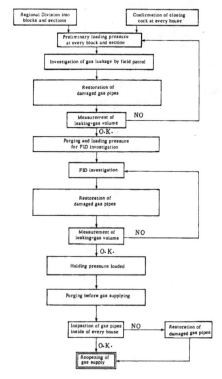

Fig. III-4   Flow Chart of the Restoration of Gas Supply Facilities from
Regional Division to Reopening of Gas Supply
(Gas Supply Bureau of Sendai Municipal Government)

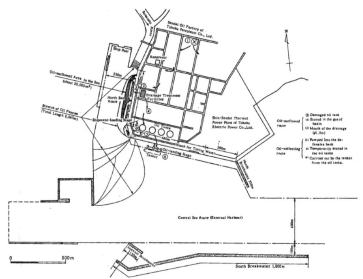

Fig. III-5 Arrangement of the Facilities and Outflow of Oil at Sendai
Oil Factory of Tohoku Petroleum Co., Ltd.

Table IV-1 Budgets of Reconstruction Works of Public Utilities in Miyagi Prefecture

Classification	Prefectural Budget				Municipal Budget				Total			
	Applied		Assessed		Applied		Assessed		Applied		Assessed	
	Sites	Budget	Sites	Budget	Sites	Budget	Sites	Budget	Sites	Budget	Sites	Budget
River Works	138	1,912,146	139	1,573,696	24	113,303	24	91,112	162	2,025,449	163	1,664,808
Coastal Works	14	433,656	14	350,360					14	433,656	14	350,360
Sabo Works	1	129,256	1	98,783					1	129,256	1	98,783
Highways	339	1,683,673	335	1,401,613	219	1,526,148	214	1,260,573	558	3,209,821	549	2,662,186
Bridges	40	319,668	39	272,111	26	414,027	26	346,570	66	733,695	65	618,681
Total	532	4,478,399	528	3,696,563	269	2,053,478	264	1,698,255	801	6,531,877	792	5,394,818

Table IV-2   Budget of Reconstruction Works of Water Supply Facilities

Classification	Budget in Thousand Yen
Receiving and Storing Reservoirs	565
Filtration Plants	9,057
Distribution Stations	14,126
Conduits	79,303
Others	74,321
Total	177,372

SOCIAL ASPECTS OF EARTHQUAKE MITIGATION

AND PLANNING IN THE UNITED STATES

William A. Anderson

National Science Foundation

Washington, D.C.

ABSTRACT

A number of social scientists in the United States have recently turned their attention
to the pre-disaster responses of individuals and social units and are investigating the
socioeconomic factors related to mitigation and preparedness. This research is important
because it promises to result in the identification of the principal factors which influence
the utilization of known and developing social and technological adjustments to earthquakes,
including building codes, land use regulations, public education and earthquake prediction.
As a result of on-going and recently completed social science studies, increased knowledge is
developing on a number of important topics. The emerging findings on the following questions
are discussed in this paper: How is the earthquake hazard perceived by officials and the
public? What types of mitigation and preparedness measures are citizens and officials willing to accept? What is the impact of some existing hazard mitigation programs? What are
some of the social factors to consider in trying to enhance the benefits of future earthquake
predictions?

KEYWORDS: Mitigation and preparedness measures; socioeconomic factors; hazard awareness.

INTRODUCTION

Systematic research on disasters and natural hazards by social scientists in the United States is relatively new. It was not until the 1950s that a significant program of disaster research began emerging in the social science field in the United States. For a long time the field was hampered by a lacking of funding, and even today social and behavioral scientists working in the area of disaster and natural hazards research receive only a fraction of the financial support received by colleagues in the physical sciences and engineering.

During the last decade, however, there has been a manifold increase in disaster and natural hazards studies conducted by American social scientists. The principal catalyst for this has been the availability of more dependable funding and an increase in well-trained social and behavioral scientists interested in conducting research in the field. Nevertheless, there is still considerable unevenness in existing knowledge. [1] Much of the accumulated knowledge is on the responses of individuals, groups and institutions during the immediate post-impact period of disaster. [2] On the other hand, one of the most neglected periods has been the pre-disaster period, the time when mitigation and preparedness actions can be undertaken by threatened populations.

Fortunately, since the mid-1970s, a small number of social scientists in the United States have turned their attention to the pre-disaster responses of individuals and social units and are investigating the socioeconomic factors related to mitigation and preparedness for earthquakes and other hazards. Social science research on earthquake mitigation and planning can be a vital complement to research conducted by engineers and geophysicists. This research is important because it promises to result in the identification of the principal factors which influence the utilization of known and developing social and technological adjustments to earthquakes, including building codes, land use regulations, educational programs, and earthquake prediction.

In the rest of this paper, I will briefly discuss some of the findings, in some instances tentative ones, of a few recent and on-going studies supported by the National Science Foundation on the socioeconomic aspects of mitigation and preparedness and various issues such investigations have thus far raised. I will particularly note emerging answers to the following questions: How is the earthquake hazard perceived by officials and the public? What types of mitigation and preparedness measures are citizens and officals willing to accept? What is the impact of certain existing hazard mitigation programs? What are some of the social factors to consider in trying to enhance the benefits of future earthquake predictions? It should be stressed that some of the research findings reported below are the result of preliminary analyses of data by researchers involved in on-going projects and are subject to some modification once the final analyses have been completed.

HAZARD PERCEPTION

Awareness is a key factor in reducing the impact of earthquakes and other hazards on society. [3] If individuals and groups have little awareness of, or concern for, the earthquake hazard, they cannot be expected to give serious consideration to developing mitigation and preparedness measures. Thus, it is important to know how the earthquake hazard is perceived by various segments of society. Such information would provide insight into the types of education efforts that might be required and also offer a more reaslistic basis for predicting support for particular kinds of hazard reduction policies and programs.

Information is currently emerging from the social sciences in the United States on how both public officials and private citizens perceive the threat of earthquakes and other natural hazards. Among the findings which have surfaced thus far is that such problems have a minor place on the political agenda of most states and local communities in the country. Put simply, most influential groups exhibit little real concern for such issues as hazard mitigation and preparedness at this time.

For example, sociologists at the University of Massachusetts conducted a national survey in 1977 concerned with the local and state politics of natural hazards policies. [4] Interviews were conducted with over 2300 key local and state persons. The researchers found that these political influentials do not perceive natural hazards as a very important problem in comparison to the many other issues they must consider.

A similar conclusion has been reached by political scientists at the University of California, Santa Barbara following a preliminary analysis of data from their study on seismic safety preparedness of local governments in California. [5] This study focuses upon 14 local jurisdictions, including 12 cities and 2 counties. The investigators found that a concern for seismic safety was almost universally low among community leaders.

There is also evidence that concern for earthquakes in the United States is very low among private citizens. For example, sociologists at the University of California, Los

Angeles have completed a survey of 1450 residents of Los Angeles County as part of an
on-going study of the social response to the southern California Uplift, an anomaly which
some scientists have interpreted as a possible precursor to a major earthquake in that
area. [6] Though respondents in the study had several opportunities to mention earthquakes
as one of southern California's serious problems, the investigators report that few did so.
For example, only 35 persons in the sample mentioned earthquakes as one of the 3 most important problems facing residents in southern California today.

Findings from such studies, then, clearly indicate the difficulties that lie ahead for
creating earthquake hazard awareness in the United States. With so little concern for this
hazard, widespread acceptance of mitigation and preparedness measures in endangered areas
may be unlikely at this time.

ACCEPTANCE OF MITIGATION AND PREPAREDNESS MEASURES

A few of the recent social science studies conducted in the United States have provided
a clearer understanding of the extent to which people find particular mitigation and preparedness measures acceptable. For example, one form of financial protection available to
individuals in threatened areas is hazard insurance. A study was recently completed by
economists at the University of Pennsylvania to determine the extent to which people took
advantage of this type of protection. [7] Part of the study involved a survey of homeowners
in 18 earthquake-susceptible areas of California. What they learned was not particularly
encouraging. It was found that a lack of concern for earthquakes, combined with inadequate
information on the availability and terms of earthquake insurance, resulted in few sales of
this form of protection. Similar results were found regarding flood insurance in the study.

In the study of the southern California Uplift previously mentioned, the principal
investigator, Ralph Turner, and his associates asked respondents if they had suggestions
regarding what the government should be doing to prepare for future earthquakes. The
researchers learned that most people had some concern for this issue and had specific suggestions to make. Most respondents offered suggestions on hazard reduction activities, such as
taking steps to enforce building codes, rather than on improving the state of emergency preparedness to handle disaster-generated problems. The investigators noted that it was

reassuring to detect this potential public support for mitigation programs, but they were also quick to point out they could not be certain of the amount of such support in real situations.

The Turner study also shows that while most of the respondents indicated a willingness to see government action to mitigate earthquake hazards, few had taken actions to prepare their own households for an earthquake; for example, reported having taken such steps as storing supplies of water and food and rearranging shelves to prevent objects from falling on occupants. However, there was a tendency for persons with more suggestions for government officials to be better prepared for earthquake themselves.

IMPACT OF EXISTING PROGRAMS

Little is known about the impact of many existing programs in the United States that attempt to achieve greater earthquake hazard reduction and awareness. Evaluations of such programs are highly desirable in order to make them more effective in the long run.

Two on-going studies are assessing the impact of programs in California on individual and community concern for seismic safety and actions taken in this regard. The previously mentioned University of California, Santa Barbara study has looked at the impact of legislation requiring cities and counties in California to develop a seismic safety element as part of their general plans. This element is to consist of an identification and appraisal of earthquake hazards in the local jurisdiction and is expected to serve as part of the basis for mitigation in the area. The 1971 San Fernando earthquake provided the impetus for this and some other new laws related to earthquake safety. After a preliminary analysis of their data, the researchers have concluded that the seismic safety elements have had little policy impact in the 14 jurisdictions they studied. They report that the development of the seismic elements or plans has had some consciousness-raising impact, but that with few exceptions little has been done to implement them and jurisdictions are not better off than before in terms of emergency preparedness, in dealing with old hazardous buildings, and in public education.

In 1975 California also passed legislation requiring a seller to inform prospective buyers if a property is located in an earthquake hazard area, referred to as a special study zone. An on-going study by geographers at the University of Colorado is evaluating this disclosure measure. [8] As part of the study, a survey has been undertaken in two

California communities to determine the salience of earthquake hazards in the house selection process and to measure the impact of disclosure on the search for and purchase of a house. A preliminary analysis of the data by the researchers shows that over half of the persons contacted knew their homes were in an earthquake hazard area or special study zone. And almost all the persons who knew about the zone said that this knowledge had no effect on their decision to purchase their home. Data from the study initially supports the conclusion then that the earthquake hazard has little or no importance to most buyers. This study will not be completed until 1980. The researchers hope that at its completion a number of recommendations can be offered regarding this particular disclosure measure and the role that such approaches in general can play in hazard mitigation and planning. This is particularly important since other states have considered such measures.

SOCIAL FACTORS TO CONSIDER IN PREDICTION

Prediction is one of the most promising means to mitigate the impacts of earthquakes. Future earthquake predictions promise to provide long lead times, thus enabling significant hazard reduction and emergency preparedness activities to be undertaken. For example, a study on the possible socioeconomic effects of earthquake predictions conducted by sociologists at the University of Colorado suggests that a credible prediction of a major earthquake in the United States will serve as the impetus for a number of mitigation and preparedness activities that will reduce loss of life and property. [9] This conclusion was based on results from a survey of private citizens, and government and private officials in two California communities. However, this study also suggests that a long-term earthquake prediction has the potential to create serious socioeconomic disruption in a community. For example, businesses could limit mortgage loans, the availability of insurance and investment in a prediction area. Additionally, the researchers note that there could be a possible net out-migration and reduced tourist trade, leading to increased unemployment, falling property values and reduced tax revenue.

One of the important tasks social scientists can perform in the hazards field, then, is conducting research that identifies socioeconomic, political and legal factors which officials should take into account in order to maximize the benefits of prediction and minimize its negative consequences. This role is exemplified in the case of a recent study completed at the Association of Bay Area Governments of California on the legal liability of government in connection with earthquake hazards reduction. The researchers note that local and state governments are potentially liable in earthquake prediction and warning situations. [10]

In their view liability could result from problems growing out of the issuance of a warning or the failure to issue one. The study concludes that there is a need for a form of immunity to cover public forecast situations when officials take reasonable actions. The researchers feel that by dealing with immunity in this fashion on an obstacle to prompt and effective government utilization of earthquake prediction and warning would be removed.

The southern California Uplift situation which Turner and his associates are studying, while not involving an actual prediction, does serve as a prototype of the possible initial stage of a prediction. Attention has been given to the situation by the media, the California Seismic Safety Commission has expressed concern about it to officials in the state, and the U.S. Geological Survey has been conducting special studies on the anomaly. The Uplift situation, then, can be viewed as a near prediction and Turner and his colleagues have been investigating it for insight it might offer on the likely social responses to an actual prediction in the United States. The researchers have learned that social groups differ in their awareness of the Uplift and in understanding its potential significance. Thus one of the cautions they offer is that groups will vary in their awareness and appreciation of an actual earthquake prediction and therefore responsible officials must be prepared to launch special efforts to insure that certain groups, such as the less educated and the poor, are made fully aware of any future earthquake prediction.

CONCLUSION

Some progress is now being made in the United States in developing knowledge on the socioeconomic factors related to earthquake mitigation and planning. Work by social scientists in this field will hopefully make the research conducted by geophysicists and engineers even more useful to policy makers.

There is now a need for social scientists in the United States to become involved in research on mitigation and planning in other societies. The real challenge is for American social scientists to engage in cooperative research efforts with colleagues in other countries, including Japan. Such joint research ventures would tell us what extent our accumulated knowledge can be generalized across societies. They would also facilitate the diffusion of innovative hazards reduction policies and programs among societies.

REFERENCES

[1] Taylor, V. A., "Future Directions for Study," in E. L. Quarantelli, ed., Disasters: Theory and Research (Sage, Beverly Hills, 1978).

[2] Mileti, D. S., Drabek, T. E., and Haas, J. E., Human Systems in Extreme Environments: A Sociological Perspective (University of Colorado, Boulder, 1975).

[3] Working Group on Earthquake Hazards Reduction, Earthquake Hazards Reduction: Issues for an Implementation Plan (Office Science and Technology Policy, Washington, D.C., 1978).

[4] Rossi, P. H., Wright, J. D., Wright, S. R., "Social Science and Natural Hazards, Social and Demographic Research Institute, University of Massachusetts, May, 1979.

[5] Mann, D., Post-Disaster Response, Natural Hazards Research Applications Workshop, Boulder, Colorado, July 1979.

[6] Turner, R. H., Nigg, J. M., Paz, D. H., and Young, B. S., Earthquake Threat: The Human Response in Southern California (Institute for Social Science Research, University of California, Los Angeles, 1979).

[7] Kunreuther, H., Disaster Insurance Protection: Public Policy Lessons (John Wiley and Sons, New York, 1978).

[8] Palm, R., "Real Estate Agents and Dissemination of Information on Natural Hazards in the Urban Area," Natural Hazards Research Applications Workshop, Boulder, Colorado, July 1979.

[9] Haas, J. E., and Mileti, D. S., Socioeconomic Impact of Earthquake Prediction (Institute of Behavioral Science, University of Colorado, Boulder, Colorado, n.d.).

[10] Margerum, T., Will Local Government Be Liable for Earthquake Losses? (Association of Bay Area Governments, Berkeley, 1979).

ORIENTATION OF TSUNAMI RESEARCH IN JAPAN

Hiroshi Takahashi

Yukio Fujinawa

National Research Center for Disaster Prevention

Science and Technology Agency

ABSTRACT

We have been struck by large tsunamis which were induced by earthquakes occurring near the Tonankai area of Japan on an average of every thirteen years in a 350 year period. It is conjectured by some seismologists that a large earthquake may occur in the area of Tonankai. We have experienced nine earthquakes which are considered to be grouped in the Tonankai earthquake zone. All of these earthquakes were accompanied by tsunamis. Victims rose to 31,000 in 1498 (Meio), 3,906 in 1605 (Keicho), 4,924 in 1707 (Hoei), and 3,427 in 1854 (Ansei), respectively. We urgently need to plan countermeasures against tsunamis, especially to develop a warning system for tsunamis which take less than about ten minutes to reach our coasts. Tsunami research in Japan is briefly reviewed from the standpoint of developing an effective warning system.

KEYWORDS: Disaster warning; earthquake detection; flood warning; Tsunami prediction; Tsunami research.

1. DISASTROUS TSUNAMIS IN JAPAN

Some 130 tsunamis have been recorded in this country over a period of some 1,500 years. A scale of tsunamis is sometimes defined by wave height along the coast as shown in Table 1. The number m is denoted as tsunami magnitude. Large tsunamis in this country are listed in Table 2. The biggest three are the 1896 Sanriku tsunami, the 1792 Shimabara tsunami and the 1771 Okinawa tsunami.

The epicenters of earthquakes which caused tsunamis are shown in Figure 1. The coastal area along the Pacific Ocean were frequently struck by huge tsunamis. Return period of tsunamis with magnitude larger than 2 is about 10 years, and larger than 3 is about 30 years. Thus, an improved tsunami warning system ranks high in regard to coastal disaster prevention needs.

2. PRESENT METHOD OF TSUNAMI PREDICTION

Various measures are taken to prevent disasters by tsunamis. The tsunami warning system plays an important role in evacuation measures. As a result of the great damage by the 1946 tsunami in the Hawaii Islands, an international tsunami warning system named the Seismic Sea Wave Warning System (SSWS) was instituted to provide information on tsunamis to countries on the Pacific Ocean. Further improvements to such a warning system based on tsunami monitoring data are needed in Japan.

3. ADOPTED METHOD OF TSUNAMI MAGNITUDE PREDICTION

The tsunami magnitude is predicted by the Japanese Meteorological Society (JMA) immediately after the occurrence of earthquakes. The prediction is based on tsunami statistics gathered in this country by Iida [1]. His result is reproduced in Figure 2, which is a graph of simultaneous plots of earthquake magnitude M, hypocenter depth H, and tsunami magnitude m. As shown in the figure, the magnitude of an earthquake accompanying a tsunami contains some lower limit, Mc, and Mc is related to the hypocenter depth H as,

$$Mc = 6.3 + 0.01 H$$

In the case of a large tsunami with magnitude larger than 2 the relation is,

$$Mc' = 7.75 + 0.008 \; H$$

On the base of these empirical results, various types of warning are given by the JMA.

It is a problem that some tsunamis do not follow this empirical relation with the result that erroneous warnings are sometimes issued. For instance, the diagram predicts only a minor tsunami for the 1896 Sanriku tsunami. Some improved interpretations can be made using the results of Hatori [2] who analyzed tsunami statistics limiting himself to those which occurred near the Sanriku district. His results are shown in Figure 3. Correspondence between tsunami magnitude m and tsunami source area is very good, which is considered to be due to the existence of a type of fracture in a fixed area.

4. QUANTITATIVE EVALUATION OF TSUNAMI MAGNITUDE

Evaluation of tsunami wave height requires knowledge of the following four stages of a tsunami:

1) Generation

2) Propagation

3) Resonance

4) Run-up

i) Generation of Tsunami

If we can obtain the size of the deformation at the ocean bottom, the ocean surface deformation can be evaluated, at least in principle. Such a calculation was done in the realm of the long wave approximation by Kajiura. However, the most difficult problem is that the actual ocean bottom deformation area is rarely observed. We are compelled to estimate the area and position of ocean bottom movement by the use of various indirect methods. Representative methods are:

1) reverse refraction method ·········· (deformation area) Sr

2) aftershock activity ······································ Sa

3) surface projection of fault plane ··· (geodetic data) ·· Sg

4) surface projection of fault plane ··· (seismic data) ··· Ss

In the case of the 1946 Nankai tsunami, the tsunami source area, Sr, is estimated to be much larger than the aftershock area, Sa,

Sr > Sa,

On the other hand, the tsunami source area, Sr, is nearly equal to the area Sg of surface projection of fault plane by the use of geodetic data,

Sr ~ Sg

and the aftershock activity area, Sa, is nearly equal to the area, Ss, of the surface projection of fault plane by the use of seismic data,

Sa ~ Ss

We are led to a conclusion that there exists a situation where the ocean bottom deformation accompanying an earthquake has a time constant too small to cause earthquakes, but large enough to cause tsunamis. Therefore, a tsunami prediction method on the basis of just the earthquake magnitude M and the hypocenter depth h is incomplete.

Kanamori [3] made a new contribution in approach. Frequency spectrum of effective moment Mo(f) was introduced (Figure 4). Observed values are illustrated with several symbols, and full lines are estimated moments with the parameter being time constant. If we can obtain the frequency spectrum of moment by the use of several types of seismograph, we can obtain more reliable estimates.

ii) Propagation and Resonance

When tsunamis approach a shallow ocean or a bay, they undergo deformation and give rise to a new type of motion, such as an edge wave, or seiche. Seiches along the Japanese coast have long been observed. We set six tsunami gauges along the coast of the Sagami Bay and observed long period ocean waves. Analysis of the data revealed the spectral structure of a seiche at those areas and the evolution of representative modes. Oscillation of the Sagami Bay itself is usually one of the three most important modes, as illustrated in Figure 5 [4].

Evolution of an edge wave along the curved coast has been studied by Fujinawa [5] who calculated the spatial change of wave characteristics. Experimental and observational studies were performed by the Earthquake Research Institute (ERI) group.

iii) Numerical Simulation of Tsunami

The ocean bottom movement which accompanies an earthquake can be estimated somewhat more accurately as a result of recent progress in seismology. Fault plan characteristics may be evaluated from seismic data. Once ocean bottom deformation size is known, calculation of tsunami magnitude can be made at least in principle. Aida [6] made a numerical calculation in the case of the Tokachi-Oki tsunami where good tidal records were obtained along the Pacific coast of this country. Some of his results are reproduced in Figure 6. The correspondence between the observed and the calculated wave forms is good. Numerical simulation of tsunamis will become more practical and useful as details of such problems as sea bottom dissipation, or nonlinear deformation in a shallow region, become better known.

5. SYSTEM OF MONITORING FOR TSUNAMIS OFF THE JAPANESE COAST

We have a plan to develop a more reliable monitoring system for those tsunamis which occur in the sea not far from our coast. The system consists of both tsunami observation and tsunami prediction. The monitoring is performed by pressure gauges and current meters placed on the sea bottom. Data are telemetered to land stations through marine cable, or by radio signals from buoys which are connected to the gauges and meters. The input data for the tsunami prediction algorithm are the direction and wave height of a tsunami which is received at several stations. The resultant predicted value is transmitted to the authorities concerned.

REFERENCES

[1] Iida, K., Proceedings Tsunami Meeting, 10th Pacific Sa Congr. 1969, IUGG Monograph, 24, 7, (1963).

[2] Hatori, T., Bulletin of Earthquake Research Institute, 47, 185, (1969).

[3] Kanamori, H., Phys. Earthquake Planet Inter., 6, 346 (1972).

[4] Fujinawa, Y. et al., Rep. Nat. Res. Cent. Dis. Prev., 19, 117, (1978).

[5] Fujinawa, Y., Rep. Nat. Res. Cent. Dis. Prev., 21, 75, (1979).

[6] Aida, I., J. Phys. Earth, 26, 57, (1978).

Table 1 Definition of tsunami magnitude (Iida,1963)

magnitude	wave height	energy $(\times 10^{22} erg)$
-1	< 50cm	0.06
0	about 1m	0.25
1	about 2m	1.0
2	4 ～ 6m	4.0
3	10～ 20m	16.0
4	>30m	64.0

Table 2 Large tsunami in Japan (Hatori,1977)

year	source	mag.M	mag.m	death		year	source	mag.M	mag.m	death
684	Tokai Nankaido	8.4	3			1703	Boso	8.2	3	5,233
869	Sanriku	8.6	4	1,000		1707	Tokai Nankaido	8.4	4	4,900
887	Kii	8.6	3			1741	Hokkaido	7.5?	3	1,467
1096	Tokai	8.4	3			1771	Okinawa	7.4	4	11,861
1099	Tokaido	8.0	3?			1792	Shimabara	6.4	2?	15,030
1361	Kii	8.4	3			1854	Tokai	8.4	3	900
1498	Tokaido	8.6	3	5,000		1854	Tokaido	8.4	3	3,000
1605	Boso Nankaido	8	3	3,800		1896	Sanriku	7.6	3-4	27,122
						1933	Sanriku	8.3	3	3,008
1611	Sanriku	8.1	4	6,800		1944	Tokai	8.0	2.5	998
1677	Boso	8	3	500		1946	Tokai	8.1	2.5	1,330

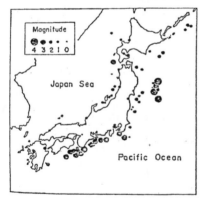

Figure 1 Epicenters which caused large tsunami
(from Kajiura)

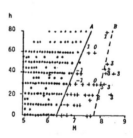

Figure 2 Relation between tsunami
magnitude and earthquake
magnitude M , hypocenter
depth H (Iida,1963)

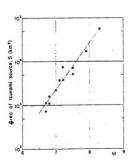

Figure 3   Relation between earthquake
magnitude M and tsunami
source area in the case of
Sanriku earthquake
(Hatori, 1969)

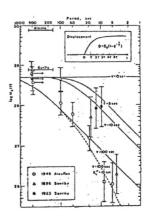

Figure 4   Spectrum of effective moment
of earthquake (Kanamori, 1972)

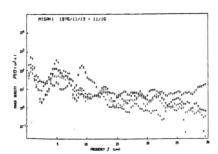

Figure 5   Seiche spectrum at Misaki along the
Bay of Sagami (Fujinawa et al, 1978)

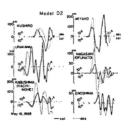

Figure 6   Result of tsunami numerical
simulation in the case of
the 1968 Tokachi-oki
tsunami (Aida, 1978)

Hiroshi Hashimoto

Taka-aki Uda

Public Works Research Institute

ABSTRACT

Typhoon 7010 caused heavy damage to Kochi city brought on by flooding from the storm surge. Numerical computations were carried out but they do not explain fully the abnormall high rise of the sea level. As one explanation, it is suggested that wave setup contributes to an abnormally high rise since a typhoon is usually accompanied by high waves.

Experimental investigation and numerical calculations were carried out to make clear th magnitude of wave setup at the mouth of Kochi harbor. Generally, on a straight coast, the normal component of radiation stress caused by breaking waves generates wave setup and the tangential component generates along-shore current. Investigations, however, show that the along-shore current is obstructed by a breakwater at Kochi harbor and wave setup is generate by the tangential component together with the normal one. The wave setup contributes to the anomaly, and its height is estimated to range from 0.5 m to 1.0 m. Numerical calculations confirm the experimental results though they need further improvement.

KEYWORDS: Storm surge; typhoon damage; wave setup.

1. INTRODUCTION

A storm surge is a meteorological tide caused by strong wind and an abrupt atmospheric pressure depression. Numerical estimations of sea levels have been carried out by many investigators. Since a tropical low pressure system such as a typhoon is usually accompanied by high waves, wave setup becomes an additional reason for abnormal sea level rise. Especially on a coast facing the open sea, the wave setup becomes comparable to a storm surge.

The typhoon 7010 which landed on the Kochi Prefecture of Shikoku Island in 1970 caused heavy damage by storms and floods in Kochi city. At Kochi harbour, the sea level was higher than 3 m and the anomaly reached 2.4 m. The numerical calculation of storm surge carried out by the Meteorological Agency does not fully explain the anomaly. As one of the causes, we think that it is necessary to consider the wave setup. In this paper, the pattern and the magnitude of the wave setup is examined by using a hydraulic model and numerical computation.

Typhoon 7010 was generated about 300 km northeast of the Saipan Islands on August 19 and landed on the Shikoku Island on August 21. At one time, the minimum depression became 910 mb and the maximum wind speed exceeded 55 m/sec near its center. The typhoon caused the storm surge along the coast of Tosa Bay as shown in Figure 1 and brought heavy damage by flooding in Kochi city. The sea level became high at the foot of Tosa Bay and the anomaly was high at those bays where the direction of the mouth is eastward, such as Katsurahama at Kochi harbour or Usa at Uranouchi Bay. The anomaly at Katsurahama was extremely different from the previously observed data as shown in Figure 2. The magnitude of the typhoon was not extremely large but its route was different from others. It approached from the southeast.

At Kochi city, the maximum wind speed was 29.2 m/sec and the direction was east. Waves also reached their maximum value of 6.6 m and a period of 12.2 sec at 10 a.m. The measurements were taken 260 m offshore of a breakwater at Kochi harbour at a depth of 12.5 by a pressure-type wave gauge.

2. EXPERIMENTAL METHOD

The wave setup at the mouth of Kochi harbour was investigated using a 1/250 scale model. The model was 3 km wide and 3 km long in prototype as shown in Figure 3. Waves propagate in widths of 1.5 km between two guide walls.

Applying the Froude scale law to the model experiment, the length scale is 1/250 and the time scale is 1/15.8. Waves were generated by a pendulum-type wave generator and measured with resistance-type wave gauges at 30 points. The wave setup was also measured with pitot tubes and pipes settled at the bottom. For experimental conditions, we selected three wave heights (2, 3, 4 cm), three directions (S22.5°E, S45°E, S67°E), and two period conditions (0.8, 1.0 sec).

3. EXPERIMENTAL RESULTS

Incident waves break offshore and regenerate new waves. Diffraction occurs behind the breakwater. In Figure 4 wave height distribution is shown as an example. The height is almost comparable to the water depth and is small behind the breakwater.

The distribution of the wave setup is shown in Figure 5. In the case of S45°E incident waves, mean sea level rises high at the shore line and behind the breakwater. In the case of S67.5°E incident waves, waves approach obliquely the shoreline and generate longshore currents. The current is blocked by the breakwater and water piles up. Then wave setup occurs in both the onshore and alongshore directions.

The effects of tidal level, wave height, period, and direction on the wave setup were investigated experimentally. First, the effect of tide is negligible because the wave height distribution is not influenced by the tidal level. The wave period is related to the breaking conditions. However, in this case, because of a wide breaker zone, the change of wave period brings only a small change of the width. Therefore, the period is not an important factor. Incident wave height and direction are strongly related to the wave setup. The relation between wave height and setup height is shown in Figure 6. The setup increases as the incident wave height increases.

In this experiment, waves were bounded by two guide walls. Thus, the formation of longshore current was not fully developed. Since it was expected that the distribution of wave setup height would be influenced by the side wall, numerical computations were carried out to make the effect clear.

4. NUMERICAL ANALYSIS

As a method to predict wave setup and current, Longuet-Higgins proposed an equation based on radiation stress concept. Basic equations are:

$$\frac{\partial M}{\partial t} + \frac{\partial M}{\partial (h + \eta)} \frac{\partial M}{\partial x} + \frac{N}{h + \eta} \frac{\partial M}{\partial y} + \frac{M}{h + \eta} \frac{\partial N}{\partial y} = -g(h + \eta) \frac{\partial \eta}{\partial x} - \gamma_b U_o \frac{M}{h + \eta} + LV^2 M$$
$$- \frac{1}{\rho} (\frac{\partial S_{xx}}{\partial x} + \frac{\partial S_{xy}}{\partial y}) \quad , \tag{1}$$

$$\frac{\partial N}{\partial t} + \frac{M}{h + \eta} \frac{\partial N}{\partial x} + \frac{N}{h + \eta} \frac{\partial M}{\partial x} + \frac{2N}{h + \eta} \frac{\partial N}{\partial y} = -g(h + \eta) \frac{\partial \eta}{\partial y} - \gamma_b U_o \frac{N}{h + \eta} + LV^2 N$$
$$- \frac{1}{\rho} (\frac{\partial S_{xy}}{\partial x} + \frac{\partial S_{yy}}{\partial y}) \quad , \tag{2}$$

$$\frac{\partial \eta}{\partial t} + \frac{\partial M}{\partial x} + \frac{\partial N}{\partial y} = 0 \tag{3}$$

where the x and y axes are taken horizontally, t is the time, M and N represent the flux flows in the x and y directions respectively, η the water surface elevation, h the water depth, g the acceleration of gravity, γ_b the friction coefficient at the bottom, L the horizontal eddy viscosity, U_o the amplitude of the orbital velocity at the bottom induced by wave action, and S_{xx}, S_{xy} and S_{yy} the radiation stress tensors.

Numerical computation was carried out using the ADI method [1, 2]. In order to find the distribution of radiation stress tensors, the wave height and the wave direction angle must be given beforehand. Equations suggested by Noda [3] were used. They are:

$$\cos\theta \frac{\partial \theta}{\partial x} + \sin\theta \frac{\partial \theta}{\partial y} = \frac{1}{C} (\sin\theta \frac{\partial c}{\partial x} - \cos\theta \frac{\partial c}{\partial y}) \tag{4}$$

$$\frac{\partial E}{\partial t} + \frac{\partial}{\partial x} (EC_g \cos\theta) + \frac{\partial}{\partial y} (EC_g \sin\theta) = 0 \tag{5}$$

where θ is the wave direction angle, C the wave velocity, C_g the group velocity, and E the wave energy.

In a breaker zone, the following equation is used,

$$H = 0.7 h \tag{6}$$

where H is the wave height in the breaking zone, and h the water depth.

Numerical computation was carried out in the case of a S67.5°E incident wave. The domains were taken as shown in Figure 7 and covered with a mesh of discrete points. The spacing was uniform and 75 m in prototype. The friction coefficient and the horizontal eddy viscosity were taken as 0.0064 and 20 cm^2/sec respectively.

5. NUMERICAL RESULTS

Computations were carried out to make clear the effect of side walls in the model. Two cases were analysed. One has the same domain as in the model and the other has a wider domain in the direction of longshore as shown in Figure 7.

A comparison of the two cases are shown in Figure 8. The wave setup in the narrow one is about 10% lower than that of wide one. Though the difference is not large, this demonstrates that the longshore component of radiation stress brings the wave setup at the harbor mouth.

The experimental results shown in Figure 5 are a little lower than the computed one shown in Figure 8. There are several reasons that produce the difference. For simplicity, in the calculation, equation (6) was used. In the experiment, the coefficient varies from 0.5 to 0.7 according to breaking condition and the formation of regenerated waves. The topography is also different in each case. In the calculation the change of topography is averaged over one mesh size.

Current patterns are shown in Figures 9 and 10. Longshore currents are generated by the obliquely incident waves and they turn offshore near the breakwater. In the narrow domain case as shown in Figure 9, the longshore current is suppressed by the side wall. This produces a low wave setup. While the current patterns were not measured in the experiment, they were different from those in the calculated ones. In the experiment, the offshore current was predominant at both sides of the breakwater. On the other hand, in the calculation circulation developed at the tip of the breakwater. The mesh pattern seems to be inappropriate.

6. CONCLUSIONS

Experiments and numerical analysis were carried out to better account for the abnormal rise of sea level which was caused by typhoon 7010 at Kochi harbor. Experimental results show that the wave setup becomes especially high when the waves approach the shoreline obliquely. The longshore component of radiation stress, together with the onshore component, play an important role in the formation of water piling up near a breakwater.

Numerical calculations were made to assess the effect of side walls in the model. When the width of side walls was extended, the wave setup becomes higher. Actual current patterns

etup heights do not agree well with those of the experiments. It is important to make

ate estimates of wave heights in a breaker zone. In a complex topography the arrange-

of the experimental mesh pattern is also important.

REFERENCES

Hashimoto, H. and Uda, T., "Method for Numerical Computation of Nearshore Currents and its Application," 21st Coastal Engineering Conference in Japan, pp. 355-361, 1974.

Hashimoto, H. and Uda, T., "Simulation of Nearshore Currents," 22nd Hydraulics Conference in Japan, pp. 147-154, 1978.

Noda, E. K., Sonu, C. J., Rupert, V. C. and Collions, J. I., "Nearshore Circulation Under Sea Breeze Conditions and Wave-Current Interaction in the Surf Zone, TETRA TECH. Report, TC-P-72-149-4, 1974.

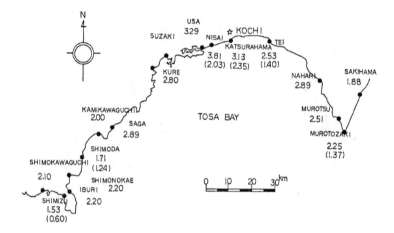

Fig. 1   Maximum sea level above T.P. (Tokyo Peil) and anomaly along the
coast of Tosa Bay caused by Typhoon 7010 (unit: m)

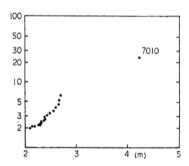

Fig. 2   Return periods of the maximum tidal level at Katsurahama (From
1950 to 1970)

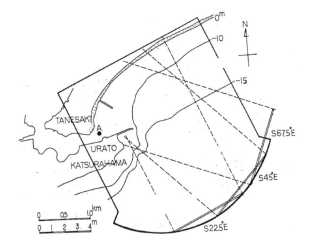

Fig. 3   Range of the model

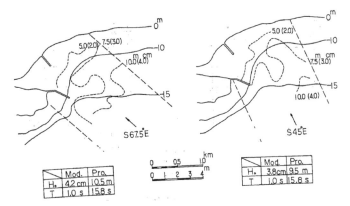

	Mod.	Pro.
H₀	4.2 cm	10.5 m
T	1.0 s	15.8 s

	Mod.	Pro.
H₀	3.8 cm	9.5 m
T	1.0 s	15.8 s

Fig. 4   Wave height distribution

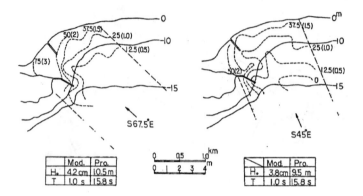

Fig. 5   Wave setup distribution

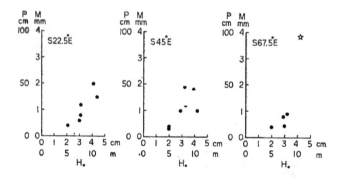

Fig. 6   Relation between wave height and wave setup height

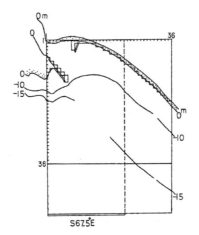

Fig. 7   Domain of numerical calculation

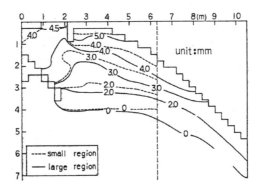

Fig. 8   Computed wave setup distribution

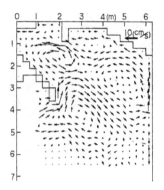

Fig. 9 Current pattern (narrow domain)

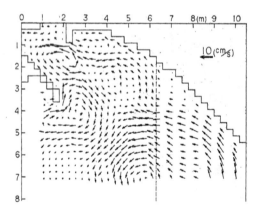

Fig. 10 Current pattern (wide domain)

SPECIFICATION AND PREDICTION OF SURFACE WIND FORCING

FOR OCEAN CURRENT AND STORM SURGE MODELS

Celso S. Barrientos

Kurt W. Hess

Techniques Development Laboratory

Systems Development Office

National Weather Service

National Oceanic and Atmospheric Administration

ABSTRACT

Forecast methods to predict movements of oil spills in the ocean are being developed in the Techniques Development Laboratory of the National Weather Service (NWS). An operational model for oil movement forecast is being implemented in NWS. The model will be available for routine use in the event of oil spills and for assessment studies of probable impacts of oil spills.

The most important component in the movement of an oil spill is due to the surface wind forcing. Surface wind stress acts on the spilled oil in two ways: (1) to generate ocean surface currents, and (2) to drag directly the oil on the surface. We examined different boundary layer wind formulations and assessed the resulting surface currents response.

A dynamical storm surge model has been developed in NWS and is being used in real time forecasting of surges when a hurricane is approaching land. The model has proved useful in routine use during the last 10 years. Recently, a model has been developed that forecast surges in bays and estuaries.

Storm surge is generated by the action of the wind and low atmospheric pressure in the storm (inverted barometer effect). The wind forcing is the dominant factor in surge genera- tion. The spatial distribution of the wind in a storm or typhoon determines the resulting characteristics of the surge, such as, maximum surge and location on the coast, extent of the coastline affected by the surge and the height variation of the surge along the coast. We tested different storm wind models and determined the response of the storm surge model.

KEYWORDS: Ocean current; oil spill trajectory; wind forcing; hurricane; storm surge; wind models.

INTRODUCTION

The National Weather Service (NWS), NOAA, has been involved in monitoring and prediction in the marine environment. The Techniques Development Laboratory, NWS, is developing forecast methods to predict movement of oil spills in the ocean and storm surges in the coastal regions. We are implementing an operational model to forecast oil spill movements. The model will be available for routine use in the event of oil spills and for assessment studies of probable impacts of oil spills.

NWS has an operational model to predict storm surges in the coastal regions. The storm surge model has been working well during the last 10 years and has been tested in many actual storm situations. Recently, a dynamical model has been developed to forecast storm surges and overland flooding in bays and estuaries.

In predicting movements of oil spill and storm surges, surface wind forcing is the most important element. The wind generates the surface current and drags the oil, while storm surge is due to the wind stress and to the low atmospheric pressure in the storm center. Specification and prediction of surface wind forcing are therefore most important in oil spill movement and storm surge.

The development of ocean current and storm surge models are two independent studies, but wind is common to both. The range of wind speed is different for the two models. Storm surge is associated with strong winds, while for oil spill trajectory, the wind speed is for the more normal range including the strong winds of course.

In our research in NWS, we have to consider that results will have to be implemented for operational use. We have to design our forecast techniques so that the input data is readily available either from other forecast models or provided by the forecasters. We require, therefore, that input must be simple or minimally required. We also design that model output is also simple. This way the users can easily interpret the results.

WIND FORCING FOR OCEAN CURRENT MODEL

NWS routinely runs several multi-level atmospheric models on an operational basis. The LFM II is the most likely candidate for use in the oil spill movement model because of its finer grid size with a large area of coverage.

The predicted winds at the earth's surface, however, are based on simple formulations, preventing their direct use in the oil spill model. However, by resorting to models of the planetary boundary layer (PBL), we can get more accurate forecasts of surface winds and wind stresses. We do this by first calculating the surface geostrophic wind from load horizontal temperature and pressure gradients. The actual wind, which incorporates the effects of turbulent friction and the earth's rotation in the lower kilometer of the atmosphere, is assumed to match the geostrophic wind above the PBL. Three models of the PBL tested are similar to each other in that a logarithmic layer near the surface (where velocity increases semi-logarithmically but maintains constant direction) is assumed, because fluxes of momentum and heat are easier to approximate. By semi-logarithmically we mean proportional to the sum of (a) the log of the altitude plus (b) another function which accounts for variations due to atmospheric stability.

Above the log-layer, in the upper layer of the PBL, the assumptions vary. For example, the marine boundary layer model (Cardone, 1978) takes a constant eddy viscosity above the log-layer. There the velocity deviation from geostrophic is described by the Ekman spiral, which is then matched with winds in the lower layer. By contrast, the Rossby Number similarity model proposed by researchers at the University of Texas (Wagner and Fredricksen, 1979) uses empirical functions for conditions above the log-layer to give the problem closure. These functions of stability and horizontal temperature gradient are used to get the surface angle between, and ratio of, friction velocity and geostrophic wind. The surface Rossby number employs variable roughness height (the roughness follows a quadratic law on the friction velocity for all three models). Both these models assume steady-state conditions.

Finally, the one-dimensional dynamic model, uses an analytic expression for the eddy viscosity profile to compute dynamic upper-layer fluxes, and hence the velocity and temperature profiles, on a finite-difference grid. A statistical evaluation, using observed winds at the Argo Merchant, was necessary to compare each method.

We are fortunate to have wind data at the wreck site, from NOAA (1977), to test each of these PBL schemes. Observed pressure distributions, each 6 hours, allows surface geostrophic winds to be computed. Similarly, the observed surface temperature fields give the thermal winds. When neutral atmospheric conditions were assumed, each PBL model was run for the interval, 17 days, for which observations were available. The first two models were run as a series of steady state conditions, based on observations each 6 hours. The dynamic

IX-21

model ran as one continuous case, with a 1-hr timestep, driven by linear interpolations on the same observations. The statistics (Table 1) show that the Rossby number similarity scheme gives the best results. The empirical functions apparently represent the upper portion of the PBL better than the Ekman spiral does. The dynamic method suffers because of the large inertial oscillations introduced during the simulation.

WIND MODEL FOR STORM SURGE

In the storm surge model, we derive the wind in the following manner. We formulated a storm model. In the Ninth Proceedings of our UJNR Panel, Fujiwhara and Kurashige (1978) have presented distribution models of pressure and wind for stationary typhoons. We formulated a similar model. We assumed the pressure profile is similar. That is, in nondimensional plot, all the pressure distribution will fall in one curve.

We begin with the stationary symmetric storm. The pressure, wind speed, and wind direction for the stationary symmetric storm are not quite arbitrarily chosen like most storm surge models. The pressure and direction are determined from simplified equations of motion by balancing of forces. The equations are adapted from Myers and Malkin (1961):

$$\frac{1}{Pa}\frac{dp}{dr} = \frac{K_s V^2}{\sin \phi} - V\frac{dV}{dr} \tag{1}$$

$$\frac{1}{Pa}\frac{dp}{dr}\cos \phi = fV + \frac{V^2}{r}\cos \phi - V^2\frac{d}{dr}\sin \phi + K_n V^2. \tag{2}$$

Here Pa is the surface atmospheric density considered constant, $p(r)$ is the pressure, $\phi(r)$ the inflow angle or angle toward the storm center, and $V(r)$ is the wind speed; all are functions of r, the distance from the storm center. The terms K_s and K_n are empirically determined coefficients, representing stress coefficients in the directions opposite and to the right of the wind respectively. These stresses are given by the coefficient times the square of the wind speed, and f denotes the coriolis parameter.

Equations (1) and (2) are to be solved for $p(r)$ and $\phi(r)$, but the form of wind speed profile, $V(r)$, must be known first. We examined several wind profiles and has adapted the form

$$V(r) = V_r \frac{2Rr}{R^2 + r^2} \tag{3}$$

The term, R, the "radius of max winds," is the distance from the center at which $V(r)$ is greatest, and $V_r = V(R)$ is the value of that maximum wind speed. Of course, similar other profiles maybe chosen. Equation (3) has the property of increasing wind speed from $r = 0$ to a maximum at $r = R$, then decreasing back to zero at r very large.

.The. procedure for solving (1) and (2) for $p(r)$ and $\phi(r)$, with a given wind profile has been a Runge-Kutta type integration scheme applied to the transformation of (1) and (2). The transformation reduced problems of instability associated with a straightforward application of the Runga-Kutta method. This procedure was stable for the wind profiles given by (3) and for commonly encountered values of V_r, R, and latitude, but not for extreme values. That is, it is only marginally stable.

In our storm surge model, we add the vector $\vec{U}_{sm}$ to the stationary storm wind vector as a gross correction for storm motion $\vec{U}_s$. The form of the correction vector $\vec{U}_{sm}$ is

$$\vec{U}_{sm} = \frac{Rr}{R^2 + r^2} \tag{4}$$

where $\vec{U}_s$ the storm directed translation speed and the other symbols were previously defined. Only storms moving with uniform rectilinear motion are considered. Accordingly, the wind stress used in the numerical storm surge model includes the storm motion effects in a gross way.

We need specification and prediction of surface wind stresses in our storm surge model in great detail. This data is not available either in observations or forecasts. Our storm model approach provides for this information.

We investigated other storm models and tested the response of our storm surge model. We found the best agreement with the observed surge data with the storm model we adapted. Other storm models were developed for different applications. A model represents the purpose best for which it was developed. For example, the Standard Project Hurricane (SPH) model in the U.S. was developed for the use of the U.S. Army Corps of Engineers for the design of structures. SPH represents a wide range of averaged situations and very good for its purpose; but not for forecasting storm surge. We believe, our storm model for SPLASH is excellent for our purpose.

REFERENCES

[1] Cardone, V. J., "Specification and Prediction of the Vector Wind on the United States Continental Shelf for Application ot an Oil Slick Trajectory Forecast Program." Final Report T-35430, Institute of Marine and Atmospheric Sciences, The City College, City University of New York, 63 pp and Appendices.

[2] Fujiwhara, S. and Kurashige, K., "Distribution Models of Pressure and Wind over Stationary Typhoon Fields." Proceedings of the Ninth Joint UJNR Panel Conference on Wind and Seismic Effects," NBS Special Publication 523, pp. I-1 to I-10, 1978.

[3] Myers, V. A. and Malkin, W., "Some Properties of Hurricane Wind Fields as Deduced from Trajectories." National Hurricane Research Report No. 49, U.S. Weather Bureau, Washington, D.C., 45 pp, 1961.

[4] National Oceanic and Atmospheric Administration, "The Argo Merchant Oil Spill: A Preliminary Scientific Report." Grose, P. L. and Matson, J. S., eds., 133 pp, 1977.

[5] Wagner, N. K. and Fredricksen, E. J., "Similarity Theory Applied to Surface Winds and Oil Slick Transport." Preprints Workshop on the Physical Behavior of Oil in the Marine Environment, May 1979, Princeton, N.J. sponsored by Princeton University and National Weather Service.

Table 1. Comparison of errors (observed wind minus computed wind) from three planetary boundary models of the Argo Merchant winds.

Model	RMS Error		AVG Error	
	Speed $\left(\frac{M}{S}\right)$	Direction (°)	Speed $\left(\frac{M}{S}\right)$	Direction (°)
Marine B.L.	4.21	41.4	1.03	11.98
Rossby No. Similarity	3.81	37.6	-0.10	6.81
One-Dimensional Dynamic	5.09	53.5	2.15	-0.44

A DYNAMIC MODEL TO PREDICT STORM SURGES

AND OVERLAND FLOODING IN BAYS AND ESTUARIES

Chester P. Jelesnianski

Celso S. Barrientos

Jye Chen

Techniques Development Laboratory

Systems Development Office

National Weather Service

National Oceanic and Atmospheric Administration

ABSTRACT

In the United States, most of the damages caused by hurricanes are attributed to storm surges. NOAA provides forecast and warning services to prevent loss of lives and property. A dynamical model was developed to forecast storm surges in bays and estuaries. The development of the model and the operational application for forecasting will be described.

KEYWORDS: Storm surge; hurricane; overland flooding; forecast and warning.

INTRODUCTION

The National Weather Service (NWS), NOAA provides forecast and warning services for possible natural disasters. A hurricane can affect a broad area along the 3500 miles of coast along the Gulf and East coasts of the U.S. Most of the damages, which is considerabl caused by hurricanes are due to storm surges. Damages are caused by the pounding action of storm surge flood.

We define storm surge as the abnormal high water level due to meteorological forces. This is to differentiate with the high water due to tidal forces. Storm surge is due to th action of the wind and the low atmospheric pressure (inverted barometer effect). In hurricanes, the wind forcing and low pressure produce a rotating mound of water. When this mound of water reaches the coastal area, it will feel the bottom of the continential shelf and the storm surge is the result.

NWS has an operational model to predict outer coast hurricane storm surge since 1969. This model, SPLASH — Special Program to List Amplitudes of Surges from Hurricanes, has been used in several actual hurricanes and has been proved useful. SPLASH was also used to compute surges for historical hurricanes for flood insurance applications. Last year, during our Tenth Panel Conference, I discussed storm surge in general including SPLASH.

The main limitation in our SPLASH model is that we assumed vertical wall along the coast. Therefore, SPLASH is not able to forecast overland flooding or overtopping of barrier islands. The limitations leads us to the development of a new dynamic surge model, SLOSH — Sea, Lake, and Overland Surges from Hurricanes. SHOSH is capable of predicting storm surges and overland flooding in bays and estuaries.

THE SLOSH MODEL

The basic dynamics of SLOSH is the same with SPLASH. We used a vertically integrated equation of motion and compute surges generated by tropical storm via finite differencing. The model is solved numerically.

We call the input simulated geographic areas to the model as the basin. The geographical area of the basins vary considerably, but usually extends about 100 miles seaward from the coast, shoreward into high terrain, and with about 200 miles of coastline. For such a large area of the basin, there is no requirement to couple a bay model onto an open coast

IX-26

model. Input boundary values are not required, except for an initial and quiescent sea level before a storm affects a basin region, that is, the initial sea height unaffected by a storm or astronomical tide. The sea level datum used is "National Geodetic Vertical Datum (NGVD)." Published bathymetry on marine charts for seas and inland water bodies are corrected to NOVD. Land topography and vertical barriers such as roads, levees, spoil banks, dunes, etc., are simulated for each basin.

The model allows overtopping vertical barriers, inland inundation from sea or bay waters, and flow through channels. The astronomical tide, river discharge, and rain are not considered; under the assumption, their effects are small compared to a transient storm surge. The astronomical tide is superimposed on the surge after computations, assuming negligible interaction with the surge. The model allows for inland inundation; water can inundate across low-lying river basins.

The equations of motion of the model are identical to an earlier open coast surge model, SPLASH, developed by NWS. The one exception is the finite amplitude effect which can be important in shallow bay waters and especially so with inland inundation. A time-history bottom stress is used rather than a standard Chezy or Manning type formulation. No local calibration or tuning are effected to locally set values for undetermined coefficients in the equations of motion; instead, best fit type constants for the coefficients were determined in a universal sense from a multitide of basins, historical storms, and surge observations using the coastal surge model SPLASH.

The storm model is the same as used by SPLASH. That is, a simple balance of surface forces to compute wind and inflow angle across isobars; the inflow angle is dynamically computed and is not empirically derived. One exception in the SLOSH model is higher surface friction across inland bodies of water, but not the sea. The friction is not a function of a particular bay, and is applied the same way for all bays.

We use very simple input meteorological data. This consists of two storm parameters, central pressure of the storm and storm size (distance from storm center to maximum winds), as well as storm track; the two storm parameters, and speed of storm along track, can be variant with time. Maximum wind is not an input parameter in the storm model. A useful property of the storm model allows high winds but small inflow angles across isobars for weak friction, and otherwise for strong friction. The feedback between winds and inflow angles in the storm model, with alternate friction values is a compensating effect for surge generation. Therefore, high accuracy for wind speed is not a requirement, however, central pressure, storm size, speed along track, and track orientation relative to a geographical area are important parameters and these must be accurate.

Coarse and find grids are used in the model for numerical computations. If a bay is large and uncomplicated, then a large, invariant, grid size (four miles grid spacing) is used. If a bay is small or complicated, then an expanding grid scheme (1/2 miles in bay regions to 4 miles in deep sea waters) is used. The expanding grid is a polar coordinate system, transform conformally onto a Clark Ellipsoid.

The SLOSH model, particularly the storm model, has been developed with the data for Lake Okeechobee, Florida during a hurricane in 1949. The model has been tested with real surge data for several historical storms affecting Lake Okeechobee and Tampa Bay, Florida and Lake Pontchartrain, Louisiana. Most of the results appear realistic, that is, usually within 20% of observed surge values. SLOSH was adapted and is used for real time forecasting of surges for the above three areas. We are now in the development stage for a total of 20 Bays and estuaries.

TOPICS ON TSUNAMI PROTECTION ALONG THE PORT AREAS IN JAPAN

Yoshimi Goda

Marine Hydrodynamics Division

Port and Harbour Research Institute

ABSTRACT

The major Japanese tsunamis since the 15th century and a general description of counter-measures against them are briefly introduced. Particular reference is made on tsunami break-waters used to protect port areas. The performance of the Ofunato tsunami breakwater in the instance of the Tokachi-oki Earthquake Tsunami in 1968 is discussed, and a new plan for a Kamaishi tsunami breakwater is introduced. Also mentioned are several tsunami prediction plans in progress.

KEYWORDS: Disaster planning; seawall protection; tsunami breakwaters; tsunami prediction.

1. MAJOR TSUNAMIS IN JAPAN AND COUNTERMEASURES

Tsunami records in Japan date back to the seventh century. Suffering from tsunamis has increased with the passage of the centuries because the population greatly increased. Table 1 lists the major tsunamis since the 15th century. Names of areas are indicated in Figure 1. Tsunami damage has been severe along the coasts of the Sanriku Area (from Aomori to Miyagi Prefectures), the Tokaido Area (from Chiba to Mihe Prefectures), and the Nankaido Area (from Mihe to Kochi Prefectures). Other areas including the Japanese Sea coast have also suffered injury from tsunamis though to a lesser extent.

Countermeasures against tsunami attacks adopted in Japan are:

1) The establishment of a tsunami-alert system and evacuation drills,

2) construction of refuge roads,

3) relocation of houses to neighboring highlands,

4) planting counter-tsunami groves and forests,

5) construction of frontage sheds and other facilities in port sections as solid, permanent structures to shelter the resident areas behind them,

6) construction of seawalls, and

7) construction of tsunami breakwaters.

The popularity of television and portable radios nowadays greatly assist spreading the warnings of possible tsunamis which are issued by the Japanese Meteorological Agency. Relocation of houses to highlands is difficult due to the scarcity of housing lots. Because people continue to live along shoreside areas even with the knowledge of tsunami danger, seawalls against tsunamis have been built for centuries and continue to be constructed at many places. Some seawalls have a crest height of 16 meters above the mean sea level. Seawalls require movable gates at their junctions with roads. Proper maintenance of these gates is not easy.

2. TSUNAMI BREAKWATERS

Construction of seawalls along the port area waterline are a hindrance and cause inconvenience to daily port operation. Reduction of tsunami height by means of offshore breakwaters provide a solution for the protection of some port areas against tsunamis. The ports of Ofunato, Onagawa, and a few others have been provided with such breakwaters, and the

Kamaishi Port will soon have one. These breakwaters restrict the entrance of tsunami waves into harbours by narrowing the entrances. The possibility of harbour resonance does not materialize because of the energy loss at the entrance by wakes and eddies.

The Ofunato tsunami breakwater shown in Figure 2 was built in the period from 1963 to 1967 after the suffering of the lives of 53 people and 8 billion yen by the Chilean Earthquake Tsunami in 1960. The City of Ofunato has the population of about forty thousand and a port cargo of 5 million tons annually. The breakwater site is 38 meters deep (maximum) and 738 meters wide. The section was reduced to the width of 200 meters and the depth of 16.3 meters below the low water level to permit the navigation of 100,000 DWT tankers. The effect of this breakwater against monocromatic tsunami of sinusoidal wave form (amplitude of 0.5 meter) is exhibited in Figure 3 in the form of the resonance curve of Ofunato Bay with and without the breakwater. The analysis was done by Ito [1, 2] with the introduction of the head loss at the entrance, which was taken as 1.5 times the velocity head.

The Tokachi-oki Earthquake Tsunami of May 16, 1968, provided an opportunity to demonstrate the effectiveness of this breakwater. The observed time histories of water level variation due to the 1968 Tsunami are shown in Figure 4. These tsunami profiles were successfully reproduced in numerical computation by providing 30 harmonic components of the tsunami waves recorded outside the breakwater and adjusted for the respective amplitude amplification there. Computation was extended to the case when the breakwater had not been constructed. Figure 5 is the predicted time history of water level variation at the innermost part of Ofunato Bay. It is clear that the breakwater reduced the tsunami height by one half. Figure 6 is the comparison of highest water level elevation with and without the breakwater at the arrival of the first crest of the 1968 Tsunami.

A new tsunami breakwater is planned at Kamaishi Port as shown in Figure 7. The City of Kamaishi has a population of about seventy thousand, and the port handles 4.6 million tons of cargo annually. The breakwater site is 60 meters deep (maximum) and about 2100 meters wide. The breakwater will have a central opening 300 meters wide and 19 meters deep below the low water level as well as two side openings each 50 meters wide. The layout of the breakwater is the best among several alternatives for tsunami reduction capability and minimization of eddy currents around entrances.

The breakwater is expected to reduce tsunami heights as listed in Table 2 according to numerical analysis and hydraulic model tests, both of which were commissioned by the Second

District Port Construction Bureau, Ministry of Transport [3]; the testing was carried out by Professor Iwasaki of Tohoku University. The construction work will begin in 1980 and be completed in ten years with an estimated cost of 53 billion yen.

3. TSUNAMI PREDICTION SCHEMES

Sanriku and Tokaido are the areas most liable to tsunami damage. To adequately plan for tsunami protection at the ports in these areas, a tsunami prediction program has been undertaken by the Ministry of Transport with the cooperation of several professors and researchers. The program consists of:

1) An assumption of the spatial distribution of sea bottom dislocation in the source area,

2) the computation of tsunami propagation from the source to the coastline,

3) the computation of tsunami runup on land topography, and

4) the evaluation of tsunami protection work against predicted tsunamis.

The project of the Kamaishi tsunami breakwater is a result of the tsunami prediction scheme along the Sanriku coast. Another scheme is being carried out for the coast of the Tokaido area against a possible Tokai Earthquake Tsunami in the near future.

4. CONCLUDING REMARKS

The protection of coastal areas from powerful tsunamis is not an easy task, because tsunami runup is quite high and the protection work is quite costly; especially when the frequency of extreme tsunamis is taken into consideration. Nevertheless, the severity of tsunami damage answers any doubt about the necessity for tsunami protection. Port areas with dense populations and busy daily operation are one of the most difficult places to provide effective tsunami protection. Though the tsunami breakwater is a good solution, we need to explore other solutions. Numerical analysis, as well as hydraulic model tests, will further help us find such solutions for tsunami protection.

REFERENCES

[1] Ito, Y., "On the Effect of Tsunami-Breakwater," Coastal Engineering in Japan, Vol. 13, Japan Society of Civil Engineers, 1970, pp. 89-102.

[2] Ito, Y., "Head Loss at Tsunami-Breakwater Opening," Proceedings of the 12th Conference on Coastal Engineering, Washington, D.C., 1970.

[3] Yokohama Investigation and Design Office, Second District Port Construction Bureau, Ministry of Transport, "On Tsunami Protection Project in Kamaishi Bay," Reference No. 52-13, 1978 (in Japanese).

Table 1 Major Tsunamis in Japan

Name	Date	Earthquake Magnitude	Area Most Suffered	Max. Runup Height	Nos. of Persons Killed or Missing	Remarks
Meio Tsunami	1498- 9-20	M = 8.6	Tokaido	over 10 m	several tens of thousands	Building of Great Buddha in Kamakura was washed away.
Keich 9th Tsunami	1605- 1-31	M = 7.9	Nankaido Tokaido	over 10 m	over 5,000	
Keicho 16th Tsunami	1611-12- 2	M = 8.1	Sanriku Hokkaido	over 30 m	over 5,000	
Hoei Tsunami	1707-10-28	M = 8.4	Nankaido	over 25 m	over 10,000	
Meiwa Tsunami	1771- 4-24	M = 7.4	Okinawa	85 m	12,000	Miyako island was washed over.
Tempo Tsunami	1843- 4-25	M = 8.4	Hokkaido	no record	small	Area was little inhabited.
Ansei Tsunami	1854-12-23 1854-12-24	M = 8.4 M = 8.4	Tokaido Nankaido	over 20 m	about 5,000	Russian warship "Diana" was damaged at Shimoda Port.
Meiji 29th Tsunami	1896- 6-15	M = 7.9	Sanriku	30 m	27,122	
Showa 8th Tsunami	1933- 3- 3	M = 8.1	Sanriku	29 m	3,644	
Chilean Tsunami	1960- 5-24	M = 8.75	All the Pacific coast of Japan	8 m	123	Tsunami period was some 70 minutes.

Table 2 Tsunami Height Reduction by Kamaishi Breakwater
- ratio of tsunami heights with and without breakwater -

Tsunami period (minutes)	10	15	16	20	30	45
Ratio by computation	0.61	-	0.37	0.39	-	0.97
Ratio by model test	0.69	0.42	0.32	0.21	0.39	0.33

Fig. 1 Location Map of Japan

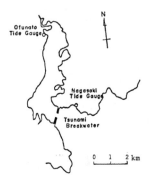

Fig. 2 Map of Ofunato Bay

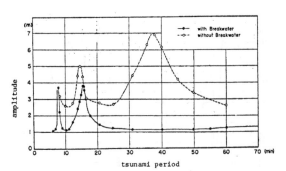

Fig. 3 Amplitude of tsunami waves at Ofunato Harbour
for sinusoidal waves with the amplitude of 0.5 m
(after Ito[1],[2])

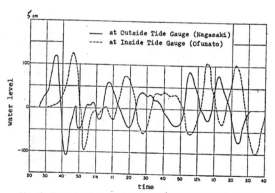

Fig. 4 Records of water level variation due to the 1968 Tsunami

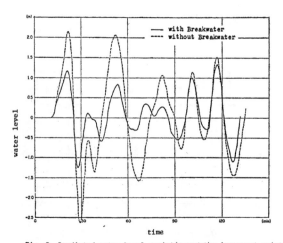

Fig. 5 Predicted water level variations at the innermost point
of Ofunato Bay with and without the breakwater
(after Ito[1],[2])

Fig. 6 Distribution of water level in centimeters at the first peak of the 1968 Tsunami (after Ito[1],[2])

Fig. 7 Map of Kamaishi Bay

SEISMIC RISK MAPS (MAXIMUM ACCELERATION AND MAXIMUM PARTICLE VELOCITY)

IN THE SOUTHEAST ASIAN COUNTRIES OF THE PHILIPPINES, INDONESIA, AND INDO-CHINA

Sadaiku Hattori

International Institute of Seismology and Earthquake Engineering

Building Research Institute

ABSTRACT

Seismic risk maps in the Southeast Asian Countries of the Philippines, Indonesia, and Indo-China were made using existing seismic data, attenuation models, and the method of extreme value fitting. The maps consist of the following two kinds: (i) The maximum particle velocity (kine) on the base rock, and (ii) The maximum acceleration (gal) on the ground. The return periods of these maps are 50, 100 and 200 years, respectively.

This paper is a modification of the report which the author is going to present at the International Conference on Engineering for Protection from Natural Hazard, which was held in the Conference Center of the Asian Institute of Technology in Thailand January 7-10, 1980.

KEYWORDS: Disaster prediction; earthquake motions; hypocenters; seismic risk maps; seismic zoning.

1. INTRODUCTION

Needless to say, accurate earthquake prediction techniques are extremely valuable. However, from the standpoint of preventing earthquake disaster, it is not enough only to predict
hypocenters, magnitudes, and occurrence times of future earthquakes. It is also important to
estimate earthquake motions for earthquakes which may occur in the future and to forecast the
disasters they might cause. Furthermore, it is necessary to consider counterplans to prevent
them. Ideally speaking, the occurrence time of earthquakes (tn), earthquake motions on the
base rock and on the ground ($f(tsn)_{BP}$, $f(tsn)_{SP}$), and the ground characteristics (Gp) should
be exactly estimated (Figure 1).

However, it is impossible at present to estimate all of the above related terms with
great accuracy. Much research has been conducted in approximating the above ideal one. The
most typical approach in this research is to define the expectations of the maximum earthquake motions for a number of return periods by using (i) the seismic data of the past, (ii)
attenuation models, and (iii) a statistical method. There are the following two problems
which must be heeded in using this approach: (A) The return periods are used, instead of
denoting directly the occurrence time of earthquakes, and only the maximum earthquake motions
are given in place of defining concretely the earthquake motions. (B) When seismic data are
not reliable or the data period is too short, it is meaningless to define the expectations
because of the large errors which will be caused.

Much research of this kind has been performed for the vicinity of Japan. [1] This
research is concerned with only the regional characteristics of the seismicity, and no consideration is given to the ground characteristics or to the temporal variations of the seismicity.

Hattori [2] showed approximately the ground characteristics of the long period range
(2-5 sec), in the whole vicinity of Japan, using the seismic data of the maximum amplitudes
and the period observed by strong motion seismographs at JMA stations. He provided a definition of base rock and proposed a new seismic risk map. The map is based on the maximum
earthquake motions which are estimated from the spatial characteristics of the seismicity,
and from the ground characteristics [3]. Further, Hattori [4] defined the temporal variations of the seismicity of the whole vicinity of Japan using the accumulative energy curve

and refined the seismic risk map. This is a solution for the above described problem (A), which means that a term indicating occurrence times of earthquakes was added to the seismic risk.

There are some examples of seismic risk analysis for a few regions in the world [5, 6, 7, 8, 9]. They, also, concerned the maximum earthquake motions based on only the spatial characteristics of the seismicity. It goes without saying that many have recognized the necessity for studies of the ground characteristics and the temporal characteristics of the seismicity, but there seem to be none which have actually adopted such an approach.

The analyses of Oliveira [9, 10] are as follows: A source area of earthquake, and a point where the maximum earthquake motions are to be estimated were supposed to be S and P. In general, there are plural source areas, and P is regarded as not one point but an area. Furthermore, Oliveira supposed that occurrences of earthquake in the source areas depend on the Poisson process and Gutenberg-Richter's law. The maximum earthquake motion at an arbitrary point for an earthquake was calculated by an attenuation model. The expectations of the maximum earthquake motion for some return periods were estimated by the method of extreme value fitting.

As mentioned above, seismic risk maps have been made for some regions in the world. There is even a case where the ground characteristics, the temporal variations of the seismicity, and the modifications of the seismic data were taken into consideration. [4] The first step in this kind of research, however, is to make a map based on the expectations of the maximum earthquake motions for some return periods being based on the spatial characteristics of the seismicity.

The working group E2, Ground Motion and Seismic Risk, the U.S.-Southeast Asia Symposium on Engineering for Natural Hazards Protection recommended cooperation in six research areas among the southeast Asian countries, Japan, and the United States. One of the six research areas was the development of a Seismic Zoning Map for the Philippines, Indonesia, and other Southeast Asian countries. This Seismic Zoning Map is a seismic risk map which is based only on the spatial characteristics of the seismicity.

2. SEISMIC DATA

The following data are available for areas concerning: world seismic data (magnetic tape) which were compiled by the National Oceanic and Atmospheric Administration Environmental Data Service (NOAA). Figure 2 shows the temporal variations of seismic data in Philippines area.

The seismic data of the NOAA magnetic tape include four kinds of magnitude, namely, (i) the body wave magnitude M_b, (ii) the surface wave magnitude M_s, (iii) the magnitude M, which was determined by each observational organization, and (iv) what is termed the local magnitude M_L. In order to estimate maximum earthquake motions, it is necessary to use a unified magnitude. So, using all the data in NOAA magnetic tape, the relations among these magnitudes were researched (Figure 3). The results obtained are as follows: The correlation coefficients are about 0.7 or more.

$$M_s = 1.081(\pm 0.014)M_b - 0.390(\pm 0.077); \; \rho = 0.699 \tag{1}$$

$$M = 1.161(\pm 0.019)M_b - 0.662(\pm 0.104); \; \rho = 0.723 \tag{2}$$

$$M_L = 0.862(\pm 0.017)M_b + 0.565(\pm 0.079); \; \rho = 0.739 \tag{3}$$

$$M = 0.853(\pm 0.015)M_s + 0.929(\pm 0.090); \; \rho = 0.847 \tag{4}$$

$$M_L = 0.679(\pm 0.041)M_s + 1.826(\pm 0.216); \; \rho = 0.751 \tag{5}$$

$$M_L = 0.860(\pm 0.024)M + 0.677(\pm 0.123); \; \rho = 0.913 \tag{6}$$

In this research, the magnitudes M_b, M, and M_L were transformed into M_s by Equations 1, 4, and 5, respectively.

The longer the data period, the better the data for statistical analysis. However, the further back we go in time, the sparser the data becomes, especially for data with magnitudes and epicenters defined. For Japan and China, areas in which the author had already made the seismic risk maps [1, 11], we could obtain the seismic data for fairly long periods. For other Asian countries, however, only the NOAA magnetic tape is available due to a lack of historical seismic data.

Figure 4 shows the epicentral distribution in the Indonesia area. It is founded on the data with the above-mentioned unification of magnitude.

X-4

3. ATTENUATION MODELS

Some of the attenuation models which have been proposed are shown in Table 1.

Oliveira [10] gave Equations 1, 2, and 3 (y maximum earthquake motion; M(m) magnitude; R hypocentral distance) as attenuation models on firm ground. Further, he suggested another one (Equation 4), which gave maximum particle velocity on firm ground in California. McGuire [12] proposed equations similar to those of Oliveira (Equations 5, 6 and 7). Esteva and Rosenblueth proposed Equation 8 for maximum acceleration on the ground [10].

Kanai [13] recognized that velocity spectra of earthquake motions on base rock are flat in a certain period range, from the results of observations under the ground (depth 300 m) of the Hitachi Mine. Further, he suggested (Equation 10) for the maximum particle velocity on the base rock, using Tsuboi's equation which defines magnitudes.

Figures 5 (a, b) are shown to compare the attenuation models referred to above. In addition to these, many attenuation models have been proposed and recently, some models based on the dislocation theory of the earthquake fault have also been suggested.

The author [1] made seismic risk maps in the vicinity of Japan using Kanai's equation (Equation 10 in Figure 5 (b)) so seismic risk maps for many areas may be compared directly.

It is reasonable that maximum earthquake motions on the ground are to be obtained by considering the ground characteristics of each site for the input on the base rock. However, the risk maps expressed by the maximum acceleration on the ground and disregard ground characteristics, are also useful in earthquake engineering. For this purpose, the attenuation model (A) in Figure 5 (a) was used. This is the mean of the attenuation models by Oliveira (Equation 1) and McGuire (Equation No. 5). It is very close to the one given by Katayama (Equation 13). It can be seen from the figure that attenuation model (A) shows nearly the same mean value as many other ones.

4. ESTIMATION OF MAXIMUM EARTHQUAKE MOTIONS

Using the probability that the variable x is equal to or larger than any of X_1, X_2, $\cdots$ X_n,

$$\Phi_\eta(x) = P(x_1 \leq x, \ x_2 \leq x, \ \cdots\cdots\cdots \ x_n \leq x), \tag{7}$$

the return period T(X) and the reduced variable z are defined by the following.

$$T(x) = 1/(1 - \phi_\eta(x)) \tag{8}$$

$$z = -\ell_\eta\{-\ell_\eta\phi_\eta(x)\} \tag{9}$$

If the maximum earthquake motion at a certain point in the m-th unit period (for example, one year) is assumed to be $A_{max,m}$ and the variable X_m is defined by the equation:

$$X_m = \log A_{max,m} \tag{10}$$

the distribution of x may be approximated by Gumbel's third asymptotic distribution. The third distribution is denoted as follows:

$$\phi_n^{(3)}(x) = \exp\left\{-\left(\frac{W - x}{W - V}\right)^K\right\} \tag{11}$$

where the parameters (W, K, V) are the upper limit of maximum value, the shape parameter, and the characteristic maximum value. The number of the unit period is put to be n. Furthermore, the reduced variable for the observation, and the theoretical one of the third asymptotic distribution, are put to be $y_m = m/(n + 1)$ and Z_m, respectively. The third asymptotic distribution most fit for n values of x may be obtained by determining the values of W, V and K so as to minimize

$$\sigma = \sum_{m=1}^{n} (z_m - y_m)^2$$

5. SEISMIC RISK MAPS IN THE PHILIPPINES, INDONESIA, AND INDO-CHINA

For statistical analysis, the longer the period of data the better, and the larger the quantity of the data the better. However, this holds true only under the following conditions: (a) The accuracy of all the data should be on the same level. (b) The data should be obtained homogeneously from all the regions concerned. As stated before, however, the older the data becomes, the more unsatisfactory it is, not only in accuracy and in quanity, but also in the regional homogeneity.

The unit period of one year was taken and as the yearly maximum value $A_{max,m}$ for the variable $X_m = \log A_{max,m}$, the biggest one was taken among maximum earthquake motions which were calculated for each earthquake occurred during the m-th one year.

X-6

Figure 6 shows an example of analyses for the vicinity of Manila (15.0°N, 121.0°E). It is read from Figure 6 that the maximum particle velocities on the base rock for the return periods of 50, 100 and 200 years are 4.05, 6.38, and 9.61 kine and the maximum accelerations on the ground are 59.7, 98.9, and 120.7 gal, respectively.

Analyses were made at every degree in the latitude and longitude for each area. Figures 7 through 9 show the regional distributions of the maximum particle velocity (kine) on the base rock and the maximum acceleration (gal) on the ground for Philippines (Return period Tr ; 100 year), Indonesia (Tr ; 100 year), and Indo-China (Tr ; 100 year), respectively.

The above may be regarded as a standard of the regional characteristics of the seismic risk in the southeast Asian countries and the vicinity.

Figures 10, 11, and 12 show the simplified seismic zoning maps for areas concerned. These maps should be modified by the countries concerned in the future according to specific purposes, because these are only tentative ones.

6. CONCLUSIONS

It is very desirable that destructive earthquake motions, which will be experienced at an arbitrary point in the future, be exactly estimated. However, it is impossible at present to do it with great accuracy. Nevertheless, seismic design for buildings or civil engineering structures which will remain for decades and centuries must now be made based on the knowledge which currently exists. As a substitute for the above ideal situation, the author proposes the use of the seismic risk map made using the following approach for the whole of Japan [10]: (i) The seismic inputs on the base rock were estimated from the spatial characteristics of seismicity in the past (S(S)); (ii) The ground characteristics were presumed by utilizing strong motion seismograms (M(S)); (iii) The temporal variations of the seismicity were estimated from the accumulative energy curve (S(T)); (iv) The more reasonable seismic risk map was expressed by combining the above S(S), M(S) and S(T).

The above first step (S(S)) is the most basic one to produce a more reliable seismic risk map. This research approach was used to produce the above mentioned first step (S(S)) type seismic risk map for the Asian countries.

The primary seismic data were those of the NOAA magnetic tape. The relations among magnitudes using different definitions were studied and a unified magnitude was used in this research.

X-7

Many attenuation models were used: (i) Kanai's attenuation model which gives the maximum particle velocity on the base rock (iii). Oliveira-McGuire's attenuation model which gives the maximum acceleration on the ground. The former one had been used by the author for the seismic risk map in the vicinity of Japan. Therefore, this one was used because when it is used, we can compare directly the present seismic risk maps in the Southeast Asian countries with the ones of Japan. From another viewpoint, the maximum acceleration on the ground is convenient for the practical use of earthquake engineering. The latter attenuation model, which was made by averaging attenuation models by Oliveira [10] and McGuire [11], is comparatively close to the other ones and it seems to give usable values. This is the reason for adopting the latter one.

Adding to the seismic data and attenuation models, the method of extreme value fitting (Gumbel's third asymptotic distribution) was used to extract the maximum earthquake motions.

The final results of this research are the seismic risk maps for many Asian countries, which are shown in Figures 7 through 12.

ACKNOWLEDGMENTS

In the course of this research, Mr. Hiroshi Tanaka of International Institute of Seismology and Earthquake Engineering (IISEE) helped the author in calculating and making diagrams. The author would like to express his thanks to him.

REFERENCES

[1] Hattori, S., Regional Distribution of Presumable Maximum Earthquake Motions at the Base Rock in the Whole Vicinity of Japan, Bulletin IISEE Vol. 14, 1976, pp. 47-86.

[2] Hattori, S., Regional Peculiarities on the Maximum Amplitudes of Earthquake Motion in Japan, Bulletin IISEE, Vol. 15, 1977, pp. 1-21.

[3] Hattori, S., A Proposal of the Earthquake Danger Man Based on Seismicity and Ground Characteristics, Bulletin IISEE, Vol. 15, 1977, pp. 23-32.

[4] Hattori, S., Temporal Variations of Seismicity and Seismic Risk in and around Japan, bulletin IISEE, Vol. 16, 1978, pp. 105-118.

[5] Borges, J. F., and Ravara, A., "Recent Portuguese Research on Earthquake Engineering," III European Conference on Earthquake Engineering, Sofia, 1970.

[6] Lomnitz, C., "Earthquake Risk Map of Chile," 4th World Conference on Earthquake Engineering, 1969.

[7] Esteva, L., Seismicity (Seismic Risk and Engineering Decisions; Developments in Geotechnical Engineering; 15) Elsevier Scientific Publishing Company, 1976, pp. 179-224.

[8]   Vit Karnik and Algermissen, S. T., Seismic Zoning (Seismic Risk Evaluation of Balkan Region), Report for UNDP/Unesco Survey of the Seismicity of the Balkan Region, Denver, U.S. Geological Survey.

[9]   Oliveira, C., Seismic Risk Analysis, EERC 74-1, Earthquake Engineering Research Center, University of California, Berkeley, 1974, pp. 1-102.

[10] Oliveira, C., Seismic Risk Analysis for a Site and a Metropolitan Area, EERC 75-3, Earthquake Engineering Research Center, University of California, Berkeley, 1975, pp. 1-182.

[11] Hattori, S., Seismic Risk Maps in the World (Maximum Acceleration and Maximum Particle Velocity) (I) - China and its Vicinity, Bulletin IISEE, Vol. 16, 1978, pp. 119-150.

[12] McGuire, R. K., Seismic Structural Response Risk Analysis Incorporating Peak Response Regressions on Earthquake Magnitude and Distance, Massachusetts Institute of Technology Department of Civil Engineering, R74-51, 1974.

[13] Kanai, K. and Suzuki, I., Expectancy of the Maximum Velocity Amplitude of Earthquake Motions at Bed Rock, Bulletin Earthquake Research Institute Vol. 46, 1968, pp. 663-666.

Table 1. Attenuation models which have
been proposed until the present.
y: Maximum earthquake motion, M(m);
Magnitude, R,r(X,x): Hypocentral
(epicentral) distance and I: Intensity.

NO	Researcher	Formula
1.	Oliveira, C. (Acc.; gal)	$y = b_1 e^{b_2 m} (R+25)^{-b_3}$ $b_1 = 1230$ $b_2 = 0.8$ $b_3 = 2$
2	Oliveira, C. (Vel.; kine)	$y = b_1 e^{b_2 m} (R+25)^{-b_3}$ $b_1 = 16$ $b_2 = 1$ $b_3 = 1.7$
3	Oliveira, C. (Dis.; cm)	$y = b_1 e^{b_2 m} (R+25)^{-b_3}$ $b_1 = 0.128$ $b_2 = 1.2$ $b_3 = 1$
4	Oliveira, C. (Vel.; kine)	$v = 15e^m (R + 0.17e^{0.59})^{-1.7}$
5	McGuire, R.K. (Acc.; gal)	$y = b_1 10^{b_2 M} (R+25)^{-b_3}$ $b_1 = 472.3$ $b_2 = 0.278$ $b_3 = 1.301$
6	McGuire, R.K. (Vel.; kine)	$y = b_1 10^{b_2 M} (R+25)^{-b_3}$ $b_1 = 5.64$ $b_2 = 0.401$ $b_3 = 1.202$
7	McGuire, R.K. (Dis.; cm)	$y = b_1 10^{b_2 M} (R+25)^{-b_3}$ $b_1 = 0.393$ $b_2 = 0.434$ $b_3 = 0.885$
8	Esteva,L. Rosenblueth,E. (Acc.; gal)	$a_D = 110 \, e^{0.8M} R^{-1.6}$
9	Kawasumi, H. (Acc.; gal)	$a = 0.253 \cdot 10^{\frac{1}{2}I}$ (gal) $I = Mk + I'$ $I' = \begin{cases} 2 \ln(\frac{100}{\Delta}) - 0.00183 (\Delta - 100) & \Delta \geq 100 \\ 2 \log(\frac{r}{r_0}) - \frac{2l}{\ln 10}(r - r_0) & \Delta < 100 \end{cases}$ $L = 0.0192/\text{km}, \quad r_0 = 100^2 + 18^2$
10	Kanai, K. (Vel.; kine)	$\log V(\text{kine}) = 0.61M - (1.66 + \frac{1.60}{X})\log X - (0.631 + 1.83/X)$

NO	Researcher	Formula
11	Muramatsu,I. (Vel.; kine)	$V (\text{kine}) = C(M)r^{-1} e^{-r/r_c(M)}$ $\log C(M) = 0.655M - 1.719$ $\log r_c(M) = 0.145M + 1.353$
12	Katayama, T. (Acc.; gal)	$\log A = 0.982 - 1.290 \log\Delta + 0.466M$
13	Katayama, T. (Acc.; gal)	$\log A = 2.308 - 1.637 \log(R+30) + 0.411M$
14	Watanabe, H. (Acc.; gal)	$\log A = M - 2.31 \log r - 1.38 \qquad r < 40\text{km}$
15	Watanabe, H. (Vel.; kine)	$\log A_v = 0.85M - 1.73\log r - 2.50 \quad r \leq 200\text{km}$ $\log A_v = 0.85M - 1.73\log r + 0.0015(r-200) - 2.50 \quad r > 200\text{km}$
16	Watabe, M. (Vel.; kine)	$V (\text{kine}) = 10^{0.607M - 1.191\log X - 1.4}$
17	Watabe, M. (Acc.; gal)	$A (\text{gal}) = 10^{0.472M - (1.97 - 1.8/X) \log X + (2.2 - 11.1/X)}$

Table 1. Attenuation models which have
been proposed until the present.
y: Maximum earthquake motion, M(m);
Magnitude, R,r(X,x): Hypocentral
(epicentral) distance and I: Intensity.

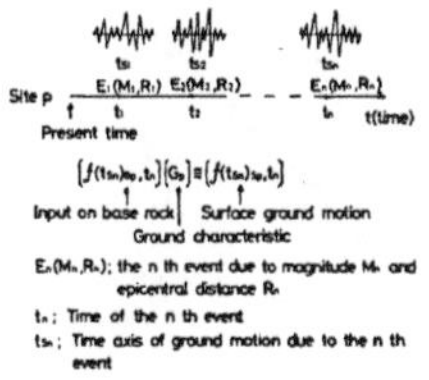

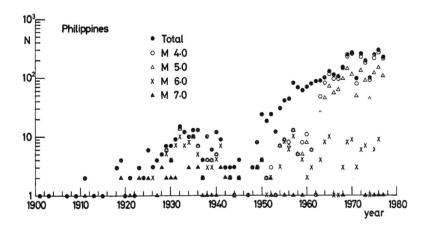

Fig. 1. Ideal way of estimation of the earthquake motions which will be experienced in the future.

Fig.2 Temporal variations of seismic data in Philippines area.

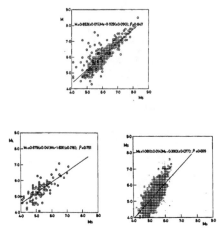

Fig. 3. Examples of relationships among
several kinds of magnitude.

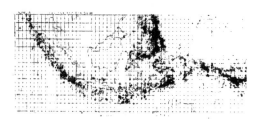

Fig. 4. Epicentral distributions of earthquakes in and
around Indonesia.

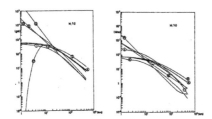

Fig. 5. Comparisons of various attenuation
models(M=7.0). (a) Maximum acceleration,
(b) Maximum particle velocity.

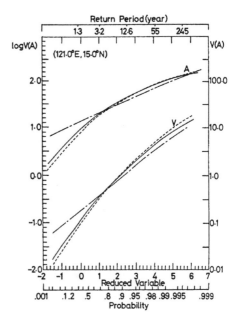

Fig. 6. An example of analysis for one point near Manila.

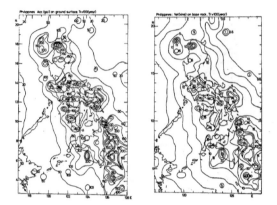

Fig. 7. Seismic risk maps (Maximum acceleration and
Maximum particle velocity) in and around Philippines.

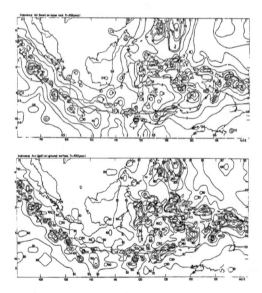

Fig. 8. Seismic risk maps (Maximum acceleration and
Maximum particle velocity) in and around Indonesia.

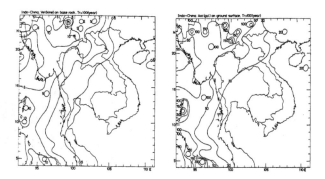

Fig. 9. Seismic risk maps (Maximum acceleration and
Maximum particle velocity) in and around Burma and
Thailand.

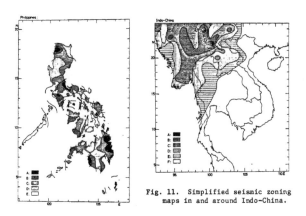

Fig. 11. Simplified seismic zoning
maps in and around Indo-China.

Fig. 10. Simplified seismic
zoning maps in and around
Philippines.

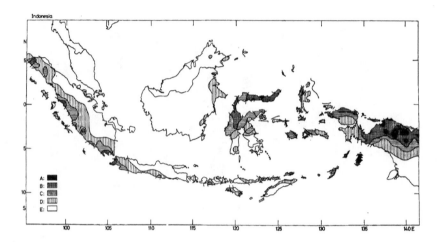

Fig. 12. Simplified seismic zoning maps in and around Indónesia.

CALIFORNIA SCHOOL AND HOSPITAL CEILINGS

John F. Meehan

Structural Safety Section

Office of the State Architect

State of California

Sacramento, California

ABSTRACT

A description of the methods of anchoring plaster, panel applied and T-bar ceilings against seismic forces is presented which is acceptable to the Structural Safety Section of the Office of the State Architect for public school and hospital buildings in California.

KEYWORDS: Seismic forces; anchorage; plaster ceilings; applied ceilings; T-bar ceilings.

INTRODUCTION

Traditionally the structural design of ceilings has been given only nominal attention by designers. Since ceilings are primarily an appearance item, they have received much more architectural input than structural. During quiescent periods ceilings perform quite adequately. However, during seismic disturbances, if proper precautions are not taken, they can become a serious life hazard because they are immediately overhead and their damage can add greatly to property loss.

Furred ceilings are constructed to conceal the underside of the overhead structure as well as the myriad of unsightly service ducts for heating and ventilating and the conduits for energy, water, sewage, gases, communications, etc. which are placed below the structure. Ceilings are usually constructed of applied cementatious materials or panels of various types of boards either applied directly to the underside of the ceiling framing or hung in light-weight metal frames. Each type of ceiling has certain deficiencies when considering lateral forces from seismic events.

PLASTER CEILINGS

The cementatious types of ceilings are those which are made from any one of the many types of plasters applied by hand or mechanical means to lath. The lath may be metal, gypsum board of even wood strips. The soft plaster bonds or is keyed to the lath for its support. The lath, in turn, must be properly anchored to its framing, and it follows that the framing must be properly anchored to the structure for its support.

Normally, metal lath is nailed or screwed to the underside of its framing to obtain its support. However, the fasteners in this case are acting solely in withdrawal. Due to deterioration withdrawal anchorage offers no second line of defense and is not dependable over long time periods and in particular during earthquakes. A second line of defense can be developed with the use of well anchored metal lath and stripping to the framing. This construction is required in California school buildings constructed pursuant to the Field Act and is provided in Appendix A which is taken from Title 21 California Administrative Code.

Briefly, these regulations require wood stripping to be fastened to the bottom of each wood joist with one slant and one straight common nail, and, in addition, the stripping must be wire tied to nails placed in the sides of the joists, or the stripping may be fastened to

panel falls out of place it does not continue to pull the adjacent panel out. On the other hand, because of the continuous metal lath, plaster ceilings will continue to peel from the supports. Gypsum board panel ceilings have in general, performed well during earthquakes. The only gypsum board ceiling failure known to this author is shown in Figure 5 which is in the corridor of a high school in Anchorage, Alaska following the 1964 earthquake. It is believed that several loosened panels were removed by hand following the earthquake for the convenience of the occupants.

T-Bar Ceilings

T-bar ceilings consist of panels of lightweight panels, usually 2' x 4' (.6 x 1.2 m) supported on all four edges on the lower flange of metal T-shaped members which, in turn, are supported by the structure.

The ceilings have the advantage that they are economical and provide unlimited access to the space above from the floor below. However, they have had the disadvantage of producing poor performance during past earthquakes as shown in Figures 6, 7 and 8. The ceiling industry has done considerable work in developing features of more competence in seismic design over the recent past several years. The T-bar ceiling regulations adopted for schools and hospitals in California are presented in Appendix B.

The T-bars are suspended vertically from the structure with wires placed through the stem of the T-bar web. The wire is usually soft annealed 12 ga. wire and is twisted around itself three times within 1 1/2" (3.8 cm) of the T-bar. The opposite end of the wire is fastened to the side of wood joists with 1 1/2" x 9 ga. stronghold fence staples placed at the 10 and 2 o'clock positions of the wire loop and another staple placed over the twisted portion of wire or through the eye of a tested-in-place shot-in concrete anchor. Where possible the connector is designed to resist shear loads rather than withdrawal loads.

The vertical hanger wires are connected to the main T-bar members near the cross T-bar member intersection. Depending upon the design, the hanger wires are usually spaced 4' x 4' (1.2 x 1.2 m) on centers. The vertical load span is therefore 4' (1.2 m). The main T-bars are spliced, usually 12' (3.67 m) oc. The cross T-bars span between the main T-bars and are spliced through slots in the web of the main T-bar. These cross T-bar splices have frequently failed during earthquakes which allowed the panels to fall to the floor together with the cross T-bars. It was quite obvious that this connection needed improvement as the failures were widespread in innumerable ceilings throughout the earthquake stricken area.

Another common location for T-bar ceiling damage from earthquakes was around the perimeter of the room at and near the interaction of the walls and ceiling. Here the main and cross T-bar members would buckle, come off the wall angle and allow the ceiling panel to fall and the T-bar to hang down precariously or fall.

These failures disclosed two primary considerations. The fact that the T-bar buckled indicated that the whole ceiling was loose and that the joints of the T-bars permitted too much free motion. The other was that the perimeter T-bars must be properly supported by a second means of defense to prevent their falling.

Several things were done to prevent these failures. It was proposed that points within the ceiling be partially held in placed thus making smaller tributary areas of ceilings rather than the whole ceiling of the room being braced by the perimeter walls. This was done by providing 45° sway bracing wires in each direction at 12' (3.67 m) oc. These wires provide a positive means to resist the horizontal component of the earthquake force but because there is no vertical strut provided at this point, the vertical component resistance is limited by the local tributary area dead load. However this concept does provide a compound pendulum effect. Some enforcing agencies require a vertical strut at the braced points.

Currently the C_p factor for California schools is 0.30 and for hospitals it is 0.50. For a 4 psf dead load ceiling, assuming the above indicated brace points at 12' oc each way (see Section 47.1812(c) of Appendix B), the tributary horizontal force in schools is 4 x 0.3 x 12 x 12 or 172.8 pounds. For hospitals with a spacing of 8' x 12' the horizontal force is 4 x 0.5 x 8 x 12 or 192 pounds.

The assumption is made that half of the horizontal load is delivered to the braced point in tension and half the load is delivered to the braced point in compression. Therefore the design load to be resisted by the members is rounded to 90 pounds. With a factor of safety of two the minimum load to be transferred is 180 pounds. This is the minimum ultimate load required for all splices and is given in Section 47.1811(b) as shown in Appendix B. Appendix B is a compilation of the California school requirements together with the 1976 UBC Standards.

It is recognized that the ceiling will probably undulate during the earthquake and that the joints should be capable of some hinged motion. The code (Section 47.1811(b) further requires that the joints be capable of resisting at least 180 pounds acting with a 5° or 1:24 offset in any direction. Several manufacturers have demonstrated that their product can meet these requirements. This provision required redesign of their equipment in many instances.

To reduce the ceiling perimeter damage, the regulation in Section 47.1812(b) requires that a hanger wire be placed within 8" (20 cm) of the wall on all T-bars abutting the walls. For ease of construction, the code, Section 47.1812(e) will permit the T-bars to be pop riveted or otherwise connected to the wall closure angle on two adjacent sides of the room and may be free at the opposite sides of the room. If the T-bars are not connected to the wall angle then a continuous supplementary member or tie must be provided in accordance with Section 47.1812(d) at the ends of the T-bars to prevent their spreading and thus to prevent the panels from falling through.

Details of several of the above-mentioned requirements are shown in Figure 9.

APPENDIX A

The following regulations are exceptions or modifications to the 1976 Uniform Building Code sections. That is, the following Section T21-4704(b) entirely replaces Section 4704(b) of the 1976 UBC. Also since T21-4704(a) is not included here, then Section 4704(a) of the 1976 UBC applies.

T21-4704. Horizontal Assemblies. (b) **Wood Framing.** Wood stripping or suspended wood systems, where used, shall be not less than 2 inches nominal thickness in the least dimension except that furring strips not less than 1-inch by 2-inch nominal dimension may be used over solid backing.

Wood furring strips for ceilings fastened to floor or ceiling joists shall be nailed at each bearing with 2 common wire nails, one of which shall be a slant nail and the other a face nail, or by one nail having spirally grooved or annular grooved shanks approved by the Office of the State Architect for this purpose. All stripping nails shall penetrate not less than 1¾ inches into the member receiving the point. Holes in stripping at joints shall be subdrilled to prevent splitting.

Where common wire nails are used to support horizontal wood stripping for plaster ceilings, such stripping shall be wire tied to the joists 4 feet on center with two strands of No. 18 W&M gage galvanized annealed wire to an 8d common wire nail driven into each side of the joist 2 inches above the bottom of the joist or to each end of a 16d common wire nail driven horizontally through the joist 2 inches above the bottom of the joist, and the ends of the wire secured together with 3 twists of the wire.

T21-4705. Interior Lath. (c) **Application of Metal Plaster Bases.** The type and weight of metal lath, and the gage and spacing of wire in welded or woven lath, the spacing of supports, and the methods of attachment to wood supports shall be as set forth in UBC Tables No. 47-B and No. 47-C.

Where interior lath is attached to horizontal wood supports, either of the following attachments shall be used in addition to the methods of attachment set forth in UBC Table No. 47-C.

(1) Secure lath to alternate supports with ties consisting of a double strand of No. 18 W&M gage galvanized annealed wire at one edge of each sheet of lath. Wire ties shall be installed not less than 3 inches back from the edge of each sheet and shall be looped around stripping, or attached to an 8d common wire nail driven into each side of the joist 2 inches above the bottom of the joist or to each end of a 16d common wire nail driven horizontally through the joist 2 inches above the bottom of the joist and the ends of the wire secured together with 3 twists of the wire.

(2) Secure lath to each support with ½ inch wide, 1½ inches long, No. 9 W&M gage, ring shank, hook staple placed around a 10d common nail laid flat under the surface of the lath not more than 3 inches from edge of each sheet. Such staples may be placed over ribs of ¾ inch rib lath or over back wire of welded wire fabric or other approved lath, omitting the 10d nails.

Metal lath shall be attached to metal supports with not less than No. 18 U.S. gage tie wire spaced not more than 6 inches apart or with approved equivalent attachments.

Metal lath or wire fabric lath shall be applied with the long dimension of the sheets perpendicular to supports.

Metal lath shall be lapped not less than ½ inch at sides and 1 inch at ends. Wire fabric lath shall be lapped not less than one mesh at sides and ends, but not less than 1 inch. Rib metal lath with edge ribs greater than ⅛ inch, shall be lapped at sides by nesting outside ribs. When edge ribs are ⅛ inch or less, rib metal lath may be lapped ½ inch at sides, or outside ribs may be nested. Where end laps of sheets do not occur over supports, they shall be securely tied together with not less than No. 18 U.S. gage wire.

Cornerite shall be installed in all internal corners to retain position during plastering. Cornerite may be omitted when lath is continuous or when plaster is not continuous from one plane to an adjacent plane.

T21-4706. **Exterior Lath.** (e) Application of Metal Plaster Bases. The application of metal lath or wire fabric lath shall be specified in Section T21-4705 (c) and they shall be furred out from vertical supports or backing not less than ¼ inch except as set forth in footnote No. 2, UBC Table No. 47-B.

Where exterior lath is attached to horizontal wood supports, either of the following attachments shall be used in addition to the methods of attachment set forth in UBC Table No. 47-C.

(1) Secure lath to alternate supports with ties consisting of a double strand of No. 18 W&M gage galvanized annealed wire at one edge of each sheet of lath. Wire ties shall be installed not less than 3 inches back from the edge of each sheet and shall be looped around stripping, or attached to an 8d common wire nail driven into each side of the joist 2 inches above the bottom of the joist or to each end of a 16d common wire nail driven horizontally through the joist 2 inches above the bottom of the joist and the ends of the wire secured together with 3 twists of the wire.

(2) Secure lath to each support with ½ inch wide, 1½ inches long, No. 9 W&M gage ring shank, hook staple placed around a 10d common nail laid flat under the surface of the lath not more than 3 inches from edge of each sheet. Such staples may be placed over ribs of ¾ inch rib lath or over back wire of welded wire fabric or other approved lath, omitting the 10d nails.

Where no external corner reinforcement is used, lath shall be furred out and carried around corners at least one support on frame construction.

A weep screed shall be provided at the foundation plate line on all exterior stud walls. The screed shall be of a type which will allow trapped water to drain to the exterior of the building.

T21-4713. **Shear-resisting Construction with Wood Frame.**

(a) General. Portland cement plaster, gypsum lath and plaster, gypsum sheathing board and gypsum wallboard shall not be used as design load resisting materials.

TABLE NO. T21-47-I

Allowable Shear for Wind or Seismic Forces in Pounds per Foot for Vertical Diaphragms of Lath and Plaster, Gypsum Sheathing Board, and Gypsum Wallboard Wood Framed Wall Assemblies

(UBC Table No. 47-I is not applicable to public school buildings.)

Suspended Acoustical Ceiling Systems. Metal ceiling suspension systems used primarily to support acoustical tile or other types of lay-in-panels shall be designed and installed in accordance with UBC Standard No. 47-18 except that the Sections noted below shall be applied in place of like Sections in the UBC Standards.

The member sizes, connections, support systems, light fixture attachments, partition supports and installation of bracing to resist lateral forces shall be fully detailed on the approved plans and/or specifications.

Sec. 47.1802. Classification. (a) The structural performance required from a ceiling suspension system shall be defined in terms of a suspension system structural classification.

The load-carrying capacity shall be the maximum uniformly distributed load (pounds per linear foot) that a simply supported main runner section having a span length of 4 feet, 0 inch is capable of supporting without a mid-span deflection exceeding 0.133 inch or 1/360 of the 4-foot, 0-inch span length.

The structural classification listed in Table No. 47-18-A shall be determined by the capability of main runners or nailing bars to support a uniformly distributed load. These classifications shall be:

TABLE NO. 47-18-A
Minimum Load-Carrying Capabilities of
Main Runner Members

Main Runner Members	Suspension System		
	pounds per linear foot		
	Direct Hung	Indirect Hung	Furring Bar
Intermediate-duty	12.0	3.5	6.5
Heavy-duty	16.0	8.0	...

Ceiling suspension systems used in public schools shall be either Intermediate-Duty Systems or Heavy-Duty Systems. Ceilings that support light fixtures, air ventilation grills or partitions shall have a classification of Heavy-Duty Systems.

Cross runners shall be capable of carrying the design load as dictated by job conditions without exceeding the maximum allowable deflection equal to 1/360 of its span. A cross runner that supports another cross runner is a main runner for the purpose of structural classification and shall be capable of supporting a uniformly distributed load at least equal to the "Intermediate" classification.

TABLE No. 47-18-B
Straightness Tolerances of Structural Members of Suspension Systems

Deformation	Straightness Tolerances
Bow	1/32 in. in any 2 ft. (1.30 mm/m), or 1/32 in. × (total length, ft)/2
Camber	1/32 in. in any 2 ft (1.30 mm/m), or 1/32 in. × (total length, ft)/2
Twist	1 deg in any 2 ft (1.64 deg/m), or 1 deg × (total length, ft)/2

TABLE NO. 47-18-B—STRAIGHTNESS TOLERANCES OF
STRUCTURAL MEMBERS OF SUSPENSION SYSTEMS

Deformation	Straightness Tolerances
Bow	1/32 in. in any 2 ft. (1.30 mm/m), or 1/32 in. X (total length, ft)/2
Camber	1/32 in. in any 2 ft (1.30 mm/m), or 1/32 in. X (total length, ft)/2
Twist	1 deg in any 2 ft (1.64 deg/m), or 1 deg X (total length, ft)/2

Dimensional Tolerance

Sec. 47.1803. (a) Straightness. The amount of bow, camber, or twist in main runners, cross runners, wall molding, splines, or nailing bars of various lengths shall not exceed the values shown in Table No. 47-18-B.

Main runners, cross runners, wall moldings, splines, or nailing bars of ceiling suspension systems shall not contain local kinks or bends.

Straightness of structural members shall be measured with the member suspended vertically from one end.

(b) **Length.** The variation in the specified length of main runner sections or cross runner sections that are part of an interlocking grid system shall not exceed ± 0.010 inch/4 feet.

The variation in the specified spacing of slots or other cutouts in the webs of main runners or cross runners that are employed in assembling a ceiling suspension grid system shall not exceed ± 0.010 inch.

(c) **Overall Cross-section Dimensions.** For steel systems, the overall height of the cross section of main runners, cross runners, wall molding, or nailing bar shall be the specified dimensions ± 0.030 inch. The width of the cross section of exposed main runners or cross runners shall be the specified dimension ± 0.008 inch.

(d) **Section Squareness.** Intersecting webs and flanges of structural members ("I," "T," or "Z" sections) shall form angles between them of 90 degrees ± 2 degrees. If deviations from squareness at more than one such intersection are additive with respect to their use in a ceiling, the total angle shall not be greater than 2 degrees.

The ends of structural members that abut or intersect other members in exposed grid systems shall be cut perpendicular to the exposed face, 90 degrees + 0, - 2 degrees.

(e) **Suspension System Devices.** Suspension system assembly devices shall satisfy the following requirements and tolerances.

A joint connection shall be judged suitable both before and after ceiling loads are imposed if the joint provides sufficient alignment so that:

The horizontal and the vertical displacement of the exposed surfaces of two abutting main runners does not exceed 0.015 inch.

There shall be no visually apparent angular displacement of the longitudinal axis of one runner with respect to the other.

Assembly devices shall provide sufficient spacing control so that horizontal gaps between exposed surfaces of either abutting or intersecting members shall not exceed 0.020 inch.

Spring wire clips used for supporting main runners shall maintain tight contact between the main runners and the carrying channels when the ceiling loads are imposed on the runners.

Coatings and Finishes for Suspension System Components

Sec. 47.1804. (a) **Protective Coatings.** Component materials that oxidize or corrode when exposed to normal use environments shall be provided with protective coatings as selected by the manufacturer except for cut or punched edges fabricated after the coating is applied.

Components fabricated from sheet steel shall be given an electrogalvanized, hot dipped galvanized, cadmium, or equal protective coating.

Components fabricated from aluminum alloys shall be anodized or protected by other approved techniques.

Components formed from other candidate materials shall be provided with an approved protective coating.

(b) **Adhesion and Resilience.** Finishes shall exhibit good adhesion properties and resilience so that chipping or flaking does not occur as a result of the manufacturing process.

(c) **Coating Classification for Severe Environment Performance.** Protected components for acoustical ceilings that are subject to the severe environmental conditions of high humidity and salt spray (fog), or both, shall be ranked according to their ability to protect the components of suspension systems from deterioration. A salt spray (fog) test conducted in accordance with the following test conditions shall be performed.

1. **Salt Solution.** Five parts by weight NaCl to 95 parts distilled water.

2. **Humidity in Chamber.** Ninety percent relative humidity.

3. **Temperature in Chamber.** Ninety degrees F.

4. **Exposure Period**—96 hours continuous.

5. **Report.** Upon request the manufacturer shall provide photographs showing worst corrosion conditions on components and shall provide comments regarding corrosion that occurs on cut metal edges, on galvanized surfaces without paint, on galvanized and painted surfaces, at edges rolled after being painted, and on any change of paint color or gloss that is apparent at the conclusion of the test. Color and gloss inspection of the component shall be made after washing in a mild soap solution.

(d) **High-Humidity Test.** The test and inspection shall be identical to that of the salt spray test, except that distilled water instead of salt solution shall be used.

Structural Members

Sec. 47.1805. The manufacturer shall determine the load-deflection performance.

The structural members tested shall be identical to the sections used in the final system design. All cutouts, slots, etc., as exist in the system component shall be included in the sections evaluated.

Load-deflection studies of structural members shall utilize sections fabricated in accordance with the system manufacturers' published metal thicknesses and dimensions.

Section Performance

Sec. 47.1806. The performance of structural members of suspension systems shall be represented by individual load-deflection plots obtained from tests performed at each different span length used in service.

The results of replicate tests of three individual sections, each tested on the same span length, shall be plotted and averaged to obtain a characteristic load-deflection curve for the structural member.

The average load deflection curve shall be used to establish the maximum uniformly distributed load that the structural member can successfully sustain prior to reaching the deflection limit of 1/360 of the span length in inches.

The load deflection curve shall be used to establish the maximum loading intensity beyond which the structural member begins to yield.

Suspension System Performance

Sec. 47.1807. Published performance data for individual suspension systems shall be developed by the manufacturer upon the basis of results obtained from load-deflection tests of its principal structural members. Where a ceiling design incorporated a number of components, each of which experiences some deflection as used in the system, the additive nature of these displacements shall be recognized in setting an allowable system deflection criteria.

Part II—Installation

Scope

Sec. 47.1808. This Standard describes procedures for the installation of suspension systems for acoustical tile and lay-in panels.

Sec. 47.1809. Installation of Components. (a) **Hangers.** Hangers shall be attached to the bottom edge of the wood joists or to the vertical face of the wood joists near the bottom edge. Bottom edge attachment devices shall be an approved type.

In concrete construction, mount hangers using cast-in-place hanger wires, hanger inserts, or other hanger attachment devices shall be an approved type. If greater center-to-center distances than 4 feet, 0 inch are used for the hangers, reduce the load-carrying capacity of the ceiling suspension system commensurate with the actual center-to-center hanger distances used.

Hangers shall be plumb and shall not press against insulation covering ducts or pipes. Hangers shall be spaced a minimum of 6 inches from all horizontal piping or duct work that is not provided with bracing restraints for horizontal forces. If some hangers must be splayed, offset the resulting horizontal force by bracing, countersplaying, or other acceptable means.

Hangers formed from galvanized sheet metal stock shall be suitable for suspending carrying channels or main runners from an existing structure provided that the hangers do not yield, twist, or undergo other objectionable movement.

Wire hangers for suspending carrying channels or main runners from an existing structure shall be prepared from a minimum of No. 12 gage, galvanized, soft-annealed, mild steel wire.

Special attachment devices that support the carrying channels or main runners shall be approved to support five times the design load.

(b) **Carrying Channels.** The carrying channels shall be installed so that they are all level to within ⅛ inch in 12 feet.

Leveling shall be performed with the supporting hangers taut.

Local kinks or bends shall not be made in hanger wires as a means of leveling carrying channels.

In installations where hanger wires are wrapped around carrying channels, the wire loops shall be tightly formed to prevent any vertical movement or rotation of the member within the loop.

(c) **Main Runners.** Main runners shall be installed so that they are all level to within ⅛ inch in 12 feet.

Where main runners are supported directly by hangers, leveling shall be performed with the supporting hanger taut.

Local kinks or bends shall not be made in hanger wires as a means of leveling main runners.

In installations where hanger wires are wrapped through or around main runners, the wire loops shall be tightly wrapped and sharply bent.

(d) **Cross Runners.** Cross runners shall be supported by either main runners or by other cross runners to within ½ inch of the required center distances. This tolerance shall be noncumulative beyond 12 feet.

Intersecting runners shall form a right angle.

The exposed surfaces of two intersecting runners shall lie within a vertical distance of 0.015 inch of each other with the abutting (cross) member always above the continuous (main) member.

(e) **Splines.** Splines used to form a concealed mechanical joint seal between adjacent tiles shall be compatible with the tile kerf design so that the adjacent tile will be horizontal when installed. Where splines are longer than the dimension between edges of supporting members running perpendicular to the splines, place the splines so that they rest either all above or all below the main running members.

(f) Assembly Devices. Abutting sections of main runner shall be joined by means of suitable connections such as splices, interlocking ends, tab locks, pin locks, etc. A joint connection shall be judged suitable both before and after ceiling loads are imposed if the joint provides sufficient alignment so that the exposed surfaces of two abutting main runners lie within a vertical distance of 0.015 inch of each other and within a horizontal distance of 0.015 inch of each other.

There shall be no visually apparent angular displacement of the longitudinal axis of one runner with respect to the other.

Assembly devices shall provide sufficient spacing control so that horizontal gaps between exposed surfaces of either abutting or intersecting members shall not exceed 0.020 inch.

Spring wire clips used for supporting main runners shall maintain tight contact between the main runners and the carrying channels when the ceiling loads are imposed on the runners.

(g) Ceiling Fixtures. Fixtures installed in acoustical tile or lay-in panel ceilings shall be mounted in a manner that will not compromise ceiling performance.

All recessed or drop-in light fixtures and grilles shall be supported directly from the fixture housing to the structure above with a minimum of two 12 gage wires located at diagonally opposite corners. Leveling and positioning of fixtures may be provided by the ceiling grid. Fixture support wires may be slightly loose to allow the fixture to seat in the grid system.

Fixtures shall not be supported from main runners or cross runners if the weight of the fixtures causes the total dead load to exceed the deflection capability of the ceiling suspension system.

Fixtures shall not be installed so that main runners or cross runners will be eccentrically loaded.

Surface mounted fixtures shall be attached to the main runner with at least two positive clamping devices made of material with a minimum of 14 gage. Rotational spring catches do not comply. A 12 gage suspension wire shall be attached to each clamping device and to the structure above.

(h) Lay-in-panels. Metal panels and panels weighing more than ⅓ pound per square foot other than acoustical tile shall be positively attached to the ceiling suspension runners.

Part III—Lateral Design Requirements

Scope

Sec. 47.1810. Suspended ceiling systems which are designed and constructed to support ceiling panels or tiles, with or without lighting fixtures, ceiling mounted air terminals, or other ceiling mounted services shall comply with the requirements of this Standard.

> **EXCEPTIONS:** 1. Ceiling area of 144 square feet or less surrounded by walls which connect directly to the structure above shall be exempt from the lateral load design requirements of these standards.
> 2. Ceilings constructed of lath and plaster or gypsum board.

Minimum Design Loads

Sec. 47.1811. (a) Lateral Forces. Such ceiling systems and their connections to the building structure shall be designed and constructed to resist a lateral force specified in Chapter 23 of the Uniform Building Code.

Where the ceiling system provides lateral support for nonbearing partitions, it shall be designed for the prescribed lateral force reaction from the partitions as specified in Section 47.1815.

Connection of lighting fixtures to the ceiling system shall be designed for a lateral force of 100 percent of the weight of the fixture in addition to the prescribed vertical loading as specified in Section 47.1813.

(b) Grid Members, Connectors and Expansion Devices. The main runners and crossrunners of the ceiling system and their splices, intersection connectors and expansion devices shall be designed and constructed to carry a mean ultimate test load of not less than 180 pounds or twice the actual load, whichever is greater, in tension with a 5° misalignment of the members in any direction and in compression. In lieu of a 5° misalignment, the load may be applied with a 1 inch eccentricity on a sample not more than 24 inches long each side of the splice. The connection at splices and intersections shall be of the mechanical interlocking type.

Evaluation of test results shall be made on the basis of the mean values resulting from tests of not fewer than three identical specimens, provided the deviation of any individual test result from the mean value does not exceed plus or minus 10 percent. The allowable load-carrying capacity as determined by test shall not exceed one-half of the mean ultimate test value.

(c) Substantiation. Each ceiling systems manufacturer shall furnish lateral loading capacity and displacement or elongation characteristics for his systems indicating the following:

1. Maximum bracing pattern and minimum wire sizes.
2. Tension and compression force capabilities of main runner splices, cross runner connections and expansion devices.

All tests shall be conducted by an approved testing agency.

Sec. 47.1812. Installation. (a) **Vertical Hangers.** Suspension wires shall be not smaller than No. 12 gage spaced at 4 feet on center or No. 10 gage spaced at 5 feet on center along each main runner unless calculations justifying the increased spacing are provided.

Each vertical wire shall be attached to the ceiling suspension member and to the support above with a minimum of three turns. Any connection device at the supporting construction shall be capable of carrying not less than 100 pounds.

Suspension wires shall not hang more than 1 in 6 out-of-plumb unless countersloping wires are provided.

Wires shall not attach to or bend around interfering material or equipment. A trapeze or equivalent device shall be used where obstructions preclude direct suspension. Trapeze suspensions shall be a minimum of back-to-back 1¼-inch cold rolled channels for spans exceeding 48 inches.

(b) **Perimeter Hangers.** The terminal ends of each cross runner and main runner shall be supported independently a maximum of 8 inches from each wall or ceiling discontinuity with No. 12 gauge wire or approved wall support.

(c) **Lateral Force Bracing.** Where substantiating design calculations are not provided, horizontal restraints shall be effected by four No. 12 gage wires secured to the main runner within 2 inches of the cross runner intersection and splayed 90° from each other at an angle not exceeding 45° from the plane of the ceiling. These horizontal restraint points shall be placed within 4 feet of each surrounding wall and not over 12 feet on center. Attachment of the restraint wires to the structure above shall be adequate for the load imposed.

Lateral force bracing members shall be spaced a minimum of 6 inches from all horizontal piping or duct work that is not provided with bracing restraints for horizontal forces. Bracing wires shall be attached to the grid and to the structure in such a manner that they can support a design load of not less than 200 pounds or the actual design load, whichever is greater, with a safety factor of 2.

(d) **Perimeter Members.** Unless a structural part of the approved system, wall angles or channels shall be considered as aesthetic closers and shall have no structural value assessed to themselves or their method of attachment to the walls. Ends of main runners and cross members shall be tied together to prevent their spreading.

(e) Attachment of Members to the Perimeter. To facilitate installation, main runners and cross runners may be attached to the perimeter member at two adjacent walls with clearance between the wall and the runners maintained at the other two walls or as otherwise shown or described for the approved system.

Sec. 47.1813. Lighting Fixtures. All lighting fixtures shall be positively attached to the suspended ceiling system. The attachment device shall have a capacity of 100 percent of the lighting fixture weight acting in any direction. (See Sec. 47.1809 (g) for hanger wires to the fixture.)

Lighting fixtures weighing 56 pounds or more shall be supported directly from the structure above by approved hangers. In such cases the slack wires required by Sec. 47.1809 (g) may be omitted.

Sec. 47.1814. Mechanical Services. Ceiling mounted air terminals or services shall be positively attached to the ceiling suspension main runners or to cross runners with the same carrying capacity as the main runners.

Terminals or services weighing not more than 56 pounds shall have two No. 12 gage hangers connected from the terminal or service to the structure above. These wires may be slack.

Terminals or services weighing more than 56 pounds shall be supported directly from the structure above by approved hangers.

Partitions

Sec. 47.1815. Where the suspended ceiling system is required to provide lateral support for permanent or relocatable partitions, the connection of the partition to the ceiling system, the ceiling system members and their connections, and the lateral force bracing shall be designed to support the reaction force of the partition from prescribed loads applied perpendicular to the face of the partition. These partition reaction forces shall be in addition to the loads described in Section 47.1811. Partition connectors, the suspended ceiling system and the lateral force bracing shall all be engineered to suit the individual partition application and shall be shown or defined in the drawings or specifications.

Drawings and Specifications

Sec. 47.1816. The drawings shall clearly identify all systems and shall define or show all supporting details, lighting fixture attachment, lateral force bracing, partition bracing, etc. Such definition may be by reference to this Standard, or approved system, in whole or in part. Deviations or variations must be shown or defined in detail.

Sec. 47.1801. This Standard covers metal ceiling suspension systems used primarily to support acoustical tile or acoustical lay-in panels.

Classification

Sec. 47.1802. (a) The structural performance required from a ceiling suspension system shall be defined in terms of a suspension system structural classification.

The load-carrying capacity shall be the maximum uniformly distributed load (pounds per linear foot) that a simply supported main runner section having a span length of 4 feet, 0 inch is capable of supporting without a mid-span deflection exceeding 0.133 inch or 1/360 of the 4-foot, 0-inch span length.

The structural classification listed in Table No. 47-18-A shall be determined by the capability of main runners or nailing bars to support a uniformly distributed load. These classifications shall be:

TABLE NO. 47-18-A—MINIMUM LOAD-CARRYING CAPABILITIES OF MAIN RUNNER MEMBERS

Main Runner Members	Suspension System		
	lb/linear ft (kg/m)		
	Direct Hung	Indirect Hung	Furring Bar
Light-duty	5.0	2.0	4.5
Intermediate-duty	12.0	3.5	6.5
Heavy-duty	16.0	8.0	...

1. **Light-Duty Systems.** Used primarily for residential and light commercial structures where ceiling loads other than acoustical tile or lay-in panels are not anticipated.

2. **Intermediate-Duty Systems.** Used primarily for ordinary commercial structures where some ceiling loads, due to light fixtures and air diffusers, are anticipated.

3. **Heavy-Duty Systems.** Used primarily for commercial structures in which the quantities and weights of ceiling fixtures (lights, air diffusers, etc.) are greater than those for an ordinary commercial structure.

Cross runners shall be capable of carrying the design load as dictated by job conditions without exceeding the maximum allowable deflection equal to 1/360 of its span. A cross runner that supports another cross runner is a main runner for the purpose of structural classification and shall be capable of supporting a uniformly distributed load at least equal to the "Intermediate" classification.

TABLE NO. 47-18-B—STRAIGHTNESS TOLERANCES OF STRUCTURAL MEMBERS OF SUSPENSION SYSTEMS

Deformation	Straightness Tolerances
Bow	$\frac{1}{32}$ in. in any 2 ft. (1.30 mm/m), or $\frac{1}{32}$ in. X (total length, ft)/2
Camber	$\frac{1}{32}$ in. in any 2 ft (1.30 mm/m), or $\frac{1}{32}$ in. X (total length, ft)/2
Twist	1 deg in any 2 ft (1.64 deg/m), or 1 deg X (total length, ft)/2

Dimensional Tolerance

Sec. 47.1803. (a) Straightness. The amount of bow, camber, or twist in main runners, cross runners, wall molding, splines, or nailing bars of various lengths shall not exceed the values shown in Table No. 47-18-B.

Main runners, cross runners, wall moldings, splines, or nailing bars of ceiling suspension systems shall not contain local kinks or bends.

Straightness of structural members shall be measured with the member suspended vertically from one end.

(b) Length. The variation in the specified length of main runner sections or cross runner sections that are part of an interlocking grid system shall not exceed ± 0.010 inch/4 feet.

The variation in the specified spacing of slots or other cutouts in the webs of main runners or cross runners that are employed in assembling a ceiling suspension grid system shall not exceed ± 0.010 inch.

(c) Overall Cross-section Dimensions. For steel systems, the overall height of the cross section of main runners, cross runners, wall molding, or nailing bar shall be the specified dimensions ± 0.030 inch. The width of the cross section of exposed main runners or cross runners shall be the specified dimension ± 0.008 inch.

(d) Section Squareness. Intersecting webs and flanges of structural members ("I," "T," or "Z" sections) shall form angles between them of 90 degrees ± 2 degrees. If deviations from squareness at more than one such intersection are additive with respect to their use in a ceiling, the total angle shall not be greater than 2 degrees.

The ends of structural members that abut or intersect other members in exposed grid systems shall be cut perpendicular to the exposed face, 90 degrees + 0, - 2 degrees.

(e) Suspension System Devices. Suspension system assembly devices shall satisfy the following requirements and tolerances.

A joint connection shall be judged suitable both before and after ceiling loads are imposed if the joint provides sufficient alignment so that:

The horizontal and the vertical displacement of the exposed surfaces of two abutting main runners does not exceed 0.015 inch.

There shall be no visually apparent angular displacement of the longitudinal axis of one runner with respect to the other.

Assembly devices shall provide sufficient spacing control so that horizontal gaps between exposed surfaces of either abutting or intersecting members shall not exceed 0.020 inch.

Spring wire clips used for supporting main runners shall maintain tight contact between the main runners and the carrying channels when the ceiling loads are imposed on the runners.

Coatings and Finishes for Suspension System Components

Sec. 47.1804. (a) Protective Coatings. Component materials that oxidize or corrode when exposed to normal use environments shall be provided with protective coatings as selected by the manufacturer except for cut or punched edges fabricated after the coating is applied.

Components fabricated from sheet steel shall be given an electrogalvanized, hot dipped galvanized, cadmium, or equal protective coating.

Components fabricated from aluminum alloys shall be anodized or protected by other approved techniques.

Components formed from other candidate materials shall be provided with an approved protective coating.

(b) Adhesion and Resilience. Finishes shall exhibit good adhesion properties and resilience so that chipping or flaking does not occur as a result of the manufacturing process.

(c) Coating Classification for Severe Environment Performance. Protected components for acoustical ceilings that are subject to the severe environmental conditions of high humidity and salt spray (fog), or both, shall be ranked according to their ability to protect the components of suspension systems from deterioration. A salt spray (fog) test conducted in accordance with the following test conditions shall be performed.

1. Salt Solution. Five parts by weight NaCl to 95 parts distilled water.

2. Humidity in Chamber. Ninety percent relative humidity.

3. Temperature in Chamber. Ninety degrees F.

4. Exposure Period—96 hours continuous.

5. Report. Upon request the manufacturer shall provide photographs showing worst corrosion conditions on components and shall provide comments regarding corrosion that occurs on cut metal edges, on galvanized surfaces without paint, on galvanized and painted surfaces, at edges rolled after being painted, and on any change of paint color or gloss that is apparent at the conclusion of the test. Color and gloss inspection of the component shall be made after washing in a mild soap solution.

(d) High-Humidity Test. The test and inspection shall be identical to that of the salt spray test, except that distilled water instead of salt solution shall be used.

Structural Members

Sec. 47.1805. The manufacturer shall determine the load-deflection performance.

The structural members tested shall be identical to the sections used in the final system design. All cutouts, slots, etc., as exist in the system component shall be included in the sections evaluated.

Load-deflection studies of structural members shall utilize sections fabricated in accordance with the system manufacturers' published metal thicknesses and dimensions.

Section Performance

Sec. 47.1806. The performance of structural members of suspension systems shall be represented by individual load-deflection plots obtained from tests performed at each different span length used in service.

The results of replicate tests of three individual sections, each tested on the same span length, shall be plotted and averaged to obtain a characteristic load-deflection curve for the structural member.

The average load deflection curve shall be used to establish the maximum uniformly distributed load that the structural member can successfully sustain prior to reaching the deflection limit of 1/360 of the span length in inches.

The load deflection curve shall be used to establish the maximum loading intensity beyond which the structural member begins to yield.

Suspension System Performance

Sec. 47.1807. Published performance data for individual suspension systems shall be developed by the manufacturer upon the basis of results obtained from load-deflection tests of its principal structural members. Where a ceiling design incorporated a number of components, each of which experiences some deflection as used in the system, the additive nature of these displacements shall be recognized in setting an allowable system deflection criteria.

Part II—Installation

Scope

Sec. 47.1808. This Standard describes procedures for the installation of suspension systems for acoustical tile and lay-in panels.

Installation of Components

Sec. 47.1809. (a) Hangers. Hangers shall be attached to the bottom edge of the wood joists or to the vertical face of the wood joists near the bottom edge. Bottom edge attachment devices shall be an approved type.

In concrete construction, mount hangers using cast-in-place hanger wires, hanger inserts, or other hanger attachment devices shall be an approved type. If greater center-to-center distances than 4 feet, 0 inch are used for the hangers, reduce the load-carrying capacity of the ceiling suspension system commensurate with the actual center-to-center hanger distances used.

Hangers shall be plumb and shall not press against insulation covering ducts or pipes. If some hangers must be splayed, offset the resulting horizontal force by bracing, countersplaying, or other acceptable means.

Hangers formed from galvanized sheet metal stock shall be suitable for suspending carrying channels or main runners from an existing structure provided that the hangers do not yield, twist, or undergo other objectionable movement.

Wire hangers for suspending carrying channels or main runners from an existing structure shall be prepared from a minimum of No. 12 gauge, galvanized, soft-annealed, mild steel wire.

Special attachment devices that support the carrying channels or main runners shall be approved to support five times the design load.

(b) Carrying Channels. The carrying channels shall be installed so that they are all level to within ⅛ inch in 12 feet.

Leveling shall be performed with the supporting hangers taut.

Local kinks or bends shall not be made in hanger wires as a means of leveling carrying channels.

In installations where hanger wires are wrapped around carrying channels, the wire loops shall be tightly formed to prevent any vertical movement or rotation of the member within the loop.

(c) Main Runners. Main runners shall be installed so that they are all level to within ⅛ inch in 12 feet.

Where main runners are supported directly by hangers, leveling shall be performed with the supporting hanger taut.

Local kinks or bends shall not be made in hanger wires as a means of leveling main runners.

In installations where hanger wires are wrapped through or around main runners, the wire loops shall be tightly wrapped and sharply bent.

(d) Cross Runners. Cross runners shall be supported by either main runners or by other cross runners to within ⅜₂ inch of the required center distances. This tolerance shall be noncumulative beyond 12 feet.

Intersecting runners shall form a right angle.

The exposed surfaces of two intersecting runners shall lie within a vertical distance of 0.015 inch of each other with the abutting (cross) member always above the continuous (main) member.

(e) Splines. Splines used to form a concealed mechanical joint seal between adjacent tiles shall be compatible with the tile kerf design so that the adjacent tile will be horizontal when installed. Where splines are longer than the dimension between edges of supporting members running perpendicular to the splines, place the splines so that they rest either all above or all below the main running members.

(f) Assembly Devices. Abutting sections of main runner shall be joined by means of suitable connections such as splices, interlocking ends, tab locks, pin locks, etc. A joint connection shall be judged suitable both before and after ceiling loads are imposed if the joint provides sufficient alignment so that the exposed surfaces of two abutting main runners lie within a vertical distance of 0.015 inch of each other and within a horizontal distance of 0.015 inch of each other.

There shall be no visually apparent angular displacement of the longitudinal axis of one runner with respect to the other.

Assembly devices shall provide sufficient spacing control so that horizontal gaps between exposed surfaces of either abutting or intersecting members shall not exceed 0.020 inch.

Spring wire clips used for supporting main runners shall maintain tight contact between the main runners and the carrying channels when the ceiling loads are imposed on the runners.

(g) Ceiling Fixtures. Fixtures installed in acoustical tile or lay-in panel ceilings shall be mounted in a manner that will not compromise ceiling performance.

Fixtures shall not be supported from main runners or cross runners if the weight of the fixtures causes the total dead load to exceed the deflection capability of the ceiling suspension system. In such cases, the fixture load shall be supported by supplemental hangers within 6 inches of each corner, or the fixture shall be separately supported.

Fixtures shall not be installed so that main runners or cross runners will be eccentrically loaded except where provision is inherent in the system (or is separately provided for) to prevent undesirable section rotation or displacement, or both. In any case, runners supporting ceiling fixtures shall not rotate more than 2 degrees after the fixture loads are imposed.

Where fixture installation would produce rotation of runners in excess of 2 degrees, install fixtures with the use of suitable accessory devices. These devices shall support the fixture in such a manner that the main runners and cross runners will be loaded symmetrically rather than eccentrically.

X-31

Part III—Lateral Design Requirements

Scope

Sec. 47.1810. Suspended ceiling systems which are designed and constructed to support ceiling panels or tiles, with or without lighting fixtures, ceiling mounted air terminals, or other ceiling mounted services shall comply with the requirements of this Standard.

> **EXCEPTIONS:** 1. Ceiling area of 144 square feet or less surrounded by walls which connect directly to the structure above shall be exempt from the lateral load design requirements of these standards.
>
> 2. Ceilings constructed of lath and plaster or gypsum board.

Minimum Design Loads

Sec. 47.1811. (a) Lateral Forces. Such ceiling systems and their connections to the building structure shall be designed and constructed to resist a lateral force specified in Chapter 23 of the Uniform Building Code.

Where the ceiling system provides lateral support for nonbearing partitions, it shall be designed for the prescribed lateral force reaction from the partitions as specified in Section 47.1815.

Connection of lighting fixtures to the ceiling system shall be designed for a lateral force of 100 percent of the weight of the fixture in addition to the prescribed vertical loading as specified in Section 47.1813.

(b) Grid Members, Connectors and Expansion Devices. The main runners and crossrunners of the ceiling system and their splices, intersection connectors and expansion devices shall be designed and constructed to carry an axial design load of not less than 60 pounds or twice the actual load, whichever is greater, in tension and compression. The connection at splices and intersections shall be of the mechanical interlocking type.

Where the composition or configuration of ceiling systems members or assemblies and their connections are such that calculations of their allowable load-carrying capacity cannot be made in accordance with established methods of analysis, their performance shall be established by test.

Evaluation of test results shall be made on the basis of the mean values resulting from tests of not fewer than three identical specimens, provided the deviation of any individual test result from the mean value does not exceed plus or minus 10 percent. The allowable load-carrying capacity as determined by test shall not exceed one-half of the mean ultimate test value.

(c) Substantiation. Each ceiling systems manufacturer shall furnish lateral loading capacity and displacement or elongation characteristics for his systems indicating the following:

1. Maximum bracing pattern and minimum wire sizes.
2. Tension and compression force capabilities of main runner splices, cross runner connections and expansion devices.

All tests shall be conducted by an approved testing agency.

Installation

Sec. 47.1812. (a) Vertical Hangers. Suspension wires shall be not smaller than No. 12 gauge spaced at 4 feet O.C. along each main runner unless calculations justifying the increased spacing are provided.

Each vertical wire shall be attached to the ceiling suspension member and to the support above with a minimum of three turns. Any connection device at the supporting construction shall be capable of carrying not less than 100 pounds.

Suspension wires shall not hang more than 1 in 6 out-of-plumb unless countersloping wires are provided.

Wires shall not attach to or bend around interfering material or equipment. A trapeze or equivalent device shall be used where obstructions preclude direct suspension. Trapeze suspensions shall be a minimum of back-to-back 1½-inch cold rolled channels for spans exceeding 48 inches.

(b) Perimeter Hangers. The terminal ends of each cross runner and main runner shall be supported independently a maximum of 8 inches from each wall or ceiling discontinuity with No. 12 gauge wire or approved wall support.

(c) Lateral Force Bracing. Where substantiating design calculations are not provided, horizontal restraints shall be effected by four No. 12 gauge wires secured to the main runner within 2 inches of the cross runner intersection and splayed 90° from each other at an angle not exceeding 45° from the plane of the ceiling. These horizontal restraint points shall be placed 12 feet O.C. in both directions with the first point within 4 feet from each wall. Attachment of the restraint wires to the structure above shall be adequate for the load imposed.

(d) Perimeter Members. Unless a structural part of the approved system, wall angles or channels shall be considered as aesthetic closers and shall have no structural value assessed to themselves or their method of attachment to the walls. For tile ceilings, ends of main runners and cross members shall be tied together to prevent their spreading.

(e) Attachment of Members to the Perimeter. To facilitate installation, main runners and cross runners may be attached to the perimeter member at two adjacent walls with clearance between the wall and the runners maintained at the other two walls or as otherwise shown or described for the approved system.

Lighting Fixtures

Sec. 47.1813. Only "intermediate" and "heavy duty" ceiling systems as defined in Section 47.1802 (a) may be used for the support of lighting fixtures.

All lighting fixtures shall be positively attached to the suspended ceiling system. The attachment device shall have a capacity of 100 percent of the lighting fixture weight acting in any direction.

When "intermediate" systems are used, No. 12 gauge hangers shall be attached to the grid members within 3 inches of each corner of each fixture. Tandem fixtures may utilize common wires.

Where "heavy duty" systems are used, supplemental hangers are not required if a 48-inch modular hanger pattern is followed. When cross runners are used without supplemental hangers to support lighting fixtures, these cross runners must provide the same carrying capacity as the main runner.

Lighting fixtures weighing more than 20 pounds but less than 56 pounds shall have, in addition to the requirements outlined above, two No. 12 gauge hangers connected from the fixture housing to the ceiling system hangers or to the structure above. These wires may be slack.

Lighting fixtures weighing 56 pounds or more shall be supported directly from the structure above by approved hangers.

Pendant hung lighting fixtures shall be supported directly from the structure above using No. 9 gauge wire or approved alternate support without using the ceiling suspension system for direct support.

Mechanical Services

Sec. 47.1814. Ceiling mounted air terminals or services weighing less than 20 pounds shall be positively attached to the ceiling suspension main runners or to cross runners with the same carrying capacity as the main runners.

Terminals or services weighing 20 pounds but not more than 56 pounds, in addition to the above, shall have two No. 12 gauge hangers connected from the terminal or service to the ceiling system hangers or to the structure above. These wires may be slack.

Terminals or services weighing more than 56 pounds shall be supported directly from the structure above by approved hangers.

Partitions

Sec. 47.1815. Where the suspended ceiling system is required to provide lateral support for permanent or relocatable partitions, the connection of the partition to the ceiling system, the ceiling system members and their connections, and the lateral force bracing shall be designed to support the reaction force of the partition from prescribed loads applied perpendicular to the face of the partition. These partition reaction forces shall be in addition to the loads described in Section 47.1811. Partition connectors, the suspended ceiling system and the lateral force bracing shall all be engineered to suit the individual partition application and shall be shown or defined in the drawings or specifications.

Drawings and Specifications

Sec. 47.1816. The drawings shall clearly identify all systems and shall define or show all supporting details, lighting fixture attachment, lateral force bracing, partition bracing, etc. Such definition may be by reference to this Standard, or approved system, in whole or in part. Deviations or variations must be shown or defined in detail.

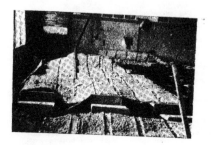

Figure 1.   Pre-Field Act School following the
1940 El Centro earthquake shows the need for
positive anchorage of the metal lath.

Figure 2.   Metal lath anchorage pulled out in the
1971 San Fernando earthquake.

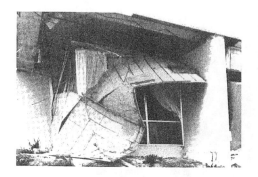

Figure 3. Ceiling framing must be properly anchored.

Figure 4. Each portion of this ceiling responded to
the 1971 San Fernando earthquake. Note the damage to
the plaster at the end of the ceiling strip ventilator
openings.

Figure 5. Gypsum board panels became dislodged in the
1964 earthquake in Anchorage, Alaska.

Figure 6. Ceiling damage to the T-bar ceiling system in
a post-Field Act school building in the 1969 Santa Rosa
earthquake.

Figure 7. View of exit corridor in a hospital in the 1971 San Fernando earthquake.

Figure 8. View of a corridor in a hospital in the 1971 San Fernando earthquake.

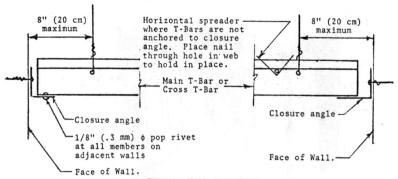

8" (20 cm) maximum

Horizontal spreader where T-Bars are not anchored to closure angle. Place nail through hole in web to hold in place.

8" (20 cm) maximum

Main T-Bar or Cross T-Bar

Closure angle

Closure angle

1/8" (.3 mm) φ pop rivet at all members on adjacent walls

Face of Wall.

Face of Wall.

TYPICAL WALL CONDITION

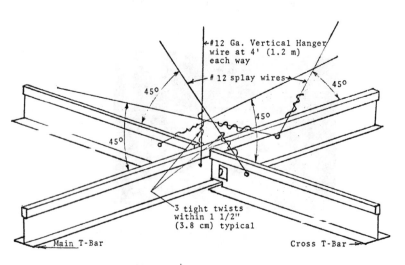

#12 Ga. Vertical Hanger wire at 4' (1.2 m) each way

12 splay wires

45°

45°

45°

45°

45°

3 tight twists within 1 1/2" (3.8 cm) typical

Main T-Bar

Cross T-Bar

TYPICAL T-BAR INTERSECTION

Splay wires typical 12' (3.67 m) oc each way, 6' (1.82 m) from walls

Figure 9. Typical T-bar ceiling details.

X-38

☆ U.S. GOVERNMENT PRINTING OFFICE: 1983—380-997/5122

PERFORMING ORGANIZATION *(If joint or other than NBS, see instructions)*

NATIONAL BUREAU OF STANDARDS
DEPARTMENT OF COMMERCE
WASHINGTON, D.C. 20234

SPONSORING ORGANIZATION NAME AND COMPLETE ADDRESS *(Street, City, State, ZIP)*

Same as item 6.

. SUPPLEMENTARY NOTES

Library of Congress Catalog Card Number: 83-600550

☐ Document describes a computer program; SF-185, FIPS Software Summary, is attached.

1. ABSTRACT *(A 200-word or less factual summary of most significant information. If document includes a significant bibliography or literature survey, mention it here)*

The Eleventh Joint Meeting of the U.S. - Japan Panel on Wind and Seismic Effect was held in Tsukuba, Japan on September 4-7, 1979. The proceedings of the Joir Meeting include the program, the formal resolution and the technical papers. subjects covered in the paper include (1) the engineering characteristics of wi (2) the characteristics of earthquake ground motions, (3) the earthquake respor of structures, (4) the wind response of structures, (5) recent design criteria against wind and earthquake disturbances, (6) the design and analysis of specia structures, (7) the evaluation, repairing, and retrofitting for wind and earthquake disaster, (8) earthquake disaster prevention planning, (9) storm surge and tsunamis, and (10) technical cooperation with developing countries.

2. KEY WORDS *(Six to twelve entries; alphabetical order; capitalize only proper names; and separate key word* Accelerograph; codes; design criteria; disaster; earthquakes; hazards; g failures; seismicity; solids; standards; structural engineering; structu responses; tsunamis; wind loads; winds.

3. AVAILABILITY

☒ Unlimited
☐ For Official Distribution. Do Not Release to NTIS
☒ Order From Superintendent of Documents, U.S. Government Printing Office, Washington, D.C. 20402.

☐ Order From National Technical Information Service (NTIS), Springfield, VA. 22161

1. Price

$12.00

USCOMM-DC

NBS TECHNICAL PUBLICATIONS

PERIODICALS

JOURNAL OF RESEARCH—The Journal of Research of the National Bureau of Standards reports NBS research and development in those disciplines of the physical and engineering sciences in which the Bureau is active. These include physics, chemistry, engineering, mathematics, and computer sciences. Papers cover a broad range of subjects, with major emphasis on measurement methodology and the basic technology underlying standardization. Also included from time to time are survey articles on topics closely related to the Bureau's technical and scientific programs. As a special service to subscribers each issue contains complete citations to all recent Bureau publications in both NBS and non-NBS media. Issued six times a year. Annual subscription: domestic $18; foreign $22.50. Single copy, $5.50 domestic; $6.90 foreign.

NONPERIODICALS

Monographs—Major contributions to the technical literature on various subjects related to the Bureau's scientific and technical activities.

Handbooks—Recommended codes of engineering and industrial practice (including safety codes) developed in cooperation with interested industries, professional organizations, and regulatory bodies.

Special Publications—Include proceedings of conferences sponsored by NBS, NBS annual reports, and other special publications appropriate to this grouping such as wall charts, pocket cards, and bibliographies.

Applied Mathematics Series—Mathematical tables, manuals, and studies of special interest to physicists, engineers, chemists, biologists, mathematicians, computer programmers, and others engaged in scientific and technical work.

National Standard Reference Data Series—Provides quantitative data on the physical and chemical properties of materials, compiled from the world's literature and critically evaluated. Developed under a worldwide program coordinated by NBS under the authority of the National Standard Data Act (Public Law 90-396).

NOTE: The principal publication outlet for the foregoing data is the Journal of Physical and Chemical Reference Data (JPCRD) published quarterly for NBS by the American Chemical Society (ACS) and the American Institute of Physics (AIP). Subscriptions, reprints, and supplements available from ACS, 1155 Sixteenth St., NW, Washington, DC 20056.

Building Science Series—Disseminates technical information developed at the Bureau on building materials, components, systems, and whole structures. The series presents research results, test methods, and performance criteria related to the structural and environmental functions and the durability and safety characteristics of building elements and systems.

Technical Notes—Studies or reports which are complete in themselves but restrictive in their treatment of a subject. Analogous to monographs but not so comprehensive in scope or definitive in treatment of the subject area. Often serve as a vehicle for final reports of work performed at NBS under the sponsorship of other government agencies.

Voluntary Product Standards—Developed under procedures published by the Department of Commerce in Part 10, Title 15, of the Code of Federal Regulations. The standards establish nationally recognized requirements for products, and provide all concerned interests with a basis for common understanding of the characteristics of the products. NBS administers this program as a supplement to the activities of the private sector standardizing organizations.

Consumer Information Series—Practical information, based on NBS research and experience, covering areas of interest to the consumer. Easily understandable language and illustrations provide useful background knowledge for shopping in today's technological marketplace.

Order the above NBS publications from: Superintendent of Documents, Government Printing Office, Washington, DC 20402.

Order the following NBS publications—FIPS and NBSIR's—from the National Technical Information Service, Springfield, VA 22161.

Federal Information Processing Standards Publications (FIPS PUB)—Publications in this series collectively constitute the Federal Information Processing Standards Register. The Register serves as the official source of information in the Federal Government regarding standards issued by NBS pursuant to the Federal Property and Administrative Services Act of 1949 as amended, Public Law 89-306 (79 Stat. 1127), and as implemented by Executive Order 11717 (38 FR 12315, dated May 11, 1973) and Part 6 of Title 15 CFR (Code of Federal Regulations).

NBS Interagency Reports (NBSIR)—A special series of interim or final reports on work performed by NBS for outside sponsors (both government and non-government). In general, initial distribution is handled by the sponsor; public distribution is by the National Technical Information Service, Springfield, VA 22161, in paper copy or microfiche form.

U.S.

SP

CPSIA information can be obtained
at www.ICGtesting.com
Printed in the USA
BVHW061039081118
532529BV00016B/709/P